MATHEMATICS OF DATA FUSION

THEORY AND DECISION LIBRARY

General Editors: W. Leinfellner (*Vienna*) and G. Eberlein (*Munich*)

Series A: Philosophy and Methodology of the Social Sciences

Series B: Mathematical and Statistical Methods

Series C: Game Theory, Mathematical Programming and Operations Research

Series D: System Theory, Knowledge Engineering and Problem Solving

SERIES B: MATHEMATICAL AND STATISTICAL METHODS

VOLUME 37

Scope: The series focuses on the application of methods and ideas of logic, mathematics and statistics to the social sciences. In particular, formal treatment of social phenomena, the analysis of decision making, information theory and problems of inference will be central themes of this part of the library. Besides theoretical results, empirical investigations and the testing of theoretical models of real world problems will be subjects of interest. In addition to emphasizing interdisciplinary communication, the series will seek to support the rapid dissemination of recent results.

The titles published in this series are listed at the end of this volume.

MATHEMATICS
OF
DATA FUSION

by

I. R. GOODMAN

NCCOSC RDTE DIV,
San Diego, California, U.S.A.

RONALD P. S. MAHLER

Lockheed Martin Tactical Defences Systems,
Saint Paul, Minnesota, U.S.A.

and

HUNG T. NGUYEN

Department of Mathematical Sciences,
New Mexico State University,
Las Cruces, New Mexico, U.S.A.

KLUWER ACADEMIC PUBLISHERS
DORDRECHT / BOSTON / LONDON

A C.I.P. Catalogue record for this book is available from the Library of Congress.

ISBN 978-90-481-4887-5

Published by Kluwer Academic Publishers,
P.O. Box 17, 3300 AA Dordrecht, The Netherlands.

Sold and distributed in the U.S.A. and Canada
by Kluwer Academic Publishers,
101 Philip Drive, Norwell, MA 02061, U.S.A.

In all other countries, sold and distributed
by Kluwer Academic Publishers Group,
P.O. Box 322, 3300 AH Dordrecht, The Netherlands.

Printed on acid-free paper

Contents

III   Use of Conditional and Relational Events in Data Fusion   339

Preface

Data fusion is the name which has been given to variety of interrelated expert-system problems which have arisen primarily in military applications. However, in a more general sense data fusion refers ro a broad range of problems which require the combination of quite diverse types of information, provided by a wide variety of sensors, in order to make decisions and initiate actions. This information is typically uncertain in at least one of the many possible senses of the word: probabilistic, imprecise, vague/fuzzy, contingent, and so on. This problem domain includes not only military applications but also statistics, decision theory, decision-support systems, expert systems, engineering (e.g, robotics, decision and control), and so on. At the current time, a sometimes bewildering variety of techniques have and are being applied to data fusion/expert system problems: statistical estimation and decision theory, Bayesian nets, fuzzy logic, Dempster-Shafer evidence theory, rule-based approaches, neural nets, genetic algorithms, wavalet transform theory, and so on.

Three book-level introductions to the basic mathematics, techniques, and heuristics of data fusion have appeared to date: *Mathematical Techniques in Multisensor Data Fusion* (1992) by D. L. Hall, *Multisensor Data Fusion* (1990) by E. Waltz and J. Llinas, and *Principles of Data Fusion Automation*(1995) by R. T. Antony (all by Artech House). These books present the basic mathematical and methodological tools for current data fusion practice, based primarily on classical and Bayesian statistics and Dempster-Shafer theory. However, Waltz and Llinas observe (p. 11-12) that current theoretical developments in the field have been moving towards:

(1) Development of representation and management calculi for reasoning in the presence of uncertainty.

(2) Development of spatial and contextual reasoning processes suitable for assessment of activity and intent on the basis of temporal and spatial behavior.

(3) Development of integrated, smart sensors with soft-decision signal processing of use in systems that perform numeric and symbolic reasoning in uncertainty.

In light of current advances in expert-system techniques, cognitive science, and so on, it seems desirable to incorporate into a single volume recent research work involving the representation and combination of disparate information types (statistical, linguistic, imprecise, contingent, etc.). In particular, our own researches in *fuzzy logic, random set theory,* and *conditional event algebra* have come to a sufficient degree of fruition that a book-length treatment seems warranted. Thus, this book is intended to be both an update on research progress in data fusion and an introduction to potentially powerful new techniques. As such, it consists of three parts: Introduction to Data Fusion and Standard Techniques (Part I); The Random Set Approach to Data Fusion (Part II); and Use of Conditional and Relational Event Algebra in Data Fusion (Part III). It is intended as a reference book for researchers and practioners in data fusion or expert systems theory, or for graduate students as a text for a research seminar or a graduate-level course.

Ackowledgements

The basic ideas underlying the work described in Chapters 4 through 8 - global density functions, the set derivative and set integral, global estimators - were developed in 1993 and early 1994 under internal research and development funding at Lockheed Martin Tactical Defense Systems, Eagan, Minnesota (then part of the Government Systems Group of Unisys Corp.) The writing of the book has been supported since then by the Electronics Division of the U. S. Army Research Office under contract DAAH04-94-C-0011. (*Note:* The content does not necessarily reflect the position or the policy of the Government. No official endorsement should be inferred.) In this regard, the second-named author, Ronald Mahler, wishes to express his gratitude to the following people. First, Dr. William Sander, Associate Director of the Electronics Division of US-ARO, whose financial support and personal encouragement has made this book possible. Second, Dr. Kenneth Matysik, former R& D Director at LMTDS-E, whose advocacy of advanced development , and financial and managerial support at critical times, made this work possible. Third, Mr. T. Craig Poling, manager of the Sensor and Data Fusion group at LMTDS-E ,for his years of reliable advice and encouragement. Fourth, Mr. Charles Mills, Director of Engineering, LMTDS-E, for his frequent support and patience in managing the unmanageable (i.e., advanced development). Fifth, I. R. Goodman and H. T. Nguyen, whose pioneering efforts in random set theory and conditional event algebra have been a constant inspiration. Sixth and finally, my wife Sue.

I. R. Goodman wishes to acknowledge the support by Independent Research Program, Code D14, NCCOSC RDTE Div, San Diego, California, under Project Number ZU074R7AO1.

We would like to thank Professor H. Skala for giving us the opportunity to publish our research in his Series Mathematical and Statistical Methods. We are grateful to Dr. Tonghui Wang of New Mexico State University for proof reading of the Text. The design of the book is due to him.

I. R. Goodman,
Ronald P. S. Mahler,
Hung T. Nguyen.

Chapter 1

Introduction

One of the long-time goals in both engineering and everyday life activities has been the efficient use and combination of all available and relevant information. Typically, such evidence involves various types of uncertainty which must be utilized in order to make good decisions. Technically speaking, the methodologies involved in accomplishing the above goal has been given a variety of names including: "data fusion", "information fusion", "combination of evidence", and "synthesis of observations'. Data fusion or information fusion are names which have been primarily assigned to military-oriented problems. In military applications, typical data fusion problems are: multisensor, multitarget detection, object identification, tracking, threat assessment, mission assessment and mission planning, among many others. However, it is clear that the basic underlying concepts underlying such fusion procedures can often be used in nonmilitary applications as well. The purpose of this book is twofold: First, to point out present gaps in the way data fusion problems are conceptually treated. Second, to address this issue, by exhibiting mathematical tools which treat combination of evidence in the presence of uncertainty in a more systematic and comprehensive way. These techniques are based essentially on two novel ideas relating to probability theory: the newly developed fields of *random set theory* and *conditional and relational event algebra*.

This book is concerned with both military and non-military applications–as well as technology transfer between the two. Therefore, with this in mind, we will use the above terms interchangeably, with emphasis on the two labels "data fusion" and " combination of evidence". Indeed, the modeling and numerical combination, or pooling, of observed data and prior information by use of standard mathematical statistics procedures is certainly a special form of data fusion. Another special case of data fusion is that of medical diagnosis based, in general, upon both statistical and linguistic information. Yet another example of data fusion involving use of both statistical and natural language information is in the diagnoses of faults in large-scale systems such as business organizational structures, manufacturing facilities, and Command, Control, Communication (C3) Systems. In view of this society which in many other ways is quite technologically advanced, it is time that similar advances be made with the treatment and combining of information.

Information relevant to a given problem can arise in various forms, including: numerical observations, from automated sensors and linguistic reports and expert opinions from purely human-based sources. To reiterate: All of this information, including models of the problem domain, are known only approximately, because of the presence of uncertainty. With the advance of Cognitive Science, Artificial Intelligence, and Expert and Rule-Based Systems, we know that there are various types of uncertainty besides classical probabilistic randomness, such as imprecision in the Dempster (or selection function, or interval analysis) sense, i.e., uncertain values are replaced by known sets of values and vagueness (or fuzziness), i.e., uncertain values are represented in the form of generalized or fuzzy set membership functions. Intimately connected with the various types of uncertainty, there is also the issue of causality or contingency of relations. It is obvious that in order to use these various types of uncertainty in the most effective manner, we need to implement the following scheme: (1) Understand better the mathematical and logical issues underlying uncertainty in general; (2) Develop reasoning processes and models capable of dealing with various sorts of uncertainty in a unified, rigorous manner; (3) Design operator-assisted or fully-automated computer algorithms which are both computationally feasible and faithful to the sound principles provided in steps (1) and (2).

Again referring to steps (1) and (2), the full processing of information can be decomposed into different aspects. First, one is faced with the retrieval-of-data problem, where one decides what type and amount of data should be obtained. Second, one partitions the problem domain into tractable subproblems and develops techniques for reasoning with the data. At present, these two aspects have been treated, more or less, by a combination of heuristics and techniques which are more situation (or subproblem)-specific, rather than generic or global, in their inception. Up until recently, this philosophy of approach would be considered sufficient. However, it should be re-affirmed because of the complexity of the issues involved, the combination of evidence problem may need to be addressed from a more systematic viewpoint, rather than from a localized stance.

In developing these tools and foundations at times we can be guided by intuition, but at other times we may need deeper mathematical and logical concepts. For example, it was noted that the well-established theory of mathematical statistics had a gap in it concerning admissibility of the standard least squares (or best linear unbiased) estimator. It was shown by Stein and James (1955) that, in problems of simultaneous estimation involving three or more parameters, the standard estimator, indeed, was inadmissible and they constructed, based on general theory, a class of biased estimators providing significantly lower Bayesian risk than the standard least-squares estimator under Gaussian-data assumptions. As another example, one can recall that the now-familiar Kalman filter was in the early 1960's regarded only as a theoretical curiosity! It was developed because a relative handful of researchers (Kalman, Bucy, Swerling, Stratonovitch, etc.), in effect, stepped back and pointed out that all prior mathematical models for filtering (such as the Wiener and alpha-beta filters)had one thing in common: they were all essentially conceptually sound, but either were entirely non-recursive or non-efficiently recursive in form. In summary, the spirit of our approach to data fusion is to develop a rational basis which is as global and free from ad hoc techniques as possible.

Present mathematical techniques for data fusion are very well surveyed in the two recent Artech House monographs by Hall *(Mathematical Techniques in Multisensor Data Fusion, 1992)*, by Waltz and Llinas *(Multisensor Data Fusion, 1990)*, and by Antony *(Principles of Data Fusion Automation, 1995)*. However, with the new tools presented in our monograph, the ideas presented in these two texts can be supplemented, and unified to handle even more realistic and complicated situations involving linguistic or natural language-based – as well as conditioned – evidence, among other factor. This treatise is written explicitly for that purpose.

Let us now turn from philosophical generalities to an overview of the two general concepts which will concern us for the remainder of the book: random sets and conditional (or, more generally, relational) events. The goal of this introductory chapter is to answer the following three basic questions: What is data fusion? What are random sets? What are conditional/relational event algebras? To answer these, we must also be concerned with related issues, such as: Why are they useful for solving data fusion problems? Given the existence of more familiar approaches, why do we need such (seemingly) esoteric concepts? Before addressing these questions, however, it would be well to emphasize at the outset that neither random set theory nor conditional event algebra are "esoteric." Neither claims to supplant existing techniques with some exotic competitor beyond mere mortal ken, but rather to extend well-understood probabilistic reasoning to new realms. Conditional event algebra, for example, seeks to bring conditional knowledge–e.g., AI production rules, evidence contingent upon evidence–under the umbrella of standard probability theory. Likewise, the random set approach to data fusion seeks to unify a broad range of data fusion techniques, ranging from multisensor, multitaget estimation to "fuzzy" inference, within a single, unified, probabilistic paradigm.

1.1   What is Data Fusion?

Very broadly speaking, data fusion is the name which has been given to a variety of interrelated problems which have arisen primarily in military applications. Condensed into a single statement, information fusion might be defined thusly:

Locate and identify many unknown objects of many different types on the basis of different kinds of evidence. This evidence is collected on an ongoing basis by many possibly allocatable sensors having varying capabilities. Analyze the results in such a way as to supply local and over-all assessments of the significance of a scenario.

However, and as earlier noted, we could be just as well be describing any number of application problems in "artificial intelligence" which involve both the finding and identifying of objects: computer processing for autonomous robots, to name but one example. Indeed, in terms of basic mathematical methodologies, the distinction between "information fusion" and "knowledge-based systems' or "expert systems" serves mainly to distinguish military from other applications of expert systems theory.

The figure "The Information Fusion Problem" in Chapter 2 illustrates one typical such military application: A large-area air-sea surveillance system. At the right of the figure, a variety of platforms, some of which may be of military interest and some of these posing a possible threat immediately or in the future, are under observation by a force consisting of many military vehicles. At the lower-left-hand corner of their figure, a headquarters operations center, responsible for overall assessment of the tactical situation, communicates with remote mobile units. In the center third of the figure, four different types of reconnaissance platforms collect data: an antisubmarine patrol aircraft; an antisubmarine helicopter; a missile cruiser; and a reconnaissance aircraft. At the top left of the figure, additional data is collected by satellites and by local reconnaissance aircraft.

These platforms collect different kinds of data using different types of sensors. For example, they may be equipped with a variety of radars: search radars, tracking radars, doppler radars, fire control radars, or OTH (Over the Horizon) radars. They may also be equipped with electronic detection devices which passively collect transmitter emissions from platforms under surveilllance (e.g., communications systems, fire control radars, and other radars). Local reconnaissance aircraft may supply active imaging data using perhaps radar and infra-red imaging sensors. Other passive-surveillance sensors may collect radar and communications. Satellites may supply imaging and other data. Remote locations may supply additional information via standardized coded messages transmitted on military data links. Information regarding surface and subsurface platforms may be supplied by passive and active sonar equipment and magnetic-detection equipment carried on or deployed by, aircraft. Again, it should be noted that much of the above information may be derived from human-based sources or experts who supply additional data in the form of experience and deductive capabilities honed by training.

Much of the data supplied by many of these sensors may be statistical in form: location estimates with estimated uncertainties; direction-to-target estimates; detected platform "attributes" or "features" (e.g., base frequencies, bandwidths, scanrates, and pulse repetition rates of radars; acoustic lines emitted by machineries on surface ships and submarines). However, data may also be highly non-statistical–for example, natural-language descriptions of platforms observed by a human. Data may also take the form of prior information: ground or sea-bottom topography, known maneuver characteristics of platforms, costs of possible actions, predetermined tactical goals and priority levels, etc. Especially when the process of analyzing tactical situations is considered, data may also consist of statistical or heuristic rules describing such things as rules of engagement, tactical doctrine, and so on.

From this data, information fusion systems must assemble facsimiles of (hypotheses concerning the nature of) "ground truth" which describe the platforms being observed: numbers. positions, velocities, accelerations, and identities; military status (hostile, allied, neutral, unknown); threat level; likely future actions of hostile platforms; and recommendations concerning appropriate friendly-force action. However, the sheer volume and variety of data that must be taken into account often exceeds the abilities of human beings to analyze and digest it and to deliver digested results in a timely fashion to military personnel in the field. This reality has been the driving force behind a range of attempts to automate the data fusion process, whether in part or in whole. One common taxonomy has been devised to break the general data fusion problem domain into smaller and possibly more tractable pieces. These pieces are as follows:

Sensor fusion.: In this kind of fusion, evidence from two or more usually closely located sensors–e.g., pictures from imaging sensors–is combined in order to get information which is more precise than that obtainable from each separate piece of evidence. Fusion of this type can involve the combination of raw signals supplied by sensors (sometimes called "pre-detection fusion), or of elementary "features" or "segments" extracted from the signals by pre-processors. So-called "registration" problems (e.g., the necessity of correctly aligning different images of the same object) are of great importance here.

Level 1 fusion.: This type of fusion is concerned with the gross problem of tracking and identifying multiple targets using data supplied by multiple sensors. It is the most well-developed part of information fusion, both in terms of theory and practice. Level 1 fusion is further subdivided into four interrelated parts: detection, correlation, tracking, and classification. Detection refers to the process of estimating the number of targets present in a scenario. Classification is the process of computing estimates of the identities of targets. In tracking, one must determine estimates of the positions, velocities, and accelerations of targets, as well as predictions of their future positions, velocities, and accelerations. Correlation involves the process of associating sensed data (called "reports" or "observations") with running estimates of target positions, velocities, accelerations, and identities (these estimates are also called "tracks"). This latter process is also sometimes called "detection-to-track fusion"

Situation/mission assessment: This type of fusion tends to rely more heavily on heuristic methodologies than the previous types. The purpose of situation assessment is to provide an over-all picture of the military significance of the data collected by the previous levels of fusion. This includes the military status of target (friendly, hostile, etc.), the threat level of hostile targets, and prediction of the probable intentions and courses of action of threatening targets. Situation information must then be fused into an over-all picture– a mission assessment–of the probable courses of action that should be followed by friendly forces, on the basis of preordained goals and priorities. Such courses of actions can include the proper use of sensors. If sensors collect data purely on the basis of happenstance, then data on some targets may be overcollected, resulting in little added information, while data on other targets may be undercollected, resulting in poor information about these targets. Many sensors cannot be easily redirected at poorly observed regions, whereas others–e.g., electronically-scanned radars–can

rapidly dwell as required on a number of different "observation cells." Sensor
management refers to the process of adaptively allocating the dwells of each
member of a suite of sensors more intelligently among existing objects of interest. The result is a more accurate Level 1-fusion picture.

A great deal of research in data fusion has, and continues to be, focused
on the "lower" levels: sensor fusion, detection, correlation, classification, and
tracking. The techniques which have been applied to sensor fusion, particularly
in the image processing and automatic target recognition area, include neural
nets, wavelet theory, Markov random field theory, and random set theory. A
host of techniques and algorithms have been devised to deal with the multitarget tracking problem alone: multihypothesis tracking (MHT), joint probabilistic
data association (JPDA) filters, symmetric measurement equation (SME) filters,
Gaussian-sum algorithms, jump-diffusion algorithms, etc. Classification problems have inspired an even more greater diversity of techniques: Bayesian nets
(Pearl trees), fuzzy logic, Dempster-Shafer theory, nonmonotonic logics, rule-
based approaches, and others. In recent years, however, much of the research
emphasis in data fusion has shifted to the "higher" levels of the taxonomy,
that is towards sensor management, threat and situation assessment, and mission assessment. The last two levels rely even more heavily on "contingent"
or rule-based evidence than the other levels. For example, the discovery that
a platform possesses certain characteristic identifying attributes may tell us
something about the threat status of a platform. As another example, the detection of a characteristic radar emission may identify the country-of-origin of
the platform and thereby perhaps its "friendly, unfriendly, or neutral' status.
Or, the detection of an actively emitting fire control radar may allow us to
conclude that the platform constitutes an immediate threat and that defensive
measures of some kind must be undertaken. Rules of engagement provide yet
another example: knowledge of the tactical doctrine for a given platform can
allow one to deduce the probable intent, or most likely future course of action,
of a potentially hostile platform.

In summary, data fusion tends to break down into a collection of weakly
connected subdisciplines, some of which are more mature than others. In the
more mature ones, such as multisensor, multitarget tracking, a powerful theoretical apparatus exists to aid application. On the other hand, the less mature
ones continue to be dominated by heuristic and ad hoc methodologies. We have
paid a price for the lack of a unifying paradigm. Subsystems tend to be poorly
integrated, since they tend to be designed separately from each other, often
using incongruent methodologies. Evidence which is designated "ambiguous"
is often poorly utilized, and utilized in an ad hoc manner when use is made of
it. Given that algorithms tend to be "stitched together" from disparate subsystems, performance optimization is often just a dream: improving one aspect
of performance can result in degradations in other aspects for reasons that are
difficult to pin down. Finally, it is difficult merely to compare two different
algorithms in a meaningful way, given that no theoretical "level playing field"
for doing so exists at the current time.

1.2   Random Set Theory

Random set theory offers a potential means of integrating most of the different aspects of data fusion together. At its current stage of development, it cannot claim to do so completely. Nevertheless, it seems capable at this point of putting under a single probabilistic umbrella the following aspects of data fusion: Level 1 fusion; sensor management, performance evaluations, and target identification evidence which is imprecise, vague, or otherwise ambiguous in form. The reason for this is that recent research, much of it conducted by the present authors, has revealed the following:

1) Random set theory provides a means of extending classical single-sensor, single-target point-variate statistics to a multisensor, multitarget statistics. Despite the rather complex random-set mathematics which undergirds it, this "finite-set statistics" is almost identical in form and behavior to the kind already routinely used by engineers and theorists in data fusion. By finite set statistics we mean data consisting of finite sets of observations, rather than just individual point or vector observations. The main conceptual hurdle which must be surmounted is the fact that, in this sort of statistics, probability density function (PDFs) describe the comprehensive statistical behavior of an entire sensor suite, rather than the internally-generated noise statistics of a single sensor and the fact that observations are finite sets of ordinary observation. Given this, data fusion algorithms may be interpreted as statistical estimators of a finite-set parameter.

2) Random set theory provides a common theoretical framework for several aspects of uncertainty on expert systems, which in the past were either considered unrelated or, in effect, treated apart from probability theory. This is illustrated by the following examples:

A. It is now known that fuzzy sets are equivalent to a weak specification of random sets, analogous to the limited description of a ordinary probability distribution by only its expectation. More specifically, it has been shown that the membership function of a fuzzy set can be considered, in effect, the probability of a random set covering a generic point. In addition, it has been demonstrated that the most familiar fuzzy logic–the Zadeh max-min logic–can be embedded into random subsets in such a way that the fuzzy AND, OR connectives correspond (homomorphically) with random-set intersection, union. It is also known that many other fuzzy logics can be regarded as approximations to these random-set operations, given different assumptions about the statistical correlation between random sets.

B. It is well-known now that the Dempster-Shafer theory of evidence can be expressed formally within the context of random set theory. As such, various evidential operations have their counterparts in probability theory. For example, the concept of independent pieces of evidence can be defined in terms of stochastic independence of random sets. Also, the Dempster-Shafer rule of combination of evidence corresponds to the intersection of non-empty, independent random sets.

C. More recently, it has been shown that certain "Type I" conditional event algebras can be represented as random sets, thus leading to the possibility that rule-based evidence can be incorporated into the random set formalism advocated in this book. In addition, there is reason to believe that other types of conditional event algebras can also be similarly represented.

Given the above, it is not surprising that random set theory provides a potential unifying structure for many aspects of data fusion. One purpose of this book is to verify that this is indeed the case. First, however, let us now go into more detail concerning the approach that is advocated in this book. The basic idea is this. Several sensors, supplying varying kinds of evidence, interrogate an unknown number of targets and transmit their reports over data links to a central data fusion site for processing. The basic trick is to:

1) model the entire sensor suite as a single "global" sensor,

2) model the unknown target-set as a single "global" target,

3) notice that the "global report"–the actual observable which is "seen" by the sensor suite as a while– is a randomly varying finite subset (of randomly varying size) whose elements are ordinary measurements.

To illustrate the concept of a global report, consider the situation where three targets are observed by a single sensor. Three separate observation sets are taken: Z_1, consisting of four observations(including a false alarm) each denoted by the symbol $*$; Z_2, consisting of two observations (implying a missed detection) each denoted by the symbol $\square$; and Z_3, consisting of three observations each denoted by $\bigcirc$. In general, the global report seen by entire sensor suite at a given time will have the form

$$Z = Z_{sensor\#1} \cup \cdots \cup Z_{sensor\#s},$$

Where $Z_{sensor\#j}$ denotes the reports delivered by the jth sensor, and which can consist not only of reports due to actual targets, but also of spurious reports caused by clutter generators. This means, in turn, that $Z_{sensor\#j}$ has the general form

$$Z_{sensor\#j} = Z_{j;target\#1} \cup \cdots \cup Z_{j;target\#t} \cup Z_{j;clutter\#1} \cup \cdots \cup Z_{j;clutter\#M(j)},$$

where $Z_{j;target\#i}$ denotes the reports collected from the ith target by the jth sensor, and $Z_{j;clutter\#k}$ denotes the reports collected from the kth clutter generator by the jth sensor. The observation-set Z is actually just one instance of a randomly varying observation-set, that is, a random finite set, denoted by Σ. The statisics of the entire sensor suite can be constructed from this random set Σ as follows. Recall that in ordinary statistics that the randomly varying observation, as seen by a single position-reporting sensor when it observes a single target, is a random vector $\mathbf{Z}$. (note: boldface Z indicates a random vector, whereas non-boldface Z indicates a nonrandom set of observations.) The complete probabilistic information concerning the random vector $\mathbf{Z}$ is

$$P_{\mathbf{Z}} = P(\mathbf{Z} \in S),$$

which measures the degree to which the random vector $\mathbf{Z}$ is concentrated in the set S. In like manner, in the multisensor, multitarget case we will see that the comprehensive statistics of the randomly varying observation-set Σ are summarized by the probability assignment

$$\beta_\Sigma(S) = P(\Sigma \subseteq S),$$

which measures the degree to which all of the points in the finite random set S are concentrated within the set S. The set function is called a belief function. Unlike probability measures, which satisfy the additivity property $P_\mathbf{Z}(S \cup T) = P_\mathbf{Z}(S) + P_\mathbf{Z}(T)$, whenever S and T are disjoint, the belief function is nonadditive. In conventional single-sensor, single-target statistics, one usually works not with the probability measure $P_\mathbf{Z}$ of a random vector $\mathbf{Z}$ but rather with its associated probability density function $f_\mathbf{Z}$. This density function can be constructed from $P_\mathbf{Z}$ by iteratively differentiating the cumulative probability function $F_\mathbf{Z}(z) = P(\mathbf{Z} \leq z)$ with respect to its scalar argument. The construction of the density requires, of course, the existence of the differential and integral calculus. In the multisensor, multitarget case we will show that there is a generalized differential and integral calculus of set functions, in particular a generalized Radon-Nikodym derivative. Using this generalized differentiation, it is possible to construct a generalized or "global" probability density function, denoted f_Σ, which summarizes the comprehensive statistics of the entire multisensor suite. In the case of a conventional PDF, the quantity $f_\mathbf{Z}(z)$ is a measure of the likelihood with which the vector z will occur as a value of the random vector $\mathbf{Z}$. In like fashion, the value $f_\Sigma(S)$ is a measure of the likelihood with which the finite set S will occur as a value of the random finite set Σ. However, the global PDF of a random finite set differs from the PDF of a random vector. The PDF of a single sensor describes only the self-noise of the statistical behavior of all of the sensors in the sensor suite. This includes: 1)the self-noise of each sensor; 2)the probabilities of detection and probabilities of false alarm for each sensor; 3)clutter models for each sensor; 4) the detection profiles of each sensor (as functions of range, aspect, target type, etc.); 5) unknown parameters describing each target; and 6) the known parameters which describe the states (modes, look angles, dwell times, etc.) of each of the sensors.

One can likewise extend the concept of prior and posterior PDFs to the multisensor, multitarget realm. In particular, prior PDFs encapsulate within the mathematical object all prior knowledge in regard to parameters such as numbers of targets, identities of targets, and kinematics of targets.

Once we have the concept of a global PDF in hand, it then is easy to see that there is an almost perfect collection of parallels between conventional single-sensor, single-target point-variate statistics, and a new multisensor, multitarget set-variate statistics. For example, one can construct multitarget, multisensor analogy of the maximum likelihood (ML) and maximum a posterior (MAP) estimator,–resulting in fully integrated, statistically consistent data fusion algorithms. Or, one can construct analogues of the familiar entropy metrics of Shannon information theory, thus opening the way for information-based Figures of Merit which measure the global or overall-all performance of data fusion algorithms. One can also construct analogues of nonparametric estimators–thus allowing us to estimate the global PDFs of arbitrary data fusion algorithms directly from empirical observations. In fact, the resulting parallelism is so nearly complete that one can state it in the form of general principle:

"Almost Parallel-Worlds Principle": To solve a general multisensor, multitarget data fusion algorithm, first solve the corresponding single-sensor, single-target tracking problem and then directly generalize the mathematics to the multisensor, multitarget case. We say "almost" here because the parallelism is not actually quite complete. Whereas it is possible to add and subtract vectors, there seems to be no analogous addition and subtraction operations for finite sets. This means, among other things, that it is not possible to compute expected values of random finite sets. Nevertheless, the parallelism is complete enough that, provided one exercises a little care, it is possible to directly generalize a century's worth of knowledge concerning point-variate statistics to multisensor, multitarget data fusion.

For example, it is possible to prove a data fusion analog of the Cramer-Rao inequality–thus potentially permitting calculations of theoretical-best-performance bounds for nontrivial multisensor, multitarget data fusion algorithms. Or, as another example, in sensor management our goal is to assign allocatable sensors to targets in as optimal a manner as possible. The corresponding single-sensor, single-target problem is simply that of a sensor tracking-and control problem—as when a missile-tracking camera must be constantly redirected so that its aperture is always directed at the current position of a rapidly moving target. One might expect, therefore, that the Almost Parallel Works Principle might be applied to generalized single-sensor tracking-and control mathematics to the multicensor, multitarget case.

This, in brief, describes how nonadditive probability measures–specifically, belief measures–arise when one adopts a systematic, rigorous, top-down analysis of the multisensor, multitarget problem. What, however, of the situation of target-identification evidence which is imprecise or vague?

1.3   Conditional and Relational Event Algebra

As already mentioned, conditional event algebra seeks to bring conditionality and contingency under the umbrella of standard probability theory. What do we mean by "standard probability theory?" Probability theory is concerned with numerically describing the frequency of occurrence of entities called "events". For example, let Ω denote the "universe" of all fifty-two playing cards in a deck. Then the observation "a king is drawn" ("king" for short) is an event. If a card is dealt and then placed back in the deck a large number of times, this event will occur with probability $P(\text{"king"}) = 4/52$. It has become common practice to represent events as subsets of the universe and Boolean logical operations & ("and"), $\vee$ ("or"), and ()' ("not") on events by the corresponding set-theoretic operations $\cap$ (intersection), $\cup$ (union) and ()' (complementation). Thus, the event "king" is represented by the subset S containing all kings: and if the event "spade" is represented by the set T of all spades, then the event "king" and "spade" is represented by the intersection $S \cap T$.

However, there are many kinds of evidence, having the general form "if T then S" or "S given T", which are not "events" in the usual sense of the word. For example, AI production rules have the form "if pressure is somewhat high then open valve slightly". Or, knowledge can come in the form of a rule base consisting of assertions such as "if Tweety is a bird then Tweety has wings". The basic idea behind conditional event algebra is twofold. The first idea, known as Stalnaker's thesis, is that the correct probability assignment of a conditional "S given T", denoted by $S|T$, is the conditional probability of its constituent events:

$$P(S|T) = P(S \cap T)/P(T),$$

whenever $P(T) > 0$. The second idea is that, if appropriately modeled, the set of conditionals should become a Boolean algebra, in such a way that conditional probability becomes a true probability measure on conditionals.

In this book we show that this is, indeed, possible and that, moreover, it leads to powerful clarifications of the meaning of contingency when ambiguity of evidence is involved. To mention just one particularly important instance, suppose that we have two conditionals "T given S" and "S given R". Then in conventional boolean logic, the meta-rule known as "syllogism" allows us to conclude that "T given R". (e.g., "Socrates is a man" and "All men are mortal".) However, it can be shown that conditional probability is not necessarily compatible with syllogism: It is possible, for example, for both $P(T|S)$ and $P(S|R)$ to be large and yet for $P(T|R)$ to be small. Unfortunately and non-intuitively, there is an apparent fundamental discontinuity between nontrivial probability extensions of the syllogism argument and the classical form itself and has been the subject of much research. However, conditional event algebra provides an alternative approach to this issue by providing the class of all deductions from the premises, without forcing the counterpart of the term "T given R": Instead, it provides a closed-form expression for the probability evaluation of the conjunction of the conditionals as a closed-form expression for the probability evaluation of the conjunction of the conditionals as a natural lower deductive bound of information content $P\left((T|S)\ \& \ (S|R)\right)$ [which is not equal to $P(T|R)$!] In addition, recently, a relatively simple criterion has been derived for testing whether or not arbitrary a large class of conditional events are deductions of this conclusion.

Let us now describe the ideas behind conditional event algebra in a little more detail, deferring a fuller treatment until Part C of the book. Whether we use the logical operators $\&$, $\vee$, $()'$ on events or the set-theoretic operators $\cap$, $\cup$, $()'$ on the subsets corresponding to them, they constitute a "Boolean algebra." That is, they satisfy the laws of symbolic logic:

$$T \cup \Omega = \Omega, \quad T \cup \emptyset = T, \quad T \cap \Omega = T, \quad T \cap \emptyset = \emptyset, \quad T \cap T = T, \quad T \cup T = T,$$

$$S \cap T = T \cap S, \quad S \cup T = T \cup S,$$

$$(R \cap S) \cap T = R \cap (S \cap T), \quad (R \cup S) \cup T = R \cup (S \cup T),$$

$$(S')' = S, \quad S \cup S' = \Omega, \quad S \cap S' = \emptyset,$$
$$(S \cup T)' = S' \cap T', \quad (S \cap T)' = S' \cup T',$$
$$R \cap (S \cup T) = (R \cap S) \cup (R \cap T), \quad R \cup (S \cap T) = (R \cup S) \cap (R \cup T).$$

In addition, the fundamental binary relation $\subseteq$ over all events representing partial, or subset, order is given as

$$S \subseteq T \text{ iff } S = S \cap T \text{ iff } T = S \cup T,$$

Probability theory begins by assuming the existence of a "probability measure" $P(\cdot)$ defined on certain class of subsets of Ω, which satisfies the following three properties:

$$P(\emptyset) = 0, \quad P(\Omega) = 1,$$

and

$$P(S \cup T) = P(S) + P(T) \quad \text{whenever} \quad S \cap T = \emptyset.$$

If Ω is infinite then things become a little more complicapted and we must assume that events are restricted to membership in a σ-algebra and assume the following additional postulate:

$$P(S_1 \cup S_2 \cup \cdots \cup S_n \cup \cdots) = P(S_1) + P(S_2) + \cdots + P(S_n) + \cdots$$

for any countable sequence of mutually disjoint events $S_1, S_2, \ldots$. From this, all of the probability and statistics familiar to data fusion engineers follows. For example, the derived laws

$$P(S') = 1 - P(S), \quad P(S \cup T) + P(S \cap T) = P(S) + P(T)$$

show the relationship between probability and the boolean operations on events. Likewise, one defines conditional probability as the probability evaluation of the contingency of S to T: $P(S|T) = P(S \cap T)/P(T)$ whenever $P(T) > 0$, where $P(S|T)$ is read "probability of: if T, then S" or as "probability of S given T".

Thus, the above is an encapsulated form of standard probability theory. The basic idea underlying conditional event algebra is that contingencies are "events" of a generalized kind, and that it should be possible to fit them in a framework similar to the one described above. That is, it should be possible to represent such "conditional events" as mathematical objects of the form $(S|T)$, where S and T are ordinary events. Ordinary events are special kind of conditional events, i.e., $(S|\Omega) = S$. Further, it should be possible to find logical operators &, $\vee$, $()'$, extending the corresponding ones over ordinary events or sets defined on these objects which possibly satisfy all, or most, properties of boolean algebra and probability theory (as briefly described above). In addition, other properties may be satisfied, such as *modus ponens*, i.e.,

$$(S|T) \cap (T|\Omega) = (A \cap T|\Omega), \quad P\left((S|T) \cap (T|\Omega)\right) = P(S \cap T),$$

this means that when a rule $(S|T)$ is "fired" by the presence of its antecedent T, then what results is knowledge of both the consequent S and the antecedent T of the rule.

Thus, one goal of developing conditional event algebra is to find a suitable framework in which rules can *both* be consistently modeled as conditional events and standard probability theory can be extended to these rules (e.g., rules can be logically manipulated in much the same way as ordinary events). Applying this to rule-based system, one can legitimately regard $P((S|T)) = P(S|T)$ as the validity or strength assignment to the rule $(S|T)$. Moreover, this will be true regardless of which probability distribution P on ordinary events one uses.

In Part III of this book, one type of conditional event algebra ("Type I") is shown to be quite useful, is relatively simple in form, but does not possess a full Boolean structure and does not satisfy all probabilities. Within the Type I class of conditional event algebras, mention must be made of two particularly important algebras: GNW (developed by Goodman, Nguyen, and Walker) and SAC (developed independently by Schay, Adams, and Calabrese). On the other hand, another type of conditional event algebra –Type II– is developed which does satisfy all of these properties (and more), but, in contrast with Type I, is much more complex in form. This is achieved by using a countably-infinite product probability space construction. In Type II conditional events algebras, conditional events can be of two forms: "simple" or "non-simple". Simple events refers back to the original definition provided above, typically of the form $(S|T)$, where S, T are ordinary sets or events in the given boolean algebra of (unconditional) events. Because of the product structure of Type II events, there are many more events in the overall boolean structure each of the three above-mentioned conditional event algebras possess, there is a fundamental tie-in relative to their application to deduction in general, and the syllogism problem, in particular. Following the philosophy that Type II algebra is more theoretically satisfactory, it can be utilized to model problems of deduction, by providing explicit forms for logical compounds (or combinations) of conditional premises, and consequently probability evaluations of such compounds. In turn, noting the aforementioned embedding of Type I into Type II, it can be shown that the best lower bound approximation to conjunctions of premises by simple types of conditional events is that furnished through GNW. On the other hand, the best upper bound approximations to such conjunctions by simple conditional events is determined through SAC. Most importantly, the last fact can be used also to describe and provide a test procedure for determining all those simple conditional events that are deductions, from the conjunction of premises Similar remarks hold for simultaneous or disjunctive deductions, with the roles of GNW and SAC reversed.

In summary, the three leading conditional event algebras developed can be used in an integrated natural manner to model and characterize deduction. This is especially useful in extracting information from rule-based systems

Thus far we have spoken of conditional events as models for evidential forms such as rules. Forming a conditional event, however, can also be thought of as an algebraic analog of the process of dividing two probabilities. One might ask, therefore, do there exist algebraic analogues of other arithmetic operations, such as squares, cubes, weighted sums, and so on? Such questions are important to combination of evidence problems since, for example, it is important to know how to pool evidence collected by several experts (who may have varying degrees of trustworthiness) a given event. The algebraic analog of a weighted sum would, therefore, show us how to pool events together in such a way as to reflect an a priori weighting scheme. This extension of conditional event algebra is called *relational event algebra.* A number of theoretical properties, as well as applications, of relational event algebra is provided.

Finally, we will also show that there are close connections between random set theory and conditional event algebra. Specifically, we will show that it is possible to represent the GNW conditional event algebra as a subalgebra of random subsets, thus potentially making it possible bring rule-based evidence under the random-set umbrella. We will also show that there is a "second degree" conditional event algebra, denoted GNW* and consisting of "compound" or "iterated" conditional events of the form $((S|T)|R)$, that is, the antecedent is an ordinary event R and the consequent is a rule $(S|T)$.

Part I

Introduction to Data Fusion

The Information Fusion Problem

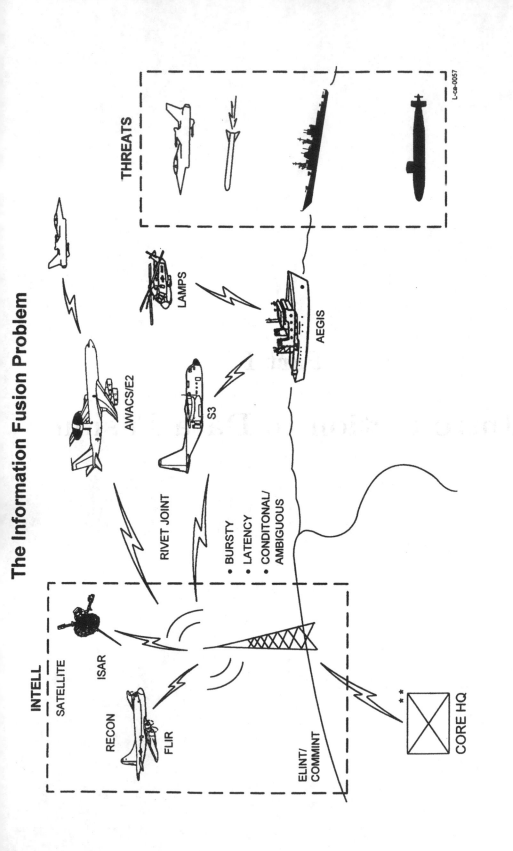

THREATS

LAMPS

AWACS/E2

S3

AEGIS

RIVET JOINT

- BURSTY
- LATENCY
- CONDITONAL/
  AMBIGUOUS

INTELL
SATELLITE
ISAR
RECON
FLIR

ELINT/
COMMINT

CORE HQ

L-ce-0057

Chapter 2

Data Fusion and Standard Techniques

2.1 Data Fusion

Data fusion, or *information fusion* as it is also known, is the name which has been attached to a variety of interrelated problems arising primarily in military applications. Condensed into a single statement, data fusion might be defined thusly:

> Locate and identify an unknown number of unknown objects of many different types on the basis of different kinds of evidence. This evidence is collected on an ongoing basis by many possibly re-allocatable sensors having varying capabilities. Analyze the results in such a way as to supply local and over-all assessments of the significance of a scenario
>
> and to determine proper responses based on those assessments.

From this description it should be clear that we could just as easily be describing any number of applications in "artificial intelligence" which involve finding and identifying objects and making consequent decisions: Computer processing for autonomous robots is the most obvious example. Indeed, in terms of basic mathematical methodologies, the distinction between "data fusion" and "knowledge-based systems" or "expert systems" serves mainly to distinguish military from other applications of artificial intelligence.

This book is *not* intended to be an introduction to the current mathematical practice of data fusion. Such a purpose has been more than adequately fulfilled already by the three excellent survey-level books on data fusion which have appeared in recent years: *Multisensor Data Fusion* by Edward Waltz and James Llinas [133], *Mathematical Techniques in Multisensor Data Fusion* by David Hall [58], and *Principles of Data Fusion Automation* by Richard Antony [6] (all published by Artech House).

17

Rather, this book is intended to be an introduction to certain new and comprehensive mathematical approaches to data fusion and expert systems theory developed by the authors in recent years. These approaches are *random set theory, conditional event algebra*, and *relational event algebra*. Briefly expressed, conditional event algebra is a rigorous, probabilistic calculus for dealing with problems of contingency in data fusion (e.g. knowledge-base rules, contingent decision-making, etc.). Relational event algebra is a generalization of conditional event algebra which provides a systematic and rigorous basis for problems involving the *pooling of evidence*. Random set theory (the study of random variates which are *sets* rather than *points*) turns out to provide a rigorous and systematic mathematical foundation for three major aspects of data fusion: multisensor-multitarget estimation; expert-systems theory; and performance evaluation.

2.1.1   Chapter Summary

In the remainder of this section we will briefly describe data fusion (Section 2.1.2), the different functional aspects of data fusion (Sections 2.1.3 and 2.1.4), the major theoretical and practical problems encountered in the field (Section 2.1.5), and introduce certain important methodological distinctions (Section 2.1.6). Sections 2.2 and 2.3 summarize some of the major mathematical approaches currently used in data fusion.

Section 2.2 is devoted to two basic perspectives in regard to *multitarget estimation*: "indirect" estimation (Section 2.2.1) and "direct" estimation (Section 2.2.2).

In Section 2.3, on *expert systems theory*, we will briefly describe *imprecise information* and the *Dempster-Shafer theory* (Section 2.3.1), *vague information* and *fuzzy logic* (Section 2.3.2), *contingent information* and *conditional event algebra* (Section 2.3.3), and *partial-probabilistic information* (Section 2.3.4).

In the remainder of the chapter we will introduce the major topics of the book. *Random set theory* is introduced in Section 2.4. *Finite-set statistics*, a special case of random set theory specifically designed to address data fusion problems, is summarized in Section 2.5. Finally, in Section 2.6, we describe the basic ideas behind *conditional event algebra* and *relational event algebra*.

2.1.2   What is Data Fusion?

The figure on page 16 illustrates one typical data fusion application, a large-area air-and-sea surveillance system. At the right of the figure a variety of platforms, some of which may be of military interest and some of which may pose a threat in the immediate or near term, are under observation by a force consisting of many military vehicles. At the lower-left-hand corner of the figure a headquarters operations center, responsible for overall assessment of the tactical situation, communicates with remote mobile units. In the center third of the figure, four different types of reconnaissance platforms collect data: an S-3 antisubmarine patrol aircraft; a LAMPS antisubmarine helicopter; an AEGIS missile cruiser;

and an AWACS reconnaissance aircraft. At the top left of the figure, additional data is collected by satellites and by local reconnaissance aircraft.

These platforms collect data from different types of "on-board" or "organic" sensors, as well as "offboard" or "nonorganic" data generated by sensors on other platforms. For example, military platforms may be equipped with a variety of radars: search radars, tracking radars, doppler MTI (moving target indicator) radars, fire control radars, or OTH (Over the Horizon) radars. They may also be equipped with ESM (Electronic Support Measures) and IFF (Identify Friend or Foe) receivers and other sensors which passively detect and collect transmitter emissions from platforms under surveillance (generated, for example, by "non-cooperative" communications systems and targeting and other radars). Local reconnaissance aircraft may supply active imaging data using SAR (Synthetic Aperture Radar), ISAR (Inverse SAR), FLIR (Forward-Looking Infrared) and LADAR (Laser Radar) sensors. Satellites may supply imaging and other data. Remote locations may supply additional information via messages transmitted on special military data links. Information in regards to surface and subsurface platforms may be supplied by passive and active sonar equipment and magnetic-detection equipment carried on or deployed by, for example, the S-3 aircraft. Last but not least, many "sensors" are human observers and military operator-analysts who supply additional data in the form of experience and expertise. Still other "sensors" are databases of rules and other information that may or may not pertain to the current situation at hand.

Much of the data supplied by many of these sensors may be statistical in form: location estimates with estimated uncertainties; direction-to-target estimates; detected platform "attributes" or "features" (e.g., base frequencies and pulse repetition rates of radars; acoustic lines emitted by machineries on surface ships and submarines). However, observations may also be highly non-statistical—for example, natural-language descriptions of platforms observed by a human. Data may also take the form of prior information: ground or sea-bottom topography, known maneuver characteristics of platforms, military costs of possible actions, predetermined tactical goals and priority levels, etc. Especially when the process of analyzing tactical situations is considered, data may also consist of statistical or heuristic rules describing such things as rules of engagement, tactical doctrine, and so on.

From such information, data fusion systems must assemble facsimiles of (hypotheses concerning the nature of) "ground truth" which describe the platforms being observed: numbers, positions, velocities, accelerations, and identities; military status (hostile, allied, neutral, unknown); degree and imminence of threat; likely future actions of hostile platforms; and recommendations concerning appropriate friendly-force action. However, the sheer volume and variety of data that must be taken into account increasingly exceeds the abilities of human beings to analyze and digest it and to deliver analyzed results in a timely and appropriate fashion to military personnel in the field. This fact has encouraged attempts to automate data fusion processes whether in part or in whole.

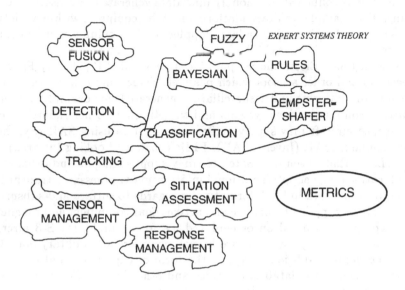

Figure 2.1: The major data fusion subfunctions are displayed, ranging from those closest to the sensor data (sensor fusion) to those closest to the end-user (situation/threat assessment and response management).

2.1.3   The Subdisciplines of Data Fusion

A standard taxonomy known as the "JDL (Joint Directors of Laboratories) model" breaks the data fusion problem into smaller and more tractable "levels" known as multisource integration (Level 1), situation assessment (Level 2), threat assessment (Level 3), and response management (Level 4). Here we will adopt a nonstandard taxonomy, illustrated in Figure 2.1, which arranges data fusion functions hierarchically from those which are closest to the sensor to those which are closest to the human end-user(s):

(1) Sensor fusion
(2) Multisource integration
(3) Sensor management
(4) Situation/threat assessment
(5) Response management

In the following paragraphs we briefly describe these subdomains.

Sensor fusion. In this kind of fusion, evidence from two or more (often but not always closely located) sensors of similar type—e.g. images from an ISAR sensor and a FLIR sensor—is combined in order to get information which is more precise than that which could be deduced from each piece of evidence alone (see, for example, [135]). This can involve the fusion of raw signals supplied by sensors (sometimes called "pre-detection fusion), or of elementary "attributes," "features" or "segments" extracted from the signals by pre-processors, see [138]. So-called "registration" problems (e.g., the necessity of correctly aligning different images of the same object) are of crucial importance here.

Multisource integration. This type of fusion is concerned with the gross problem of tracking and identifying multiple targets using data supplied by multiple sensors. It is the most well-developed part of data fusion in terms of both theory and practice. Multisource integration is further subdivided into four interrelated parts: detection, correlation, tracking, and classification. *Detection* refers to the process of estimating the number of targets present in a scenario. *Classification* is the process of computing estimates of the identities of targets. If these estimates are computed from data collected by imaging sensors, then classification is more commonly known as "automatic target recognition" (ATR). In *tracking*, one must determine estimates of the positions, velocities, and/or accelerations of targets, as well as predictions of their future positions, velocities, and/or accelerations. *Correlation* is the process of associating observed data (called "reports," "observations," or "measurements") with running estimates of target positions, velocities, accelerations, and identities (also called "tracks"). This latter process is also sometimes called "report-to-track fusion".

Sensor management. If sensors collect data purely on a happenstance basis then information concerning some targets may be overcollected (resulting in little added information) while data on other targets may be undercollected (resulting in poor information about these targets). Many sensors cannot be easily redirected at poorly observed regions, whereas others—e.g., electronically-scanned radars—can rapidly dwell as required on a number of different "observation cells." Sensor management refers to the process of adaptively allocating the dwells of each re-allocatable member of a suite of sensors more intelligently among existing targets of interest. The basis for re-allocation decisions can be purely statistical (e.g., determining what sensor configurations result in smaller target uncertainties) and purely heuristic (e.g., determining which sensors will provide information given current limiting atmospheric, weather, terrain, and/or other conditions). It also includes such processes as sensor cueing, sensor scheduling, and sensor tasking.

Of necessity, the mathematical approaches employed in the following two levels of fusion tend to be more heuristic and less probabilistic than those used for sensor fusion, multisource integration, or sensor management.

Situation/threat assessment. The purpose of situation assessment is to provide an over-all picture of the military significance of the data collected by the previous two levels of fusion. This includes the military status of targets (friendly, hostile, etc.), the threat level of hostile targets (immediate threat, imminent threat, potential threat, etc.), and prediction of the probable intentions and courses of action of threatening targets. Situation assessment depends on one's ability to model the typical behavior of targets of interest. It requires extensive knowledge of the threat characteristics of targets (e.g. on-board weapons and sensors; ranges of these weapons and sensors; etc.), their rules of engagement, tactical doctrines, etc. Ideally, it requires one to infer the amount of information that unfriendly forces have regarding the numbers, identities, and locations of one's own platforms on the basis of the presumed. It also requires knowledge of environment (weather, terrain, visibility, electromagnetic interference, etc.).

An important aspect of situation assessment is *force estimation* [28], which involves the detection, identification, and tracking of *group targets*. Group targets are military formations consisting of many individual targets operating in a coordinated fashion—e.g. brigades and regiments, tank columns, aircraft carrier battlegroups, etc.

Response management. Response management is the process of deciding upon courses of action which are appropriate responses to current and evolving military situations. Such responses can include evasive or other tactical maneuvers, deployment of countermeasures, delivery of weapons, etc. Sometimes, however, the best response is simply to collect more data in order to resolve the identified ambiguities in a situation. This, in turn, may require the tactically optimal—as opposed to statistically optimal—allocation of sensor resources.

2.1.4   Central vs. Distributed Fusion

In addition to this basic taxonomy, there is another distinction which is very important: that between central fusion and distributed fusion.

Central fusion refers to the process by which a suite of sensors supplies their observations directly to a central fusion processing site over data links. This site then processes the observations into a single, coherent estimate of overall "ground truth." However, data fusion resources tend to be widely dispersed among many autonomous, or partially autonomous, fusion sites which may in turn share their estimates with each other over data links. *Distributed fusion* (sometimes also called *track-to-track fusion*) [2], [82], [4] refers to the process of fusing together both observations *and* the target estimates supplied by remote fusion sources—*even though these sites may (and usually do) share data collection resources.* Distributed fusion of information strongly depends on the particular topology of the communications network which interconnects the fusion sites. For example, if different sites share data from some of the same sources then one may run afoul of the "rumor propagation" problem. That is, if fusion site D receives information from sites B and C that a given target is, say, a diesel submarine, then it may conclude that the evidence for that hypothesis is stronger than if it had been asserted by site B alone or by site C alone. If, however, both sites B and C reached their conclusions on the basis of an estimate supplied by a site A situated at an earlier node in the communications network, then the evidence from B and C would not actually be independent and site D's greater confidence would be misplaced. Estimates from fusion sites may, of course, be correlated in even more complex and subtle ways. Various techniques for dealing with such correlation problems have been proposed, e.g. decorrelation of estimates using inverse Kalman filters, or attaching additional

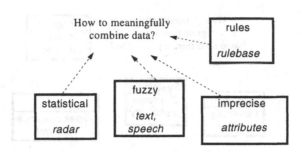

Figure 2.2: A major problem in data fusion is that of fusing an extreme variety of observations, e.g. a radar report with an English language report.

information to transmitted estimates in order to "document" their origins.

2.1.5   Some Major Problems of Data Fusion

Four basic practical problems of information fusion are of major interest in the field. The first is the *problem of incongruent data* (see Figure 2.2). Despite the fact that information can be statistical (e.g. radar reports), imprecise (e.g. target features such as sonar frequency lines), vague/fuzzy (e.g. natural language statements), or contingent (e.g. rules), one must somehow find a way to pool this information in a meaningful way so as to gain more knowledge than would be available from any one information type alone.

Second is the *problem of incongruent legacy systems* (see Figure 2.3). A great deal of investment has been made in algorithms which perform various speciality functions. These algorithms are based on a wide variety of mathematical and/or heuristic paradigms (e.g. Bayesian probability, fuzzy logic, rule-based inference). New approaches—e.g. "weak evidence accrual"—are introduced on a continuing basis in response to the pressure of real-world necessities. Integrating the knowledge produced by these "legacy" systems in a meaningful way is a major requirement of current large-scale fusion systems.

The third difficulty is the *level playing field problem* (see Figure 2.4). Comparing the performance of two data fusion systems, or determining the performance of any individual system relative to some predetermined standard, is far more daunting a prospect than at first might seem to be the case. Real-world test data is expensive to collect and often hard to come by. Algorithms are

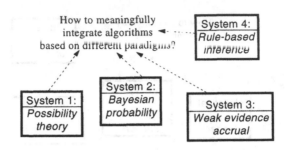

Figure 2.3: Another outstanding problem of data fusion is that of meaningfully integrating existing ("legacy") systems which may be based on very different fusion paradigms, e.g. Bayesian probability vs. fuzzy logic.

typically compared using metrics which measure some very specific aspect of performance (e.g., miss distance, probability of correct identification). Not infrequently, however, optimization of an algorithm with respect to one set of such measures results in degradation in performance with respect to other measures. Also, there is no obvious over-all "figure of merit" which permits the direct comparison of two complex multi-function algorithms. In the case of performance evaluation of individual algorithms exercised with real data, it is often not possible to know how well one is doing against real—and therefore often high-ambiguity—data unless one also has some idea of the best performance that an ideal system could expect. This, in turn, requires the existence of theoretical best-performance bounds analogous to, for example, the Cramér-Rao bound.

The fourth practical problem is *multifunction integration* (see Figure 2.5). With few exceptions, current information fusion systems are patchwork amalgams of various subsystems, each subsystem dedicated to the performance of a specific function. One algorithm may be dedicated to target detection, another to target tracking, still another to target identification, and so on. Better performance should result if these functions were tightly integrated—e.g. so that target I.D. could help resolve tracking ambiguities and vice-versa.

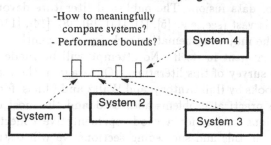

Figure 2.4: It is difficult to compare the performance of different data fusion algorithms, or compare different subsystems of the same algorithm, without the existence of some "level playing field" which allows "apples with apples" comparisons to be made

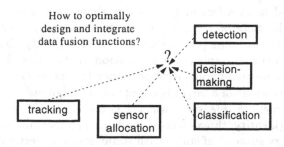

Figure 2.5: Better performance would result if the different subfunctions of data fusion—detection, classification, tracking, sensor management, etc.—could be integrated into a single, unified statistical operation.

2.1.6   How Does One Fuse Data?

Having summarized *what* data fusion is and *why* it is of interest we now turn
to the "how" of data fusion. The published literature devoted to multitarget
tracking alone is vast (see, e.g. [5], [1], [11], [9], [8], [14], [116], [122], [131]). A
survey of just the major mathematical techniques currently in use would require
book-length treatment in itself. No attempt will be made here to provide a
comprehensive survey of this literature. (Once again, the reader is directed to
the excellent books by Hall, Antony, and Waltz and Llinas, for basic information
concerning the practical problems addressed and the most commonly applied
practical solutions. Also, for a good representative collection of papers see
[113]). Rather, in this and succeeding sections we will restrict ourselves to a
description of *general mathematical methodologies*. Currently, there are four
general methodological families, represented by the four possible combinations
of the following two distinctions: *indirect estimation* vs. *direct estimation* and
point estimation vs. *set estimation*.

Indirect estimation refers to the process by which the states of multiple
targets are deduced by first attempting to determine which observations were
generated by which targets. In *direct estimation* the problem of estimating
the correct report-to-track observation is regarded as *unnecessary* (since it is
possible to estimate target states directly) and even *counterproductive*.

Point estimation is the process of determining a single point-estimate of
the state of a target or targets in some state-space, along with an estimate
of the uncertainty involved in this estimate. The maximum likelihood (ML)
and maximum *a posteriori* (MAP) estimators are the most well known point
estimators used in data fusion. In *set estimation*, by way of contrast, one tries
to constrain the state of any given target to membership within some subset
of possible states, to some prespecified degree of certainty. Interval estimation
is the most familiar example of set estimation in statistics. In data fusion, set
estimation is more commonly disguised under the name *expert systems theory*.
Here an "observation" of a target is a piece of "ambiguous" evidence—a natural
language statement, a rule, etc.—which has the effect of constraining the state
of the target (identity, threat status, etc.) to membership in some subset (or
subsets, or fuzzy subsets) of states with some degree of certainty. "Estimation"
in this context refers to the process of using various techniques of *evidential
pooling*—fuzzy logic, Dempster's rule, rule-based inference, etc.—to construct
constraints on states which are more refined than those associated with each
individual piece of evidence taken alone. Because the nature of ambiguous
evidence is often rather murky from a conceptual point of view, these pooling
techniques often mix heuristics with mathematical rigor in varying proportions.
Statistically oriented researchers and practitioners in data fusion have, for this
reason, sometimes been skeptical of many of these techniques.

2.2 Multisensor, Multitarget Estimation

In this section we will briefly describe indirect and direct estimation and some of the major techniques commonly being used. Section 2.2.1 will be devoted to indirect estimation and section 2.2.2.

2.2.1 Indirect Estimation

In this section we will describe several indirect estimation techniques which are of interest either because of their importance in current practical data fusion, or because of their pertinence to one of the major themes of this book: The potential application of random set theory to data fusion. We will describe three such techniques:

(1) Multi-Hypothesis Estimation (MHE) filter
(2) Joint Probabilistic Data Association (JPDA) filter
(3) The Mori, Chong, Tse, and Wishner (MCTW) filter.

We will devote greater attention to the first and third of these techniques since the MHE filter most strongly illustrates the distinction between indirect and direct estimation, and since the MCTW filter is apparently the first attempt to apply the random set perspective to multisensor, multitarget estimation.

Multi-Hypothesis Estimation (MHE) Filter

By far the dominant approach to multitarget estimation is MHE, proposed in 1979 by Donald B. Reid [111] of the Lockheed Palo Alto Research Laboratory and subsequently popularized in Samuel Blackman's seminal book on multitarget tracking [14]. Though MHE was originally proposed as a multitarget *tracking* technique, it is easily integrated with a wide variety of target-identification approaches and forms the basis for a variety of Level 1 fusion algorithms.

The basic motivating idea behind MHE is as follows. For the moment suppose that there are no false alarms and that all targets are perfectly detected and let $\hat{x}_1, ..., \hat{x}_t$ be the estimated parameters of the targets as computed the previous time data was collected. With probability one, t observations $z_1, ..., z_t$ will be collected during a single sensor observation. Especially when targets are close together or when the sensor noise level is high, it is not easily determined which observations were generated by which targets. In fact, there will be $t!$ possible ways in which the data could be associated with targets, each association corresponding to a permutation π on the numbers $1, ..., t$:

$$\hat{x}_1 \leftrightarrow z_{\pi 1}, \qquad ..., \qquad \hat{x}_k \leftrightarrow z_{\pi t}$$

Suppose that it were possible to determine which of these associations is the "true" one, say π_0. Then one could simply apply a Kalman filter t times, using the report $z_{\pi_0 i}$ to update the track $\hat{x}_i$, for each $i = 1, ..., t$.

In MHE one makes the following assumption: *To determine the multitarget state (i.e., numbers of targets and individual target states) it is first necessary to determine which observations were generated by which targets (if any). In this* case it is legitimate to ask, Which association is the most probable? and then proceed as before. *Single-hypothesis estimation* results if, at each time-step, the current estimate of the mulititarget state is based only on the most probable association at the previous time-step. Since the most probable association at any time-step may in actuality be erroneous, however, it would be prudent to keep a list of the *higher-probability* associations $\pi_0, ..., \pi_l$ and, for each association π from this list, update the tracks $\hat{\mathbf{x}}_1, ..., \hat{\mathbf{x}}_t$ using the track-to-report association $\hat{\mathbf{x}}_1 \leftrightarrow \mathbf{z}_{\pi 1}, ..., \hat{\mathbf{x}}_t \leftrightarrow \mathbf{z}_{\pi t}$. Since there is always a small likelihood that the "true" association is one of the $k! - l$ that have been ignored, it is desirable to make l as large as possible within the constraints imposed by computer processing speed and memory. Instead of one hypothesis about "ground truth" consisting of a set of t tracks, in this case, one instead has a list of l *hypotheses* about ground truth, each hypothesis consisting of t tracks. Applying the same procedure at the next data collection period, one gets l^2 hypotheses which must, in some manner, be reduced to a list of l hypotheses.

Note that, regardless of the hypothesis, reports are partitioned on a one-to-one basis with the tracks in the hypothesis—there are no "partial" assignments of the same report to two or more tracks.

At the extreme is so-called "ideal MHE," in which one maintains *all* possible associations at each step—a process which results in massive combinatorial explosion in the number of hypotheses as time progresses. In practice computational loads are reduced via various stratagems such as *gating* (ignoring reports that lie outside of some region surrounding any given track); *pruning* (eliminating the lowest-probability hypotheses); *merging* (consolidating similar tracks into one track and similar hypotheses into one hypothesis); *clustering* (breaking up tracks into smaller groups which are statistically unrelated to each other); and *fast-association techniques* (see [14], [109], [17], [131]).

In actual practice some observations will be false alarms, many observations will be missing because some targets did not generate returns during the report-collection period, and new targets may enter the scene. Thus the above scheme must be extended to include more general track-to-report associations. At each time-step α a set of "daughter" hypotheses is generated from each "parent" hypothesis which existed at time-step $\alpha - 1$. Given the parent hypothesis o, one knows the hypothesis probability $p(o|Z^{\alpha-1})$ and the (generally Gaussian) posterior density $f(\mathbf{x}_1, ..., \mathbf{x}_{t(o)}|Z^{\alpha-1}, o)$ of a set $\mathbf{x}_1, ..., \mathbf{x}_{t(o)}$ of tracks, given previous measurements $Z^{\alpha-1} = \{\mathbf{z}_0, ..., \mathbf{z}_{\alpha-1}\}$ and the parent hypothesis o. Target state-estimates $\hat{\mathbf{x}}_1, ..., \hat{\mathbf{x}}_{t(o)}$ are computed from $f(\mathbf{x}_1, ..., \mathbf{x}_{t(o)}|Z^{\alpha-1}, o)$. Daughter hypotheses are generated from o based on the following three possible interpretations of a new report $\mathbf{z}_\alpha$:

(1) $\mathbf{z}_\alpha$ is a false alarm;

(2) $\mathbf{z}_\alpha$ is associated with an established track $\hat{\mathbf{x}}_i$ in the parent hypothesis o;

(3) $\mathbf{z}_\alpha$ cannot be associated with any established track in o and thus must have originated with a previously unnoticed target.

Bayes' rule then leads to a set of daughter hypotheses o' with hypothesis probabilities $p(o'|Z^\alpha)$ and posteriors $f(\mathbf{x}_1, ..., \mathbf{x}_{t(o')}|Z^\alpha, o')$. (In practice it is not efficient to update hypotheses on a report-by-report basis as just described.

Rather, entire "scans" of observations are processed at the same time with various algorithms—e.g. the Munkres or the Jonker-Volgenant-Castanon (JVC) algorithms—being used to determine the "optimal" report-to-track association as well as a range of "suboptimal" associations.)

In principle one could compute an estimate of the number and states of targets by computing the *total posteriors*

$$f_{MHE}(\mathbf{x}_1, ..., \mathbf{x}_k, k | Z^\alpha) = \sum_{o':t(o')=k} f(\mathbf{x}_1, ..., \mathbf{x}_k | Z^\alpha, o') \, p(o' | Z^\alpha)$$

for all $k \geq 0$ and then deriving a single estimate of ground truth from them [124], [7]. In practice this is never done in MHE algorithms. Rather, the state-estimate $\hat{\mathbf{x}}_1, ..., \hat{\mathbf{x}}_{t(\hat{o})}$ corresponding to that daughter hypothesis $o' = \hat{o}$ which has maximal $p(o' | Z^\alpha)$ is usually what is presented to the end-user as the "MHE estimate" of ground truth.

(Notice that when the number of targets is not known *a priori*, the determination of an estimate from the total posteriors is not straightforward: $f_{MHE}(\mathbf{x}_1, ..., \mathbf{x}_k, k | Z^\alpha)$ has different units for different k and so one cannot compute a MAP estimate or conditional expectation. Instead one could determine the maximizing value $k = \hat{k}$ of the marginal distribution $f_{MHE}(k | Z^\alpha) = \sum_{o':t(o')=k} p(o' | Z^\alpha)$ and then determine the maximizing value—or conditional expectation—$\hat{\mathbf{x}}_1, ..., \hat{\mathbf{x}}_{\hat{k}}$ of $f_{MHE}(\mathbf{x}_1, ..., \mathbf{x}_{\hat{k}}, \hat{k} | Z^\alpha)$. The statistical consistency of the resulting multitarget estimator would then have to be proved.)

The dominance of the MHE technique in data fusion is due to a number of factors. First, MHE is conceptually consistent with the way military operators intuitively propagate tracks by associating them with reports. Second, it allows highly complex multitarget scenarios to be reduced to a large number of single-target problems, each of which can be addressed using standard Kalman filtering techniques.

Third, MHE is easily extended to incorporate target-identification capability. One common scheme is to attach target-I.D. propositions to tracks and then use some rule of evidential combination—most commonly, Dempster's rule—to update the propositional part of the track at the same time that a Kalman filter is used to update the geokinematic part of the track.

Fourth, and perhaps most important of all, is the widespread belief that "ideal" MHE is *the theoretically optimal solution* to multitarget tracking problems. However and as previously noted, MHE rests on the assumption that estimating the true report-to-track association adds additional information. If this assumption is erroneous then in determining the optimal association one is improperly adding information and thus introducing an *inherent bias* into the estimation process.

The following counterexample is adapted from [6] and illustrates the difficulties that can result from the use of indirect-estimation approaches. Assume the special case characterized by the following assumptions:

(1) A single sensor, with single-target measurement-model density $f(\mathbf{z} | \mathbf{x})$, observes a known number t of targets with state vectors $\mathbf{x}_1, ..., \mathbf{x}_t$ at a fixed

moment of time;

(2) There are no false alarms or missed detections;

(3) Observations are independent;

(4) We are to estimate the system-level target state vector $\mathbf{x} = (\mathbf{x}_1, ..., \mathbf{x}_t)$ on the basis of M scans $Z_\alpha = \{\mathbf{z}_{\alpha;1}, ..., \mathbf{z}_{\alpha;t}\}$ of data for $\alpha = 1, ..., M$;

(5) The joint prior distributions $f(\mathbf{z}_1, ..., \mathbf{z}_t)$, $f(\mathbf{x}_1, ..., \mathbf{x}_t, \pi)$ on data and parameters are uniform in the region of interest.

Given these assumptions, the probability density of observing Z_α given the target states $\mathbf{x}_1, ..., \mathbf{x}_t$ and the association π_α on the α'th scan is

$$f(\mathbf{z}_{\alpha;1}, ..., \mathbf{z}_{\alpha;t} | \mathbf{x}_1, ..., \mathbf{x}_t, \pi_\alpha) = f(\mathbf{z}_{\alpha;\pi_\alpha 1} | \mathbf{x}_1) \cdots f(\mathbf{z}_{\alpha;\pi_\alpha t} | \mathbf{x}_t)$$

Thus the density of observing the sequence $Z_1, ..., Z_M$ given the sequence of associations $\pi = \{\pi_1, ..., \pi_M\}$ is

$$f(\mathbf{z}_{1;1}, ..., \mathbf{z}_{1;t}, ..., \mathbf{z}_{M;1}, ..., \mathbf{z}_{M;t} | \mathbf{x}_1, ..., \mathbf{x}_t, \pi) = \prod_{\alpha=1}^{M} \prod_{i=1}^{t} f(\mathbf{z}_{\alpha;\pi_\alpha i} | \mathbf{x}_i)$$

We can compute the posterior distribution from Bayes' rule:

$$\begin{aligned} &f(\mathbf{x}_1, ..., \mathbf{x}_t, \pi | Z_1, ..., Z_M) \\ &= \frac{f(\mathbf{z}_{1;1}, ..., \mathbf{z}_{1;t}, ..., \mathbf{z}_{M;1}, ..., \mathbf{z}_{M;t} | \mathbf{x}_1, ..., \mathbf{x}_t, \pi) \; f(\mathbf{x}_1, ..., \mathbf{x}_t, \pi)}{f(Z_1, ..., Z_M)} \end{aligned}$$

where by assumption the distributions $f(\mathbf{x}_1, ..., \mathbf{x}_t, \pi)$ and $f(Z_1, ..., Z_M)$ are uniform in the region of interest. Thus

$$f(\mathbf{x}_1, ..., \mathbf{x}_t, \pi | Z_1, ..., Z_M) \propto \prod_{\alpha=1}^{M} \prod_{i=1}^{t} f(\mathbf{z}_{\alpha;\pi_\alpha i} | \mathbf{x}_i)$$

If we compute the maximum *a posteriori* (MAP) estimate of $\mathbf{x}_1, ..., \mathbf{x}_k, \pi$ then, for each sequence π of associations, we can first compute

$$(\hat{\mathbf{x}}_1(\pi), ..., \hat{\mathbf{x}}_t(\pi)) = \arg\max_{\mathbf{x}_1, ..., \mathbf{x}_t} \prod_{\alpha=1}^{M} \prod_{i=1}^{t} f(\mathbf{z}_{\alpha;\pi_\alpha i} | \mathbf{x}_i)$$

(Note that if f is Gaussian the MAP estimate of $(\mathbf{x}_1, ..., \mathbf{x}_t)$ coincides with the conditional expectation.) Defining

$$p(\pi | Z_1, ..., Z_M) \propto \prod_{\alpha=1}^{M} \prod_{i=1}^{t} f(\mathbf{z}_{\alpha;\pi_\alpha i} | \hat{\mathbf{x}}_i(\pi))$$

and $\hat{\pi} = \arg\max_\pi p(\pi | Z_1, ..., Z_M)$ it follows that $\hat{\mathbf{x}}_1(\hat{\pi}), ..., \hat{\mathbf{x}}_t(\hat{\pi}), \hat{\pi}$ is the MAP estimate of the system-level target state, given the observation sets $Z_1, ..., Z_M$.

Now consider the specific case when there are two targets in one dimension located at $x = \pm 1$ and $f(z|x) = N_{\sigma^2}(z - x)$ is the normal distribution with variance σ^2 centered at x. Then maximizing the posterior

$$f(x_1, x_2, \pi | Z_1, ..., Z_M) \propto \exp\left(-\frac{1}{2\sigma^2}\sum_{\alpha=1}^{M}\sum_{i=1}^{2}(z_{\alpha;\pi_\alpha i} - x_i)^2\right)$$

is the same as minimizing the square-error

$$\varepsilon^2(x_1, x_2, \pi) = \sum_{\alpha=1}^{M}(z_{\alpha;\pi_\alpha 1} - x_1)^2 + \sum_{\alpha=1}^{M}(z_{\alpha;\pi_\alpha 2} - x_2)^2$$

which, for fixed π, yields:

$$\hat{x}_1(\pi) = \frac{1}{M}\sum_{\alpha=1}^{M} z_{\alpha;\pi_\alpha 1}, \qquad \hat{x}_2(\pi) = \frac{1}{M}\sum_{\alpha=1}^{M} z_{\alpha;\pi_\alpha 2}$$

Substituting $x_1 = \hat{x}_1(\pi)$, $x_2 = \hat{x}_2(\pi)$ back into $\varepsilon^2(x_1, x_2, \pi)$ and expanding yields

$$\varepsilon^2(\hat{x}_1(\pi), \hat{x}_2(\pi), \pi)$$

$$= \sum_{\alpha=1}^{M}(z_{\alpha;\pi_\alpha 1}^2 + z_{\alpha;\pi_\alpha 2}^2) - \frac{M-1}{M^2}\left[\left(\sum_{\alpha=1}^{M} z_{\alpha;\pi_\alpha 1}\right)^2 + \left(\sum_{\alpha=1}^{M} z_{\alpha;\pi_\alpha 2}\right)^2\right]$$

$$= \sum_{\alpha=1}^{M}(z_{\alpha;1}^2 + z_{\alpha;2}^2) - \frac{M-1}{M^2}\left[\left(\sum_{\alpha=1}^{M} z_{\alpha;\pi_\alpha 1}\right)^2 + \left(\sum_{\alpha=1}^{M} z_{\alpha;\pi_\alpha 2}\right)^2\right]$$

This will be minimized if $\pi_1, ..., \pi_M$ are chosen to maximize

$$\left(\sum_{\alpha=1}^{M} z_{\alpha;\pi_\alpha 1}\right)^2 + \left(\sum_{\alpha=1}^{M} z_{\alpha;\pi_\alpha 2}\right)^2$$

However, let

$$c_\alpha \triangleq \min\{z_{\alpha;1}, z_{\alpha;2}\}, \; C_\alpha \triangleq \max\{z_{\alpha;1}, z_{\alpha;2}\}$$

$$\Delta_\alpha \triangleq z_{\alpha;1} - \min\{z_{\alpha;1}, z_{\alpha;2}\} = \max\{z_{\alpha;1}, z_{\alpha;2}\} - z_{\alpha;2} \geq 0$$

Then

$$\left(\sum_{\alpha=1}^{M} z_{\alpha;\pi_\alpha 1}\right)^2 + \left(\sum_{\alpha=1}^{M} z_{\alpha;\pi_\alpha 2}\right)^2$$

$$= \left(\sum_{\alpha=1}^{M} c_\alpha + \sum_{\alpha=1}^{M} \Delta_\alpha\right)^2 + \left(\sum_{\alpha=1}^{M} C_\alpha - \sum_{\alpha=1}^{M} \Delta_\alpha\right)^2$$

$$= \left(\sum_{\alpha=1}^{M} c_\alpha\right)^2 + \left(\sum_{\alpha=1}^{M} C_\alpha\right)^2 - 2\left(\sum_{\alpha=1}^{M} \Delta_\alpha\right)\left(\sum_{\alpha=1}^{M}(C_\alpha - c_\alpha - \Delta_\alpha)\right)$$

Since $\Delta_\alpha \geq 0$ and $C_\alpha - c_\alpha - \Delta_\alpha = z_{\alpha;2} - c_\alpha \geq 0$ for all α it follows that the maximum value is achieved when $\sum_{\alpha=1}^{M} \Delta_\alpha$ is minimized and thus when $\Delta_\alpha = 0$ for all $\alpha = 1, ..., M$. Accordingly $\varepsilon^2(\hat{x}_1(\pi), \hat{x}_2(\pi), \pi)$ will be minimized when, for each α, the permutation π_α is chosen so that either of the following two cases hold:

$$z_{\alpha;\pi_\alpha 1} = \min\{z_{\alpha;1}, z_{\alpha;2}\} \quad \text{and} \quad z_{\alpha;\pi_\alpha 2} = \max\{z_{\alpha;1}, z_{\alpha;2}\}$$
$$z_{\alpha;\pi_\alpha 1} = \max\{z_{\alpha;1}, z_{\alpha;2}\} \quad \text{and} \quad z_{\alpha;\pi_\alpha 2} = \min\{z_{\alpha;1}, z_{\alpha;2}\}$$

Thus the MAP (as well as the conditional expectation) estimate is

$$\hat{x}_1(\hat{\pi}) = \frac{1}{M} \sum_{\alpha=1}^{M} \min\{z_{\alpha;1}, z_{\alpha;2}\}, \qquad \hat{x}_2(\hat{\pi}) = \frac{1}{M} \sum_{\alpha=1}^{M} \max\{z_{\alpha;1}, z_{\alpha;2}\}$$

That is, the MAP estimate always assigns the largest report to one track and the smallest report to the other track. When the sensor variance $\sigma^2 << 1$ and thus the targets are well separated, one will get $\hat{x}_1(\hat{\pi}) \cong -1$ and $\hat{x}_2(\hat{\pi}) \cong +1$ for large M—i.e. the correct answer. If σ^2 is large, however, then reports tend to be assigned to the nearest target. This results in a biased estimate regardless of how large M is—i.e., how many scans of data are processed. See [6] for more details.

In work funded by the Office of Naval Research, Lawrence Stone and his associates have produced a counterexample [124, Example 2, pp. 35-36] which shows that MHE remains non-optimal even if one constructs an estimate from the total posterior $f_{MHE}(\mathbf{x}_1, ..., \mathbf{x}_k, k|Z^\alpha)$. They have also offered sufficient (but not necessary) conditions for the optimality of MHE [124, Theorem 3, p. 20], [7]. In summary, let $f(\mathbf{z}, \mathbf{z} \leftrightarrow \mathbf{x}_i | \mathbf{x}_1, ..., \mathbf{x}_t)$ denote the probability density that measurement $\mathbf{z}$ is observed and that it was generated by the i'th target, given target states $\mathbf{x}_1, ..., \mathbf{x}_t$. Assume that

(1) Each sensor response is generated by a single target and is independent of all other sensor responses;

(2) The number t of targets is known *a priori*.

(3) The likelihood functions for each sensor response are conditionally independent in the sense that

$$f(\mathbf{z}, \mathbf{z} \leftrightarrow \mathbf{x}_i | \mathbf{x}_1, ..., \mathbf{x}_t) = \prod_j g_j^i(\mathbf{z}, \mathbf{x}_j)$$

for some functions g_j^i for all $i = 1, ..., t$; and

(4) The Markov motion models for the targets are independent

Then MHE is Bayes-optimal in the sense that the MHE total posteriors $f_{MHE}(\mathbf{x}_1, ..., \mathbf{x}_k, k|Z^\alpha)$ are identical to the actual Bayesian posteriors $f(\mathbf{x}_1, ..., \mathbf{x}_k, k|Z^\alpha)$.

Joint Probabilistic Data Association (JPDA) Filter

The JPDA filter was developed by Yaakov Bar-Shalom and Thomas Fortmann at about the same time as the MHE approach (see [35], [1], [11]) as a "suboptimal" approach specifically designed for multitarget tracking in very high-clutter environments. The number of targets is assumed known on an *a priori* basis, as is a conventional Kalman measurement and state-transition model for each of the targets:

$$\mathbf{x}_{\alpha+1} = F_\alpha \mathbf{x}_\alpha + G_\alpha \mathbf{w}_\alpha, \qquad \mathbf{z}_\alpha = H_\alpha \mathbf{x}_\alpha + \mathbf{v}_\alpha$$

In what follows, for notational clarity we suppress the time-step index α. Suppose that $\mathbf{z}_1, ..., \mathbf{z}_m$ denotes a list of gated observations associated with a cluster of t targets. Let $\hat{\mathbf{x}}^i$ denote the estimate of the i'th target at time-step $\alpha - 1$. Then from the pairwise innovations $\mathbf{y}_j^i \triangleq \mathbf{z}_j - HF\hat{\mathbf{x}}^i$ construct the composite innovations $\mathbf{y}^i \triangleq \sum_{j=1}^m \beta_j^i \mathbf{y}_j^i$ for each target $i = 1, ..., t$, where β_j^i is the posterior probability that observation $\mathbf{z}_j$ was generated by the i'th target [12, pp. 132, 92]. These innovations are information-updated using one Kalman filter per target.

Strictly speaking, JPDA is not an "indirect" estimation approach in the sense originally defined and is, in fact, a sort of hybrid of the indirect-estimation and direction-estimation perspectives. On the one hand, JPDA was devised as a means of *approximating* the true mulititarget posterior distributions. In contrast to MHE, no attempt is made in JPDA to estimate the "correct" report-to-track association. The result is that—unlike the case with MHE—the actual reports are nonuniquely apportioned among the tracks. On the other hand, JDPA shares much of MHE's indirect-estimation spirit. That is, the underlying idea is to determine an association $\mathbf{y}^i \leftrightarrow \mathbf{x}_i$ ($i = 1, ..., t$) between the tracks $\mathbf{x}_i$ and a set of composite innovations $\mathbf{y}^i$, which is then used as though it were an *approximation of the optimal association* between the tracks and the actual innovations. As with MHE, this approximate association is used to reduce the multitarget filtering problem to a collection of single-target Kalman filter problems.

The computationally demanding part of the JPDA approach—determination of the posteriors β_j^i—is equivalent to computing the permanents of certain matrices and can be accomplished exactly with computational complexity $O(2^t)$ [105]. A variation of JPDA known as "integrated probabilistic data association" [100] allows the number of targets to be either one or zero, thus integrating detection with single-target tracking in a single algorithm.

Mori-Chong-Tse-Wishner (MCTW) Filter

Shozo Mori, Chee-Yee Chong, Edison Tse, and Richard Wishner seem to have been the first researchers to advocate a random set perspective for multisensor, multitarget estimation and filtering, in an unpublished 1984 Advanced Information and Decision Systems, Inc. technical report [22]. Under Defense Advanced

Research Projects Administration (DARPA) funding they used their formalism to develop a multihypothesis-type, fully integrated multisensor, multitarget detection, association, classification, and tracking algorithm [23]. They also demonstrated that the Reid and JPDA algorithms, as well as algorithms due to Morefield and Goodman, are special cases of their approach (hereafter called the "MCTW filter"). Given that the random-set aspect of this work is largely unpublished and therefore not easily accessible, this section is devoted to a fairly lengthy discussion of the MCTW filter.

Mori et al. began by observing that the multitarget tracking problem is "non-classical" in the sense that:

> (1) The number of the objects to be estimated is, in general, random and unknown. The number of measurements in each sensor output is random and a part of observation information. (2) Generally, there is no a priori labeling of targets and the order of measurements in any sensor output does not contain any useful information. For example, a measurement couple $(\mathbf{y}_1, \mathbf{y}_2)$ from a sensor is totally equivalent to $(\mathbf{y}_2, \mathbf{y}_1)$. When a target is detected for the first time and we know it is one of n targets which have never been seen before, the probability that the measurement originat[ed] from a particular target is the same for any such target, i.e., it is $1/n$. The above properties (1) and (2) are properly reflected when both targets and sensor measurements are considered as random sets as defined in [Mathéron]....The uncertainty of the origin of each measurement in every sensor output should then be embedded in a sensor model as a stochastic mechanism which converts a random set (set of targets) into another random set (sensor outputs)...

Rather than making use of Mathéron's detailed random-set formalism, however, they argued that:

> ...a random finite set X of reals can be probabilistically completely described by specifying probability $Prob.\{|X| = n\}$ for each nonnegative n and joint probability distribution with density $p_n(\mathbf{x}_1, ..., \mathbf{x}_n)$ of elements of the set for each positive n...[which is completely symmetric in the variables $\mathbf{x}_1, ..., \mathbf{x}_n$] [22, pp. 1-2].

They then went on to construct a representation of random finite-set variates in terms of conventional stochastic quantities (i.e., discrete variates and continuous vector variates).

In more detail: Mori et al. represent target states as elements of a "hybrid" continuous-discrete space of the form $\mathcal{R} = \mathbb{R}^n \times U$ endowed with the "hybrid" (i.e., product Lebesgue-counting) measure. (Note: We have simplified the definition of $\mathcal{R}$ somewhat to keep things clearer.) Finite subsets of state space are modeled as vectors in disjoint-union spaces of the form

$$\mathcal{R}^{(n)} = \bigcup_{k=0}^{\infty} \mathcal{R}^k \times \{k\}$$

Thus a typical element of $\mathcal{R}^{(n)}$ is $(\mathbf{x}, k)$ where $\mathbf{x} = (x_1, ..., x_k)$; and a random finite subset of $\mathcal{R}$ is represented as a random variable $(\mathbf{X}, N_T)$ on $\mathcal{R}^{(n)}$. More generally, a finite stochastic subset of $\mathcal{R}$ is represented as a discrete-time stochastic process $(\mathbf{X}(\alpha), N_T)$ where N_T is a random variable which is constant with respect to time. The basic model is completed by specifying the measurement space. Various assumptions are made with permit the reduction of simultaneous multisensor, multitarget observations to a discrete time-series of multitarget observations. Given a sensor suite with sensor tags $U_{\odot} = \{1, ..., s\}$, the observation space has the general form

$$\mathcal{R}_{\odot}^{(m)} = \mathcal{R}^{(m)} \times \mathbb{N} \times U_{\odot}$$

where $\mathbb{N} = \{1, 2, ...\}$ is the set of possible discrete time tags beginning with the time tag 1 of the initial time-instant. Any given observation thus has the form $(\mathbf{y}, m, \alpha, s)$ where $\mathbf{y} = (y_1, ..., y_m)$, where α is a time tag, and where $s \in U_{\odot}$ is a sensor tag. The time-series of observations is a discrete-time stochastic process of the form $(\mathbf{Y}(\alpha), N_M(\alpha), \alpha, s_{\alpha})$. Here, for each time tag α the random integer $s_{\alpha} \in U_{\odot}$ is the identifying sensor tag for the sensor active at that moment. The random integer $N_M(\alpha)$ is the number of measurements collected by that sensor. The vector $\mathbf{Y}(\alpha)$ represents the (random) subset of actual measurements.

In the MCTW formalism state estimation (i.e., number, identities, and geodynamics of targets) is contingent on multihypothesis track estimation which, in turn, requires an intricate Bayesian bookkeeping system. A "track" is a time-history of reports all of which are believed to have been generated by the same target, with missed detections and false alarms taken into account. Any collection of such tracks is a "data-to-data hypothesis" H concerning the historical generation of reports by all targets, including a history of missed detections and false alarms. Missed and spurious observations are modeled by a random "assignment function" (a one-to-one correspondence)

$$A_{\alpha} : I_{DT}(\alpha) \to J_{DT}(\alpha)$$

between the elements of $I_{DT}(\alpha)$ and the elements of $J_{DT}(\alpha)$. Here, the domain $I_{DT}(\alpha)$ of the transformation A_{α} is a randomly varying subset of the randomly varying set $I_T = \{1, ..., N_T\}$ of target indices; and the image-set $J_{DT}(\alpha)$ of A_{α} is a randomly varying subset of the randomly varying finite set $J_T(\alpha) = \{1, ..., N_M(\alpha)\}$ of measurement indices. The integer $A_{\alpha}(i) = j$ is interpreted as the index of the observation $\xi_j \in \mathcal{R}^{(m)}$ which is hypothetically generated by the i'th target (assuming that no missed detection has occurred). The elements of $I_{DT}(\alpha)$ are interpreted as the set of hypothesized detections, and the difference set $I_T - I_{DT}(\alpha)$ is interpreted as the set of hypothesized missed detections. The set $J_{FA}(\alpha) = J_M(\alpha) - J_{DT}(\alpha)$, which consists of measurements not associated with parameters, is interpreted as the hypothesized set of false alarms. For fixed $I_{DT}(\alpha)$ and $J_{DT}(\alpha)$ the random function A_{α} is assumed to range uniformly over all the possible hypothesized associations between the parameters in $I_{DT}(\alpha)$ and the measurements in $J_{DT}(\alpha)$.

Hypothesized tracks are modeled as follows. Let us be given a time tag $\alpha' \geq 1$, a target index $i \in I_T$, and a random association $A_{\alpha'} : I_{DT}(\alpha') \to J_{DT}(\alpha')$. Define the random finite set $T_{\alpha'}(i)$ as

$$T_{\alpha'}(i) = \{(A_1(i), 1), ..., (A_{\alpha'}, \alpha')\}$$

That is, for any time-sequence of associations $A_1, ..., A_{\alpha'}$ the set $T_{\alpha'}(i)$ collects together the time-sequence of all indices $A_1(i), ..., A_{\alpha'}(i) \subset J_M(\alpha)$ of measurements which have hypothetically been generated by the target which has index $i \in I_T$. In other words, $T_{\alpha'}(i)$ is a mathematical model of a track over the time range $\alpha = 1, ..., \alpha'$. Accordingly, the random set of random subsets

$$H_{\alpha'} = \{T_{\alpha'}(1), ..., T_{\alpha'}(N_T)\}$$

models a randomly varying data-to-data hypothesis. Every such hypothesis has a unique "parent" hypothesis defined by:

$$T_{\alpha'-1}(i) = \{(A_1(i), 1), ..., (A_{\alpha'-1}(i), \alpha' - 1)\}$$
$$H_{\alpha'-1} = \{T_{\alpha'-1}(1), ..., T_{\alpha'-1}(N_T)\}$$

Finally, this vector formalism is rendered consistent with finite-set behavior by requiring that the following five probability distributions be invariant under all possible permutations of elements of $\mathcal{R}^k$ for all k:

$$p(\mathbf{X}(t_0) \in d\mathbf{X}| N_T = k\} \quad \text{(initial distribution)}$$
$$p(\mathbf{X}(\alpha + 1) \in d\mathbf{X}| \mathbf{X}(\alpha) = \mathbf{X}, N_T = k) \quad \text{(state transition)}$$
$$p(I_{DT}(\alpha) = D| \mathbf{X}(\alpha) = \mathbf{x}, N_T = k) \quad \text{(detections)}$$
$$p(\mathbf{y} \in d\mathbf{Y}| A_\alpha = a, N_M(\alpha) = m, I_{DT} = D, \mathbf{X}(\alpha) = \mathbf{x}, N_T = k)$$
$$\text{(measurements)}$$
$$p(N_{FA}(\alpha) = m| I_{DT}(\alpha) = D, \mathbf{X}(\alpha) = \mathbf{x}, N_T = k) \quad \text{(false alarms)}$$

(Here, $N_{DT}(\alpha) = |I_{DT}(\alpha)|$ is the number of detections and $N_{FA}(\alpha) = |J_{FA}(\alpha)|$ is the number of false alarms. Hereafter we will simplify notation by abusing it, i.e. by suppressing random variables. For example we will write $p(D|\mathbf{x}, k, \alpha)$ instead of $p(I_{DT}(\alpha) = D|\mathbf{X}(\alpha) = \mathbf{x}, N_T = k)$, etc.)

The MCTW filter is then constructed as follows. Given a time-sequence $z_1, ..., z_\alpha$ of measurements we may write a Bayesian updating equation

$$p(h_{\alpha'}|z_1, ..., z_{\alpha'}) = \frac{p(z_1, ..., z_{\alpha'}, h_{\alpha'}| z_1, ..., z_{\alpha'-1}, h_{\alpha'-1})}{p(z_1, ..., z_{\alpha'}|z_1, ..., z_{\alpha'-1})} p(h_{\alpha'-1}|z_1, ..., z_{\alpha'-1})$$

where $h_{\alpha'}$ any specific hypothesis and where $h_{\alpha'-1}$ is the unique parent hypothesis of $h_{\alpha'}$. Similar recursive updating formulas hold for the number-of-targets (i.e., detection) distribution $p(n|h_{\alpha'-1}, z_1, ..., z_{\alpha'})$ and the state distribution $p(\mathbf{x}|n, h_{\alpha'-1}, z_1, ..., z_{\alpha'})$.

The main result of [22] is a collection of formulas which permit the computationally explicit implementation of these recursion formulas. Given the recursively computed estimate from the previous time step, therefore, one may first compute the estimate $H_{\alpha'} = \hat{h}_{\alpha'}$ of the most probable data-to-data association at the current time-step α'. This, in turn, is used to compute the estimate $N_T = \hat{k}$ of the most probable number of targets at the current time-step. Finally, $\hat{h}_{\alpha'}$ and $\hat{k}$ are used to compute the estimate of the most probable identities and geodynamics of all $\hat{k}$ estimated targets at the current time-step.

By making additional independence, Markov, and Poisson assumptions Mori et. al. simplify their equations as follows:

1) The number-of-targets distribution $p(N_T = n)$ is Poisson

2) The number-of-false-alarms distribution $p(m|D, \mathbf{x}, k, \alpha)$ is independent of $D, \mathbf{x}$ and k and thus has the form $p_{NFA}(m|\alpha)$

3) The detection distribution has the form

$$p_D(D|\mathbf{x}, k, \alpha) = \left(\prod_{i \in D} p_D(\mathbf{x}_i|\alpha) \right) \left(\prod_{j \in \{1,...,m\} - D} (1 - p_D(\mathbf{x}_j|\alpha)) \right)$$

4) The measurement distribution has the form

$$p_M(\mathbf{y}|a, m, D, \mathbf{x}, k, \alpha) = \left(\prod_{i \in D} p_M(\mathbf{z}_{a(i)}|\mathbf{x}_i, \alpha) \right) \left(\prod_{j \in \{1,...,m\} - \mathrm{Im}(a)} p_{FA}(\mathbf{z}_j|\alpha) \right)$$

As we shall see, these models are similar to those derived from a "pure" random set analysis.

2.2.2   Direct Estimation

As already noted, *direct estimation* techniques treat the problem of estimating the true report-to-track association as unnecessary and instead attempt to directly estimate the multitarget state $(\mathbf{x}_1, ..., \mathbf{x}_t)$ as a whole. A handful of direct-estimation techniques have been proposed but, for the most part, are considerably less familiar than the MHE and JPDA techniques. In what follows we will briefly summarize the following direct-estimation filters:

(1) Point-Process Filter
(2) Symmetric Measurement Equation (SME) Filter
(3) Jump Diffusion Filter
(4) Event-Averaged Maximum Likelihood (EAMLE) Filter
(5) Hidden Markov Model (HMM) Filters

Many of these approaches share some strong affinities with the random set direct-estimation techniques proposed in this book. With the exception of the jump-diffusion filter, all assume that the number of targets is known *a priori*.

Point-Process Filter

In 1987 and under Office of Naval Research funding, Robert Washburn of Alphatech, Inc. proposed a theoretical approach (see [134] and also [33]) to multitarget tracking which bears some similarities to the random set approach since it is based on *point process theory* [5]. Washburn notes that the basis of the point process approach

> ...is to consider the observations occurring in one time period or scan of data as an image of points rather than as a list of measurements. The approach is intuitively appealing because it corresponds to one's natural idea of radar and sonar displays—devices that provide two-dimensional images of each scan of data.[134, p. 1846]

Washburn makes use of the fact that randomly varying finite sets of data can be mathematically represented as *random integer-valued ("counting") measures.* (For example, if Σ is a random finite subset of measurement space then $N_\Sigma(S) = |\Sigma \cap S|$ defines a random counting measure N_Σ.) In this formalism, therefore, the "measurement" collected at time-instant α by a single sensor in a multitarget environment is an integer-valued measure μ_α.

Washburn assumes that the number n of targets is known *a priori* and specifies a measurement model of the form

$$\mu_\alpha = \tau_\alpha + \nu_\alpha$$

The random counting measure ν_α, which models clutter, is assumed to be a Poisson measure. (Poisson measures have the following properties: For any two disjoint measurable subsets B_1, B_2 of observation space the random integers $\nu_\alpha(B_1), \nu_\alpha(B_2)$ are statistically independent, and $\nu_\alpha(B_1)$ is Poisson-distributed with mean $\iota(B_1)$, where ι is some nonnegative, real-valued measure on observation space called the "intensity measure.") The random counting measure τ_α models detected targets, and is assumed to have the form

$$\tau_\alpha = \sum_{i=1}^{n} l_{i;\alpha}\, \delta_{\mathbf{y}_{i;\alpha}}$$

Here, (1) for any measurement $\mathbf{y}$ the Dirac measure $\delta_\mathbf{y}$ is defined by $\delta_\mathbf{y}(S) = 1$ if $\mathbf{y} \in S$ and $\delta_\mathbf{y}(S) = 0$ otherwise; (2) $\mathbf{y}_{i;\alpha}$ is the (random) observation at time-step α generated by the i'th target $\mathbf{x}_i$ if detected, and whose statistics are governed by a probability measure $h_\alpha(B|\mathbf{x}_i)$; (3) $l_{i;\alpha}$ is a random integer which takes only the values 1 or 0 depending on whether the i'th target is detected at time-step α or not, and whose statistics are governed by the probability distribution $q_\alpha(\mathbf{x}_i)$. Various independence relationships are assumed to exist between $\nu_\alpha, l_{i;\alpha}, \mathbf{y}_{i;\alpha}, \mathbf{x}_i$.

Washburn develops a Bayesian nonlinear filtering procedure, which in turn necessitates the reformulation of the measurement model $\mu_\alpha = \tau_\alpha + \nu_\alpha$ as a measurement-model likelihood $f(\mu_\alpha|\mathbf{x})$. In the single-target case this turns out to be

$$f(\mu_\alpha|\mathbf{x}) = 1 - q_\alpha(\mathbf{x}) + q_\alpha(\mathbf{x}) \int k_\alpha(\mathbf{y}|\mathbf{x})\, d\mu_\alpha(\mathbf{y})$$

where k_α is the Radon-Nikodým derivative of $h_\alpha(-|\mathbf{x}_i)$ with respect to the intensity measure ι. In the multitarget case with perfect detections it turns out to be

$$f(\mu_\alpha | \mathbf{x}_1, ..., \mathbf{x}_n) = \int k_\alpha(\mathbf{y}_1 | \mathbf{x}_1) \cdots k_\alpha(\mathbf{y}_n | \mathbf{x}_n) \, d\mu_\alpha^{(n)}(\mathbf{y}_1, ..., \mathbf{y}_n)$$

where $k_\alpha(-|\mathbf{x}_i)$ is the Radon-Nikodým derivative of $h_\alpha(-|\mathbf{x}_i)$ with respect to ι, for all $i = 1, ..., n$ and where $\mu_\alpha^{(n)}$ is a certain random counting measure on the n'th Cartesian product of observation space. Formulas are derived for the imperfect detection case as well.

Symmetric Measurement Equation (SME) Filter

The symmetric measurement equation filter was introduced by Edward Kamen and his associates at the Georgia Institute of Technology in the early 1990s as a means of estimating the states of multiple targets while avoiding the problem of explicit report-to-track association [64], [66], [67], [99], [65]. The approach has also attracted attention because of its potentially attractive computational characteristics. The key idea behind the SME filter is to

> "...convert the measurement data (with associations not known) into a measurement equation by defining new measurements that are symmetric functionals [sic] of the original measurements. In this way the target states can be estimated without ever considering the associations between targets and measurements." [66, p. 476]

In essence, the SME approach is a method of converting the finite random set of observations generated by multiple targets into a random vector which is a nonlinear function of the set $X = \{\mathbf{x}_1, .., \mathbf{x}_t\}$ of state vectors of the targets.

For simplicity, assume that we are tracking a known number t of targets on the real line $\mathbb{R}$ and that there are no missed detections or false alarms. For any $1 \leq j \leq t$ define the *canonical symmetric functions* [76, p. 132-134]

$$s_{t;j}(z_1, ..., z_t) = \sum_{1 \leq i_1 < ... < i_j \leq t} z_{i_1} \cdots z_{i_j}$$

Thus, for example,

$$
\begin{aligned}
s_{3;1}(z_1, z_2, z_3) &= z_1 + z_2 + z_3 \\
s_{3;2}(z_1, z_2, z_3) &= z_1 z_2 + z_2 z_3 + z_1 z_3 \\
s_{3;3}(z_1, z_2, z_3) &= z_1 z_2 z_3
\end{aligned}
$$

Given an observation-set $Z = \{z_1, ..., z_t\}$ with $|Z| = t$ define $s_{t;j}(Z) = s_{t;j}(z_1, ..., z_t)$ and the t-dimensional vector $\mathbf{s}_t(Z)$ by

$$\mathbf{s}_t(Z) = (s_{t;1}(Z), ..., s_{t;t}(Z))$$

The transformation $(z_1, ..., z_t) \mapsto \mathbf{s}_t(\{z_1, ..., z_t\})$ is nonsingular.

Next, the system-level state vector is assumed to be governed by a standard Gauss-Markov state transition equation. The measurements $z_1, ..., z_t$ are assumed to satisfy the measurement equations $z_i = x_i + u_i$ where u_i are zero-mean white Gaussian noise processes. The randomly varying observation-set Z is replaced by the random vector $\mathbf{y} = \mathbf{s}_t(Z)$ which, it can be shown, can be written in the form

$$\mathbf{y} = \mathbf{s}_t(Z) = \mathbf{s}_t(X) + \mathbf{v}$$

where $\mathbf{v}$ is a vector zero-mean white noise process having no dependence on the report-to-track associations [66]. A Levenberg-Marquardt algorithm is used to estimate the set of parameters $X = \{x_1, ..., x_t\}$. (Note that $x_1, ..., x_t$ can be determined only up to permutation, so that the set X is actually what is being estimated.)

The SME approach has potentially attractive computational characteristics. Define new coordinates $y_i = c_i s_{t;i}(z_1, ..., z_t)$ where $c_i > 0$ for all $i = 1, ..., t$ and define $y_{i,j} = s_{j;i}(z_1, ..., z_j)$ for $k \geq j \geq i \geq 1$ and $y_{i,j} = 0$ otherwise. Then the y_i can be efficiently computed from the $z_1, ..., z_t$ using the recursive formulas

$$
\begin{aligned}
y_{1,j} &= z_1 + ... + z_j & (j = 1, ..., t) \\
y_{i,j} &= y_{i,j-1} + z_j y_{i-1,j-1} & (j = 2, ..., t; \; i = 2, ..., j)
\end{aligned}
$$

since $y_i = c_i y_{i,t}$ for $i = 1, ..., t$. Thus the number of additions and multiplications required to compute the coordinates $y_1, ..., y_t$ is only of order t^2.

The SME filter has been extended to multiple dimensions and to the case of missed detections and false alarms [65], [99]. A variation on the approach in [99] is to embed measurement-sets $Z = \{\mathbf{z}_1, ..., \mathbf{z}_k\}$ in the multinomial ring $\mathbb{R}[\theta_0, \theta_1, ..., \theta_n]$ in $n + 1$ indeterminates $\theta_0, \theta_1, ..., \theta_n$ where n is the dimension of observation space $\mathbb{R}^n$. Given a measurement $\mathbf{z} = (z_1, ..., z_n)$ define $L(\mathbf{z}) = z_1\theta_1 + ... + z_n\theta_n$ and, given a set $Z = \{\mathbf{z}_1, ..., \mathbf{z}_k\}$ of observations with $|Z| = k$ define

$$\mathbf{s}(Z) = (1 + L(\mathbf{z}_1)) \; \cdots \; (1 + L(\mathbf{z}_k)) \; \theta_0^k$$

In the one-dimensional case with $k = t$ this reduces to

$$\mathbf{s}(Z) = (1 + z_1\theta_1) \; \cdots \; (1 + z_k\theta_1) \; \theta_0^k = (1 + s_{t;1}(Z)\theta_1 + ... + s_{t;t}(Z)\theta_1^t) \; \theta_0^t$$

which is an alternate vector representation of the function $\mathbf{s}_t(Z)$ defined previously.

Jump-Diffusion Filter

The jump-diffusion approach was developed by Joseph O'Sullivan, Dennis Miller and their associates at Washington University in St. Louis [79]. It integrates detection, tracking, aspect-angle estimation, and automatic target recognition (ATR) into a single algorithm. The ATR aspect of the problem is addressed using parametrized deformable templates and by representing the aspect variables of target states as rotational transformations of the templates in the special orthogonal group $SO(3)$ (rather than Euler angles or pitch, roll, and yaw angles).

Number, kinematics, aspect, and I.D.s of targets are inferred simultaneously via a random sampling algorithm which estimates the joint posterior density. This "jump diffusion" sampling algorithm is so named because the discrete random variables (number of targets, target type) are dealt with by randomly jumping among multiple diffusion processes, each diffusion being used to follow the gradient of the posterior with respect to the continuous state variables. Since the multitarget state space includes multiple copies of the non-Euclidean manifold $SO(3)$, a theory of diffusion equations on Lie groups must be used. The jump-diffusion process is very computationally expensive and so the algorithm is typically run on a supercomputer.

Event-Averaged Maximum Likelihood Estimator (EAMLE) Filter

EAMLE is a multitarget extended Kalman filter developed in 1993 by Keith Kastella of Lockheed Martin Tactical Defense Systems to estimate multitarget parameters directly. According to Kastella, techniques such as MHE take the view that

> ...the correct permutation $\hat{\pi}$ can be determined from the data and that determining $\hat{\pi}$ is part of the job of a data fusion system...In contrast...we abandon the notion that the correct report-to-track assignment $\hat{\pi}$ is an observable of the system. Instead we evaluate [the multitarget state] directly, summing over [all] report-to-track assignments... [6, pp. 387, 388]

EAMLE assumes that: (1) there is a single Gaussian kinematics-reporting sensor; (2) the number t of targets is known *a priori*; (3) observations are independent; and (4) clutter is spacially dense and uniform with Poisson arrivals. The basic idea is as follows. Suppose for convenience that the number of targets is known, observations are independent, and that there are no missed detections or false alarms. Then in our earlier discussion of the MHE approach we observed that the density of observing the sequence $Z_1, ..., Z_M$ of observation-sets, given the sequence of associations $\pi = \{\pi_1, ..., \pi_M\}$, is

$$f(\mathbf{z}_{1;1}, ..., \mathbf{z}_{1;t}, ..., \mathbf{z}_{M;1}, ..., \mathbf{z}_{M;t} | \mathbf{x}_1, ..., \mathbf{x}_t, \pi) = \prod_{\alpha=1}^{M} \prod_{i=1}^{t} f(\mathbf{z}_{\alpha;\pi_\alpha i} | \mathbf{x}_i)$$

Since finding the true sequence of report-to-track associations is considered irrelevant, one should *average the likelihood* over all association events π, thus obtaining

$$f_{ass}(\mathbf{z}_{1;1}, ..., \mathbf{z}_{1;t}, ..., \mathbf{z}_{M;1}, ..., \mathbf{z}_{M;t} | \mathbf{x}_1, ..., \mathbf{x}_t)$$

$$= \frac{1}{(t!)^M} \sum_{\pi} f(\mathbf{z}_{1;1}, ..., \mathbf{z}_{1;t}, ..., \mathbf{z}_{M;1}, ..., \mathbf{z}_{M;t} | \mathbf{x}_1, ..., \mathbf{x}_t, \pi)$$

$$= \frac{1}{(t!)^M} \sum_{\pi} \prod_{\alpha=1}^{M} \prod_{i=1}^{t} f(\mathbf{z}_{\alpha;\pi_\alpha i} | \mathbf{x}_i)$$

A maximum likelihood estimate can then be deduced from this average-likelihood function.

The formula for f_{ass} is generalized to deal with missed detections and (spatially uniform, Poisson) clutter. In general f_{ass} is highly non-Gaussian. Since the number of targets is known, however, the system state vector $(\mathbf{x}_1, ..., \mathbf{x}_t)$ is an element of $\mathbb{R}^{tn}$, a vector space of fixed known dimensionality. One can therefore assume a standard Kalman motion model for the system state $(\mathbf{x}_1, ..., \mathbf{x}_t)$ and linearize the measurement density by replacing it with its Gaussian approximation. A conventional EKF is used to propagate the estimate of the system state and the error covariance matrix for this estimate. This "system" covariance describes not only individual track covariances but also cross-correlations between tracks. These cross-correlations arise from the fact that, as with any direct-estimation filter, reports are implicitly assigned to multiple tracks to some degree of proportion. Because tracks are constructed from the same reports, they are no longer independent.

Another notable feature of the EAMLE technique is the use of so-called "mean-field" methods borrowed from physics to drastically reduce computational complexity to order $\max\{t^3, mt^2\}$, where t is the number of targets and m is the number of observations per scan. This is accomplished by rewriting the summations in the equation for f_{ass} above as an integral of a function of Dirac delta functions, approximating the delta functions by very small-variance Gaussians, and then approximating the resulting integral using the saddle-point method. See [7], [6], [5] for details.

As will noted in Remark 3 of Section 6.2.3 of Chapter 6, the single-scan version of the EAMLE filter is a special case of the random set maximal likelihood estimation approach proposed in this book.

Hidden Markov Model Filters

In 1991 Xianya Xie and Robin J. Evans of the University of Melbourne developed a hidden Markov model (HMM) filter for tracking multiple slowly varying acoustic tones in highly cluttered frequency data. Though not couched in random set language, this direct-estimation filter bears strong similarities with the random set approach. Xie and Evans argue that

> The usual approach to this problem is to maintain lock on each individual track, projecting it forward to the next time interval so that measured data can be associated with the correct track. This is often quite difficult when the tracks come close or cross....[The HMM] approach avoids the difficult problem of data association and provides MAP estimates for multiple crossing tracks in clutter....Our aim is to estimate the "mixed" Viterbi track $I(K)$, i.e. there is no explicit notion of data association and the tracks are not considered separately....[A]ll the state vectors are viewed as nonordered arrays rather than ordered ones. In fact, they are symbols rather than vectors."[15, pp. 2659, 2660, 2661]

What Xie and Evans call a "mixed track" or "nonordered array" is in fact a *finite parameter-set*. Discretizing the problem, they use the symmetric motion model

$$p_{\{i,j\}}(\{k,l\}) = p_{il} \cdot p_{jk} + p_{ik} \cdot p_{jl}$$

to describe the state-transition from the mixed track $\{i,j\}$ to the mixed track $\{k,l\}$. Here, p_{il} denotes the transition probability of an individual target moving from state a_i at time α to state a_l at time $\alpha + 1$. In other words, this motion model asserts that complete confusion in regard to individual target identities/tags can occur between measurements. (As we will see later, this is a special case of a random set Markov motion model in which the possibility of complete between-measurements confusion between target I.D.s is modeled.)

The computational strategy for this filter is based on the Viterbi and Baum-Welch algorithms for hidden Markov models.

Another HMM filter, due to V. Krishnamurthy and R.J. Evans, assumes that the state space is discrete and finite and that the number t of targets is known *a priori* [73]. The basis of the approach is a measurement model of the form

$$\mathbf{z} = A_\sigma \mathbf{x} + \mathbf{v}$$

where $\mathbf{v}$ is the additive noise term, where the matrix A_σ is defined by $A_\sigma(\mathbf{x}_1, ..., \mathbf{x}_t) = (\mathbf{x}_{\sigma 1}, ..., \mathbf{x}_{\sigma t})$, and where the nonadditive noise variable σ is a uniformly distributed random permutation on the numbers $1, ..., t$.

2.3 Expert Systems

So far it has been assumed that observations collected by individual sensors are "unambiguous" in the sense that they are *mutually exclusive* and *exhaustive*. In other words, it has been assumed that one has at hand a comprehensive list of all of the observations that can possibly be collected and that each such observation is distinct and easily distinguished from any other such observation. However, many observations collected in real-world applications are *not* unambiguous in this sense. Major reasons for this are:

(1) It is often not possible to exhaustively denumerate all possible observations beforehand even if one could know them with precision;

(2) Not all of the observations that occur may be known to us ahead of time;

(3) The possible observations may be too variable or complex to be analyzed and compiled in a reasonable amount of time or with reasonable expense; and

(4) The nature of the observations themselves may be so conceptually murky that it is difficult to clearly demarcate one observation from another or even to mathematically model individual observations.

Unambiguous observations are unique and isolated—for example, a radar detection which gives us the range and azimuth of a target, but perhaps not its elevation. Other kinds of observations, however, have the effect of *constraining* the diversity of unambiguous observations that would otherwise be possible.

For example, one might learn that a target cannot possibly be in some particular region. As a consequence, any (unambiguous) observations subsequently generated in that region should be treated as most probably erroneous.

Another typical example of an ambiguous observation is a natural-language statement such as *"The truck is near the woods near the town of Smallville."* A concept such as *NEAR* is highly subjective in the sense that its interpretation varies considerably with the individual (the "sensor") who collects it and in the sense that it is not easy to "calibrate" this "sensor" to determine whether or not any particular event would be regarded as an instance of *NEAR*. Moreover, the concept *NEAR* is highly context-dependent: The implied distance between the truck and the woods is considerably different than the implied distance between the woods and the town, even though both distances are expressed by the concept *NEAR*. Such a concept is said to be *vague* or, more commonly, *fuzzy*.

Still another example of an ambiguous observation is a *rule* such as *"If a fire-control radar of type X is detected then it is likely that a missile of type Y will be launched."* This rule can be regarded as an observation and the *knowledge-base* which contains the rule can be regarded as the "sensor" that generated it. Conventional observations are *events*—they are assertions that something happened *unconditionally* (e.g., a fire control radar was detected; or a missile was launched). Rules, however, are *contingent* or *conditional events*: They express the fact that some event occurs only contingently upon the occurrence of some other event. Such "observations" exhibit a third kind of ambiguousness that is difficult to express using conventional means—that of *contingency*.

A fourth type of ambiguity in observations is due to what Pawlak [108] has called *indiscernibility* or *roughness*—that is, to the "granularity of knowledge." According to Pawlak, knowledge is based on the ability to classify things according to pre-existing knowledge bases which specify the *a priori* "concepts" or "categories" which form our cognitive models of the world. Given a universe of discourse U a "concept" is the subset of U of all objects which embody the concept), a "dimension of classification" is an exhaustive list of distinct concepts of similar type, and a "knowledge base" is a family of classification dimensions. (For example, one classification dimension might be "wingedness"— "fixed-wing aircraft", "rotor-winged aircraft," etc.—and another might be "propulsion type"—"turboprop", "jet turbine", etc. Finer discriminations, e.g. "fixed-wing jet aircraft", are formed by intersecting concepts.) Any two objects which are contained in the same elementary concepts cannot be distinguished from each other and are therefore indiscernible in this sense.

Expert-systems theory is the name given to various techniques developed to deal with ambiguous evidence (see, for example, [75], [36], [101], [133], [58]). In what follows we will briefly summarize some of the major expert-system approaches being used in data fusion to address the problem of inference using ambiguous observations, and briefly describe how such problems can be addressed using *random set theory* and *conditional event algebra*. Universes of discourse U will be assumed to be *finite* throughout. Some readers may prefer to consult Section 2.4, an introduction to random set theory, before continuing.

2.3.1 Imprecise Evidence

The Dempster-Shafer Theory of Evidence

Two events $X, Y \subseteq U$ are said to be *contradictory* if they have no overlap: $X \cap Y = \emptyset$, and X is *consistent* with Y if $X \subseteq Y$. Given a mass assignment m, the weight or mass $m(S)$ is the amount of belief which accrues to the hypothesis S alone and not to any strictly smaller subset of S. The *belief* in S is the total belief which is consistent with S:

$$Bel_m(S) = \sum_{T \subseteq S} m(T)$$

The *plausibility* of the body of evidence is the total weight of all events which do not contradict S:

$$Pl_m(S) = \sum_{S \cap T \neq \emptyset} m(T)$$

Clearly $Bel_m \leq Pl_m$ identically and $Bel_m(S) = 1 - Pl_m(S^c)$ for all $S \subseteq U$. The interval $[Bel_m(S), Pl_m(S)]$ is called the *interval of uncertainty* of the event S given the body of evidence m. The mass assignment can be recovered from the belief function using the Möbius transform: $m(S) = \sum_{T \subseteq S} (-1)^{|S-T|} Bel_m(T)$. So-called "fast Möbius transforms" have been developed by Thoma [127], [128], Black [13], and Smets and Kennes [120].

The *conflict* between two bodies of evidence m, m' is the total weight of contradiction between the events of m and the events of m':

$$K(m, m') = \sum_{X \cap Y = \emptyset} m(X) \, m'(Y)$$

The quantity $1 - K$ is the cumulative degree to which the two bodies of evidence do not contradict each other, and is called the *agreement* between m and m'. *Dempster's rule of combination* results when we form a new body of evidence whose focal subsets are all non-contradictory intersections $X \cap Y$. Given any $S \subseteq U$ there are many pairs $X, Y \subseteq Y$ such that $X \cap Y = S$ and so the total weight of agreement assignable to the focal subset $X \cap Y$ is $\sum_{X \cap Y = S} m(X) \, m'(Y)$. However, in general $\sum_{S \subseteq U} \left(\sum_{X \cap Y = S} m(X) \, m'(Y) \right) < 1$ since it may be the case that $K \neq 0$. Normalizing, we get Dempster's rule of combination of imprecise evidence:

$$(m * m')(S) = \frac{1}{1 - K} \sum_{X \cap Y = S} m(X) \, m'(Y)$$

for all $\emptyset \neq S \subseteq U$.

Dempster's rule of conditioning is a generalization of conditional probability. Given a probability measure q and an event T such that $q(T) \neq 0$, the conditioning of q given T is just $q(S|T) = q(S \cap T)/q(T)$ for all $S \subseteq U$. Similarly, Dempster's rule of conditioning of a mass assignment m with respect to T is the (unique) body of evidence m' such that

$$Pl_{m'}(S) = \frac{Pl_m(S \cap T)}{Pl_m(T)}$$

for all $S \subseteq U$. The dual relationship

$$Bel_{m''}(S) = \frac{Bel_m(S \cap T)}{Bel_m(T)}$$

produces what is known as *geometric conditioning* [118]. See also [31] for still another approach to conditioning. For more details concerning the Dempster-Shafer theory in theory and in practice, see [75], [32], [59], [133], [58]. Various methods have been proposed for integrating Dempster-Shafer theory with Bayesian probability theory [129], [137], [84], [34].

Random Set Models of Imprecise Evidence

Suppose that we are given a mass assignment m on the subsets of the finite universe U. Because U is finite the set $\mathcal{P}(U)$ of all subsets of U is also finite and so it is always possible to construct a random subset Σ of U such that $p(\Sigma = S) = m(S)$ for all $S \subseteq U$. In this case we write $m_\Sigma = m$ Furthermore, suppose that Σ, Λ are two random subsets of U which are statistically independent. Then it is easy to show that the mass assignment $m_{\Sigma \cap \Lambda}$ of $\Sigma \cap \Lambda$ is:

$$
\begin{aligned}
m_{\Sigma \cap \Lambda}(S) &= p(\Sigma \cap \Lambda = S) = \sum_{X \cap Y = S} p(\Sigma = X, \Lambda = Y) \\
&= \sum_{X \cap Y = S} p(\Sigma = X)\, p(\Lambda = Y) = \sum_{X \cap Y = S} m_\Sigma(X)\, m_\Lambda(Y)
\end{aligned}
$$

for all $\emptyset \neq S \subseteq U$ and thus

$$
\begin{aligned}
p(\Sigma \cap \Lambda \;\neq\; \emptyset) &= 1 - p(\Sigma \cap \Lambda = \emptyset) = 1 - \sum_{X \cap Y = \emptyset} m_\Sigma(X)\, m_\Lambda(Y) \\
&= \sum_{X \cap Y \neq \emptyset} m_\Sigma(X)\, m_\Lambda(Y)
\end{aligned}
$$

Since it may be the case that $\Sigma \cap \Lambda = \emptyset$, the assignment $S \mapsto p(\Sigma \cap \Lambda = S)$ cannot be the mass function of a body of evidence. If we force algebraic closure on the intersection operator '$\cap$' by assuming $\Sigma \cap \Lambda \neq \emptyset$ then it immediately follows that, for any $\emptyset \neq S \subseteq U$,

$$p(\Sigma \cap \Lambda = S \mid \Sigma \cap \Lambda \neq \emptyset) = \frac{\sum_{X \cap Y = S} m_\Sigma(X)\, m_\Lambda(Y)}{\sum_{X \cap Y \neq \emptyset} m_\Sigma(X)\, m_\Lambda(Y)} = (m_\Sigma * m_\Lambda)(S)$$

That is, Dempster's rule of combination can be regarded as an algebraic representation of the intersection of nonempty, independent random subsets of U when algebraic closure is imposed. See [75], [60], [104], [110] for more details.

Imprecision in *a Priori* Knowledge: Indiscernibility

The Dempster-Shafer formalism provides a methodology for dealing with problems in which evidence about unknown targets is imprecise. However, it is often the case that not only evidence but also *a priori* knowledge can be imprecise. Specifically, the *a priori* classification categories or concepts used to identify targets can be ambiguous: They can overlap; there may be more than one plausible interpretation of any given concept; the relative frequencies with which the categories are expected to occur can be ambiguous; etc. The reason for this is the additional kind of ambiguousness associated with *a priori* knowledge called *roughness* or *indiscernibility* [108]. More formally, a system of knowledge consists of a *partition* of some (finite) universe U into mutually disjoint cells $K_1, ..., K_m \subseteq U$ which exhaust U. Each K_i is a concept and the elements of any one concept are indiscernible with respect to each other. If C is some other concept that we encounter, then the best that we can do is try to approximate it using our existing concepts, e.g. the *lower approximation*

$$\underline{C} = \text{union of all } a \text{ priori concepts } K_i \text{ such that } K_i \subseteq C$$

Concepts which cannot be exactly represented as some combination of existing concepts are said to be *rough*. If we introduce an *a priori* probability measure q on U and define the set function β_q by

$$\beta_q(S) = q(\underline{S}) = \sum_{K_i \subseteq S} q(K_i)$$

then it is clear [117] that β_q is the belief measure of some random subset Γ of U such that $p(\Gamma = K_i) = q(K_i)$ for all i.

More generally, however, *a priori* concepts may not be mutually exclusive or exhaustive because of deficiencies in our knowledge. Concepts may exhibit some smaller or greater degree of overlap. There may be two or more plausible alternatives for the same concept, with some seemingly more plausible than others. For example, some centuries ago the categories $K_1 = ANIMAL$ and $K_2 = BIRD$ could have been regarded as overlapping, with $u = PLATYPUS$ belonging to both. $ANIMAL$ would then be an ambiguous *a priori* concept, with both K_1 and $K_1 - \{u\}$ as plausible interpretations.

In the most general case, therefore, *a priori* knowledge about targets can be imprecise in the same way that evidence about targets is ambiguous. We can therefore think of *a priori* knowledge in a more general sense as a random subset Γ of U, with each C such that $p(\Gamma = C) \neq 0$ being one aspect of an ambiguous concept. In more detail, recall that in the conventional Dempster-Shafer approach the quantity $m_\Sigma(S) = p(\Sigma = S)$ is that part of belief which adheres only to the hypothesis S and not to any proper subset of S. An analogous interpretation applies to *a priori* random subsets Γ: The quantity $p(\Gamma = C)$ is the degree of belief adhering to S alone (and not to any proper subset) *as a prior classification concept*. The elements of S are indiscernible *only as elements of that concept* and not as elements of any proper concept. Thus $p(\Gamma = \{u, v\}) = 0.5$ and $p(\Gamma = \{v, w\}) = 0.5$ asserts that there are exactly two (partially overlapping) concepts $\{u, v\}$ and $\{v, w\}$. The elements u, v are indiscernible as specific instances of the concept $\{u, v\}$ and the elements v, w are indiscernible as specific instances of the concept $\{v, w\}$. Yet $p(\{u, v, w\}) = 0$ offers no contradiction because $\{u, v, w\}$ is not a concept and, therefore, we are asserting on *a priori* grounds that there is no reason to believe that u, v, w are indiscernible with respect to each other as members of *this* concept. The extreme case of indiscernibility occurs when $p(\Gamma = U)$, when we assert that, from an *a priori* conceptual point of view, the elements of U cannot be distinguished from each other at all.

In the paper [84] it was shown that the theory of imprecise evidence can be extended so as to include the influence of *a priori* knowledge, even when that knowledge itself exhibits considerable imprecision. Given evidence modeled by a random subset Σ of U, we introduced the belief measure

$$\beta_\Gamma(S|\Sigma) = \frac{\beta_\Gamma(S \cap \Sigma)}{\beta_\Gamma(\Sigma)} = \frac{p(\Gamma \subseteq S \cap \Sigma)}{p(\Gamma \subseteq \Sigma)}$$

and showed that, in the event that Γ, Σ are independent, its Möbius transform is

$$m_\Gamma(S|\Sigma) = \frac{\delta_\Sigma(S)\, m_\Gamma(S)}{\beta_\Gamma(\Sigma)}$$

where $\delta_\Sigma(S) = p(\Sigma \supseteq S)$ is a commonality measure. The mass assignment $S \mapsto m_\Gamma(S|\Sigma)$ was called the *concordance* of the evidence Σ with respect to *a priori* knowledge Γ. It was also shown there the concordance is a *generalized pignistic probability distribution*, in the sense that it generalizes the pignistic probability distributions of Smets [119], [120].

2.3.2 Vague Evidence: Fuzzy Logic

Fuzzy Logic

Fuzzy set theory, or fuzzy logic as it is also known, generalizes conventional set theory. If U is some universe of objects then any ordinary subset $S \subseteq U$ is uniquely determined by its indicator function 1_S, which is defined by $1_S(u) = 1$ if $u \in S$ and $1_S(u) = 0$ otherwise. That is, any element $u \in U$ is either entirely contained within S or entirely outside of it. A fuzzy subset F of U is a set which can *partially* contain elements, in the sense that the indicator function $1_F(u)$ is allowed to take arbitrary values in the unit interval $[0, 1]$. Such an indicator function is called a *fuzzy membership function*. The conventional intersection, union, and complement operators of ordinary set theory are generalized using Zadeh's *max-min fuzzy logic*. Specifically, let f, g be fuzzy membership functions. Define

$$(f \wedge g)(u) \triangleq \min\{f(u), g(u)\}, \qquad (f \vee g)(u) \triangleq \max\{f(u), g(u)\}$$

$$f^c(u) \triangleq 1 - f(u)$$

for all $u \in U$. Fuzzy set theory differs from ordinary set theory in that it is not Boolean: In general,

$$f \wedge f^c \neq 1_\emptyset = 0, \qquad f \vee f^c \neq 1_U = 1$$

The Zadeh max-min logic is not the only possible fuzzy logic and, in fact, there are an infinite number of such logics [55]. For example,

$$(f \wedge g)(u) = f(u)g(u), \qquad (f \vee g)(u) = f(u) + g(u) - f(u)g(u)$$

$$f^c(u) = 1 - f(u)$$

defines the so-called *product-sum* or *prodsum* logic.

Fuzzy sets can be modeled as *consonant* Dempster-Shafer mass functions, i.e. mass functions whose focal subsets are linearly ordered under set inclusion [1], [27], [75].

Natural-language statements are often mathematically modeled by fuzzy statements [50], as is vague or approximate data [74]. Kramer and Goodman have developed a multitarget tracking algorithm called PACT, which fuses both statistical and natural-language data using the theory of random set representations of fuzzy sets described below [41].

Random Set Models of Fuzzy Evidence

The basic relationships between fuzzy logic and random set theory are due to Goodman [38], Orlov [106], [107], and Höhle [61], [62]. See also [50], [75].

Fuzzy evidence, as well as the operations of fuzzy logic, can be modeled as random sets as follows. Let Σ be a random subset of some (finite or infinite) universe. The *one-point covering function* μ_Σ of Σ (see [38], [50]) is defined as

$$\mu_\Sigma(u) \triangleq p(u \in \Sigma)$$

It defines a fuzzy subset of the universe. Conversely, let $f : U \rightarrow [0,1]$ be a fuzzy membership function of the universe and let A be a *uniformly distributed* random number on the unit interval $[0,1]$. Then the random subset $\Sigma_A(f)$ defined by

$$\Sigma_A(f) = f^{-1}[A, 1] = \{u \in U | A \leq f(u)\}$$

is a random subset of U, called the *canonical random set representation* of the fuzzy subset f. Because A is uniform it is easy to show that

$$\mu_{\Sigma_A(f)} = f$$

That is, the fuzzy subset corresponding to the canonical random set representation of a fuzzy subset is just the original fuzzy subset. Moreover, if '$\wedge$', '$\vee$', and '^{c}' denote the standard min/max fuzzy logic then the following relationships are true

$$\Sigma_A(f) \cap \Sigma_A(g) = \Sigma_A(f \wedge g), \qquad \Sigma_A(f) \cup \Sigma_A(g) = \Sigma_A(f \vee g)$$

However, it is not true that $\Sigma_A(f)^c = \Sigma_A(f^c)$ since the space of random sets of U is a Boolean algebra whereas the space of fuzzy subsets is not. Indeed,

$$\Sigma_A(f)^c = \Sigma_{1-A}^*(f^c)$$

where $\Sigma_A^*(f) \triangleq \{u \in U \mid A < f(u)\}$. In other words, the conventional fuzzy logic complementation operator can be regarded as the result of forcing algebraic closure on the canonical random set representation of the complement. Thus the celebrated non-Boolean behavior of fuzzy logic can be interpreted as the result of ignoring the statistical correlations between fuzzy subsets.

More generally, let B be another uniformly distributed random number on $[0, 1]$. Then it can be shown (see [38], [50]) that

$$\mu_{\Sigma_A(f) \cap \Sigma_B(g)}(u) = F_{A,B}(f(u), g(u))$$
$$\mu_{\Sigma_A(f) \cup \Sigma_B(g)}(u) = G_{A,B}(f(u), g(u))$$

where $F_{x,y}(a, b) = p(A \leq a, B \leq b)$ and $G_{A,B}(a, b) = 1 - F_{A,B}(1 - a, 1 - b)$ are called *copulas* and *cocopulas*, respectively (see [115]). The binary operators $F_{A,B}$ and $G_{A,B}$ often define fuzzy logics different from the standard one. For example, if A, B are independent then

$$F_{A,B}(a, b) = ab$$
$$G_{A,B}(a, b) = a + b - ab$$

thus yielding the prodsum logic. Likewise,

$$F_{A,1-A}(a, b) = \max\{0, a + b - 1\}$$
$$G_{A,1-A}(a, b) = \min\{1, a + b\}$$

which yields a non-distributive fuzzy logic often called the "nilpotent logic." It is well known that any copula $F_{A,B}$ must satisfy the inequalities

$$\max\{0, a + b - 1\} \leq F_{A,B}(a, b) \leq \min\{a, b\}$$

The more general relationships which exist between fuzzy logic and their canonical representations suggest the following: The reason that there are an infinite number of possible fuzzy logics is that many fuzzy logics correspond to different assumptions about the statistical correlations between fuzzy sets. (Indeed, this fact is often implicitly exploited in practical applications of fuzzy logic to data fusion—see, for example, [15]).

The canonical representation is by no means the only way of representing fuzzy sets using random sets. In the finite-universe case, a random set can be modeled as a field A_u, $u \in U$ of random integers A_u which take only the values 0 or 1. Using this and other formulations, Goodman has developed more general homomorphism-like relationships between fuzzy logic and random sets (see [38], [40]) and has produced a characterization of the random sets which have a given fuzzy membership function as their one-point covering function. Another probabilistic approach to fuzzy set theory has been proposed by Li [81].

2.3.3   Contingent Evidence: Conditional Event Algebra

Conditional Event Algebra

Knowledge-base rules have the form

$$X \Rightarrow S = \text{``if target is in } X \text{ then it is in } S\text{''}$$

where S, X are subsets of a (finite) universe U. The ordinary events $S \subseteq U$ of U form a Boolean algebra. The following questions can be asked: Does there exist a Boolean algebra $\hat{U}$ which has the rules of U as its elements and which satisfies the following elementary properties:

$$(U \;\Rightarrow\; S) = S$$
$$(X \;\Rightarrow\; S) \cap X = X \cap S$$

(where the first equation simply states that ordinary events are also conditional events, and the second is a mathematical formulation of the logical inference rule known as *modus ponens*)? If so, given any probability measure q on U is there a probability measure $\hat{q}$ on $\hat{U}$ such that

$$\hat{q}(X \Rightarrow S) = q(S|X) \triangleq \frac{q(S \cap X)}{q(X)}$$

for all $S, X \subseteq U$ with $q(X) \neq 0$? That is, is conditional probability a *true probability measure* on some Boolean algebra of rules in a manner that is "measure free" manner—i.e., regardless of the particular *a priori* probability measure q being assumed? The answer to both questions turns out to be "yes."

It can be shown that the usual approach to modeling conditionals used in symbolic logic—the "material implication" defined as $X \Rightarrow S = X \to S \triangleq (S \cap X) \cup X^c$ —is inconsistent with probability theory in the sense that

$$q(S \cap X) < q(X \to S) < q(S|X)$$

except under trivial circumstances. Lewis [80], [1, p. 14] showed that there is *no* binary operator " $\Rightarrow$ " on U which will solve the problem, except under trivial circumstances. Accordingly, the object $X \Rightarrow S$ must belong to some logical system which strictly extends the Boolean logic $\mathcal{P}(U)$. Any such extension— Boolean or otherwise—is called a *conditional event algebra*.

Several conditional event algebras have been discovered [1], [102]. For example, the so-called Goodman-Nguyen-Walker (GNW) algebra [1], [42] has logical operations defined by

$$
\begin{aligned}
(a|b)' &= (a'|b) \\
(a|b) \wedge (c|d) &= (ac|a'b \vee c'd \vee bd) \\
(a|b) \vee (c|d) &= (a \vee c|ab \vee cd \vee bd)
\end{aligned}
$$

The set of conditional events is thus algebraically closed, but at a price. If one defined $\hat{q}((a|b)) \triangleq q(a|b)$ then this is well-defined (since $(a|b) = (ab|b)$) but $\hat{q}$ is not a probability measure and, in fact, the set of GNW events is not even a Boolean algebra! The same is true of another algebraically closed algebra known as the Shay-Adams-Calabrese (SAC) algebra, which is defined by

$$
\begin{aligned}
(a|b)' &= (a'|b) \\
(a|b) \wedge (c|d) &= ((b' \vee a)(d' \vee c)|b \vee d) \\
(a|b) \vee (c|d) &= (ab \vee cd|b \vee d)
\end{aligned}
$$

Finally, it has been shown [91] that there is an algebraically closed *Boolean* conditional event algebra of "compound rules" of the form $((a|b)||c)$ such that $((a|b)||U) = (a|b)$ and whose conjunction and disjunction rules extend those of the GNW algebra. The rules of this algebra, called GNW*, are too complex to describe here. Though GNW* is Boolean, the extension

$$\hat{q}(((a|b)||c)) \triangleq \frac{q((a|b) \wedge (c|U))}{q(c|U)} - \frac{q(ac|c' \vee b)}{q(c)} = \frac{q(abc)}{(q(b) - q(bc)) q(c)}$$

of the probability measure q is not a probability measure on the set of compound conditional events.

Nevertheless, there indeed exists a Boolean algebra of rules, first discovered by van Fraassen [132], forgotten, and independently rediscovered fifteen years later by Goodman and Nguyen [42], which has the desired relationship with conditional probability.

Specifically, let U be a finite universe. Then conditional events are modeled as subsets of the countably infinite product measurable space

$$\hat{U} \triangleq U \times ... \times U \times ...$$

whose corresponding sigma-algebra $\sigma(\hat{U})$ is generated by "cylinders" of the form

$$a_1 \times ... \times a_k \times U \times U \times ...$$

for all $a_1, ..., a_k \in \mathcal{P}(U)$. Given two events $a, b \in \mathcal{P}(U)$ abbreviate $ab \triangleq a \cap b$, $a \vee b \triangleq a \cup b$ and $a' \triangleq U - a$. Then [42] the conditional event $(a|b)$ is defined as the infinite disjunction

$$(a|b) \triangleq (ab \times b' \times U \times ...) \vee (b' \times ab \times U \times ...) \vee ... \vee (b' \times ... \times b' \times ab \times U...) \vee ...$$

The terms of this disjunction are mutually disjoint. If q is an *a priori* probability measure on $\hat{U}$ let $\hat{q}$ denote the unique product probability measure on $\hat{U}$ which satisfies

$$\hat{q}(a_1 \times ... \times a_k \times U \times ...) \triangleq q(a_1) \cdots q(a_k)$$

for all $a_1, ..., a_k \in \mathcal{P}(U)$. If $q(b) \neq 0$ then $q(b') < 1$ and so

$$\begin{aligned}
\hat{q}((a|b)) &= \hat{q}(ab \times U \times ...) + \hat{q}(b' \times ab \times U \times ...) \\
&\quad + ... + \hat{q}(b' \times ... \times b' \times ab \times U...) + ... \\
&= q(ab) \sum_{i=0}^{\infty} (b')^i = \frac{q(ab)}{1 - q(b')} = \frac{q(ab)}{q(b)} = q(a|b)
\end{aligned}$$

for all $a, b \in \mathcal{P}(U)$ with $q(b) \neq 0$.

Conditional events defined in this way are not algebraically closed in $\hat{U}$: $(a|b) \wedge (c|d)$ and $(a|b) \vee (c|d)$ are not necessarily conditional events of the form $(e|f)$. Nevertheless, one can compute formulas for $\hat{q}((a|b) \wedge (c|d))$ and $\hat{q}((a|b) \vee$

$(c|d))$ which agree with formulas independently proposed by Bayesian theorists such as McGee [95]. For example,

$$\hat{q}((a|b) \wedge (c|d)) = \frac{q(abcd) + q(abd')q(c|d) + q(cdb')q(a|b)}{q(b \vee d)}$$

(see [42, Equ's. 41,42, p. 1693]).

Both the GNW and SAC logics play important roles in interpreting the product-space algebra. For example, suppose that we want to determine all of the possible deductions which follow from the "partial firing" of a rule: $(a|b)\tilde{b}$. The GNW and SAC algebras provide the means for computing these deductions. Also, the GNW algebra is the closest fit (in a lattice-ordering sense) to the product space algebra by an algebraically closed conditional event algebra.

Random Set Models of Rules

It has been shown (see [90], [91]) that there is at least one way to represent knowledge-base rules in random set form. This is accomplished by first representing rules as elements of the Goodman-Nguyen-Walker (GNW) conditional event algebra, and then—in analogy with the canonical representation of a fuzzy set by a random subset—representing the elements of this algebra as random subsets of the original universe. Specifically, let $(S|X)$ be a GNW conditional event in a finite universe U where $S, X \subseteq U$ Let Φ be a *uniformly distributed* random subset of U—that is, a random subset of U whose probability distribution is $p(\Phi = S) = 2^{-|U|}$ for all $S \subseteq U$. Then define the random subset $\Sigma_\Phi(S|X)$ of U by

$$\Sigma_\Phi(S|X) \triangleq (S \cap X) \cup (X^c \cap \Phi)$$

Notice that this random subset is uniformly distributed on the elements of the set interval

$$[S \cap X, X \to S] = \{T \subseteq U | S \cap X \subseteq T \subseteq X \to S\}$$

where $X \to S = (S \cap X) \cup X^c$ denotes the material implication. Ordinary events arise by setting $X = U$: $(S|U) = S$. Moreover, the following relationships are true:

$$\Sigma_\Phi(S|X) \cap \Sigma_\Phi(T|Y) = \Sigma_\Phi((S|X) \wedge_{GNW} (T|Y))$$
$$\Sigma_\Phi(S|X) \cup \Sigma_\Phi(T|Y) = \Sigma_\Phi((S|X) \vee_{GNW} (T|Y))$$
$$\Sigma_\Phi(S|X)^c = \Sigma_{\Phi^c}((S|X)^{c,GNW})$$

where '$\wedge$', '$\vee$', and 'c' represent the conjunction, disjunction, and complement operators of the GNW logic.

The Boolean algebra GNW* described earlier also has a random set interpretation, namely

$$\Sigma_\Phi((T|Y)\|S) = (T \cap Y \cap S) \cup (S \cap Y^c \cap \Phi) \cup (S^c \cap \Phi^c)$$

2.3.4　Partial-Probabilistic Evidence

Suppose that A is a random variable on a finite universe U and that we know only that the probability distribution $p_A(u) = p(A = u)$ of A belongs to some family of distributions, e.g., we know only that

$$p_A(a) \geq .3, \qquad p_A(b) \geq .1, \qquad p_A(c) \leq .8$$

for some elements $a, b, c \in U$. Let Q denote the set of all probability distributions which satisfy these constraints. Then we can define the *lower probability measure* p_* and *upper probability measure* p^* of Q by

$$p_*(S) = \inf_{q \in Q} q(S)$$

$$p^*(S) = \sup_{q \in Q} q(S)$$

Thus the true distribution p_A is always bounded between the upper and lower probabilities:

$$p_*(S) \leq p_A(S) \leq p^*(S)$$

for all $S \subseteq U$.

Because U is a *finite* universe it can be shown that there is a random subset Σ of U such that

$$p_*(S) = p(\Sigma \subseteq S) = \beta_\Sigma(S)$$

$$p^*(S) = p(\Sigma \cap S \neq \emptyset) = \rho_\Sigma(S)$$

for all $S \subseteq U$. (This fact is not true for non-finite universes [104], [9].) In this sense the information encapsulated in the family Q can be represented as a random subset Σ_Q, and so partial-probabilistic evidence on finite spaces can be represented in random set form as well.

2.4　Random Sets

Random sets are random elements whose values are sets. Random sets are generalizations of the familiar concept of random variables (or random vectors) in probability theory. Confidence intervals (or regions in euclidean spaces) in statistics are simple examples of random (closed) sets.

The simplest example is this. Let U be a finite set. Suppose elements of the power set $\mathcal{P}(U)$ of U are selected according to some specified probability law. Then we obtain a finite random set on U. Specifically, let

$$f : \mathcal{P}(U) \longrightarrow [0,1] \quad \text{with} \quad \sum_{A \in \mathcal{P}(U)} f(A) = 1.$$

The function f plays the role of a probability density function, but unlike the case of random variables, it is defined on sets rather than on points of U. The probability that the subset A of U is selected is $f(A)$. This situation can be formulated as follows.

Let $(\Omega, \mathcal{A}, P)$ be a probability space, and $X : \Omega \to \mathcal{P}(U)$ be a random element. The σ-field on $\mathcal{P}(U)$ is of course $\mathcal{P}(\mathcal{P}(U))$. The density of the random set X is

$$f(A) = P(\omega : X(\omega) = A).$$

As in the case of random variables, the density f determines a probability measure P_X on $\mathcal{P}(\mathcal{P}(U))$. The knowledge of f (or P_X) is sufficient for studying X.

Unlike the case of random variables whose range is the real line $\mathbb{R}$ (which is a totally ordered set, with rich algebraic and topological structures), the situation in random sets is more complicated since we are dealing with sets. In the case where U is finite, we have a similar concept of distribution functions. Indeed, let

$$F : \mathcal{P}(U) \longrightarrow [0,1], \quad F(A) = P(\omega : X(\omega) \subseteq A) = \sum_{B \subseteq A} f(B).$$

Then, by Möbius inversion formula (see, e.g. Aigner [3]), we can recover f from F via

$$f(A) = \sum_{B \subseteq A} (-1)^{|A \setminus B|} F(B),$$

where $A \setminus B = \{u \in U : u \in A, u \notin B\}$ and $|A|$ denotes the cardinality of A.

The distribution function F of the random set X satisfies the following conditions.

(i) $F(\emptyset) = 0, \qquad F(U) = 1,$

(ii) F is infinitely monotone, i.e., for any $A_1, \ldots, A_n \subseteq U, n \geq 1$,

$$F\left(\bigcup_{i=1}^{n} A_i\right) = \sum_{\emptyset \neq I \subseteq \{1, \ldots, n\}} (-1)^{|I|+1} F\left(\bigcap_{i \in I} A_i\right).$$

Recall that the usual method for introducing probability measures on $\mathbb{R}^d$ (probability laws of random vectors) is via distribution functions. A distribution function is defined axiomatically in such a way that it determines uniquely the probability law of a random vector. Since sets share basic algebraic and topological properties with $\mathbb{R}^d$, namely they form a lattice (with set inclusion as partial order relation) with some suitable topology, it is expected that probability laws of random sets on $\mathbb{R}^d$ (or more generally, on locally compact spaces) can be specified via distribution functions of random sets. In Chapter 3, we will present rigorously this characterization of probability laws of random closed sets in terms of distribution functions (called capacity functionals). But here is the idea. A random sets X is a set-valued random element. Specifically, X is a map from some space Ω to some collection Θ of subsets of U, which is $\mathcal{A}$-$\mathcal{B}$-measurable, where $\mathcal{A}$ and $\mathcal{B}$ are σ-field on Ω and Θ, respectively. If P is a probability measure on $\mathcal{A}$, then the probability law of X is the probability measure $P_X = PX^{-1}$ on $\mathcal{B}$. For statistical aspects of random sets, see e.g. Stoyan et al. [125].

When $\Theta = \mathbb{R}$ (Θ consists of singletons of $\mathbb{R}$), the distribution function of S is the real-valued function F defined on $\mathbb{R}$ by

$$F(x) = P(\{\omega : S(\omega) \leq x\}), \qquad \forall x \in \mathbb{R}.$$

By construction, F satisfies the following conditions:

(i) $\lim_{x \to -\infty} F(x) = F(-\infty) = 0$, $\lim_{x \to \infty} F(x) = F(\infty) = 1$.

(ii) F is right continuous on $\mathbb{R}$, i.e., for any $x \in \mathbb{R}$, if $y \searrow x$ then $F(y) \searrow F(x)$.

(iii) F is monotone non-decreasing: $x \leq y$ implies $F(x) \leq F(y)$.

The classical result is this (for a proof, see, e.g. Durrett [29]). Let

$$C(x) = \{y : y \in \Re, \, y \leq x\} \qquad (\text{here} = (-\infty, x])$$

and $\mathcal{B}(\mathbb{R})$ be the σ-field generated by the $C(x)$'s, $x \in \mathbb{R}$ (the so called Borel σ-field of $\mathbb{R}$). If a given function F, defined on $\mathbb{R}$, satisfies the above conditions (i), (ii), and (iii), then F uniquely determines a probability measure μ on $\mathcal{B}(\mathbb{R})$ such that

$$F(x) = \mu(C(x)) \qquad \forall x \in \mathbb{R}.$$

Thus to specify the probability law (measure) of a random variable, or to propose a model for a random quantity, it suffices to specify a distribution function.

For the case when $\Theta = \mathbb{R}^d$, $d > 1$ (singletons of $\mathbb{R}^d$), we note that $\mathbb{R}^d$ is a topological space, and a partially ordered set, where, by abuse of notation, the order relation on $\mathbb{R}^d$ is written as

$$x \leq y \qquad \text{iff} \qquad x_i \leq y_i, \quad i = 1, 2, \ldots, d,$$

$x = (x_1, x_2, \ldots, x_d)$ and $y = (y_1, y_2, \ldots, y_d)$. In fact, $\mathbb{R}^d$ is a semi-lattice (inf), where

$$\inf(x, y) = x \wedge y = (x_1 \wedge y_1, \ldots, x_d \wedge y_d).$$

By analogy with the case when $d = 1$, consider $F : \mathbb{R}^d \longrightarrow [0, 1]$ defined by

$$F(x_1, \ldots, x_d) = P(S_1 \leq x_1, \ldots, S_d \leq x_d),$$

where $S = (S_1, \ldots, S_d) : \Omega \longrightarrow \mathbb{R}^d$ is a d-dimensional random vector. Obviously, by construction, F satisfies the counter-parts of (i) and (ii), namely,

(i$'$) $\lim_{x_j \to -\infty} F(x_1, \ldots, x_d) = 0$ for each j, $\quad \lim_{x_1 \to \infty, \ldots, x_d \to \infty} F(x_1, \ldots, x_d) = 1$.

(ii$'$) F is right continuous on $\mathbb{R}^d$, i.e. for any $x \in \mathbb{R}^d$, if $y \in \mathbb{R}^d$ and $y \searrow x$ (i.e. $y_i \searrow x_i$, $i = 1, \ldots, d$), then $F(y) \searrow F(x)$.

As for the counter-part of (iii), first note that F is monotone non-decreasing, i.e., if $x \leq y$ (on $\mathbb{R}^d$), then $F(x) \leq F(y)$, by construction of F in terms of P

and S. Now, also by construction of F, we see that for $x = (x_1, \ldots, x_d) \le y = (y_1, \ldots, y_d)$,

$$
\begin{aligned}
P(x < S \le y) &= F(y) - \sum_{i=1}^{d} F_i \\
&\quad + \sum_{i<j} F_{ij} \pm \cdots \pm \cdots + (-1)^d F(x) \ge 0,
\end{aligned}
$$

where $F_{ij\cdots k}$ denotes the value $F(z_1, z_2, \ldots, z_d)$ with $z_i = x_i, \ldots, z_k = x_k$ and the others $z_t = y_t$.

Note that $P(x < S \le y)$ can be written in a more compact form as follows (see Durrett [29]). Let the set of vertices of $(x_1, y_1] \times \cdots \times (x_d, y_d]$ be

$$
V = \{x_1, y_1\} \times \cdots \times \{x_d, y_d\},
$$

and for $v \in V$, let

$$
sgn(v) = (-1)^{\alpha(v)},
$$

where $\alpha(v)$ is the number of the x_i's in v, then

$$
P(x < S \le y) = \sum_{v \in V} sgn(v) F(v).
$$

Of course, being a probability value, $P(x < S \le y) \ge 0$ for any $x \le y$.

For a probability measure μ on $(\mathbb{R}^d, \mathcal{B}(\mathbb{R}^d))$ to be uniquely determined through a given function $F : \mathbb{R}^d \to [0, 1]$, by

$$
\mu(C(x)) = F(x), \qquad \forall x \in \mathbb{R}^d,
$$

where $C(x) = \{y \in \mathbb{R}^d : y \le x\}$, we not only need that $F(x)$ satisfies (i'), (ii') and be monotone non-decreasing, but also the additional condition

(iii') For any $x \le y$ in $\mathbb{R}^d$,

$$
\sum_{v \in V} sgn(v) F(v) \ge 0.
$$

Again for a proof, see, e.g., Durrett [29]. Here is an example where F satisfies (i'), (ii') and is monotone non-decreasing, but fails to satisfy (iii'). Consider

$$
F(x, y) = \begin{cases} 1 & \text{if } x \ge 0, y \ge 0, \text{ and } x + y \ge 1 \\ 0 & \text{otherwise,} \end{cases}
$$

then for $(1/4, 1/2) \le (1, 1)$, we have

$$
F(1, 1) - F\left(\frac{1}{4}, 1\right) - F\left(1, \frac{1}{2}\right) + F\left(\frac{1}{4}, \frac{1}{2}\right) < 0.
$$

We can group the monotone non-decreasing property and (iii′) into a single condition:

(iv) For $a, a_1, \ldots, a_n$ in $\mathbb{R}^d$, $n \geq 0$,

$$F(a) \;\geq\; \sum_{\emptyset \neq I \subseteq \{1,2,\ldots,n\}} (-1)^{|I|+1} F\left(a \wedge_I a_i\right) \quad \forall a, a_1, \ldots, a_n \in \mathbb{R}^d,$$

where $|I|$ denotes the cardinality of I.

The condition (iv) is stronger than monotone non-decreasing and is referred to as *monotone of infinite order* or *totally monotone* (Choquet [20] and Revuz [112]). Note that for $d = 1$, the real line $\mathbb{R}$ is a chain, so that total monotonicity is the same as monotonicity. For $d > 1$, monotone functions might not be totally monotone. For example, let $d = 2$ and

$$F(x, y) = \begin{cases} 1 & \text{for } x \geq 0,\ y \geq 0,\ \text{and } \max(x, y) \geq 1 \\ 0 & \text{otherwise.} \end{cases}$$

Let $a = (x, y)$, $a_1 = (x_1, y_1)$, $a_2 = (x_2, y_2)$ with

$$0 \leq x_1 < 1 < x < x_2, \qquad 0 \leq y_2 < 1 < y < y_1,$$

then

$$F(a) = F(a \wedge a_1) = F(a \wedge a_2) = 1, \qquad F(a \wedge a_1 \wedge a_2) = F(x_1, y_2) = 0,$$

and hence F is not totally monotone.

Now consider the case when Θ is a collection of subsets of some set U. Θ is partially ordered set-function. Suppose that Θ is stable under finite intersections, so that it is a semi-lattice, where $\inf(A, B) = A \cap B$. To carry conditions on a distribution function to this case, we need to define an appropriate topology for Θ (to formalize the concept of right continuity and to define an associated Borel σ-field $\mathcal{B}$ on Θ, so that $C(A) = \{B \in \Theta : B \subseteq A\} \in \mathcal{B}$ for all $A \in \Theta$). The condition of monotone of infinite order (iv) is expressed in terms of the lattice structure of Θ. Various situations of this setting can be found in Revuz [112].

2.5   "Finite-Set Statistics"

Our discussions of multisensor, multitarget estimation and of expert-systems theory in sections 2.2 and 2.3 suggest strong reasons why one might want to try to employ random set theory as a scientific foundation for data fusion. In this section, which summarizes the contents of Chapters 4 through 8, we propose a rigorous, fully probabilistic scientific foundation for the following aspects of information fusion:

(1) *Multisource integration* based on parametric estimation and Markov techniques[85], [89], [88]

(2) *Prior information* regarding the numbers, identities, and geokinematics of targets[87]

(3) *Sensor management* based on information theory and nonlinear control theory[86]

(4) *Performance evaluation* using information theory and nonparametric estimation[87], [93]

(5) *Expert-systems theory*: fuzzy logic, evidential theory, rule-based inference [92]

There are also reasons to believe that the approach might be further extended to many aspects of situation assessment, though will not be considered here.

This unification approach implies the existence of algorithms which fully unify, in *a single statistical process,* the following functions of information fusion: (1) target detection; (2) target identification; (3) target tracking and localization; (4) prior information with respect to detection, classification, and tracking; (5) ambiguous *evidence* and precise *data;* (6) rules of evidential combination; and (6) sensor management

Not all sensors can be *directly* subsumed within this unification, however. We assume the following sensor types: (1) point-source; (2) range-profile; (3) line-of-bearing; (4) human observers reporting in natural language; (5) rulebases; (6) imaging sensors whose target-images are point "firefly" sources; and (7) imaging sensors whose target-images consist of relatively small clusters of point energy reflectors (also sometimes called "extended targets").

The basic approach is portrayed in Figures 2.6 and 2.7. A suite of known sensors transmits to a *central data fusion site* the observations they collect regarding a group of targets whose numbers, positions, velocities, identities, threat states, etc. are unknown.. (It should be possible to extend the approach to distributed fusion problems, for example by adapting the techniques of [2]. However, we will not consider distributed fusion in this book.) We conceptually think of:

(1) The sensor suite as though it were a single "global sensor"

(2) The target set as though it were a single "global target"

(3) The observations, collected by the sensor suite at approximately the same time, as a single "global observation"

Simplifying somewhat, each individual observation will have the form

$$\xi = (z, u, i)$$

where z is a continuous variable (geokinematics, signal intensity, etc.) in $\mathbb{R}^n$, u is a discrete variable (e.g. possible target I.D.s) drawn from a finite universe U of possibilities, and i is a "sensor tag" which identifies the sensor which supplied the measurement. If the total observation-set

$$Z = \{\xi_1, ..., \xi_k\}$$

collected by the global sensor is treated as a single entity, then it is a specific value of a *randomly varying* finite observation-set Σ.

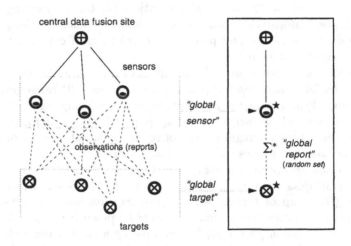

Figure 2.6: The figure illustrates the basic concept underlying "finite-set statistics." The sensor suite, target group, and observations are modeled as a single "global sensor," "global target," and "global observation," respectively. The multisensor, multitarget data fusion problem can therefore be reformulated as a single-sensor, single-target problem. However, "observations" are now *randomly varying finite sets of ordinary observations.*

Incongruent Data Types: Approach

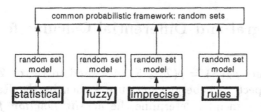

Figure 2.7: Expert systems theory can be rigorously integrated with multisensor, multitarget estimation if we first model ambiguous evidence in random set form.

We are now in a position to reformulate multisensor, multitarget estimation problems as single-sensor, single-target problems, as illustrated in Figure 2.6. Expert systems theory can also be subsumed within the same paradigm if we model "ambiguous" observations in random set form (see Figure 2.7).

2.5.1   Random Set Formulation of Data Fusion Problems

From Mathéron's random set theory we know that the class of finite subsets of measurement space has a topology, the so-called *hit-or-miss topology* [94], [96]. If O is any Borel subset of this topology then the statistics of Σ are characterized by the associated probability measure

$$p_\Sigma(O) \triangleq p(\Sigma \in O)$$

The Choquet-Mathéron capacity theorem [94, p. 30] tells us, among other things, that we need only consider Borel sets of the specific form $O_{S^c}^c$—i.e., the class whose elements are all closed subsets C of measurement space such that $C \cap S^c = \emptyset$ (i.e., $C \subseteq S$) where S is some closed subset of measurement space. In this case

$$p_\Sigma(O_{S^c}^c) = p(\Sigma \subseteq S) = \beta_\Sigma(S)$$

and thus we can use the *nonadditive* measure β_Σ defined on subsets of *ordinary observation space* instead of the *additive* measure p_Σ, which is defined on subsets of the class of finite sets of observation space.

The set function β_Σ is known as the *belief measure* of the randomly varying finite subset Σ, or alternatively as an "infinitely monotone capacity." Despite the fact that it is nonadditive, it plays the same role in multisensor, multitarget statistics that ordinary probability measures play in single-sensor, single-target statistics. The reason is that from the belief measure β_Σ we can construct a "global density function" $f_\Sigma(Z)$ which describes the *comprehensive statistical behavior of the entire sensor suite.*

2.5.2   An Integral and Differential Calculus for Data Fusion

If an *additive* measure $p_\mathbf{Z}(S) \triangleq p(\mathbf{Z} \in S)$ of a random vector $\mathbf{Z}$ is absolutely continuous with respect to Lebesgue measure λ then by the Radon-Nikodým theorem one can, in principle, determine the *density function* $f_\mathbf{Z} = \frac{dp_\mathbf{Z}}{d\lambda}$ that corresponds to it. Conversely, the measure can be recovered from the density through application of the Lebesgue integral: $\int_S f_\mathbf{Z}(\mathbf{z})d\lambda(\mathbf{z}) = p_\mathbf{Z}(S)$. We will show how to define an integral and differential calculus of functions of a set variable which obeys similar properties. Given a vector-valued function $f(Z)$ of a *finite-set* variable Z we will show how to define a "set integral" of the form $\int_S f(Z)\delta Z$. Conversely, given a vector-valued function $\Phi(S)$ of a *closed set* variable S, we will show how to *constructively* define a "set derivative" of the form $\frac{\delta\Phi}{\delta Z}$. Under certain assumptions, these operations turn out to be inverse to each other:

$$\int_S \frac{\delta\Phi}{\delta Z}(\emptyset)\, \delta Z \;=\; \Phi(S)$$

$$\left[\frac{\delta}{\delta Z}\int_S f(Z)\delta Z\right]_{S=\emptyset} \;=\; f(Z)$$

2.5.3   The Global Density of a Sensor Suite

Given certain absolute continuity assumptions which need not concern us here, if β_Σ is the belief measure of a random observation set Σ then the quantity

$$f_\Sigma(Z) \triangleq \frac{\delta\beta_\Sigma}{\delta Z}(\emptyset)$$

is the *density function* of the random finite subset Σ. The quantity $f_\Sigma(Z)$ is the likelihood that the event $\Sigma = Z$ will occur. On the other hand, it also has a completely Bayesian interpretation. Suppose that $f_\Sigma(Z|X)$ is a global density with a *set parameter* $X = \{\zeta_1, ..., \zeta_t\}$, where $\zeta_1, ..., \zeta_t$ are the unknown (discrete and continuous) parameters of the targets, and t is the unknown number of the targets. Then $f_\Sigma(Z|X)$ is the *total probability density of association* between the measurements in Z and the parameters in X.

The global density of a sensor suite differs from conventional densities in that it *encapsulates the comprehensive statistical behavior of the entire sensor suite into a single mathematical object*, and not just sensor noise statistics. Most generally speaking, a global density has the form $f_\Sigma(Z|X;Y)$ and includes the following information:

(1) The observation-set $Z = \{\xi_1, ..., \xi_k\}$

(2) The set $X = \{\zeta_1, ..., \zeta_t\}$ of unknown parameters

(3) The states $Y = \{\eta_1, ..., \eta_s\}$ of the sensors (dwells, modes, etc.)

(4) The sensor-noise distributions of the individual sensors

(5) The probabilities of detection and false alarm for the individual sensors

(6) Clutter models

(7) Detection profiles for the individual sensors (as functions of range, aspect, etc.)

Density functions can be computed explicitly from a knowledge of the sensors that belong to the sensor suite. For example, suppose that we are given a single sensor with sensor-noise density $f(\xi|\zeta)$ with no false alarms and constant probability of detection p_D, and that observations are independent. Then the global density which specifies the *multitarget measurement model* for the sensor can be shown to be

$$f_\Sigma(\{\xi_1, ..., \xi_k\}|\{\zeta_1, ..., \zeta_t\}) = p_D^k (1 - p_D)^{t-k} \sum_{1 \le i_1 \ne ... \ne i_k \le t} f(\xi_1|\zeta_{i_1}) \cdots f(\xi_k|\zeta_{i_k})$$

where the summation is taken over all distinct $i_1, ..., i_k$ such that $1 \le i_1, ..., i_k \le t$. (Here, $\xi_1, ..., \xi_k$ are assumed distinct, as are $\zeta_1, ..., \zeta_t$.)

Global densities satisfy the properties that one would expect. For example, if $\mathbf{T}(Z)$ is a measurable vector-valued transformation of a finite-set variable then the expectation $\mathrm{E}[\mathbf{T}(\Sigma)]$ of the random vector $\mathbf{T}(\Sigma)$ is

$$\mathrm{E}[\mathbf{T}(\Sigma)] = \int \mathbf{T}(Z)\, f_\Sigma(Z)\, \delta Z$$

Global densities also enjoy certain *unexpected* properties, the most interesting of which is the fact that they are *continuous analogs of the Möbius transform*. That is if $Z = \{\mathbf{z}_1 \cdots \mathbf{z}_k\}$ with $|Z| = k$ then it can be shown that

$$f_\Sigma(Z) = \lim_{i \to \infty} \frac{\sum_{Y \subseteq Z}(-1)^{|Z-Y|}\beta_{\Sigma,i}(Y)}{\lambda(E_{\mathbf{z}_1;i}) \cdots \lambda(E_{\mathbf{z}_k;i})}$$

where $\beta_{\Sigma,i}$ is defined by

$$\beta_{\Sigma,i}(\{\mathbf{z}_1, ..., \mathbf{z}_k\}) \triangleq \beta_\Sigma((E_{\mathbf{z}_1;i} \times u_1) \cup ... \cup (E_{\mathbf{z}_k;i} \times u_k))$$

where $E_{\mathbf{z}_1;i}$ denotes a closed ball of radius $1/i$ centered at $\mathbf{z} \in \mathbb{R}^n$.

2.5.4   A Simple Illustration

Assume that two identical targets $\otimes_1$ and $\otimes_2$ located on the real line are observed by a single Gaussian sensor with density $f(a|x) = N_{\sigma^2}(a - x)$ (where N_{σ^2} denotes the normal distribution with variance σ^2) and associated probability measure $p_f(S|x)$. Assume also that (1) reports are independent, (2) the probability of false alarm for the sensor is zero, (3) the probability of detection

q is not necessarily unity, and (4) that q is constant over the region of interest. The observations corresponding to target $\otimes_1$ are the specific realizations of a Gaussian random number A_1 and the observations corresponding to target $\otimes_2$ are modeled by another Gaussian random number A_2. If $q = 1$ then the observation generated by target $\otimes_1$ is a randomly-varying single-element set of the form $\Sigma_1 = \{A_1\}$ and the observation generated by target $\otimes_2$ is $\Sigma_2 = \{A_2\}$. What the sensor sees at any given instant is just an unordered two-element set $Z = \{a_1, a_2\}$ of reports generated by the randomly varying two-element set $\Sigma = \Sigma_1 \cup \Sigma_2 = \{A_1, A_2\}$. If $q < 1$, however, then the possible realizations for Σ are $\Sigma = Z$ where 1) $Z = \{a_1, a_2\}$ for any $a_1, a_2 \in \mathbb{R}$ with $a_1 \neq a_2$; 2) $Z = \{a\}$ for any $a \in \mathbb{R}$; or 3) $Z = \emptyset$.

The statistics of this random set are described by the belief measure $\beta_\Sigma(S|X) = p(\Sigma \subseteq X) = p(\Sigma_1 \subseteq S, \Sigma_2 \subseteq S)$ where X is a set parameter which can take the form $X = \emptyset$ (no tracks), $X = \{x\}$ (one track), or $X = \{x_1, x_2\}$ with $x_1 \neq x_2$ (two tracks). In fact,

$$\beta_\Sigma(S|\{x\}) = 1 - q + q\, p_f(S|x)$$
$$\beta_\Sigma(S|\{x_1, x_2\}) = [1 - q + q\, p_f(S|x_1)][1 - q + q\, p_f(S|x_2)]$$

From the belief measure it is possible to compute the global density. In the one-track case $X = \{x\}$ it is

$$f_\Sigma(\emptyset|X) = 1 - q, \qquad f(\{z\}|X) = q\, f(z|x)$$

and in the two-track case $X = \{x_1, x_2\}$ by

$$f_\Sigma(\emptyset|X) = (1 - q)^2$$
$$f_\Sigma(\{z\}|X) = q(1 - q)\, f(z|x_1) + q(1 - q)\, f(z|x_2)$$
$$f_\Sigma(\{z_1, z_2\}|X) = q^2\, f(z_1|x_1)\, f(z_2|z_1) + q^2\, f(z_2|x_1)\, f(z_1|z_1)$$

In all other cases, $f_\Sigma(Z|X) = 0$. The one-track case is just the conventional single-sensor, single-target situation with the probability of detection taken into explicit account [22, p. 30], [124, p. 15, equ. 2.4]. The two-track case is more interesting. Given any real numbers $a_1 < a_2$ the set $\{a_1, a_2\} = \{a_2, a_1\}$ does not depend on the order of a_1, a_2 and therefore can be identified with the point (a_1, a_2) on the half-plane. Likewise, when $a_1 = a_2 = a$ then the set $\{a\}$ can be identified with the diagonal boundary line $x = y$ of the half-plane. In other words, all of the possible two-observation realizations $Z = \{a_1, a_2\}$ of the random set Σ can be identified with points on the half-plane. The global density $f_\Sigma(\{a_1, a_2\}|X)$ gives the probability density that $\{a_1, a_2\}$ occurs as a value of the random set $\{A_1, A_2\}$. Its graph is just a surface over the half-plane which has a peak near (but usually not the same as) the point (x_1, x_2). The reader may also easily verify that the belief measure $\beta_\Sigma(S|X)$ can be recovered from the global density via the set integral:

$$\beta_\Sigma(S|X) = \int_S f_\Sigma(Z|X)\delta X$$
$$= f_\Sigma(\emptyset|X) + \int f_\Sigma(\{z\}|X)dz + \frac{1}{2}\int_{S\times S} f_\Sigma(\{z_1, z_2\}|X)dz_1 dz_2$$

for any closed $S \subseteq \mathbb{R}$ and any finite $X \subseteq \mathbb{R}$.

2.5.5   The Parallelism Between Point- and Finite-Set Statistics

Because of the existence of the set derivative and the set integral, one can compile the following list of *direct mathematical parallels* between the world of single-sensor, single-target statistics and the world of multisensor, multitarget statistics.

Random Vector, Z	**Finite Random Set, Σ**
sensor, $\odot$	global sensor, $\odot^*$
target, $\otimes$	global target, $\otimes^*$
observation, $\mathbf{z}$	global observation-set, Z
parameter, $\mathbf{x}$	global parameter-set, X
differentiation, $\frac{dp_{\mathbf{Z}}}{d\mathbf{z}}$	set differentiation, $\frac{\delta\beta_\Sigma}{\delta Z}$
integration, $\int_S f_{\mathbf{Z}}(\mathbf{z}|\mathbf{x})d\lambda(\mathbf{z})$	set integration, $\int_S f_\Sigma(Z|X)\delta Z$
probability measure, $p_{\mathbf{Z}}(S|\mathbf{x})$	belief measure, $\beta_\Sigma(S|X)$
density, $f_{\mathbf{Z}}(\mathbf{z}|\mathbf{x})$	global density, $f_\Sigma(Z|X)$
prior density, $f_{\mathbf{X}}(\mathbf{x})$	global prior density, $f_\Gamma(X)$
motion models, $f_{\alpha+1|\alpha}(\mathbf{x}_{\alpha+1}|\mathbf{x}_\alpha)$	global models, $f_{\alpha+1|\alpha}(X_{\alpha+1}|X_\alpha)$

The parallelism is so close that it suggests a general way of attacking data fusion problems. The above list can be thought of as a kind of "translation dictionary." That is, any theorem in conventional statistics can be thought of as a "sentence" in a language whose "words" and "grammar" consist of the basic concepts in the left-hand column above. The above "dictionary" establishes a direct correspondence between the words and grammar of the random-vector language and the cognate words and grammar of the finite-set language. Consequently, nearly any "sentence"—any theorem or mathematical algorithm—phrased in terms of the random vector language can, in principle, be directly translated into a corresponding "sentence"—theorem or algorithm—in the random-set language. The reason that we say "nearly" is because, as with any translation process, the correspondence is not one-to-one. For example, there seems to be no natural way to add and subtract finite sets as one does vectors.

Nevertheless, the parallelism is complete enough that, with the exercise of some prudence, a hundred years of conventional statistics can be *directly* brought to bear on multisensor, multitarget information fusion problems. We offer the following specific examples of the potential utility of this parallelism.

Data Fusion Information Metrics

Suppose that we wish to attack the problem of *performance evaluation* of information fusion algorithms in a scientifically defensible manner. One can define a

"global" version of the conventional Kullback-Leibler metric which is applicable
to performance evaluation problems in multisensor, multitarget data fusion:

$$I(f_\Sigma; f_\Gamma) \triangleq \int f_\Sigma(X) \ln \left(\frac{f_\Sigma(X)}{f_\Gamma(X)} \right) \delta X$$

(See Section 5 3 1 of Chapter 5 and Section 8.1 of Chapter 8.)

Multisensor, Multitarget ROC Curves

The *Receiver Operating Characteristic (ROC) curve* of an entire multisensor,
multitarget system can be defined as the parametrized curve $\tau \mapsto (p_{FA}(\tau), p_D(\tau))$
where

$$p_{FA}(\tau) \triangleq \int_{L(Z_1,...,Z_k)>\tau} f_{\Sigma_1,...,\Sigma_k|H_1}(Z_1, ..., Z_k|H_0) \, \delta Z_1 \cdots \delta Z_k$$

$$p_D(\tau) \triangleq 1 - \int_{L(Z_1,...,Z_k)>\tau} f_{\Sigma_1,...,\Sigma_k|H_0}(Z_1, ..., Z_k|H_1) \, \delta Z_1 \cdots \delta Z_k$$

and where

$$L(Z_1, ..., Z_k) \triangleq \frac{f_{\Sigma_1,...,\Sigma_k|H_0}(Z_1, ..., Z_k|H_0)}{f_{\Sigma_1,...,\Sigma_k|H_1}(Z_1, ..., Z_k|H_1)}$$

is the "global" likelihood ratio for the problem of deciding between hypotheses
H_0 and H_1 given global observations $Z_1, ..., Z_m$. (Problems of this type will not
be considered further in this book, however.)

Multisensor, Multitarget Estimation

One can define multisensor, multitarget analogs of conventional statistical esti-
mators. A *global estimator* of the set parameter of a global density $f_\Sigma(Z|X)$
is a function $\hat{X} = J(Z_1, ..., Z_m)$ of global measurements $Z_1, ..., Z_m$. One can
define a multisensor, multitarget version of the maximum likelihood estimator
(MLE) by

$$J_{GMLE}(Z_1, ..., Z_m) \triangleq \arg \sup_X L(X|Z_1, ..., Z_m)$$

where $L(X|Z_1, ..., Z_m) \triangleq f_\Sigma(Z_1|X) \cdots f_\Sigma(Z_m|X)$ is the "global" likelihood func-
tion. Notice that in determining X, one is estimating not only the geokinematics
and identities of targets, but also their number as well. Thus detection, localization, and identification are unified into a single statistical operation. This
operation is a *direct* multisource, multitarget estimation technique in the sense
defined in section 2.2.2. The definition of a multisensor, multitarget analog of
the MAP estimator is less straightforward but also possible. See Section 5.2 of
Chapter 5 for more details.

Cramér-Rao Inequalities

One can show, under assumptions analogous to those used in the proof of the conventional Cramér-Rao inequality, that the following holds for a *vector-valued* multisensor, multitarget estimator **J**:

$$\langle \mathbf{v},\, C_{\mathbf{J},\mathbf{x}}\rangle \cdot \langle \mathbf{w},\, L_{X,\mathbf{x}}(\mathbf{w})\rangle \geq \langle \mathbf{v},\, \frac{\partial}{\partial_{\mathbf{x}}\mathbf{w}} \mathbf{E}_X[\mathbf{X}]\rangle$$

for all $\mathbf{v}, \mathbf{w}$, where $\mathbf{X} \triangleq \mathbf{J}(\Sigma_1, ..., \Sigma_m)$ and where $L_{X,\mathbf{x}}$ is defined by

$$\langle \mathbf{v},\, L_{X,\mathbf{x}}(\mathbf{w})\rangle = \mathbf{E}_X\left[\left(\frac{\partial \ln f}{\partial_{\mathbf{x}}\mathbf{v}}\right)\left(\frac{\partial \ln f}{\partial_{\mathbf{x}}\mathbf{w}}\right)\right]$$

for all $\mathbf{v}, \mathbf{w}$, where $f = f_{\Sigma_1,...,\Sigma_m|\Gamma}$, and where the directional derivative $\partial f / \partial_{\mathbf{x}}\mathbf{v}$ of the function $f(X)$ of a finite-set variable X, if it exists, is defined by

$$\frac{\partial f}{\partial_{\mathbf{x}}\mathbf{v}}(X) \triangleq \lim_{\varepsilon \to 0} \frac{f((X - \{\mathbf{x}\}) \cup \{\mathbf{x} + \varepsilon\mathbf{v}\}) - f(X)}{\varepsilon} \qquad \text{(if } \mathbf{x} \in X)$$

$$\frac{\partial f}{\partial_{\mathbf{x}}\mathbf{v}}(X) \triangleq 0 \qquad \text{(if } \mathbf{x} \notin X)$$

See Section 5.3 of Chapter 5 for more details.

Multisensor, Multitarget Nonlinear Filtering

One can establish the existence of multisensor, multitarget analogs of the conventional Bayes-Markov discrete-time nonlinear filtering equations:

$$f_{\alpha+1|\alpha+1}(X_{\alpha+1}|Z^{(\alpha+1)}) = \frac{f(Z_{\alpha+1}|X_{\alpha+1})\, f_{\alpha+1|\alpha}(X_{\alpha+1}|Z^{(\alpha)})}{\int f(Z_{\alpha+1}|Y_{\alpha+1})\, f_{\alpha+1|\alpha}(Y_{\alpha+1}|Z^{(\alpha)})\, \delta Y_{\alpha+1}}$$

$$f_{\alpha+1|\alpha}(X_{\alpha+1}|Z^{(\alpha)}) = \int f_{\alpha+1|\alpha}(X_{\alpha+1}|X_\alpha)\, f_{\alpha|\alpha}(X_\alpha|Z^{(\alpha)})\, \delta X_\alpha$$

The global density $f(Z_{\alpha+1}|X_{\alpha+1})$ is the measurement model for the multisensor, multitarget problem, and the global density $f_{\alpha+1|\alpha}(X_{\alpha+1}|X_\alpha)$ is the motion model *for the entire multitarget system.* The second equation is the multitarget analog of the Markov prediction integral, and the first equation is the analog of the Bayesian information-update equation. See Section 6.4 of Chapter 6 for more details.

Sensor Management: Multisensor Nonlinear Control Theory

Recall that *sensor management* is the problem of controlling the re-directable sensors in a sensor suite in order to resolve ambiguities in our knowledge about multiple sensors. The parallelism between point-variate and finite-set-variate statistics suggests that one way of attacking this problem is by first examining the *single-sensor, single-target* case: E.g., a missile-tracking camera as it attempts to follow a missile. The camera must adjust its azimuth, elevation, and focal length in such a way as to anticipate the location of the missile at the time the next image of the missile is recorded. This is a standard problem in *optimal control theory*. Such problems are solved by defining a *controlled vector*, associated with the camera, and a *reference vector*, associated with the target, and attempting to keep the distance between these two vectors as small as possible.

An approach to the multisensor, multitarget *sensor management* problem becomes evident if we use the random set approach to reformulate such problems as a single-sensor, single-target problem. In this case the "global" sensor follows a "global" target (some of whose individual targets may not even be detected yet). The motion of the multitarget system is modeled using a global Markov transition density. The only undetermined aspect of the problem is how to define analogs of the controlled and reference vectors. This is done by determining the Kullback-Leibler information distance between two suitable global densities. Problems of this kind will not be considered further in this book (for more details see [86]).

2.5.6 Data Fusion Using Ambiguous Evidence

In section 2.3 we summarized research which shows that many kinds of ambiguous evidence can be modeled as random sets. It is possible to show, as we will do in Chapter 7, that the two sides of data fusion—multisensor, multitarget estimation on the one hand and expert-systems theory on the other—can be fully integrated.

Specifically, one begins by modeling ambiguous observations as *discrete closed random subsets of observation space* and then specifying *measurement models* for ambiguous evidence. Figures 2.8 and 2.9 illustrate how ambiguous evidence can be modeled using discrete closed random subsets. Suppose that we have a statement such as

"target 1 is *NEAR* location A and target 2 is *NEAR* location B"

Suppose that *NEAR* in these two cases can be interpreted to mean that target 1 (and therefore observations of target 1) will always be found in "cookie cutter" region G_1 and, likewise, that observations of target 2 will always be found in "cookie cutter" region G_2. Then Figure 2.8 illustrates the constraint that is imposed by this evidence. In general, however, evidence will consist not of a simple cookie-cutter constraint G_1 but rather of a range of constraints $G_1^{(1)} \subseteq ... \subseteq G_1^{(d)}$, each being an interpretation of the concept *NEAR* to some degrees $r_1, ..., r_d$ of likelihood. This evidence can be modeled as a random subset Θ_1 of observation space such that $p(\Theta_1 = G_1^{(i)}) = r_i$ for all $i = 1, ..., d$, as illustrated in Figure 2.9.

Measurement models for ambiguous observations consist of *global measurement densities* which have the form

$$f(Z_1, ..., Z_m, \Theta_1, ..., \Theta_{m'}|X)$$

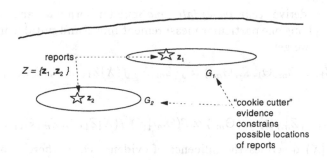

Figure 2.8: The statement "target 1 is *NEAR* location A and target 2 is *NEAR* location B" is illustrated when the concept *NEAR* is interpreted as constraining observations within "cookie cutter" regions G_1 and G_2 of locations A and B, respectively.

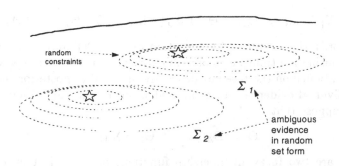

Figure 2.9: The statement "target 1 is *NEAR* location A and target 2 is *NEAR* location B" illustrated when the concept *NEAR* is interpreted as a variable constraint by *random subsets*, Θ_1 and Θ_2.

where $Z_1, ..., Z_m$ denote precise global observations, and where $\Theta_1, ..., \Theta_{m'}$ are discrete random closed subsets of observation space which model ambiguous evidence. From these measurement models one can then define global posterior densities *conditioned on both data and evidence:*

$$f(X|Z_1, ..., Z_m, \Theta_1, ..., \Theta_{m'})$$

Then one can derive recursive update equations for both data and evidence. For example, applying one particular measurement model and suitable independence assumptions, we get

$$f(X|Z_1, ..., Z_m, \Theta_1, ..., \Theta_{m'}) \propto f(Z_m|X)\, f(X|Z_1, ..., Z_{m-1}, \Theta_1, ..., \Theta_{m'})$$

and

$$f(X|Z_1, ..., Z_m, \Theta_1, ..., \Theta_{m'}) \propto \beta(\Theta_{m'}|X)\, f(X|Z_1, ..., Z_m, \Theta_1, ..., \Theta_{m'-1})$$

where $\beta(\Theta|X)$ describes the influence of evidence Θ. It thereby becomes possible to extend the Bayesian nonlinear filtering equations so that both precise and ambiguous observations can be accommodated into dynamic multisensor, multitarget estimation. For example, if one assumes one possible measurement model for evidence, one gets the following update equation:

$$f_{\alpha+1|\alpha+1}(X_{\alpha+1}|Z^{(\alpha)}, \Theta^{(\alpha+1)}) = \frac{\beta(\Theta_{\alpha+1}|X_{\alpha+1})\, f_{\alpha+1|\alpha}(X_{\alpha+1}|Z^{(\alpha)}, \Theta^{(\alpha)})}{\int \beta(\Theta_{\alpha+1}|Y_{\alpha+1})\, f_{\alpha+1|\alpha}(Y_{\alpha+1}|Z^{(\alpha)}, \Theta^{(\alpha)})\, \delta Y_{\alpha+1}}$$

where $\Theta^{(\alpha)} \triangleq \{\Theta_1, ..., \Theta_\alpha\}$.

A Bayesian Interpretation of Rules of Evidential Combination

If one specific measurement model for ambiguous evidence is assumed—the so-called "data-dependent" model—then posterior distributions will have the property that

$$f(X|Z_1, ..., Z_m, \Theta_1, ..., \Theta_{m'}) = f(X|Z_1, ..., Z_m, \Theta_1 \cap ... \cap \Theta_{m'})$$

In other words, the random-set intersection operator '$\cap$' may be interpreted as a means of fusing multiple pieces of ambiguous evidence in such a way that posteriors conditioned on the fused evidence are identical to posteriors conditioned on the individual evidence and computed using Bayes' rule alone. Thus, for example, suppose that

$$\Theta_1 = \Sigma_A(f), \qquad \Theta_2 = \Sigma_A(g)$$

where f, g are two fuzzy membership functions on U and A is a uniformly distributed random number on $[0, 1]$. Let '$\wedge$' denote the Zadeh "min" fuzzy *AND* operation and *define* the posteriors $f_{\Gamma|\Sigma}(X|Z, f, g)$ and $f_{\Gamma|\Sigma}(X|Z, f \wedge g)$ by

$$f(X|Z, f, g) = f(X|Z, \Sigma_x(f), \Sigma_x(g))$$
$$f(X|Z, f \wedge g) = f(X|Z, \Sigma_x(f \wedge g))$$

We know from section 2.3.2 that $\Sigma_A(f) \cap \Sigma_A(g)) = \Sigma_A(f \wedge g)$ and thus that

$$
\begin{aligned}
f(X|Z, f, g) &= f(X|Z, \Sigma_A(f), \Sigma_A(g)) = f(X|Z, \Sigma_A(f) \cap \Sigma_A(g)) \\
&= f(X|Z, \Sigma_A(f \wedge g)) = f(X|Z, f \wedge g)
\end{aligned}
$$

and so

$$
f(X|Z, f, g) = f(X|Z, f \wedge g)
$$

That is: the fuzzy *AND* is a means of fusing fuzzy evidence in such a way that posteriors conditioned on the fused evidence are identical to posteriors conditioned on the fuzzy evidence individually and computed using Bayes' rule alone. Thus fuzzy logic is entirely consistent with Bayesian probability—*provided that it is first represented in random set form*, and *provided that we use a specific measurement model for ambiguous observations*. Similar observations apply to *any* rule of evidential combination which bears a homomorphic relationship with the random set intersection operator.

2.5.7   Possible Objections to "Finite-Set Statistics"

A number of objections to the use of random set theory in data fusion can be envisaged. We address some of these in turn.

Isn't It Just "Bookkeeping"?

The reformulation of multisource, multitarget problems as single-sensor, single-target problems is not just a mathematical "bookkeeping" device. Generally speaking, *any group of targets observed by imperfect sensors must be analyzed as a single indivisible entity rather than as a collection of unrelated individuals*. When measurement uncertainties are large in comparison to target separations there will always be a significant likelihood that any given measurement was generated by any given target. This means that *every* measurement can be associated, partially or in some degree of proportion, to *every* target. The more irresolvable the targets are, the more our estimates of them will be statistically correlated and thus the more that they will seem as though they are a *single* target. Observations can no longer be regarded as separate entities generated by individual targets but rather as collective phenomena *generated by the entire multitarget system*. This remains true even when target separations are large in comparison to sensor uncertainties. Though in this case the likelihood is very small that a given observation was generated by any other target than the one it is intuitively associated with, nevertheless this likelihood is nonvanishing.

Why Not Vectors or Point Processes?

Other skeptics might ask: *Why not simply use vector models or point process models?* It could be objected, for example, that what we will later call the "global density" $f_\Sigma(Z)$ of a finite random subset Σ is just a new name and notation for the so-called "Janossy densities" $j_n(z_1, ..., z_n)$, $n \geq 0$ [5, pp. 122-123] of the corresponding simple finite point process defined by $N_\Sigma(S) \triangleq |\Sigma \cap S|$ for all measurable S.

In response to possible such objections we offer the following responses.

First, vector approaches *encourage carelessness in regard to basic questions.* For example, to apply the theorems of conventional estimation theory one must clearly identify a measurement space and a state space and specify their topological and metrical properties. Wald's classic proof of the consistency of the maximal likelihood and maximum *a posteriori* estimators, for example, assumes that state space is a *metric space.* As another instance suppose that we want to determine whether small deviations in the input data to a data fusion algorithm can result in large deviations in output data. To answer this question one must first have some idea of what *distance* means in both measurement space and state space. The standard Euclidean metric is clearly not adequate: If we represent an observation set $\{z_1, z_2\}$ as a vector (z_1, z_2) then $\|(z_1, z_2) - (z_2, z_1)\| \neq 0$ even though the order of measurements should not matter. Likewise one might ask, What is the distance between (z_1, z_2) and z_3? Whereas both finite set theory and point process theory have rigorous metrical concepts, attempts to define metrics for vector models can quickly degenerate into *ad hoc* invention.

More generally, the use of vector models has resulted in piecemeal solutions to information fusion problems (most typically, the assumption that the number of targets is known *a priori*). Lastly, any attempt to incorporate expert systems theory into the vector approach results in extremely awkward attempts to make vectors behave as though they were finite sets.

Second, the random set approach is *explicitly geometric* in that the random variates in question are actual sets of observations—rather than, say, abstract integer-valued measures.

Third, and as we saw in section 2.5.5, systematic adherence to a random set perspective results in a series of direct parallels between single-sensor, single-target statistics and multisensor, multitarget statistics which results in a methodology for information fusion that is nearly identical in general behavior to the "Statistics 101" formalism with which engineering practitioners and theorists are already familiar. More importantly, it leads to a systematic approach to solving information fusion problems that allows standard single-sensor, single-target statistical techniques to be directly generalized to the multisensor, multitarget case.

Fourth, because the random set approach provides a systematic foundation for both expert systems theory *and* multisensor, multitarget estimation (section 2.5.6), it permits a systematic and mathematically rigorous integration of these two quite different aspects of information fusion—a question left unaddressed by either the vector or point-process models.

Fifth, an analogous situation holds in the case of random subsets of $\mathbb{R}^n$ which are *convex* and *bounded.* Given a bounded convex subset $T \subseteq \mathbb{R}^n$, the *support function* of T is defined by $s_T(e) \triangleq \sup_{x \in T} \langle e, x \rangle$ for all vectors e on the unit hypersphere in $\mathbb{R}^n$, where $'\langle -, - \rangle'$ denotes the inner product on $\mathbb{R}^n$. The assignment $\Sigma \to s_\Sigma$ establishes a very faithful embedding of random bounded convex sets Σ into the random functions on the unit hypersphere, in the sense that it encodes the behavior of bounded convex sets into vector mathematics [19], [16]. Nevertheless, it does not follow that the theory of random bounded convex subsets is a special case of random function theory. Rather, random functions provide a useful tool for studying the behavior of random bounded convex sets.

In like manner, finite point processes are best understood as specific—and by no means the only or the most useful—representations of random finite subsets as elements of some abstract vector space.

2.6   Conditional and Relational Event Algebra

When we speak of probability we really mean two related, but distinct, concepts: an algebraic or logical part in the form of the structure of relevant events considered for evaluating; and the probability evaluation itself. In the way probability and statistics have evolved, emphasis has been placed on the latter, rather than the former. Conditional and relational event algebra are new mathematical techniques for dealing with problems where the underlying event structure for certain probabilistic concepts have not been considered in the mainline development of probability, and consequently a number of issues – especially concerning similarity or closeness of events – have not been adequately addressed.

Essentially two basic cases arise concerning the relationship between logical structure and probability evaluation proper in the modeling of the uncertainty of complex real-world phenomena – especially for C3 (Command, Control, and Communication) and data fusion problems.

Case 1. Logical structure of overall model initially known.

Often in modeling the uncertainty of the occurrences of complex real world phenomena – in particular, data fusion problems – one knows how to decompose the relevant compound events in question into finite boolean logical combinations – together with possibly related operations such as cartesian products – of simpler contributing events, whose probabilities are obtainable. In turn, such contributing event probabilities may also be obtained by possibly further logical decomposition of even simpler events, etc. Then, the standard laws of probability can be applied, showing how the overall model uncertainty is a computable function of the probabilities of the contributing event probabilities. Thus, once the logical structures of the models as single compound events are known, one can then compare, contrast, and combine them when appropriate. For example, the model uncertainty may represent a sum of probabilities, each probability corresponding to a disjoint event. Thus, the overall model uncertainty can also be considered to be the probability evaluation of a single (compound) event, namely the (disjoint) union of the two contributing events. Or, the uncertainty perhaps may be in a simple product of probabilities form, whereupon it may also be considered the product probability evaluation of a single event, namely, the cartesian product of the contributing events. Still more complicated situations can arise, where various combinations of arithmetic sums, differences, and products are involved in the overall model uncertainty description but which also can be considered as the probability evaluation of a single compound event consisting of a well-defined naturally corresponding finite boolean logical or cartesian product-sum combination of simpler events.

When two relatively complex models have a known logical decomposition, using the standard laws of boolean logic we can determine any binary boolean operator acting on them; in particular, their conjunction, and in turn, the probability of their conjunction. Next, using the separate overall model probabilities, as well as the above conjunction probability, we can compute various probability distances between the two models, such as the natural probability distance, a well-known standard measure of similarity between events [8], or the relative probability distance, among others [47]. For example, one could use simply the absolute difference of probabilities as a legitimate distance measure among events, but this measure is for various situations insensitive to the degree of disjunction or conjunction – and hence distinctness – of events. In fact, one could have two events with probability measures, say around 0.4, so that the absolute difference of probabilities is close to zero, indicating similarity, yet the events can be chosen to be completely disjoint ! Another measure of similarity, apparently superior to both the natural distance one and the absolute difference one is obtained by normalizing the natural probability distance by the disjunction probability. (See again [47] for a discussion of these and related concepts.) Furthermore, by higher order probability assumptions, this measure can also be actually used to test hypotheses of similarity [48]. But, essentially, the implementation of the nontrivial measures of similarity crucially depend on knowledge of the conjunctive probabilities of the overall model events.

Case 2. Logical structure of overall model initially unknown.

At times, the situation opposite to Case 1 occurs: the overall model uncertainty is represented as a known function of the probabilities of (known) contributing event probabilities, but the underlying logical and related structure of the model is not known initially. A basic example where the overall model event logical structure is not initially known, but where the function of probability of the contributing simpler events is known (in this case, arithmetic division) is when conditional probabilities are used to describe models. That is, there are no obvious single "conditional events" lying in the same probability space whose probability evaluation (by the same probability measure) yields back the two given conditional probabilities. Another example is subjective probability evaluation in the form of a weighting or importance combination of contributing event probabilities, where the actual logical relation among the latter events is not known, but the individual probabilities – or even the probabilities of their joint occurrences – are known. This can happen, e.g., when experts model the probability of enemy attack tomorrow as a weighted linear combination of contributing events b_i , such as the probabilities of true enemy buildup exceeding a certain threshold, and weather probabilities including ranges of expected temperature, precipitation, wind, and visibility. When the contributing event probabilities bi are mutually disjoint and exhaustive, these relations appear formally the same as the standard total probability expansions of the probability of compound events, say a, into sums of $w_i P(b_i)$, $w_i = P(a|b_i)$, but where the w_i are held fixed for a reasonably large class of probabilities P. In the standard development of probability since conditional events are not considered, there is no way to represent each such a in purely logical structure relating to the w_i and the b_i. Hence, again, here is no way to obtain the conjunction – and conjunction probability – of such models. And this implies there is no way to obtain in a sound manner, using standard probability techniques, measures of similarity between such models. (But, when conditional events are introduced, it can be shown that a certain class of such events can be used to represent the weights w_i, independent of the choice of P. See later comments.) The same issue also applies to the situation when the contributing events are not necessarily mutually disjoint. In a similar vein, the experts may instead of restricting themselves to linear combinations, may also utilize nonlinear functions of contributing probabilities, including exponentials, polynomials, or various functional forms. Such linear or nonlinear functions of probabilities representing the overall model uncertainties may also arise from the use of natural language descriptions evaluated through fuzzy logic, where modifiers can be in the form of exponentials and fuzzy partitioning may be employed in the modeling [46]. This is because it has now been established (though, unfortunately, not as well-known in the fuzzy logic and AI communities as it should be) that many basic fuzzy logic concepts have homomorphic-like representations via the one-point coverage events of corresponding random sets (in general, the correspondence between fuzzy logic concepts and boolean logic ones applied to corresponding random sets is infinitely many to one). All of this revolves around the key idea that a given fuzzy set can be identified as the one-point coverage function of any of a certain class of corresponding random subsets of the domain of that fuzzy set [40], [49].

The following basic statement and question then arises: Clearly, Case 1 allows us to establish direct computable measures and tests of similarity between competing models. Can we analogously obtain (in a sound and comprehensive manner) computable measures of similarity between models when their uncertainties satisfy Case 2?

In brief, the answer to the above question remains open, so long as the standard development of probability and statistics is pursued. The overwhelming body of literature connected with the latter fields has ignored this basic issue. In fact, even for the simple case where the uncertainties of two models are each provided through single conditional probabilities – such as the case for the models being inference rules of the form "if b, then a", "if d, then c", with natural corresponding conditional probability evaluations $P(a|b)$, $P(c|d)$, respectively, when antecedent events b and d are distinct, exhaustive literature searches have shown that standard probability theory has provided no means for determining the meaning of the possible underlying "conditional events", symbolically denoted without the probability operator as $(a|b)$, $(c|d)$, respectively. Thus, since there is no standard way to obtain the conjunction of $(a|b)$ and $(c|d)$, there is no standard means to obtain the natural absolute probability distance between $(a|b)$ and $(c|d)$, and hence, there is no standard approach for determining how similar or dissimilar the two models are, unlike Case 1. Of course, when the antecedents are identical, standard probability techniques can be use, via the normalization of the common antecedent probability and the trace space resulting by conjoining all relevant events with the antecedent, etc. But, even when the antecedents are approximately the same in some sense – but truly not identical – the same problem described above maintains. Another simplifying case where conditional events need not be introduced explicitly occurs when it is reasonable to conclude that the conditional expressions are independent of each other, in which case the usual factoring properties could be employed. A related situation which avoids the introduction of explicit conditional events is when a sequence of events are in a chaining or modus ponens form, best illustrated by the standard identity

$$P(a|bcd)P(b|cd)P(c|d) = P(abc|d),$$

where in effect, the "conditional events" represented by the left-hand conditional probabilities are all formally independent and their conjunction probability is the single conditional probability on the right-hand side.

As a result of the above difficulties connected with Case 2, and in particular for conditional probabilities, conditional event (probability) algebra was developed. Briefly, a conditional event algebra is a mathematical entity consisting of a the set of all possible conditional events formed out of an initial boolean (or sigma-) algebra of (relatively) unconditional events – which include an imbedding of all of the initial unconditional events as a special subclass – and which do not depend structurally on the choice of any specific probability measure, together with an extension of the original unconditional probability measure over the unconditional events, so that the evaluation by this extension on each such conditional event is the natural conditional probability measure, for any non-zero probability antecedent employed. Furthermore, the conditional event algebra is only useful when one can compute through a reasonably finite number of simple operations, the probability of any finite logical combination of conditional events. When the conditional event algebra can be considered a legitimate probability space, we call such an algebra a boolean conditional event algebra. This relatively new field actually has origins traceable to the nineteenth century to Boole himself [16] who basically reasoned that if arithmetic addition, subtraction and multiplication have naturally corresponding event operations compatible with probability evaluations, why should not the division of events (restricted of course to the antecedent being a subevent of the consequent)? Boole then proceeded to attempt to provide meaning to such conditional events and even developed formal canonical expansions and some limited use for them, but did not consider the problem of combining them for differing antecedents. However, the main body of logic and probability ignored this. Even over a hundred years later when Hailperin attempted to rigorize Boole's limited experimenting with conditionals [56], [57], the important general case of non-identical conditionals was not treated. However, the independent work of DeFinetti in 1935 [24], [25], and Schay [19] in 1968 offered for the first time full conditional event algebras. (In fact Schay developed two different conditional event algebras and even produced a Stone-like representation theorem. But, for the most part, his work, even though published in a leading journal was overlooked – see any citation and review listing of the period.) This was followed by independent work of Adams [1], [2], and Calabrese [18] (both offering almost identical algebras), followed by Goodman [12] and later Goodman, Nguyen and Walker [1] (both works producing an algebra similar in basic ways to DeFinetti's) , Dubois and Prade [28], and a number of others. (See Goodman [5] for a detailed survey of the history of the conditional event algebra problem up to 1990.) All of the above efforts led to non-boolean conditional event algebras. However, independently, Van Fraasen [132], McGee [95], and Goodman and Nguyen [52], from three different approaches obtained the same conditional event algebra, which turned out also to be boolean! We designate this algebra as the product space conditional event algebra, due to its structural form as the countable infinite cartesian product of the given probability space of unconditional events. Moreover, Goodman and Nguyen later showed that this algebra possesses many desirable properties and demonstrated certain characterizing theorems and explicit relations with the previously proposed nonboolean algebras of Adams-Calabrese and Goodman and Nguyen [53]. Interest in conditional event algebra has grown (see, e.g. the December 1994 issue of *IEEE Transactions on Systems, Man and Cybernetics*, devoted entirely to the subject) and, in fact, the field has been expanded – now denoted relational event algebra – to address the more general problem as outlined in Case 2 above [47], [48], [49]. In conjunction with all of this mention must be made of Lewis' "triviality" theorem [80], which states that both unconditional and conditional events cannot exist in the same probability space. However, this does not preclude the construction of a probability space properly extending the original one associated with the unconditional events and imbedding the unconditional events in the bigger space – but noting that the imbeddings – a s natural as they are – are in general not identities, and Lewis' Theorem is not violated. Nevertheless, the theorem – much more known in the logical community than the probability one – has apparently caused a lack of interest in developing conditional event algebra, because of the mistaken belief that, a priori, conditional events and unconditional events should all literally lie in the same probability space (to this end, see the recent edited work of Eells and Skyrms [30]), much as the possibly misguided work of Copeland and others who attempted, in effect, to construct such conditional event algebras, called in their terminology, implicative boolean algebras [21].

Generalizing the concept of a boolean conditional event algebra, a relational event (probability) algebra is a probability space extending in a natural way an initially given probability space of unconditional events to now include finitely determinable relational events, i.e., events whose structure, analogous to conditional events, do not depend on any particular probability measure chosen, all lying in the extended probability space whose probability evaluations coincide with the particular functions of event probabilities of interest [47]. Generally speaking, the structure of a typical relational event representing an analytic function of probabilities is a disjoint disjunction of cartesian products of events, each product consisting of an integer exponent term (i.e., itself a multi-fold cartesian product) and a coefficient term — so that its probability evaluation takes the form of a polynomial or infinite series of probabilities. Each coefficient term is in the form of a certain type of conditional event – called a constant-probability event – whose probability evaluations, regardless of the choice of initial probability measure (up to certain limitations) is the corresponding constant coefficient value. Conditional events in the product space conditional event algebra can be considered special cases of relational events in the above form where each coefficient is unity. (See [49] for more details on relational events representing analytic functions of probabilities.)

Relational event algebra is not as developed as product space conditional event algebra, but a number of particular classes of analytic functions of probabilities have proven to be satisfactorily representable, including weighted sums, polynomials and power series (or analytic functions) in one variable, including fractional exponentials, and certain miscellaneous functions. On the other hand, there are (necessarily, non-analytic) functions which do not appear to be possible to represent through relational events, such as the absolute difference, or even plain unweighted sums. However, the later can also be treated in the sense that corresponding appropriately modified relational events can be constructed which now depend structurally, in part, on the particular probability chosen probability measure of interest. So far – unlike product space conditional event algebra – few characterizations and related properties have been obtained for relational event algebra, but the expectation is that such will eventually be derived. The long-range goal here is the development of an algebraic analogue to standard numerically-based decision-making theory, utilizing similarity measures between relational events.

Bibliography

[1] Adams, E. W. (1966) Probability and the logic of conditionals, in *Aspects of Inductive Logic*, J. Hintikka and P. Suppes (eds.), North-Holland, Amsterdam, pp. 265-316.

[2] Adams, E. W. (1975) *The Logic of Conditionals*, North Holland, Dordrecht.

[3] Aigner, M. (1979) *Combinatorial Theory*, Springer-Verlag, New York.

[4] Alhakeem, S. and Varshney, P. K. (1995), A Unified approach to the design of decentralized detection systems, *IEEE Transactions on Aerospace and Electronic Systems*, **31**, 9-20.

[5] Alspach, D. L. (1975) A Gaussian Sum Approach to the Multi-Target Identification-Tracking Problem, *Automatica*, **11**, 285-296.

[6] Antony, R. T. (1995) *Principles of data fusion automation*, Artech House, Boston.

[7] Barlow, C. A., Stone, L. D., and M.V. Finn (1996) Unified data fusion, *Proceedings of the Ninth National Symposium on Sensor Fusion*, **I** (Unclassified), Naval Postgraduate School, Monterey CA. To appear.

[8] Bar-Shalom, Y. ed. (1990) *Multitarget-Multisensor Tracking: Advanced Applications*, Artech House, Boston.

[9] Bar-Shalom, Y. (1978) Tracking methods in a multitarget environment, *IEEE Transactions on Automatic Control*, **AC-23**, 618-626.

[10] Bar-Shalom, Y. and Fortmann, T. E. (1988) *Tracking and Data Association*, Academic Press, New York.

[11] Bar-Shalom, Y. and Li, X. R. (1993) *Estimation and Tracking: Principles, Techniques, and Software*, Artech House, Boston.

[12] Bar-Shalom, Y., and Li, X.-R. (1995) *Multitarget-Multisensor Tracking: Principles and Techniques*, YBS (private publication)

[13] Black, P. K. (1996) An Examination of Belief Functions and Other Monotone Capacities, Doctoral Dissertation, Carnegie-Mellon University

[14] Blackman, S. S. (1988) *Multiple-Target Tracking with Radar Applications*, Artech House, Dedham MA.

[15] Bonissone P. P. and Wood, N. C. (1989) T-norm based reasoning in situation assessment applications, in L. N. Kanal, T.S. Levitt, and J.F. Lemmer, eds., *Uncertainty in Artificial Intelligence 3*, Elsevier, pp. 241-256.

[16] Boole, G. (1854) *An Investigation of The Laws of Thought*, Walton & Maberly, London; reprinted 1951, Dover Publications, New York.

[17] Chaffee, S., Poore, A. B., Rijavec, R., Gassner, R., and Vannicola, V. C. A centralized fusion multisensor/multitarget tracker based on multidimensional assignments for data association, *SPIE Proceedings*, **2484**, 114-125.

[18] Calabrese, P. G. (1987) An algebraic synthesis of the foundations of logic and probability, *Information Sciences*, **42**, 187-237.

[19] Chong, C. Y., Mori, S., and Chang, K. C. (1990) Distributed multitarget multisensor tracking, in Y. Bar-Shalom, ed., *Multitarget-Multisensor Tracking: Advanced Applications*, Artech House, Chapter 8.

[20] Choquet, G. Theory of Capacities, *Ann. Inst. Fourier*, **5** (1953/54), 131-295.

[21] Copeland, A. H. (1950) Implicative boolean algebras, *Mathematische Zeitschrift*, **53**(3), 285-290.

[22] Daley, D. J., and Vere-Jones, D. (1988) *An Introduction to the Theory of Point Processes*, Springer-Verlag.

[23] Daum, F. (1992) A system approach to multiple target tracking, in Y. Bar-Shalom (ed.), *Multitarget-Multisensor Tracking: Applications and Advances*, Vol. II, Artech House, Dedham MA.

[24] DeFinetti, B. (1936) La logique de la probabilite in *Actes du Congres International de Philosophie Scientifique*, Hermann et Cie (eds.), Paris, pp. IV1-IV9.

[25] DeFinetti, B. (1974) *Theory of Probability*, volume I,II, J. Wiley, New York.

[26] Dubois, D., and Prade, H. (1990) Consonant approximations of belief functions, *International Journal of Approximate Reasoning*, **4**, 419-449.

[27] Dubois, D., and Prade, H. (1994) Fuzzy sets–a convenient fiction for modeling vagueness and possibility, *IEEE Transactions on Fuzzy Systems*, **2** (1), 22-26.

[28] Dubois, D. and Prade, H. (1994) Conditional objects as nonmonotonic consequence relationships, *IEEE Transactions on Systems, Man and Cybernetics*, **24**(12), 1724-1740.

[29] Durrett, R. (1996) *Probability: Theory and Examples*, Second Edition, Duxbury Press, Belmont.

[30] Eells, E. and Skyrms, B. (1994) *Probability and Conditionals*, Cambridge University Press, Cambridge.

[31] Fagin, R. and Halpern, J. Y. (1991) A new approach to updating beliefs, in P.P. Bonissone, M. Henrion, and J.F. Lemmer eds., *Uncertainty in Artificial Intelligence 6*, pp. 347-374.

[32] Fagin, R. and Halpern, J. Y. (1991) Uncertainty, belief, and probability, *Computer Intelligence*, **7**, 160-173.

[33] Fishman, P. M. and Snyder, D. L. (1976) The statistical analysis of space-time point processes, *IEEE Transactions on Information Theory*, **IT-22**, 257-274.

[34] Fixsen, D. and Mahler, R. (1997) The modified Dempster-Shafer approach to classification, *IEEE Transactions on Systems, Man and Cybernetics*, **27** (1), 96-104.

[35] Fortmann, T. E., Bar-Shalom, Y., and Scheffe, M. (1983) Sonar tracking of multiple targets using joint probabilistic data association, *IEEE Journal of Oceanic Engineering*, **OE-8**, 173-183.

[36] Goicoechea, A. (1988) Expert system models for inference with imperfect knowledge: a comparative study, *Journal of Statistical Planning and Inference*, **20**, 245-277.

[37] Goodman, I. R. (1979) A general model for the multiple target correlation and tracking problem, *Proceedings of the 18th IEEE Conference on Decision and Control*, Fort Lauderdale FL, pp. 383-388.

[38] Goodman, I. R. (1982) Fuzzy sets as equivalence classes of random sets, in R. Yager, ed., *Fuzzy Sets and Possibility Theory*, Permagon, pp. 327-343.

[39] Goodman, I. R. (1987) A general theory for the fusion of data, *Proceedings of the 1987 Joint Services Data Fusion Symposium,*.

[40] Goodman, I. R. (1994) A new characterization of fuzzy logic operators producing homomorphic-like relations with one-point coverages of random sets, in P.P. Wang, ed., *Advances in Fuzzy Theory and Technology, Vol. 2*, Duke University, Durham, pp. 133-159.

[41] Goodman, I. R. (1986) Pact: an approach to combining linguistic-based and probabilistic information in correlation and tracking, Technical Document 878, March 1986, Naval Ocean Command and Control Ocean Systems Center, RDT&E Division, San Diego; and "A Revised Approach to Combining Linguistic and Probabilistic Information in Correlation," Technical Report 1386, July 1992, Naval Ocean Command and Control Ocean Systems Center, RDT&E Division, San Diego.

[42] Goodman, I. R. (1994) Toward a comprehensive theory of linguistic and probabilistic evidence: Two new approaches to conditional event algebra," *IEEE Transactions on Systems, Man and Cybernetics*, 24, 1685-1698.

[43] Goodman, I. R. (1983) A unified approach to modeling and combining of evidence through random set theory, *Proceedings of the Sixth MIT/ONR Workshop on C3 Systems*, Boston, MA, pp. 42-47.

[44] Goodman, I. R. (1987) A measure-free approach to conditioning, *Proceedings Third AAAI Workshop on Uncertainty in Artificial Intelligence*, University of Washington at Seattle.

[45] Goodman, I. R. (1991) Evaluation of combinations of conditioned information: a history,*Information Sciences*, **57-58**, 79-110.

[46] Goodman, I. R. (1991) Similarity measures of events, relational events algebra and extensions to fuzzy logic, *1996 Biennial Conference of the North American Fuzzy Information Processing Society* - NAFIPS, University of California at Berkeley, pp. 187-191.

[47] Goodman, I. R. and Kramer, G. F. (1996) Extension of relational event algebra to a general decision making setting, *Proceedings of the Conference on Intelligent Systems: A Semiotic Perspective*, National Institute of Standards and Technology (NIST), Gaithersberg, MD, to appear.

[48] Goodman, I. R. and Kramer, G. F. (1996) Applications of relational event algebra to the development of a decision aid in Command and Control, *Proceedings of 1996 Command and Control Research and Technology Symposium*, Naval Postgraduate School, Monterey, CA, pp. 415-435.

[49] Goodman, I. R. and Kramer, G. F. (1996) Extension of relational and conditional event algebra to random sets with applications to data fusion, *Proceedings of the Workshop on Applications and Theory of Random Sets*, Institute for Mathematics and Its Applications (IMA), University of Minnesota at Minneapolis, to appear.

[50] Goodman, I. R. and Nguyen, H. T. (1985) *Uncertainty Models for Knowledge Based Systems*, North-Holland, Amsterdam.

[51] Goodman, I. R., Nguyen, H. T., and Walker, E. A. (1991) *Conditional Inference and Logic for Intelligent Systems: A Theory of Measure-Free Conditioning*, North-Holland.

[52] Goodman, I. R. and Nguyen, H. T. (1993,1994) A theory of conditional information for probabilistic inference in intelligent systems: II product space approach; III mathematical appendix, *Information Sciences*, **76**(1,2), 13-42; **75**(3), 253-277.

[53] Goodman, I. R. and Nguyen, H. T. (1995) Mathematical foundations of conditionals and their probabilistic assignments, *International Journal of Uncertainty, Fuzziness and Knowledge-Based Systems*, **3**(3), 247-339.

[54] Grabisch, M., Nguyen, H. T., and Walker, E. A. (1995) *Fundamentals of Uncertainty Calculi With Applications to Fuzzy Inference*, Kluwer Academic Publishers, Dordrecht.

[55] Gupta, M. M., Qi, J. (1991) Connectives (AND, OR, NOT) and T-operators in fuzzy reasoning, in I.R. Goodman, M.M. Gupta, H.T. Nguyen, and G.I. Rogers (eds.), *Conditional Logic in Expert Systems* , pp. 211-233.

[56] Hailperin, T. (1986) *Boole's Logic and Probability*, (Second Edition), North-Holland, Amsterdam.

[57] Hailperin, T. (1996) *Sentential Probability Logic*, Lehigh University Press, Bethlehem.

[58] Hall, D. L. (1992) *Mathematical Techniques in Multisensor Data Fusion*, Artech House, Dedham, MA.

[59] Halpern, J. Y. and Fagin, R. (1992) Two views of belief: belief as generalized probability and belief as evidence, *Artificial Intelligence*, **54**, 275-317.

[60] Hestir, K., Nguyen, H. T., and Rogers, G. S. (1991) A random set formalism for evidential reasoning, in I.R. Goodman, M.M. Gupta, H.T. Nguyen and G.S. Rogers, eds., *Conditional Logic in Expert Systems* , North-Holland, Amsterdam, pp. 309-344.

[61] Höhle, U. (1981) A mathematical theory of uncertainty: fuzzy experiments and their realizations, in R.R. Yager (ed.), *Recent Developments in Fuzzy Set and Possibility Theory*, Permagon Press, 344-355.

[62] Höhle, U. (1981) Representation theorems for L-fuzzy quantities, *Fuzzy Sets and Systems*, **5**, 83-107.

[63] Johnson, R. W. (1986) Independence and Bayesian updating methods, in L.N. Kanal and J.F. Lemmer, eds., *Uncertainty in Artificial Intelligence*, Elsevier, pp. 197-201.

[64] Kamen, E. W. (1992) Multiple target tracking based on symmetric measurement equations, *IEEE Transactions on Automatic Control*, **37**, 371-374.

[65] Kamen, E. W., Lee, Y. J., and Sastry, C. R. (1994) A parallel SME filter for tracking multiple targets in three dimensions, *SPIE Proceedings*, **2235**, 417-428.

[66] Kamen, E. W. and Sastry, C. R. (1993) Multiple target tracking using products of position measurements, *IEEE Transactions on Aerospace and Electronics Systems*, **29**, 476-493.

[67] Kamen, E. W., Sastry, C. R., and Sword, C. K. (1997) The SME filter approach to sensor fusion in multiple target tracking, *Proceedings of the 1993 SPIE Conference on Signal and Data Processing of Small Targets*, to appear.

[68] Kantor, P. B. (1991) Orbit space and closely spaced targets, *Proceedings of the SDI Panels on Tracking*, No. 2.

[69] Kappos, D. A. (1969) *Probability Algebra and Stochastic Processes*, Academic Press, New York. pp. 16-17 et passim.

[70] Kastella, K. (1995) Event averaged maximum likelihood estimation and mean-field theory in multi-target tracking, *IEEE Transactions on Automatic Control*, **40** (6), 1070-1074.

[71] Kastella, K. (1993) A maximum likelihood estimator for report-to-track association, *Proceedings of the 1993 SPIE Conference on OE/Aerospace and Remote Sensing: Signal and Data Processing of Small Targets*, SPIE Proc. Vol. 1954, 386-393.

[72] Kastella, K., and Lutes, C. (1993) Coherent maximum likelihood estimation and mean-field theory in multi-target tracking, *Proceedings of the Sixth Joint Service Data Fusion Symposium*, Vol. I (Part 2), Johns Hopkins Applied Physics Laboratory, Laurel, Maryland, June 14-18, 1993, 971-982.

[73] Krishnamurthy, V., and Evans, R. J. (1995) The data association problem for hidden Markov models, *Proceedings of the 34th IEEE Conference on Decision and Control*, Dec. 1995, New Orleans, 2764-2765.

[74] Kruse, R., and Meyer, K. D. (1987) *Statistics With Vague Data*, D. Reidel/Kluwer, Dordrecht.

[75] Kruse, R., Schwencke, E., and Heinsohn, J. (1991) *Uncertainty and Vagueness in Knowledge-Based Systems*, Springer-Verlag, New York.

[76] Lang, S. (1971) *Algebra*, Addison-Wesley, London.

[77] Lanterman, A. D., Miller, M. I., Snyder, D. L., and Miceli, W. J. (1994) Jump-diffusion processes for the automated understanding of FLIR scenes, *SPIE Proceedings*, **2234**, 416-427.

[78] Lewis, D. Probabilities of conditionals and conditional probabilities, *Philosophical Review*, **85**, 297-315.

[79] Lanterman, A.D., Miller, M.I., Snyder, D.L., and Miceli, W.J. (1994) Jump-diffusion processes for the automated understanding of FLIR scenes, *SPIE Proceedings*, Vol. 2234, 1994, 416-427

[80] Lewis, D. (1976) Probabilities of conditionals and conditional probabilities, *Philosophical Review*, **85**, 297-315

[81] Li, Y. (1994) *Probabilistic Interpretations of Fuzzy Sets and Systems*, Doctoral Dissertation, Dept. of Electrical Engineering and Computer Science, Massachusetts Institute of Technology, July 1994.

[82] Lobbia, R. and Kent, M. (1994) Data fusion of decentralized local tracker outputs," *IEEE Transactions on Aerospace and Electronic Systems*, **30**, 787-798.

[83] Lyshenko, N. N. (1983) Statistics of random compact sets in Euclidean space, *Journal of Soviet Mathematics*, **21**, 76-92.

[84] Mahler, R. P. S. (1986) Combining ambiguous evidence with respect to ambiguous *a priori* Knowledge, I: Boolean Logic, *IEEE Transactions on Systems, Man and Cybernetics—Part A: Systems and Humans*, **26**, 27-41.

[85] Mahler, R. P. S. (1994) Global integrated data fusion, *Proceedings of the Seventh National Symposium on Sensor Fusion*, Vol. I (Unclassified), Sandia National Laboratories, Albuquerque, ERIM, Ann Arbor MI, pp. 187-199.

[86] Mahler, R. P. S. (1996) Global optimal sensor allocation, *Proceedings of the Ninth National Symposium on Sensor Fusion*, Vol. I (Unclassified), Naval Postgraduate School, Monterey CA, March 11-13, 1996, 347-366.

[87] Mahler, R. (1995) Information theory and data fusion, *Proceedings of the Eighth National Symposium on Sensor Fusion*, Vol. I (Unclassified), Texas Instruments, Dallas, Mar. 17-19 1995, ERIM, Ann Arbor MI, 279-292.

[88] Mahler, R. (1995) Nonadditive probablity, finite-set statistics, and information fusion, *Proceedings of the 34th IEEE Conference on Decision and Control*, New Orleans, Dec. 1995, 1947-1952.

[89] Mahler, R. (1994) The random-set approach to data fusion, *SPIE Proceedings*, **2234**, 287-295.

[90] Mahler, R. (1996) Representing rules as random sets, I: Statistical Correlations Between Rules, *Information Sciences*, **88**, 47-68.

[91] Mahler, R. (1996) Representing rules as random sets, II: Iterated rules, *International Journal of Intelligent Systems*, **11**, 583-610.

[92] Mahler (1996) Unified data fusion: fuzzy logic, evidence, and rules, *SPIE Proceedings*, **2755**, 226-237.

[93] Mahler, R. (1995) Unified nonparametric data fusion, *SPIE Proceedings*, **2484**, 66-74.

[94] Matheron, G. (1975) *Random Sets and Integral Geometry*, J. Wiley, New York.

[95] McGee, V. (1989) Conditional probabilities and compounds of conditionals, *The Philosophical Review*, **98** (4), 485-541.

[96] Molchanov, I. S. (1993) *Limit Theorems for Unions of Random Closed Sets*, Springer-Verlag Lecture Notes in Mathematics No. 1561, New York.

[97] Mori, S., Chong, C.-Y., Tse, E., and Wishner, R. P. (1984) Multitarget multisensor tracking problems, Part I: A general solution and a unified view on Bayesian approaches, Revised Version, Advanced Information and Decision Systems report TR-1048-01, Mountain View CA, August 1984. My thanks to Dr. Mori for making this report available to me (Dr. Shozo Mori, personal communication, Feb. 28 1995).

[98] Mori, S., Chong, C. Y., Tse, E., and Wishner, R. P. (1986) Tracking and classifying multiple targets without a priori identification, *IEEE Transactions on Automatic Control*, **AC-31**, 401-409.

[99] Muder, D. J. and O'Neil, S. D. (1993) The multi-dimensional SME filter for multitarget tracking, *SPIE Proceedings*, **1954**, 587-599.

[100] Musicki, D., Evans, R., and Stankovic, S. (1994) Integrated probabilistic data association, *IEEE Transactions on Automatic Control*, **39**, 1237-1241.

[101] Neapolitan, R.E. (1992) A survey of uncertain and approximate inference, in L. Zadeh, J. Kocprzyk (eds.), *Fuzzy Logic for the Management of Uncertainty*, J. Wiley, New York.

[102] Nguyen, H. T. and Walker, E. A. (1994) A history and introduction to the algebra of conditional events, *IEEE Transactions on Systems, Man and Cybernetics*, **24**, 1671-1675.

[103] Nguyen, H. T. and Walker, E. A. (1996), *A First Course in Fuzzy Logic*, CRC Press, Boca Raton, Florida.

[104] Nguyen, H. T. (1978) On random sets and belief functions,*Journal of Mathematical Analysis and Applications*, **65**, 531-542.

[105] O'Neil, S. D. and Bridgland, M. F. (1991) Fast algorithms for joint probabilistic data association, *Proceedings of the Fourth National Symposium on Sensor Fusion*, Vol. I (Unclassified), Orlando, ERIM, Ann Arbor MI, pp. 173-189.

[106] Orlov, A. I. (1977) Relationships between fuzzy and random sets: fuzzy tolerances, *Issledovania po Veroyatnostnostatishesk. Modelirovaniu Realnikh System*, Moscow.

[107] Orlov, A. I. (1978) Fuzzy and random sets, *Prikladnoi Mnogomerni Statisticheskii Analys*, Moscow.

[108] Pawlak, Z. (1991) *Rough Sets: Theoretical Aspects of Reasoning About Data*, Kluwer Academic Publishers, Dordrecht.

[109] Poore, A. B., Robertson III, A. J., and Shea, P. J. (1995) A new class of Lagrangian relaxation based algorithms for fast data association in multiple hypothesis tracking applications, *SPIE Proceedings*, **2484**, 184-194.

[110] Quinio, P. and Matsuyama, T. (1991) Random closed sets: a unified approach to the representation of imprecision and uncertainty, in R. Kruse and P. Siegel (eds.), *Symbolic and Quantitative Approaches to Uncertainty*, Springer-Verlag, New York, pp. 282-286.

[111] Reid, R. D. (1979) An algorithm for tracking multiple targets, *IEEE Transactions on Automatic Control*, **AC-24**, 843-854.

[112] Revuz, A. (1956) Fonctions croissantes et mesures sur les espaces topologiques ordonnes, *Ann. Inst. Fourier*, VI, 187-268.

[113] Sadjadi, F.A. (ed.) (1996) *Selected Papers on Sensor and Data Fusion*, SPIE Vol. MS-124.

[114] Schay, G. (1968) An algebra of conditional events, *Journal of Mathematical Analysis and Applications*, **24**, 334-344.

[115] Schweizer, B. and Sklar, A. (1983) *Probabilistic Metric Spaces*, North-Holland, Amsterdam.

[116] Singer, R. A., Sea, R. G., and Housewright, R. B. (1974) Derivation and evaluation of improved tracking filters for use in dense multi-target environments, *IEEE Transactions on Information Theory*, **IT-20**, 423-432.

[117] Skowron, A. (1989) The relationship between the rough set theory and evidence theory, *Bulletin of the Polish Academy of Sciences: Technical Sciences*, **37** (1-2), 87-90.

[118] Smets, P. (1991) About updating, in B. D'Ambrosio, Ph. Smets, and P.P. Bonissone (eds.), *Uncertainty in Artificial Intelligence*, Morgan Kaufmann, pp. 378-385.

[119] Smets, P. (1990) Constructing the pignistic probability function in a context of uncertainty, *Uncertainty in Artificial Intelligence*, **5**, 29-39.

[120] Smets, P. and Kennes, R. (1994) The transferable belief model, *Artificial Intelligence*, **66**, 191-234.

[121] Smets, P. (1992) The Transferable belief model and random sets, *International Journal of Intelligent Systems* , **7**, 37-46.

[122] Sorenson, H. W., and Alspach, D. L. (1971) Recursive Bayesian estimation using Gaussian sums, *Automatica*, **7**, 465-479.

[123] Stein, M. C. (1995) Recursive Bayesian fusion for force estimation, *Proc. 8th Nat'l Symp. on Sensor Fusion*, Vol. I (Unclassified), Dallas TX, pp. 47-66.

[124] Stone, L. D. Finn, M. V., and Barlow, C. A. (1996) Unified data fusion, Metron Corp. report to Office of Naval Research (ONR) under contract N00014-95-C-0052, Jan. 26 1996

[125] Stoyan, D., Kendall, W. S., and Mecke, J. (1995) *Stochastic Geometry and Its Applications*, Second Edition, J. Wiley, New York.

[126] Suppes, P. and Zanotti, M. (1977) On using random relations to generate upper and lower Probabilities, *Syntheses*, **36**, 427-440.

[127] Thoma, H. M. (1991) Belief function computations, in I.R. Goodman, M.M. Gupta, H.T. Nguyen and G.S. Rogers (eds.), *Conditional Logic in Expert Systems*, North-Holland, pp. 309-344.

[128] Thoma, H. M. (1990) *Factorization of Belief Functions*, Doctoral Dissertation, Harvard University.

[129] Thomopoulos, S. C. A. (1990) Sensor integration and data fusion, *Journal of Robotic Systems*, **7**, 337-372.

[130] Tsaknakis, H., Buckley, M. D., and Washburn, R. B. (1991) Tracking closely-spaced objects using multi-assignment algorithms, *Proceedings of the 1991 Joint Services Data Fusion Symposium*, Vol. I (Unclassified), Johns Hopkins APL, Laurel MD, Naval Air Development Center, Warminster PA, pp. 293-309.

[131] Uhlmann, J. K. (1992) Algorithms for multiple-target tracking, *American Scientist*, **80**, 128-141.

[132] van Fraassen, B.C. (1976) Probabilities of conditionals, in W.L. Harper and E.A. Hooker (eds.), *Foundations of Probability Theory, Statistical Inference, and Statistical Theories of Science*, Vol. I, D. Reidel, pp. 261-308.

[133] Waltz, E., and Llinas, J. (1990) *Multisensor Data Fusion*, Artech House, Dedham, MA.

[134] Washburn, R. B. (1987) A random point process approach to multiobject tracking, *Proceedings of the American Control Conference*, **3**, 1846-1852.

[135] Willett, P., Alford, M., and Vannicola, V. (1994) The case for like-sensor predetection fusion, *IEEE Transactions on Aerospace and Electronic Systems*, **30** (4), 986-1000.

[136] Xie, X. and Evans, R. J. (1991) Multiple target tracking and multiple frequency line tracking using hidden Markov models, *IEEE Transactions on Signal Processing*, **39**, 2659-2676.

[137] Yen, J. (1986) A reasoning model based on an extended Dempster-Shafer theory, *Proceedings of the AAAI-86 Conference*, Philadephia PA, pp. 125-131.

[138] Kadar, I. Abrams, B., Louas, L., Alford, M., Berry, W., and Liggins, M. (1994) Robust centralized predetection fusion for enhanced target detection, in *Signal Processing, Sensor Fusion and Target Recognition III* (I. Kadar and Z. Libby, Eds.), Proceedings S.P.I.E. 2232, pp. 37-55.

Part II

The Random Set Approach to Data Fusion

Chapter 3

Foundations of Random Sets

This chapter is devoted to theoretical aspects of random sets. It is intended to provide information about set-valued random elements within probability theory. The writing of this chapter is tutorial in nature. Readers who are interested only in applications of random sets might skip this chapter without discontinuities. The applications of random set theory to data fusion problems will be treated in subsequent chapters of this Part II of the book, in which additional materials on theoretical issues of random sets will be incorporated at appropriate places. Also, to make the reading of applied issues easier, we choose to recall the basics of random set theory when necessary. Incorporating this freedom of choice has led to some redundancy, mostly in definitions, which may be beneficial.

3.1 Distributions of Random Sets

Simple examples of random sets and their distributions have been given in Chapter 2, Section 2.4. In this Chapter, we consider random closed sets in euclidean spaces $\mathbb{R}^d$ or more generally, in locally compact, Hausdorff and separable spaces (LCHS). The reason is this. The foundations of random closed sets are based on Choquet theorem on such spaces. It should be noted that the natural domain of probability theory is Polish spaces (a topological space is Polish if it has a countable base, is metrizable, and is complete under a metric compatible with its topology). These spaces might not be locally compact, for example, infinite dimensional Banach Spaces. Also, in using capacity theory in other applications, such as optimal control of distributed systems (see, e.g. Li and Yong [20]), it is necessary to consider infinite dimensional topological spaces.

Let $(\Omega, \mathcal{A}, P)$ be a probability space, and $(U, \mathcal{U})$ be an abstract measurable

93

space. By a *random element*, we mean a map

$$X : \Omega \longrightarrow U,$$

which is measurable with respect to $\mathcal{A}$ and $\mathcal{U}$, i.e.,

$$X^{-1}(\mathcal{U}) \subseteq \mathcal{A}.$$

The probability law of X is the probability measure $P_X = PX^{-1}$ on $\mathcal{U}$. By abuse of language, a probability measure P on $\mathcal{U}$ is referred to as a random element (we take $(\Omega, \mathcal{A}, P) = (U, \mathcal{U}, P)$, with X being the identical mapping from U to itself, $X(u) = u$).

When $U = \mathbb{R}$ (resp. $\mathbb{R}^d$, $d > 1$), X is called a *random variable* (resp. *random vector*). Random variables and random vectors are special cases of *random closed sets*.

When U is a class of subsets of some set Θ, the random element X is called a *random set*. It is a set-valued random element. For example, when U is a class of closed sets of a topological space Θ, X is called a *random closed set*.

From now on, unless stated otherwise, Θ will be a LCHS space, that is

(i) Each point in Θ has at least a compact neighborhood.

(ii) For any $x \neq y$ in Θ, there are disjoint neighborhoods of x and y.

(iii) The topology of Θ has a countable base.

To be concrete, the reader might think of such a space as $\mathbb{R}^d$.

let $\mathcal{K}$, $\mathcal{F}$, and $\mathcal{G}$ be the classes of compact, closed, and open sets of Θ, respectively. The Borel σ-field of Θ is denoted as $\mathcal{B}(\Theta)$. We will focus on the space $\mathcal{F}$ as the range of random elements.

Let $(\Omega, \mathcal{A}, P)$ be a probability space. A map $X : \Omega \to \mathcal{F}$ will be formulated as a random closed set when we equip $\mathcal{F}$ with some σ-field. We are going to topologize $\mathcal{F}$ for this purpose. We follow Matheron [15]. For $A \subseteq \Theta$, let

$$\mathcal{F}^A = \{F \in \mathcal{F} : F \cap A = \emptyset\}, \quad \mathcal{F}_A = \{F \in \mathcal{F} : F \cup A \neq \emptyset\}.$$

A base for the topology of $\mathcal{F}$ is taken to be

$$\mathcal{F}^K, \quad \mathcal{F}^K \cap \mathcal{F}_{G_1} \cap \ldots \cap \mathcal{F}_{G_n} \quad n \geq 1, K \in \mathcal{K}, G_i \in \mathcal{G}.$$

The topological space $\mathcal{F}$ so obtained is compact, Hausdorff, and separable (Matheron, pp. 3-5) [15]. The Borel σ-field of $\mathcal{F}$ is denoted as $\mathcal{B}(\mathcal{F})$. Now, we equip the space $\mathcal{F}$ with the σ-field $\mathcal{B}(\mathcal{F})$, and $X : \Omega \to \mathcal{F}$ is a random closed set if X is $\mathcal{A}$-$\mathcal{B}(\mathcal{F})$-measurable. The probability law (or distribution) of X governing the random behavior of X is the probability image P_X on $\mathcal{B}(\mathcal{F})$, where

$$P_X(B) = PX^{-1}(B) = P(\{\omega : X(\omega) \in B\}), \quad B \in \mathcal{B}(\mathcal{F}).$$

For example, $B = \mathcal{F}_K$,

$$P_X(\mathcal{F}_K) = P_X(\{F \in \mathcal{F} : F \cap K \neq \emptyset\}) = P(\{\omega : X(\omega) \cap K \neq \emptyset\}).$$

As stated earlier, a probability measure on $\mathcal{B}(\mathcal{F})$ is referred to as a random closed set.

The practical method for introducing probability measures on $\mathbb{R}^d$ is via distribution functions (see Chapter 2, Section 2.4). In the case of random sets, we will seek a similar method.

Observe that $\mathcal{B}(\mathcal{F})$ is also generated by the $\mathcal{F}_K$'s, $K \in \mathcal{K}$. Thus, it is expected that a probability measure P_X on $\mathcal{B}(\mathcal{F})$ can be completely determined from its values on the $\mathcal{F}_K$'s.

Given P_X, define $T : \mathcal{K} \to [0,1]$ by

$$T(K) = P_X(\mathcal{F}_K).$$

Then the set-function T satisfies the following conditions:

(a) $T(\emptyset) = 0, 0 \leq T(\cdot) \leq 1$.
(b) If $K_n \searrow K$ in $\mathcal{K}$, then $T(K_n) \searrow T(K)$.
(c) $S_1(K; K_1) = T(K \cap K_1) - T(K) \geq 0, \quad \ldots,$

$$S_2(K; K_1, \ldots, K_n) = S_{n-1}(K; K_1, \ldots, K_{n-1}) - S_{n-1}(K \cup K_n; K_1, \ldots, K_{n-1}) \geq 0.$$

Remark. The condition (c) says the $S_n \geq 0$ for any $K, K_1, \ldots, K_n \in \mathcal{K}, n \geq 1$. This is so because

$$
\begin{aligned}
S_n(K; K_1, \ldots, K_n) &= P\left(\mathcal{F}_{K_1,\ldots,K_n}^K\right) \\
&= P_X(F \in \mathcal{F} : F \cap K = \emptyset, F \cap K_1 \neq \emptyset, \ldots, F \cap K_n \neq \emptyset),
\end{aligned}
$$

where $\mathcal{F}_{K_1,\ldots,K_n}^K = \mathcal{F}^K \cap \mathcal{F}_{K_1} \cap \ldots \cap \mathcal{F}_{K_n}$. The condition (c) is referred to as the property of *alternating of infinite order* of T.

The alternating of infinite order property of T can be expressed in other equivalent forms which are more convenient technically. Indeed, observe that

$$S_n(K; K_1, \ldots, K_n) = \sum_{\emptyset \neq I \subseteq \{1,\ldots,n\}} (-1)^{|I|+1} T\left(K \cup \left(\bigcup_{i \in I} K_i\right)\right) - T(K),$$

where $|I|$ denotes the cardinality of the set I. Thus $S_n \geq 0, n \geq 1$ is equivalent to

(d) For any $K, K_1, \ldots, K_n \in \mathcal{K}, n \geq 1$,

$$T(K) \leq \sum_{\emptyset \neq I \subseteq \{1,\ldots,n\}} (-1)^{|I|+1} T\left(K \cup \left(\bigcup_{i \in I} K_i\right)\right).$$

But (d) is equivalent to

(e) T is monotone increasing on $\mathcal{K}$ (i.e., $K_1 \subseteq K_2$ implies that $T(K_1) \leq T(K_2)$), and for $K_1, K_2, \ldots, K_n, n \geq 2$,

$$T\left(\bigcap_{i=1}^n\right) \leq \sum_{\emptyset \neq I \subseteq \{1,\ldots,n\}} (-1)^{|I|+1} T\left(\bigcup_{i \in I} K_i\right).$$

Indeed, assume (d). Then T is monotone increasing (take $n = 1$ in (d)). For $K_1, \ldots, K_n$ with $n \geq 2$, take $K = \bigcap_{i=1}^n K_i$ in (d) yielding (e). Conversely, assume (e). Then (d) holds for $n = 1$ by monotonicity of T. For $n \geq 2$, apply (e) to $K_i' = K \cup K_i$, $i = 1, \ldots, n$, yielding

$$T\left(\bigcap_{i=1}^n K_i'\right) \leq \sum_{\emptyset \neq I \subseteq \{1,\ldots,n\}} (-1)^{|I|+1} T\left(\bigcup_{i \in I} K_i'\right)$$

$$= \sum_{\emptyset \neq I \subseteq \{1,\ldots,n\}} (-1)^{|I|+1} T\left(K \cup \left(\bigcup_{i \in I} K_i\right)\right).$$

But $K \subseteq \bigcap_{i=1}^n K_i'$ so that $T(K) \leq T(\bigcap_{i=1}^n K_i')$, since T is monotone increasing.

It turns out that the necessary conditions (a), (b), and (c) are also sufficient for a set-function T defined on $\mathcal{K}$ to determine uniquely a probability measure P_X on $\mathcal{B}(\mathcal{F})$, thanks to Choquet theorem (a proof of it can be found in Matheron [15] (1975), pp. 31-35).

Choquet Theorem. *A set-function $T: \mathcal{K} \to \mathbb{R}$ satisfying conditions (a), (b), and (c) determines uniquely a probability measure P_X on $\mathcal{B}(\mathcal{F})$ such that*

$$P_X(\mathcal{F}_K) = T(K), \qquad \forall K \in \mathcal{K}.$$

The set-function T in the above Choquet Theorem is referred to as the *capacity functional* of the random closed set X.

In view of Choquet Theorem, it suffices to specify capacity functionals in practice, as far as models for random closed sets are concerned.

In the case of LCHS spaces, one can also determine probability laws of random closed sets from appropriate set-functions defined on the class $\mathcal{G}$ of the open sets of Θ. Specifically, consider the following "dual" situation.

Let $S: \mathcal{G} \to \mathbb{R}$ such that

(i) $S(\emptyset) = 0$, $0 \leq S(\cdot) \leq 1$.
(ii) If $G_n \nearrow G$ in $\mathcal{G}$, then $S(G_n) \nearrow S(G)$.
(iii) S is alternating of infinite order on $\mathcal{G}$.

Then, there exists a unique probability measure P_X on $\mathcal{B}(\mathcal{F})$ such that

$$P_X(\mathcal{F}_G) = S(G), \qquad \forall G \in \mathcal{G}.$$

This result follows from Choquet Theorem, in view of proposition 1.4.3 in Matheron [15] (pp. 14-15), namely, if we define T on $\mathcal{K}$ by

$$T(K) = \inf\{S(G) : G \in \mathcal{G}, K \subseteq G\}, \qquad \forall K \in \mathcal{K},$$

Then T satisfies (a), (b), and (c) of Choquet Theorem and

$$S(G) = \sup\{T(K) : K \in \mathcal{K}, K \subseteq G\}, \qquad \forall G \in \mathcal{G}.$$

Remark. Given T on $\mathcal{K}$, we extend T to the power set $\mathcal{P}(\Theta)$ of Θ as follows.
For $G \in \mathcal{G}$, $T(G) = \sup\{T(K) : K \in \mathcal{K}, K \subseteq G\}$, and for $A \subseteq \Theta$, $T(A) = \inf\{T(G) : G \in \mathcal{G}, A \subseteq G\}$. Then T is a (alternating of infinite order) Choquet capacity, that is, T is monotone increasing, if $K_n \searrow K$ in $\mathcal{K}$ then $T(K_n) \searrow T(K)$, and if $A_n \nearrow A$ in $\mathcal{P}(\Theta)$ then $T(A_n) \nearrow T(A)$. As such, since Θ is LCHS, every Borel set of Θ is T-capacitable, i.e.,

$$T(A) = \sup\{T(K) : K \in \mathcal{K}, K \subseteq A\}, \qquad \forall A \in \mathcal{B}(\Theta).$$

Having the probability space $(\mathcal{F}, \mathcal{B}(\mathcal{F}), P_X)$ where P_X is generated by T, if we consider the completion $(\mathcal{F}, \mathcal{B}(\mathcal{F}), \tilde{P}_X)$, then for any $A \in \mathcal{B}(\mathcal{F})$, $\mathcal{F}_A \in \tilde{\mathcal{B}}(\mathcal{F})$ and $\tilde{P}_X(\mathcal{F}_A) = T(A)$.
Note that in the context of Choquet Theorem, the empty set $\emptyset$ is considered as a possible value of random set X, that is

$$P(X = \emptyset) = P_X(\mathcal{F}) - P_X(\mathcal{F}_\Theta) \geq 0.$$

(Observe that $\mathcal{F}_\Theta = \{F \in \mathcal{F} : F \cap \Theta \neq \emptyset\} = \mathcal{F} \setminus \{\emptyset\}$). Thus if T is identically zero on $\mathcal{K}$, say when Θ is compact, then $P_X(\{\emptyset\}) = P(X = \emptyset) = 1$, i.e., X is $\emptyset$ almost surely. If *non-emty* random closed sets are considered, then an extra boundary condition on T or S needs to be imposed, such as $S(\Theta) = 1$.

Example 1 $\Theta = \mathbb{R}^d$. *Let $f : \mathbb{R}^d \to [0,1]$ be a upper semi-continuous function, i.e., for any $t \in \mathbb{R}$, $(f \geq t) = \{x \in \mathbb{R}^d : f(x) \geq t\}$ is a closed set of $\mathbb{R}^d$. Then $T : \mathcal{K} \to [0,1]$ defined by*

$$T(K) = \sup_{x \in K} f(x), \qquad K \in \mathcal{K}$$

is a capacity functional.

This can be shown either by exhibiting a random closed set X such that for any $K \in \mathcal{K}$, $T(K) = P(X \cap K \neq \emptyset)$, or by verifying the sufficient conditions in Choquet Theorem.
(i) Let $\alpha : (\Omega, \mathcal{A}, P) \to [0,1]$ be a uniformly distributed random variable. Define $X : \Omega \to \mathcal{F}$ by

$$X(\omega) = \{x \in \mathbb{R}^d; f(x) \geq \alpha(\omega)\},$$

i.e., $X(\omega)$ is the random $\alpha(\omega)$-level set of f. Let $K \in \mathcal{K}$, Then, since f is upper semi-continuous,

$$
\begin{aligned}
P(\{\omega : X(\omega) \cap K \neq \emptyset\}) &= P(\{\omega : \alpha(\omega) \leq f(x) \text{ for some } x \in K\}) \\
&= P\left(\{\omega : \alpha(\omega) \leq \sup_{x \in K} f(x)\}\right) \\
&= \sup_{x \in K} f(x) = T(K)
\end{aligned}
$$

(ii) As an exercise, let us verify directly the condition (a)-(c) in Choquet Theorem for T. (a) is obvious. For (b), let $K_n \searrow K$ in $\mathcal{K}$. We have $T(K) \leq$

$\inf_n T(K_n) = \alpha$. Let $\varepsilon > 0$ and consider $A_n = \{x \in \mathbb{R}^d : f(x) \geq \alpha - \varepsilon\} \cap K_n$. By hypothesis, $\{x \in \mathbb{R}^d : f(x) \geq \alpha - \varepsilon\}$ is closed, and so is A_n. The A_n's are contained in K_1, and hence $\cap_n A_n \neq \emptyset$ (note that for each n, $A_n \neq \emptyset$). We have $A_n \subseteq K_n$ and hence

$$T(K) \geq \sup_{\cap_n A_n} f(x) \geq \alpha - \varepsilon.$$

For (c), observe that $T(K) = \sup_{x \in K} f(x)$ is maxitive, i.e., for any $K_1, \ldots, K_n \in \mathcal{K}$,

$$T\left(\bigcup_{i=1}^n K_i\right) = \max\{T(K_i) : i = 1, \ldots, n\}.$$

Such a set-function is necessarily alternating of infinite order (see Nguyen et al. [27]). This result is interesting in its own right.

Lemma 1 *Let $\mathcal{C}$ be a class of subsets of some set Θ, containing $\emptyset$ and stable under finite intersections and unions. Let $T : \mathcal{C} \to [0, \infty)$ be maxitive, i.e.,*

$$T(A \cup B) = \max\{T(A), T(B)\}, \qquad \forall A, B \in \mathcal{C}.$$

Then T is alternating of infinite order.

Proof. First note that, for any $A_1, \ldots, A_n$, $n \geq 2$,

$$T\left(\bigcup_{i=1}^n A_i\right) = \max\{T(A_i) : i = 1, \ldots, n\}.$$

Next, such a set-function is obviously monotone increasing on $\mathcal{C}$. Thus, it suffices to show that

$$T\left(\bigcap_{i=1}^n A_i\right) \leq \sum_{\emptyset \neq I \subseteq \{1, \ldots, n\}} (-1)^{|I|+1} T\left(\bigcup_{i \in I} A_i\right).$$

Without loss of generality we may assume that

$$0 \leq \alpha_n = T(A_n) \leq \alpha_{n-1} = T(A_{n-1}) \leq \cdots \leq \alpha_1 = T(A_1).$$

For $k \in \{1, 2, \ldots, n\}$, let $\mathcal{I}(k) = \{I \subset \{1, \ldots, n\} : |I| = k\}$. For $I \in \mathcal{I}(k)$, let $m(I) = \min\{i : i \in I\}$, and for $i = 1, \ldots, n - k + 1$, let $\mathcal{I}_i(k) = \{I \in \mathcal{I}(k) : m(I) = i\}$. Then we have

$$T\left(\bigcup_{j \in I} A_j\right) = \alpha_i \quad \text{for every } I \in \mathcal{I}_i(k), \, i = 1, \ldots, n - k + 1.$$

Also $|\mathcal{I}_i(k)| = \begin{pmatrix} n - i \\ k - 1 \end{pmatrix} = (n - i)!/[(k - 1)!(n - i - k + 1)!]$ for every $i = 1, \ldots, n - k + 1$. Thus

$$\sum_{I \in \mathcal{I}(k)} T\left(\bigcup_{i \in I} A_j\right) = \sum_{i=1}^{n-k+1} \sum_{I \in \mathcal{I}_i(k)} T\left(\bigcup_{j \in I} A_j\right)$$

$$= \sum_{i=1}^{n-k+1} \sum_{I \in \mathcal{I}_i(k)} \alpha_i = \sum_{i=1}^{n-k+1} \binom{n-i}{k-1} \alpha_i.$$

and therefore

$$\sum_{\emptyset \neq I \subseteq \{1,\dots,n\}} (-1)^{|I|+1} T\left(\bigcup_{i \in I} A_i\right) = \sum_{k=1}^{n} (-1)^{k+1} \sum_{I \in \mathcal{I}(k)} T\left(\bigcup_{i \in I} A_i\right)$$

$$= \sum_{k=1}^{n} (-1)^{k+1} \sum_{i=1}^{n-k+1} \binom{n-i}{k-1} \alpha_i$$

$$= \sum_{i=1}^{n} \left[\sum_{k=0}^{n-i} \binom{n-i}{k} (-1)^k\right] \alpha_i$$

$$= \alpha_n = T(A_n) \geq T\left(\bigcap_{i=1}^{n} A_i\right)$$

by observing that

$$\sum_{k=0}^{n-i} \binom{n-i}{k} (-1)^k = (1-1)^{n-i} = 0,$$

for any $i = 1, \dots, n-1$. $\qquad\square$

The set-function T in Example 3.1 is in fact maxitive in a very strong sense, namely, for any index set I,

$$T\left(\bigcup_{i \in I} A_i\right) = \sup\{T(A_i) : i \in I\},$$

provided $\bigcup_{i \in I} A_i$ is in the domain of T. A more general class of maxitive set-functions appears in the theory of extremal stochastic processes (see Molchanov [16] and Noberg [28], [29]).

A general construction of maxitive set-functions goes as follows. Let $(U, \mathcal{U})$ be a measurable space, and $T : \mathcal{U} \to [0, \infty)$, such that $T(\emptyset) = 0$ and T is maxitive on $\mathcal{U}$. For each $t \geq 0$, $\mathcal{J}_t = \{A \in \mathcal{U} : T(A) \leq t\}$ is an ideal in $\mathcal{U}$ (i.e., $\mathcal{J}$ is a non-empty subset of $\mathcal{U}$, stable under finite unions and hereditary, i.e., if $A \in \mathcal{J}_t$, $B \in \mathcal{U}$ with $B \subseteq A$, then $B \in \mathcal{J}_t$). Obviously, $s < t$ implies $\mathcal{J}_s \subset \mathcal{J}_t$, that is, the family of ideals $\{\mathcal{J}_t, t \geq 0\}$ is increasing.

It is obvious that $T(A) = \inf\{t \geq 0 : A \in \mathcal{J}_t\}$. Conversely, let $\{\mathcal{N}_t, t \geq 0\}$ be an increasing family of ideals of $\mathcal{U}$, then the set-function T on $\mathcal{U}$ defined by

$$T(A) = \inf\{t \geq 0 : A \in \mathcal{N}_t\}$$

is maxitive. It may happen that $\{\mathcal{N}_t, t \geq 0\}$ is different from $\{\mathcal{J}_t, t \geq 0\}$. They coincide if and only if $\{\mathcal{N}_t, t \geq 0\}$ is "right continuous", i.e., $\mathcal{N}_t = \cap_{s > t} \mathcal{N}_s$.

Here are some examples. Let $f : U \to [0, \infty)$, and

$$\mathcal{N}_t = \{A \in \mathcal{U} : A \cap \{f > t\} \in \mathcal{N}\}$$

where $\mathcal{N}$ is some given ideal in $\mathcal{U}$. Then

$$T(A) = \inf\{t \geq 0 : A \cap \{f > t\} \in \mathcal{N}\}$$

is maxitive. For example, let P be a probability measure on $(U, \mathcal{U})$, f measurable, and $\mathcal{N} = \{A \in \mathcal{U}, P(A) = 0\}$. Note that $\mathcal{N}$ is σ-ideal in $\mathcal{U}$ (also stable under countable unions). Then

$$\mathcal{N}_t = \{A \in \mathcal{U} : P(A \cap \{f > t\}) = 0\},$$

and

$$\inf\{t \geq 0 : A \in \mathcal{J}_t\} = \|f.1_A\|_\infty = T(A),$$

is maxitive in $\mathcal{U}$.

In practice, if $\mathcal{N} = \{\emptyset\}$, then

$$\inf\{t \geq 0 : A \cap \{f > t\} = \} = \sup_A f.$$

Another interesting example is the Hausdorff dimension on a metric space. Let (U, d) be a metric space. For $A \subseteq U$, $d(A)$ denotes the diameter of A. For $\alpha \geq 0$, the Hausdorff α-(outer)measure μ_α (see e.g. Billingsley [3], and Rogers [32]) is defined as

$$\mu_\alpha(A) = \lim_{\varepsilon \to 0} \inf \left[\sum_n d^\alpha(A_n) \right],$$

where the inf is taken over all countable coverings of A by closed balls A_n such that $d(A_n) < \varepsilon$. The Hausdorff dimension of A is defined to be

$$D(A) = \inf\{\alpha \geq 0 : \mu_\alpha(A) = 0\}.$$

Now, $\mathcal{N}_\alpha = \{A : \mu_\alpha(A) = 0\}$ is a σ-ideal in the power set $\mathcal{P}(U)$, and $\{\mathcal{N}_\alpha, \alpha \geq 0\}$ is an increasing family of ideals. As such, the Hausdorff dimension D is maxitive, and hence alternating of infinite order on $\mathcal{P}(U)$.

For more information on capacity functionals of random sets on $\mathbb{R}^d$ and their statistics, see Stoyan et al. [36].

3.2   Radon-Nikodym Derivatives

Unless the random closed set X is single-valued (i.e. X is a random vector), its capacity functional T is non-additive. Thus, the analysis of non-additive set-functions appear naturally in the context of random sets, especially for finding ways to specify capacity functionals. This is similar to the case of random vectors: if their distribution functions are absolutely continuous, then F can be represented by its (almost everywhere) ordinary derivative f (its probability density function), which turns out to be also the Radon-Nikodym derivative of the associated probability law $dF(x)$ with respect to Lebesgue measure dx on $\mathbb{R}^d$. Since the capacity functional T plays the role of the distribution function F, it is nature to inquire about various concepts of "derivatives" for *non-additive set-functions*. To our knowledge the literature on this topic is quite small in comparison with that on Radon-Nikodym derivatives for *finitely additive measures* (see, e.g. Bell and Hagood [2]). In fact, we are only aware of the works of Huber and Strassen [17], Graf [10], and Greco [11]. By analogy with distribution functions of random vectors, two approaches are possible.

(i) Viewing sets as points in another space, one might try to consider derivatives of set-functions. This is exemplified by the finite case. Let U be a finite set. A random set X on U has distribution:

$$F : \mathcal{P}(U) \longrightarrow [0,1], \qquad F(A) = P(X \subseteq A), \quad \forall A \subseteq U.$$

The "density" function f is the Möbius transform (inverse) of F, namely $f : \mathcal{P}(U) \to [0,1]$,

$$f(A) = \sum_{B \subseteq A} (-1)^{|A \setminus B|} F(B) = P(X = A),$$

so that

$$F(A) = \sum_{B \subseteq A} f(B).$$

In practice, it is much easier to specify the density f than the distribution F.

When U is infinite, such as $\mathbb{R}^d$, the extension is not obvious at all! This is due to the fact that the concept of Möbius inverse makes sense only in the case of locally finite partial ordered sets. This will be treated in Section 3.3.

(ii) Extending the concept of Radon-Nikodym derivatives in measure theory to the case of non-additive set-functions. In this section, we will elaborate this approach in some details.

Let $(U, \mathcal{U})$ be a measurable space. We consider set-function ν, $\nu : \mathcal{U} \to [0, \infty)$, such that $\mu(\emptyset) = \nu(\emptyset) = 0$, and μ and ν are monotone increasing. A (generalized) Radon-Nikodym derivative (RND) of μ with respect to ν should be some measurable function $f : U \to [0, \infty)$, such that $d\mu = f \, d\nu$.

Now, μ and ν are not, in general, measures, we need to specify a concept of integral to give a meaning to the expression $d\mu = f d\nu$.

For $f : U \to [0, \infty)$ measurable, and $\nu : \mathcal{U} \to [0, \infty)$ increasing, the function $t \in [0, \infty) \to \nu\{x \in U : f(x) \geq t\}$ is measurable, so that the ordinary Lebesgue integral $\int_0^\infty \nu(f \geq t)dt$ is well-defined. It coincides with the ordinary integral $\int_U f(x)d\nu(x)$ where ν is a measure. Thus, it is natural to consider $\int_0^\infty \nu(f \geq t)dt$ as the integral of the function f with respect to the monotone increasing set-function ν. This integral is referred to as the *Choquet integral*. Thus by $d\mu = f d\nu$, we mean

$$\mu(A) = \int_0^\infty \nu[A \cap (f \geq t)]dt \equiv \int_A f \, d\nu, \quad \forall A \in \mathcal{U}.$$

Definition 1 *Let* μ, ν : $\mathcal{U} \to [0, \infty)$ *such that* $\mu(\emptyset) = \nu(\emptyset) = 0$, μ *and* ν *monotone increasing on* $\mathcal{U}$. *A measurable function* $f : U \to [0, \infty)$ *is called a* **Radon-Nikodym derivative** *of* μ *with respect to* ν *if*

$$\mu(A) = \int_0^\infty \nu[A \cap (f \geq t)]dt, \qquad \forall A \in \mathcal{U}.$$

(in symbol, $f = d\mu/d\nu$, $\nu(A) = \int_A f d\nu$*).*

Remark. *If* $f = d\mu/d\nu$, *then* f *is unique in the sense that: if* $g = d\mu/d\nu$, *then* $\mu(f \neq g) = 0$.

Example 2 *Let* $f : \mathbb{R}^d \to [0, 1]$, *upper semi-continuous. Let* μ, $\nu_0 : \mathcal{B}(\mathbb{R}^d) \to [0, 1]$ *be*

$$\mu(A) = \sup_{x \in A} f(x) \qquad \nu_0(A) = \begin{cases} 0 & \text{if } A = \emptyset \\ 1 & \text{if } A \neq \emptyset. \end{cases}$$

Then f *is the RND of* μ *with respect to* ν_0. *Indeed, for any* $A \in \mathcal{B}(\mathbb{R}^d)$,

$$\int_0^\infty \nu_0[A \cap (f \geq t)]dt = \sup_{x \in A} f(x).$$

Given a pair of set-functions (μ, ν), one would like to know when a RND $d\mu/d\nu$ exists. If μ and ν are measures, then it is well-known that a necessary and sufficient condition for this is that μ is absolutely continuous with respect to ν, denoted as $\mu << \nu$, i.e., $\mu(A) = 0$ whenever $\nu(A) = 0$ (see e. g. Halmos [10]). For non-additive set-functions μ and ν, observe that if $d\mu = f d\nu$, then $\mu << \nu$. However, as in the case of finitely additive measures, this condition does not characterize the Radon-Nikodym properties for non-additive set-functions. Specifically, the absolute continuity is only a necessary condition, but not sufficient.

Example 3 *Let* $U = \mathbb{R}, \mathcal{U} = \mathcal{B}(\mathbb{R})$. *Let* μ, $\nu : \mathcal{B}(\mathbb{R}) \to [0, 1]$ *be*

$$\mu(A) = \begin{cases} 0 & \text{if } A \cap I\!N = \emptyset \\ 1 & \text{if } A \cap I\!N = \emptyset \end{cases}$$

and

$$\nu(A) = \begin{cases} 0 & \text{if } A \cap I\!N = \emptyset \\ \sup\{1/x : x \in A \cap I\!N\} & \text{if } A \cap I\!N = \emptyset, \end{cases}$$

where $I\!N = \{1, 2, \ldots\}$. *Obviously,* $\mu << \nu$. *Note that both* μ *and* ν *are alternating of infinite order, since they are maxitive.*

Suppose that there is a measurable function $f : \mathbb{R} \to [0, \infty)$ *such that* $\mu(A) = \int_A f \, d\nu$ *for any* $A \in \mathcal{B}(\mathbb{R})$. *Then*

$$\begin{aligned} \mu(\{n\}) &= \int_0^\infty \nu[\{n\} \cap (f \geq t)]dt = \int_0^{f(n)} \nu(\{n\})dt \\ &= \int_0^{f(n)} \frac{1}{n}dt = \frac{1}{n}f(n). \end{aligned}$$

But by construction, $\mu(\{n\}) = 1$ for any $n \in I\!N$, thus,

$$f(n) = n \qquad \forall n \in I\!N.$$

Now

$$
\begin{aligned}
\mu(I\!N) &= \int_{I\!N} f \, d\nu = \int_0^\infty \nu(I\!N \cap \{f \geq t\}) \, dt \\
&= \sum_{n=1}^\infty \int_{n-1}^n \nu(I\!N \cap \{f \geq t\}) \, dt \geq \sum_{n=1}^\infty \int_{n-1}^n \nu(I\!N \cap \{f \geq n\}) \, dt \\
&= \sum_{n=1}^\infty \int_{n-1}^n \nu(\{k : f(k) \geq n\}) \, dt = \sum_{n=1}^\infty \int_{n-1}^n \nu(\{k : k \geq n\}) \, dt \\
&= \sum_{n=1}^\infty \sup\{\frac{1}{k} : k \geq n\} = \sum_{n=1}^\infty \frac{1}{n} = \infty.
\end{aligned}
$$

But this is in contradiction with $\mu(I\!N) = 1$, by construction of μ.

Depending upon additional properties of μ and ν, sufficient conditions for the existence of $d\mu/d\nu$ can be found. For example, suppose that μ and ν both belong to the class of "pre-capacities" ϕ of the following type (see Graf [10]): $\phi : \mathcal{U} \to [0, \infty)$,

(i) $\phi(\emptyset) = 0$,
(ii) ϕ is monotone increasing: $A, B \in \mathcal{U}$, $A \subseteq B$ implies that $\phi(A) \leq \phi(B)$,
(iii) ϕ is sub-additive: $\phi(A \cup B) \leq \phi(A) + \phi(B)$, for any $A, B \in \mathcal{U}$,
(iv) ϕ is sequentially continuous on $\mathcal{U}$: if $A_n \nearrow$ in $\mathcal{U}$, then

$$\phi\left(\bigcup_{n \geq 1} A_n\right) = \lim_{n \to \infty} \phi(A_n).$$

Then, as shown in Graf [10], a necessary and sufficient condition for μ to admit a RND with respect to ν is

(a) $\mu << \nu$ and
(b) the pair (μ, ν) has the following decomposition property (DP):
For every $t \geq 0$, there exists $A(t) \in \mathcal{U}$ such that
(i) $t[\nu(A) - \nu(B)] \leq \mu(A) - \mu(B)$ for $A, B \subseteq A(t)$ with $B \subseteq A$,
(ii) $t[\nu(A) - \nu(A \cap A(t))] \leq \mu(A) - \mu(A \cap A(t))$ for any $A \in \mathcal{U}$.

Note that the above DP is automatically satisfied when both μ and ν are measures, in view of the Hahn decomposition of a signed measure in terms of a partition of the space U (see e.g. Shilov and Gurevich [35]). Below, we give the proof for the existence of RND for a larger class of non-additive set-functions, namely those which are countably subadditive. Note that pre-capacities in Graf's sense are countably subadditive set-functions. These countably subadditive set-functions are outer measures when the σ-field $\mathcal{U}$ of U is the power set of U. Unless stated otherwise, all set-functions μ considered are monotone increasing and satisfy $\mu(\emptyset) = 0$.

Recall that if μ and ν are measures, then for any $t \geq 0$, $t\nu - \mu$ is a signed measure, so that, by the Hahn decomposition theorem for measures, there exists $A(t) \in \mathcal{U}$ such that $(t\nu - \mu)|_{\mathcal{U}_{A(t)}} \leq 0$ and $(t\nu - \mu)|_{\mathcal{U}_{A^c(t)}} \geq 0$, where A^c denotes the set-complement of A in U, $\mathcal{U}_A = \{B \in \mathcal{U} : B \subseteq A\}$, and the notation $\nu|_{\mathcal{U}_A} \preceq \mu|_{\mathcal{U}_A}$ means $\nu(D) \leq \mu(D)$ for any $B \subset \mathcal{U}_A$. Observe that, in the case of measures, the family $\{A(t), t \geq 0\}$ has the following property:

$$t\nu|_{\mathcal{U}_{A(t)}} \leq \mu|_{\mathcal{U}_{A(t)}}, \quad t\nu|_{\mathcal{U}_{A^c(t)}} \geq \mu|_{\mathcal{U}_{A^c(t)}}. \qquad (*)$$

Thus, we are led to consider the following generalized decomposition property (GDP) for arbitrary set-functions.

Definition 2 *The pair (μ, ν) has the GDP if there exists a family $\{A(t); t \geq 0\} \subseteq \mathcal{U}$, called Hahn decomposition for (μ, ν), that satisfies (*).*

Definition 3 *Let $(U, \mathcal{U})$ be a measurable space. A set-function $\mu : \mathcal{U} \to [0, \infty)$ is called a **sub-measure** if $\mu(\emptyset) = 0$, μ is monotone increasing and σ-subadditive, i.e. for any $\{A_n, n \geq 1\} \subseteq \mathcal{U}$,*

$$\mu\left(\cup_{n \geq 1} A_n\right) \leq \sum_{n \geq 1} \mu(A_n).$$

Example 4 *Unlike measures, set-functions might not admit GDP. Let*

$$\mu(A) = \begin{cases} 0 & \text{if } A \cap \mathbb{N} = \emptyset \\ \min\left\{1, \frac{1}{2}\sum\{\frac{1}{n} : n \in A \cap \mathbb{N}\}\right\} & \text{if } A \cap \mathbb{N} \neq \emptyset \end{cases}$$

and

$$\nu(A) = \begin{cases} 0 & \text{if } A \cap \mathbb{N} = \emptyset \\ \sup\left\{\frac{1}{n} : n \in A \cap \mathbb{N}\right\} & \text{if } A \cap \mathbb{N} \neq \emptyset, \end{cases}$$

where $\mathbb{N} = \{1, 2, \ldots, \}$. Suppose that (μ, ν) has the GDP. Let $\{A(t) : t \geq 0\}$ be their Hahn decomposition. Then for $t = 1$, we have $\nu(A) \leq \mu(A)$ for any $A \in \mathcal{U}_{A(1)}$, and $\nu(A) \geq \mu(A)$ for any $A \in \mathcal{U}_{A^c(1)}$. Let $G_n = \{n, n+1, \cdots, \}$, $n \geq 1$. Then $G_n \subseteq A^c(1)$ for any $n \geq 2$. Indeed, if not, then $G_n \cap A(1) \neq \emptyset$ for some $n \geq 2$, i.e. there exists $k \geq 2$ and $k \in G_n \cap A(1)$. By definition of $A(1)$, we have $\nu(k) \leq \mu(k)$. But by construction of μ and ν, $\nu(k) = 1/k$ and $\mu(k) = 1/(2k)$, a contradiction. Thus, $G_n \subseteq A^c(1)$, implying that $\nu(G_n) \geq \mu(G_n)$ for $n \geq 2$. But by construction of μ and ν again, $\nu(G_n) = 1/n$ and $\nu(G_n) = 1$ for $n \in \mathbb{N}$, we get a contradiction.

Example 5 *The pair in Example 3 does not have the GDP. Indeed, let $A(0) = \mathbb{R}$ and $A(t) = [t, \infty)$ for $t > 0$. It suffices to show that this family $\{A(t), t \geq 0\}$ is a Hahn decomposition for (μ, ν). Obviously (*) holds for any A such that $A \cap \mathbb{N} = \emptyset$. For $A \subseteq A(t)$ with $A \cap \mathbb{N} \neq \emptyset$, we have*

$$\begin{aligned} t\nu(A) &= t\sup\{x^{-1} : x \in A \cap \mathbb{N}\} \\ &\leq t\sup\{n^{-1} : n \geq t\} \leq 1 = \mu(A). \end{aligned}$$

On the other hand, for every $A \subset A^c(t)$ *with* $A \cap \mathbb{N} \neq \emptyset$, *we have*

$$
\begin{aligned}
t\nu(A) &= t \sup\{x^{-1} : x \in A \cap \mathbb{N}\} \\
&= t \sup\{n^{-1} : n < t\} > 1 = \mu(A).
\end{aligned}
$$

Note that, although (μ, ν) *has the GDP,* μ *does not admit a RND with respect to* ν *as shown in Example 3.*

The following theorem (Nguyen et al. [27]) establishes a necessary and sufficient condition for a pair of sub-measures (μ, ν) to admit a RND. This condition contains the absolute continuity condition.

Theorem 2 *Let* μ, ν *be submeasures. Then* μ *admits the RNP with respect to* ν *if and only if there exists a family* $\{A(t) : t \in [0, \infty)\} \subset \mathcal{U}$, *satisfying following conditions:*
(i) $\mu\left(\cap_{q \in Q^+} A(q)\right) = \nu\left(\cap_{q \in Q^+} A(q)\right) = 0$, *where* Q^+ *denotes the non-negative rationales.*
(ii) $\mu(B(t) \backslash A(t)) = 0$ *for every* $t \in [0, \infty)$, *where the decreasing family* $\{B(t); t \geq 0\} \subseteq \mathcal{U}$ *is defined as follows:* $B(0) = U$, *and for* $t > 0$,

$$
B(t) = \bigcap_{q \in Q(t)} A(q) \quad \text{where } Q(t) = Q \cap [0, t),
$$

Q being the set of rationals.
(iii) For any $s, t \in [0, \infty)$ *with* $s < t$ *and* $A \in \mathcal{U}$,

$$
\begin{aligned}
s\left[\nu(A \cap A(s)) - \nu(A \cap A(t))\right] &\leq \mu(A \cap A(s)) - \mu(A \cap A(t)) \\
&\leq t\left[\nu(A \cap A(s)) - \nu(A \cap A(t))\right].
\end{aligned}
$$

Remark. Conditions (i)-(iii) arise from the following observation. Let $f : U \to [0, \infty)$, measurable, such that

$$
\mu(A) == \int_A f \, d\nu = \int_0^\infty \nu[A \cap (f \geq t)] \, dt, \quad \forall A \in \mathcal{U}.
$$

If $A \subseteq \{f \leq \alpha\}$, then $\mu(A) = \int_0^\alpha \nu[A \cap (f \geq t)] \, dt$. To compute this definite integral, we divide $[0, \alpha]$ into small intervals and taken the Riemann sum. The first inequality of (iii) is used to show that $\mu(A) \leq \int_0^\alpha \nu[A \cap (f \geq t)] \, dt$ and the secon inequality in (iii) is used to show that $\mu(A) \geq \int_0^\alpha \nu[A \cap (f \geq t)] \, dt$.
Proof: Assume that (μ, ν) has the RNP. Let $f = d\mu/d\nu$. For every $t \in [0, \infty)$, denote $A(t) = \{f \geq t\}$. Then it suffices to show that the family $\{A(t) : t \in [0, \infty)\}$ satisfies the conditions (i)-(iii). Since $f(x) < \infty$ for every $x \in U$, $\cap_{q \in Q^+} A(q) = \emptyset$. Thus

$$
\mu\left(\bigcap_{q \in Q^+} A(q)\right) = \nu\left(\bigcap_{q \in Q^+} A(q)\right) = 0,
$$

proving (i). It is easy to see that $B(t) = A(t)$ for every $t \in [0, \infty)$ and hence (ii) follows. (iii) can be obtained by noting that

$$\mu(A \cap A(s)) - \mu(A \cap A(t)) = \int_{A \cap A(s)} f \, d\nu - \int_{A \cap A(t)} f \, d\nu$$

$$= \int_0^\infty [\nu(A \cap A(s) \cap \{f \geq \alpha\}) - \nu(A \cap A(t) \cap \{f \geq \alpha\})] \, d\alpha$$

$$\leq \int_0^s [\nu(A \cap A(s) \cap \{f \geq \alpha\}) - \nu(A \cap A(t) \cap \{f \geq \alpha\})] \, d\alpha$$

$$= \int_0^s [\nu(A \cap A(s)) - \nu(A \cap A(t))] \, d\alpha$$

$$= s[\nu(A \cap A(s)) - \nu(A \cap A(t))]$$

and

$$\mu(A \cap A(s)) - \mu(A \cap A(t))$$

$$= \int_0^\infty [\nu(A \cap A(s) \cap \{f \geq \alpha\}) - \nu(A \cap A(t) \cap \{f \geq \alpha\})] \, d\alpha$$

$$= \int_0^t [\nu(A \cap A(s) \cap \{f \geq \alpha\}) - \nu(A \cap A(t))] \, d\alpha$$

$$\leq \int_0^t [\nu(A \cap A(s)) - \nu(A \cap A(t))] \, d\alpha$$

$$= t[\nu(A \cap A(s)) - \nu(A \cap A(t))].$$

Now we prove that the existence of a family $\{A(t) : t \in [0, \infty)\}$, satisfying the conditions (i)-(iii), will imply the RNP for (μ, ν). First we establish some relations between the two families $\{A(t)\}$ and $\{B(t)\}$.

Claim 1. $\mu\left(\bigcap_{k=1}^\infty B(k)\right) = \nu\left(\bigcap_{k=1}^\infty B(k)\right) = 0$.

Proof: Observe that $\bigcap_{q \in Q^+} B(q) = \bigcap_{q \in Q^+} A(q)$. Thus by (i), we have

$$\mu\left(\bigcap_{q \in Q^+} B(q)\right) = \nu\left(\bigcap_{q \in Q^+} B(q)\right) = 0,$$

which implies that $\mu\left(\bigcap_{n \in \mathbb{N}} B(n)\right) = \nu\left(\bigcap_{n \in \mathbb{N}} B(n)\right) = 0$ as $B(n)$ is decreasing in n.

Claim 2. $\lim_{n \to \infty} \mu(B(n)) = \lim_{n \to \infty} \nu(B(n)) = 0$.

Proof: Since

$$B(n) = \left[B(n) \cap \left(\bigcap_{k=1}^\infty B(k) \right) \right] \cup \left[B(n) \cap \left(\bigcap_{k=1}^\infty B(k) \right)^c \right],$$

$$\mu(B(n)) \;\le\; \mu\left(\bigcap_{k=1}^{\infty} B(k)\right) + \mu\left[B(n) \cap \left(\bigcup_{k=1}^{\infty} B^c(k)\right)\right]$$

$$\le\; \sum_{k=1}^{\infty} [B(n) \setminus B(k)]. \qquad \text{(by Claim 1)}$$

Since $\{B(n)\}$ is decreasing,

$$\lim_{n\to\infty} \mu\left(B(n) \setminus B(k)\right) = 0 \quad \text{for each } k \in \mathbb{N},$$

and the result follows. A similar reasoning applies to ν.

Claim 3. *For every $A \in \mathcal{U}$ we have*

$$\mu(A) = \lim_{n\to\infty} \mu(A \cap B^c(n)), \qquad \nu(A) = \lim_{n\to\infty} \nu(A \cap B^c(n)).$$

Proof: For any $n \in \mathbb{N}$, we have

$$\mu\left(A \cap B^c(n)\right) \le \mu(A) \;\le\; \mu(A \cap B(n)) + \mu(A \cap B^c(n))$$
$$\le\; \mu(B(n)) + \mu(A \cap B^c(n)),$$

and the result follows from Claim 2. Similarly

$$\nu(A) = \lim_{n\to\infty} \nu(A \cap B^c(n)).$$

Claim 4. $\mu(A(t) \setminus B(t)) = \nu(A(t) \setminus B(t)) = 0$ *for every* $t \in [0, \infty)$.

Proof: Let $A = A(t) \setminus A(s)$ in (iii), then we obtain

$$-s\nu\,[A(t) \setminus A(s)] \le -\mu\,[A(t)) \setminus A(s)] \le -t\nu\,[A(t)) \setminus A(s))]$$

for any $s, t \in [0, \infty)$ with $s < t$. It follows that

$$\nu\,[A(t) \setminus A(s)] = \mu\,[A(t)) \setminus A(s)] = 0 \qquad (3.1)$$

for any $s, t \in [0, \infty)$ with $s < t$. Observe that

$$A(t) \setminus B(t) = A(t) \setminus \left(\bigcap_{q \in Q(t)} A(q)\right) = \bigcup_{q \in Q(t)} (A(t) \setminus A(q)).$$

Since $q < t$, we obtain from (9),

$$\mu\left(A(t) \setminus B(t)\right) \le \sum_{q \in Q(t)} \mu\left(A(t) \setminus A(q)\right) = 0 \quad \text{for every } t \in [0, \infty)$$

and

$$\nu\left(A(t) \setminus B(t)\right) \le \sum_{q \in Q(t)} \nu\left(A(t) \setminus A(q)\right) = 0 \quad \text{for every } t \in [0, \infty).$$

Note that Claim 4 and (ii) imply that

$$\mu(A(t)) = \mu(B(t)), \qquad \nu(A(t)) = \nu(B(t)) \quad \forall t \in [0, \infty).$$

Claim 5. *For any* $t \in [0, \infty)$ *and* $A \in \mathcal{U}$,

$$\mu(A \cap A(t)) = \mu(A \cap B(t)) \quad \text{and} \quad \nu(A \cap A(t)) = \nu(A \cap B(t)).$$

Proof: Note that

$$A \cap A(t) = [A \cap (A(t) \setminus B(t))] \cup [A \cap A(t) \cap B(t)].$$

By Claim 4, we obtain $\mu(A \cap A(t)) = \mu(A \cap A(t) \cap B(t))$. Similarly we can obtain $\mu(A \cap B(t)) = \mu(A \cap B(t) \cap A(t))$. Thus $\mu(A \cap A(t)) = \mu(A \cap B(t))$. A similar reasoning applies to ν.

Claim 6. *For any* $s, t \in [0, \infty)$ *with* $s < t$ *and* $A \in \mathcal{U}$,

$$
\begin{aligned}
s[\nu(A \cap B(s)) - \nu(A \cap B(t))] &\leq \mu(A \cap B(s)) - \mu(A \cap B(t)) \\
&\leq t[\nu(A \cap B(s)) - \nu(A \cap B(t))].
\end{aligned}
$$

Proof: By (iii) and Claim 5,

$$
\begin{aligned}
s[\nu(A \cap B(s)) - \nu(A \cap B(t))] &= s[\nu(A \cap A(s)) - \nu(A \cap A(t))] \\
&\leq \mu(A \cap A(s)) - \mu(A \cap A(t)) \\
&= \mu(A \cap B(s)) - \mu(A \cap B(t))
\end{aligned}
$$

and

$$
\begin{aligned}
\mu(A \cap B(s)) - \mu(A \cap B(t)) &= \mu(A \cap A(s)) - \mu(A \cap A(t)) \\
&\leq t[\nu(A \cap A(s)) - \nu(A \cap A(t))] \\
&= t[\nu(A \cap B(s)) - \nu(A \cap B(t))].
\end{aligned}
$$

Remark. Observe that conditions (i)-(iii) imply the absolute continuity of μ with respect to ν. In fact, assume that $\nu(A) = 0$. Then from Claim 6,

$$\mu(A \cap B(s)) - \mu(A \cap B(t)) = 0 \quad \text{for every } s, t \in [0, \infty) \text{ with } s < t.$$

Letting $t = n \to \infty$, we obtain from Claim 2, $\mu(A \cap A(s)) = 0$ for every $s \in [0, \infty)$. It follows that $\mu(A \cap B(0)) = \mu(A \cap U) = \mu(A) = 0$.

Now we are able to complete the proof of our main result.
Claim 7. $\mu(A) = \int_A f d\nu$ for every $A \in \mathcal{U}$.

Proof: Let $f : U \longrightarrow [0, 1]$ be given in (8) and $A_n = A \cap \{f < n\} = A \cap B^c(n)$ for every $A \in \mathcal{U}$ and $n \in \mathbb{N}$. Then as $A_n \subset \{f < n\}$,

$$\int_0^n \nu(A_n \cap \{f \geq t\}) \, dt = \int_0^\infty \nu(A_n \cap \{f \geq t\}) dt. \qquad (3.2)$$

We first prove that

$$\mu(A_n) = \int_0^\infty \nu(A_n \cap \{f \geq t\})dt \quad \text{for every } n \in \mathbb{N}.$$

Let $0 = t_0 < t_1 < \cdots < t_k = n$,

$$s_k = \sum_{i=1}^k (t_i - t_{i-1})\nu(A_n \cap \{f \geq t_i\}),$$

and

$$S_k = \sum_{i=1}^k (t_i - t_{i-1})\nu(A_n \cap \{f \geq t_{i-1}\}).$$

Then we have

$$s_k \leq \int_0^n \nu(A_n \cap \{f \geq t\})dt \leq S_k.$$

Note that as $\max\{t_i - t_{i-1} : i = 1, \ldots, k\} \to 0$,

$$s_k \longrightarrow \int_0^n \nu(A_n \cap \{f \geq t\})dt, \quad S_k \longrightarrow \int_0^n \nu(A_n \cap \{f \geq t\})dt. \quad (3.3)$$

By using the first inequality in Claim 6 and the fact that $\mu(A_n \cap \{f \geq t_k\}) = \nu(A_n \cap \{f \geq t_k\}) = 0$, we obtain

$$
\begin{aligned}
s_k &= \sum_{i=1}^k (t_i - t_{i-1})\nu(A_n \cap \{f \geq t_i\}) \\
&= \sum_{i=1}^k t_i \nu(A_n \cap \{f \geq t_i\}) - \sum_{i=1}^k t_{i-1}\nu(A_n \cap \{f \geq t_i\}) \\
&= \sum_{i=1}^{k-1} t_i [\nu(A_n \cap \{f \geq t_i\}) - \nu(A_n \cap \{f \geq t_{i+1}\})] \\
&\leq \sum_{i=1}^{k-1} [\mu(A_n \cap \{f \geq t_i\}) - \mu(A_n \cap \{f \geq t_{i+1}\})] \\
&= \mu(A_n \cap \{f \geq t_1\}) - \mu(A_n \cap \{f \geq t_k\}) \\
&= \mu(A_n \cap \{f \geq t_1\}) \leq \mu(A_n).
\end{aligned}
$$

Hence $s_k \leq \mu(A_n)$. Similarly, by using the second inequality in Claim 6, $S_k \geq \mu(A_n)$. Therefore $s_k \leq \mu(A_n) \leq S_k$. Combining (10) and (11), we obtain

$$\mu(A_n) = \int_0^n \nu(A_n \cap \{f \geq t\})dt = \int_{A_n} f d\nu.$$

By Claim 3 we have

$$\mu(A) = \lim_{n \to \infty} \mu(A_n) = \int_0^\infty \nu(A \cap \{f \geq t\})dt = \int_A f d\nu.$$

Consequently the theorem is proved.   □

The following example is a pair of set-functions which are not capacities in Graf's sense, but are submeasures. Moreover, they satisfy the conditions of Theorem 1, and hence (μ, ν) has the RNP.

Example 6 *Let U be a Banach space, $\mathcal{U} = \mathcal{B}(U)$, and $S = \{x : \|x\| < 1\}$. Define*

$$\mu(A) = \begin{cases} 0 & \text{if } A = \emptyset \\ \frac{1}{2}\sup_{x \in A} \|x\| & \text{if } A \neq \emptyset \text{ and } d(A, S^c) > 0 \\ 1 & \text{if } A \neq \emptyset \text{ and } d(A, S^c) = 0 \end{cases}$$

and

$$\nu(A) = \begin{cases} 0 & \text{if } A = \emptyset \\ \frac{1}{2} & \text{if } A \neq \emptyset \text{ and } d(A, S^c) > 0 \\ 1 & \text{if } A \neq \emptyset \text{ and } d(A, S^c) = 0, \end{cases}$$

where S^c is the set complement of S and $d(A, B) = \inf\{\|x - y\| : x \in A, y \in B\}$.

Proof: First observe that μ and ν are not capacities in the sense of Graf. In fact, for $S_n = \{x : \|x\| \leq 1 - \frac{1}{n}\}$, we have $S_n \nearrow S = \cup_{n=1}^{\infty} S_n$, but

$$\mu(S) = \nu(S) = 1 > \frac{1}{2} = \lim_{n \to \infty} \mu(S_n) = \lim_{n \to \infty} \nu(S_n).$$

Therefore the lower continuity condition of capacities is violated.

It is easy to see that μ and ν are submeasures. Moreover, applying Theorem 1, we will show that (μ, ν) has the RNP.

To verify the conditions (i)-(iii) of Theorem 1, let

$$A(t) = \begin{cases} \{x : \|x\| \geq t\} & \text{for } t \in [0, 1] \\ \emptyset & \text{for } t > 1. \end{cases}$$

Then $\cap_{q \in Q^+} A(q) = \emptyset$, and hence (i) holds. Clearly, $B(t) = \cap_{q \in Q(t)} = A(t)$, which implies (ii). Thus it suffices to verify that (iii) holds. Note that $A(t) \subset A(s)$ for all $s < t$, and (iii) holds if $A \cap A(s) = \emptyset$. So we need only consider the case where $A \cap A(s) \neq \emptyset$ (in this case, $s \leq 1$ as $A(s) = \emptyset$ for $s > 1$).

Case 1. $d(A \cap A(s), S^c) > 0$ and $A \cap A(t) = \emptyset$. We have

$$s \leq \sup_{A \cap A(s)} \|x\| \leq t$$

and $\mu(A \cap A(t)) = \nu(A \cap A(t)) = 0$. Therefore

$$\mu(A \cap A(s)) = \frac{1}{2} \sup_{A \cap A(s)} \|x\| \geq \frac{s}{2} = s[\nu(A \cap A(s))],$$

as $\nu(A \cap A(s)) = 1/2$. Also

$$\mu(A \cap A(s)) = \frac{1}{2} \sup_{A \cap A(s)} \|x\| \leq \frac{t}{2} = t[\nu(A \cap A(s))],$$

hence (iii) follows.

Case 2. $d(A \cap A(s), S^c) > 0$ and $A \cap A(t) \neq \emptyset$. We have $d(A \cap A(t), S^c) \geq d(A \cap A(s), S^c) > 0$, $\nu(A \cap A(s)) = \nu(A \cap A(t)) = 1/2$, and

$$\mu(A \cap A(s)) = \frac{1}{2} \sup_{A \cap A(s)} \|x\| = \frac{1}{2} \sup_{A \cap A(t)} \|x\| = \mu(A \cap A(t)).$$

Thus (iii) holds.

Case 3. $d(A \cap A(s), S^c) = 0$ and $A \cap A(t) = \emptyset$. We have $t \geq 1$ and $\mu(A \cap A(s)) = \nu(A \cap A(s)) = 1$. Therefore

$$s[\nu(A \cap A(s))] = s \leq 1 = \mu(A \cap A(s)) \leq t = t\nu(A \cap A(s)),$$

i.e., (iii) holds.

case 4. $d(A \cap A(s), S^c) = 0$ and $A \cap A(t) \neq \emptyset$. We have $s \leq 1, t \leq 1, \mu(A \cap A(s)) = \nu(A \cap A(s)) = 1$, and $d(A \cap A(t), S^c) = 0$. Thus $\mu(A \cap A(t)) = \nu(A \cap A(t)) = 1$ and hence (iii) follows.

Since (μ, ν) satisfies the conditions (i) - (iii) of Theorem 1, (μ, ν) has the RNP and hence $d\mu/d\nu$ exists. Moreover, the Radon-Nikodym derivative $d\mu/d\nu$ can be obtained by :

$$\frac{d\mu}{d\nu} = f(x) = \begin{cases} \|x\| & \text{if } x \in S \\ 1 & \text{if } x \in S^c. \end{cases} \qquad \Box$$

The proof of Lemma 1 (Section 3.1) is combinatorial in nature. An alternative proof is presented below based upon the concept of *local Radon Nikodym derivatives* of set-functions, a concept of independent interest. This is based on a recent preprint of M. Marinacci (personal communication, 1996).

Observe that the capacity functional T in Example 1 has the function f as its Radon-Nikodym derivative with respect to the set-function ν_0 (see Example 2). Now, obviously, ν_0 is alternating of infinite order. Thus the fact that T is alternating of infinite order seems to be a consequence of this Choquet integral representation. It is indeed the case, and more generally, arbitrary maxitive set-functions have "local" Choquet integral representations with respect to ν_0, and hence alternating of infinite order.

Let $(U, \mathcal{U})$ be a measurable space, and $\mu, \nu : \mathcal{U} \to [0, \infty)$. We always assume that all set-functions are monotone increasing and zero on empty set.

Following Marinacci, we say that the pair (μ, ν) has the *local Radon-Nikodym Property* (LRNP) if for every finite sub-algebra $\mathcal{A}$ of $\mathcal{U}$, there is a $\mathcal{A}$-measurable function $f_\mathcal{A} : U \to [0, \infty)$ such that

$$\mu(A) = \int_A f_\mathcal{A} \, d\nu, \qquad \forall A \in \mathcal{A},$$

where, of course, the integral is taken in Choquet's sense.

In measure theory, i.e. for μ and ν being countably additive, this concept of LRNP is subsumed by the usual Radom-Nikodym property (RNP), i.e. Radon-Nikodym derivative of μ with respect to ν. Indeed, if (μ, ν) has LRNP, then obviously $\mu << \nu$, and hence $d\mu/d\nu$ exists. Conversely, suppose that $d\mu = f d\nu$. Let $\mathcal{A}$ be a finite sub-algebra of $\mathcal{U}$, say, $\mathcal{A} = \{A_1, \ldots, A_n\}$. Let $\mathcal{B}$ be the set of all atoms of $\mathcal{A}$, i.e., set of the form $\cap_{k=1}^n A_k^{i_k}$, where $i_k \in \{0, 1\}$ with $A_k^1 = A_k$ and $A_k^0 = A_k^c$. It suffices to define $f_{\mathcal{A}} : \mathcal{U} \to [0, \infty)$ by

$$f_{\mathcal{A}}(x) = \frac{1}{\nu(B)} \int_B f d\nu, \qquad \text{for } x \in B \in \mathcal{B},$$

when $\nu(B) \neq 0$, and zero if $\nu(B) = 0$.

For non-additive set-functions, however, the two concepts RNP and LRNP are distinct.

It can be shown that if the pair (μ, ν) has LRNP and ν is alternating of infinite order, then so is ν. As a consequence, any maxitive set-function is alternating of infinite order.

This can be seen as follows. Suppose that μ is maxitive. Let $\mathcal{A}$ be the finite sub-algebra of $\mathcal{U}$, and $\mathcal{B}$ be its set of atoms. Define $f : U \to [0, \infty)$ by

$$f(u) = \mu(B), \qquad \text{for } u \in B \in \mathcal{B}.$$

The function f so defined is $\mathcal{A}$-measurable. For $A \in \mathcal{A}$, $A = \cup_{i=1}^n B_i$, $B_i \in \mathcal{B}$, we have

$$\mu(A) = \max\{\mu(B_i), 1 \leq i \leq n\} = \max_{u \in A} f(u),$$

which can be written as $\int_A f d\nu_0$, implying that the pair (μ, ν_0) has the LRNP, and the result follows by observing that ν_0 is alternating of infinite order.

3.3   Möbius Transforms of Set-Functions

The standard framework in the so-called theory of evidence is this. Let U be a finite set. Consider the situation where the uncertainty on U is expected as a belief function, i.e., $F : 2^U \to [0, 1]$ satisfying

(i) $F(\emptyset) = 0, \quad F(U) = 1$,
(ii) For any $n \geq 2$, $A_1, \ldots, A_n \subseteq U$,

$$F\left(\bigcup_{i=1}^n A_i\right) \geq \sum_{\emptyset \neq I \subseteq \{1, \ldots, n\}} (-1)^{|I|+1} F\left(\bigcap_{i \in I} A_i\right),$$

where $|I|$ denotes the cardinality of the set I. Note that (ii) means that F is *monotone of infinite order*.

Although the set-function F is not an ordinary measure (F is not additive, in general), it does play the role of the "distribution function", not of some random variable, but of some random set, i.e. there exists a random set S : $(\Omega, \mathcal{A}, P) \to 2^U$ such that

$$F(A) = P(S \subseteq A), \qquad \forall A \subseteq U.$$

This can be seen by exhibiting a probability measure on the power set of 2^U. Viewing $(2^U, \subseteq)$ as a locally finite ordered set, one can utilize results from Möbius inversion in combinatorial theory (see, e.g. Aigner [1]).

Let $f : 2^U \to [0, 1]$ be the Möbius transform of F, i.e.

$$f(A) = \sum_{B \subseteq A} (-1)^{|A \setminus B|} F(B)$$

(noting that the Möbius function of $(2^U, \subseteq)$ is $\mu(A, B) = (-1)^{|A \setminus B|}$ for all $A \subseteq B \subseteq U$). Now the axioms on F ((i) and (ii)) imply that $f(A) \geq 0$, for any $A \subseteq U$ and

$$\sum_{A \subseteq U} f(A) = 1,$$

i.e. f is a bona fide probability density function of some random element taking values in 2^U.

In applications, one usually specifies f as a model for uncertainty analysis. As far as we know, the theory of evidence stops there, i.e. only the case of finite frame U is considered. Perhaps, if U is infinite, so that $(2^U, \subseteq)$ needs not be locally finite, there is no Möbius transforms of set-functions? Of course, as in standard probability theory one could then deal directly with probability measures rather than with "densities". An interesting and natural question is: not only for possible applications, but also as a mathematical problem of its independent interest does there exist the counterpart of Möbius transforms in the non locally finite case? It turns out, we believe, that the answer is yes. It is hidden in the recent paper by Marinacci [21]!

In the following, we are going to elaborate on this. Let us review briefly the concept of Möbius functions (essentially due to Rota [33]). Let X be a poset, i.e., X is a set with a partial order relation $\leq$ on it. For $x, y \in X$, the segment (or "interval") $[x, y]$ is defined as $\{z \in X : x \leq z \leq y\}$. The poset $(X, \leq)$ is said to be *locally finite* if $[x, y]$ is finite for any $x, y \in X$.

Suppose that $(X, \leq)$ is locally finite. A basic question in combinatorial theory is this. If $f, g : X \to \mathbb{R}$ such that

$$g(x) = \sum_{y \leq x} f(y).$$

Is it possible (and how) to express f in terms of g? In other words, can one find an inversion formula to recover f from g? The answer is given via the Möbius function of the poset $(X, \leq)$.

Specifically, let

$$\mathcal{A}(X) = \{\phi : X^2 \to \mathbb{R} : x \not\leq y \Rightarrow \phi(x, y) = 0, \ \phi(x, x) \neq 0, \forall x\}.$$

An operation on $\mathcal{A}(X)$ is defined as

$$\phi * \psi(x, y) = \sum_{x \le z \le y} \phi(x, z)\phi(z, y),$$

The algebraic structure $(\mathcal{A}(X), *)$ is a group in which the Kronecker function δ is the identity, where

$$\delta(x, y) = \left\{ \begin{array}{ll} 1 & \text{if } x = y \\ 0 & \text{if } x \ne y. \end{array} \right.$$

The *Möbius function* μ of $(X, \le)$ is defined to be the inverse of the Zeta-function

$$\eta(x, y) = \left\{ \begin{array}{ll} 1 & \text{if } x \le y \\ 0 & \text{otherwise,} \end{array} \right.$$

or, equivalently, μ is such that

$$\mu(x, x) = 1, \qquad \mu(x, y) = - \sum_{x \le z \le y} \mu(x, z).$$

Given a "sum function" (viewed as the discrete counterpart of the indefinite integral in calculus),

$$g(x) = \sum_{y \le x} f(y),$$

the Möbius inversion is viewed as the discrete analogue of the derivative:

$$f(x) = \sum_{y \le x} \mu(y, x)g(y).$$

In view of this inversion formula, we call f the *Möbius transform* of g.

In the case of the Boolean algebra 2^U, with U finite, the Möbius function of $(2^U, \subseteq)$ is

$$\mu(A, B) = (-)^{|A \backslash B|}, \quad \text{for } A \subseteq B \subseteq U,$$

so that if $g(A) = \sum_{B \subseteq A} f(B)$, we have

$$f(A) = \sum_{B \subseteq A} (-1)^{|A \backslash B|} g(B)$$

as the Möbius transform of the set-function g.

The question we ask is this. Suppose U is infinite, and $\mathcal{U}$ is an infinite algebra of subsets of U. Let $g : \mathcal{U} \to \mathbb{R}$ be a set-function. What is a reasonable generalization of the concept of Möbius transforms for the infinite case?

The following is from Marinacci [21]. All set-functions are defined on $\mathcal{U}$ and are zero on empty set. The space of all such set-functions is denoted by V. With pointwise addition and multiplication by scalars, V is a vector space.

First, consider the case where $\mathcal{U}$ *is finite*. Then the Möbius transform of $g \in V$ plays the role of coefficients in the development of g with respect to some linear basis of V. Specifically,

$$g(A) = \sum_{\emptyset \neq B \in \mathcal{U}} \alpha_g(B) u_B(A), \qquad A \in \mathcal{U}, \qquad (*)$$

where $u_B \in V$ for any $B \in \mathcal{U}$,

$$u_B(A) = \begin{cases} 1 & \text{if } B \subseteq A \\ 0 & \text{otherwise,} \end{cases}$$

and

$$\alpha_g(B) = \sum_{D \subseteq B} (-1)^{|B \setminus D|} g(D).$$

Now, if we view the Möbius transform $\alpha_g(\cdot)$ of g as a (signed) *measure* on the finite set $\mathcal{U}$ (for $\mathcal{A} = \{A_1, \ldots, A_n\} \subseteq \mathcal{U}$, define $\alpha_g(\mathcal{A}) = \sum_{i=1}^n \alpha_g(A_i)$), then the Möbius transform of a set-function g is a (signed) measure on $\mathcal{U}$ satisfying $(*)$.

We need another identification. For each $A \in \mathcal{U}$, we identify A with the principal filter generated by A, namely

$$p(A) = \{B \in \mathcal{U} : A \subseteq B\}.$$

But, each principal filter p (generated by A) corresponds uniquely to an element u_p where $u_p(B)$ is equal to 1 if $B \in p$ and zero otherwise. Thus in the case of a finite algebra $\mathcal{U}$, the Möbius transform $\alpha_g(\cdot)$ of a set-function $g : \mathcal{U} \to \mathbb{R}$ is a (signed) measure on the *space*

$$\mathbb{F} = \{u_P(\cdot), p \text{ principal filters of } \mathcal{U}\}$$

satisfying $(*)$. Its existence is well-known from combinatorial theory. As we will see, its existence is due to the fact that the "composition norm" of g, namely $\|g\| = \sum_{A \in \mathcal{U}} |\alpha_g(A)|$, is finite.

Now, consider the case where the algebra $\mathcal{U}$ (of subsets of U) is arbitrary (finite or not). From the above identifications, it seems natural to define the Möbius transform $g : \mathcal{U} \to \mathbb{R}$ by generalizing the space $\mathbb{F}$ and the relation $(*)$ (to an integral representation), as well as specifying sufficient conditions for their existence. In other words, the Möbius transform of g, when it exists, will be a signed measure living on some appropriate space.

The space $\mathbb{F}$ is generalized as follows. Observe that an element u_p in $\mathbb{F}$ is a special case of a set-function v_p on $\mathcal{U}$ where p is a proper filter of $\mathcal{U}$ (i.e. $p \neq \mathcal{U}$), $v_p = 1$ if $A \in p$ and 0 otherwise. Thus $\mathbb{F}$ is generalized to $\mathbb{G} = \{v_p : p \text{ proper filter of } \mathcal{U}\}$. Note that the space $\mathbb{G}$ can be topologized appropriately. By a measure on $\mathbb{G}$, we mean a Borel measure on $\mathbb{G}$.

For a set-function $g : \mathcal{U} \to \mathbb{R}$, we define its composition norm $\|\cdot\|$ by

$$\|g\| = \sup\{\|g|_{\mathcal{F}}\| : \mathcal{F} \text{ finite sub-algebra of } \mathcal{U}\},$$

where $g|_{\mathcal{F}}$ denotes the restriction of g to $\mathcal{F}$.

The basic representation theorem of Marinacci [21] is this. If $\|g\| < \infty$, then there exists a unique (regular and bounded) signed measure α_g on $\mathbb{G}$ such that

$$g(A) = \int_{\mathbb{G}} v_p(A) d\,\alpha_g(v_p), \qquad \forall A \in \mathcal{U}, \qquad (**)$$

It is clear that $(**)$ is an extension of $(*)$. Thus, via identifications, the signed measure α_g on $\mathbb{G}$ can be viewed as the Möbius transform of the set-function g (with bounded norm) in the infinite case.

3.4   Random Sets in Decision-Making

In this section, we illustrate some typical situations in which random sets come to play a useful role in decision-making under uncertainty.

3.4.1   Confidence Region Estimation

The problem of computations of various numerical characteristics of random sets appears in various contexts of statistical inference, such as in the theory of coverage processes (see Hall [13]), and in the optimal choice of confidence region estimation procedures (see Robbins [31]). In many situations, one is interested in the expected measure of random sets. Specifically, let S be a random set with values in $\mathcal{C} \subseteq \mathcal{B}(\mathbb{R}^d)$. Let $\sigma(\mathcal{C})$ be a σ-field on $\mathcal{C}$. One is interested in the expected value $E(\mu(S))$, where μ denotes the Lebesgue measure on $\mathbb{R}^d$. Of course, for this to make sense, $\mu(S)$ has to be random variable. This is indeed the case, for example, when S depends upon a finite number of random variables. In general, the distribution of $\mu(S)$ seems difficult to obtain. One then proceeds directly to the computation of $E(\mu(S))$ as follows. The *one-point coverage function* of S is defined as $\pi : \mathbb{R}^d \longrightarrow [0,1]$,

$$\pi(x_1, \ldots, x_d) = P(\{\omega : (x_1, \ldots, x_d) \in S(\omega)\}).$$

The main result in Robbins [31] is this. If the map $g : \mathbb{R}^d \times \mathcal{C} \longrightarrow \{0,1\}$, defined by $g(x, A) = 1_A(x)$, is $\mathcal{B}(\mathbb{R}^d) \otimes \sigma(\mathcal{C})$–measurable, then

$$E(\mu(S)) = \int_{\mathbb{R}^d} \pi(x) d\mu(x).$$

By considering many-point coverage functions, i.e.,

$$\pi_n(u_1, \ldots, u_n) = P(\{\omega : \{u_1, \ldots, u_n\} \subseteq S(\omega)\}),$$

for $u_i \in \mathbb{R}^d$, $i = 1, 2, \ldots, n$, higher order moments of $\mu(S)$ are obtained in a similar way, under suitable measurability conditions. Specifically, for $n \geq 1$,

$$E[\mu(S)]^n = \int_{\mathbb{R}^{nd}} \pi(u_1, \ldots, u_n) \otimes_{i=1}^n \mu_i(du_1, \ldots, du_n),$$

where $\otimes_{i=1}^{n}\mu_i$ denotes the product measure $\mu \otimes \cdots \otimes \mu$ on $\mathbb{R}^{nd}$.

When $n = 1$, the measurability of $g(x, A)$ implies that $\pi(x)$ is measurable and $\mu(S)$ is a random variable. Viewing $(g(x, \cdot), x \in \mathbb{R}^d)$ as a 0-1 stochastic process on the measurable space $(\mathcal{C}, \sigma(\mathcal{C}))$, we see that the measurability condition on g is simply that this stochastic process is a measurable process. If S is a compact-valued random set on $\mathbb{R}^d$, then the process g is measurable, and hence $\mu(S)$ is a random variable. For a general condition on the measurability of the process g, see Debreu [5].

Obviously, being unaware of Robbins' result, Matheron [15] established the same measurability condition and deduced Robbins' formula, but the purpose was to obtain a tractable criterion for almost sure continuity for random closed sets. According to Matheron, any random closed set S is measurable in the following sense: the map $g : \mathbb{R}^d \times \mathcal{F} \longrightarrow \{0, 1\}$ is $\mathcal{B}(\mathbb{R}^d) \otimes \mathcal{B}(\mathcal{F})$–measurable. Thus, if μ denotes a positive measure (σ-finite) on $(\mathbb{R}^d, \mathcal{B}(\mathbb{R}^d))$ (or more generally, on a locally compact, Hausdorff, separable space), then the map

$$F \in \mathcal{F} \longrightarrow \mu(F) = \int_{\mathbb{R}^d} g(x, F) d\mu(x)$$

is a positive random variable whose expected value is $\int_{\mathbb{R}^d} P(x \in S) d\mu(x)$.

Note that for $d = 1$ and $S(\omega) = [0, X(\omega)]$, where X is a non-negative random variable, Robbins' formula becomes

$$E(X) = \int_0^\infty P(X > t) dt.$$

Also, note that in sample surveys, e.g., Hajek [12], a sampling design on a finite population $\mathcal{U} = \{u_1, u_2, \ldots, u_N\}$ is a probability measure Q on its power set. The inclusion probabilities (first and second order) are precisely the values of the coverage functions, indeed

$$\alpha(u) = \sum_{A \ni u} Q(A) = P(u \in S),$$

$$\beta(u, v) = \sum_{A \ni u, v} Q(A) = P(\{u, v\} \subseteq S),$$

where S is a random set on $2^{\mathcal{U}}$ with probability law Q. These inclusion probabilities are related to the concept of measure of spread for sampling probabilities (defined as the entropy of the random set S).

3.4.2 Imprecise Probabilities

In its simplest form, a decision problem consists of choosing an action among a collection of relevant actions $\mathbb{A}$ in such a way that "utility" is maximized. Specifically, let Θ denote the set of possible "states of nature" where the unknown, true state is denoted by θ_0. A specific utility function is a map $u : \mathbb{A} \times \Theta \longrightarrow \mathbb{R}$, where $u(a, \theta)$ is interpreted as the "payoff" when action a is taken and nature presents θ. In a Bayesian framework, the knowledge about Θ is represented as a probability measure P_0 on Θ. The expected value $E_{P_0}(u(a, \cdot))$ can be used to make a choice as to which a to take. The optimal action is the one that maximizes $E_{P_0}(u(a, \cdot))$ over $a \in \mathbb{A}$. In many situations, the probability measure P_0 is only known partially, say, $P_0 \in \mathcal{P}$, a specified class of probability measures on Θ, so that $F \leq P_0 \leq G$, where $F = \inf_{\mathcal{P}} P$ and $G = \sup_{\mathcal{P}} P$. There are situations in which the lower envelop F turns out to be a set-function, totally monotone, i.e., a distribution functional of some random set S. Decision-making in this imprecise probabilistic knowledge can be carried out by using one of the following approaches.

(a) **Choquet Integral.** From a minimax viewpoint, we choose the action that maximizes

$$\inf \{E_p(u(a, \cdot)) : p \in \mathcal{P}\}.$$

Consider the case when Θ is finite, say, $\Theta = \{\theta_1, \theta_2, \ldots, \theta_n\}$. Suppose that F is monotone of infinite order. For each a, rename the elements of Θ if necessary so that

$$u(a, \theta_1) \leq u(a, \theta_2) \leq \cdots \leq u(a, \theta_n).$$

Let $A_i = \{\theta_i, \theta_{i+1}, \ldots, \theta_n\}$ and

$$h(\theta_i) = F(A_i) - F(A_i \setminus \{\theta_i\}).$$

Then h is a probability density on Θ such that

$$\begin{aligned}
P_h(A) &= \sum_{B \subseteq A_i} m(B) - \sum_{B \subseteq A_i \setminus \{\theta_i\}} m(B) \\
&= \sum_{\theta_i \in B \subseteq A_i} m(B),
\end{aligned}$$

where m is the Moebius inverse of F, i.e., $m : 2^{\Theta} \longrightarrow [0, 1]$,

$$m(A) = \sum_{B \subseteq A} (-1)^{|A \setminus B|} F(B).$$

Now, for each $t \in \mathbb{R}$ and a density g such that $P_g \in \mathcal{P}$, we have

$$P_h(u(a, \cdot) > t) \leq P_g(u(a, \cdot) > t)$$

since $(u(a, \cdot) > t)$ is of the form $\{\theta_i, \theta_{i+1}, \ldots, \theta_n\}$. Thus

$$E[P_h(u(a, \cdot) > t)] \leq E[P_g(u(a, \cdot) > t)]$$

for all densities g such that $P_g \in \mathcal{P}$. But, by construction of h

$$\begin{aligned}
&E[P_h(u(a, \cdot) > t)] \\
&= \sum_{i=1}^{n} u(a, \theta_i) \left[F(\{\theta_i, \ldots, \theta_n\}) - F(\{\theta_{i+1}, \ldots, \theta_n\}) \right] \\
&= \int_0^{\infty} F(u(a, \cdot) > t) \, dt + \int_{-\infty}^{0} [F(u(a, \cdot) > t) - 1] \, dt,
\end{aligned}$$

which is nothing else than the *Choquet integral* of the function $u(a, \cdot)$ with respect to the monotone set-function F, denoted as $E_F(U(a, \cdot))$.

We have

$$E_F(U(a, \cdot)) = \inf\{E_P(U(a, \cdot)) : P \in \mathcal{P}\}.$$

For example, let $\Theta = \{\theta_1, \theta_2, \theta_3, \theta_4\}$ and

$$F(A) = \sum_{B \subseteq A} m(B), \qquad A \subseteq \Theta,$$

where

$$m(\{\theta_1\}) = 0.4, \qquad m(\{\theta_2\}) = f(\{\theta_3\}) = 0.2,$$

$$m(\{\theta_4\}) = 0.1, \qquad m(\{\theta_1, \theta_2, \theta_3, \theta_4\}) = 0.1,$$

and $m(A) = 0$ for any other subset A of Θ; then, we have

$$h(\theta_1) = 0.5, \quad h(\theta_2) = h(\theta_3) = 0.2, \quad h(\theta_4) = 0.1.$$

If

$$u(a, \theta_1) = 1, \quad u(a, \theta_2) = 5, \quad u(a, \theta_3) = 10, \quad u(a, \theta_4) = 20,$$

then

$$E_F[u(a, \cdot)] = \sum_{i=1}^{4} u(a, \theta_i) h(\theta_i) = 5.5.$$

When Θ is infinite and F is monotone of infinite order, we still have

$$E_F(U(a, \cdot)) = \inf\{E_P(U(a, \cdot)) : P \in \mathcal{P}\},$$

although, unlike the infinite case, the inf might not be attained (see Wasserman [37]).

From the above, we see that the concept of Choquet integral can be used as a tool for decision-making based on expected utility in situations where imprecise probabilistic knowledge can be modeled as a distribution functional of some random set. For example, consider another situation with imprecise information. Let X be a random variable, defined on $(\Omega, \mathcal{A}, P)$, and $g : \mathbb{R} \to \mathbb{R}^+$ be a measurable function. We have

$$E(g(X)) = \int_\Omega g(X(\omega)) dP(\omega) = \int_{\mathbb{R}} g(x) dP_X(x),$$

where $P_X = PX^{-1}$.

Suppose that to each random experiment ω, we can only assert that the outcome $X(\omega)$ is some interval, say $[a, b]$. This situation can be described by a multi-valued mapping $\Gamma : \Omega \longrightarrow \mathcal{F}(\mathbb{R})$, the class of closed subsets of $\mathbb{R}$. Moreover, for each $\omega \in \Omega$, $X(\omega) \in \Gamma(\omega)$. As in Wasserman [37], the computation, or approximation of $E[g(X)]$ is carried out as follows.

Since

$$g(X(\omega)) \in g(\Gamma(\omega)) = \{g(x) : x \in \Gamma(\omega)\},$$

we have

$$g_*(\omega) = \inf\{g(x) : x \in \Gamma(\omega)\} \le g(X(\omega))$$
$$\le \sup\{g(x) : x \in \Gamma(\omega)\} = g^*(\omega).$$

Thus, $g(x)$ is such that $g_* \le g(X) \le g_*$ and hence formally,

$$E(g_*) \le E(g(X)) \le E(y^*).$$

For this to make sense, we first need to see whether g_* and g^* are measurable functions.

Let $\mathcal{B}(\mathbb{R})$ be the Borel σ-field on $\mathbb{R}$. The multi-valued mapping Γ is said to be strongly measurable if

$$B_* = \{\omega : \Gamma(\omega) \subseteq B\} \in \mathcal{A}, \quad B^* = \{\omega : \Gamma(\omega) \cap B \ne \emptyset\} \in \mathcal{A},$$

for all $B \in \mathcal{B}(\mathbb{R})$. If Γ is strongly measurable, then the measurability of g implies that of g_* and g^*, and conversely. Indeed, suppose that Γ is strongly measurable and g is measurable. For $\alpha \in \mathbb{R}$,

$$\Gamma(\omega) \subseteq \{x : g(x) \ge \alpha\} = g^{-1}([\alpha, \infty)),$$

and hence $\omega \in \left[g^{-1}([\alpha, \infty))\right]_*$. If $\omega \in \left[g^{-1}([\alpha, \infty))\right]_*$, then $\Gamma(\omega) \subseteq g^{-1}([\alpha, \infty))$, that is, for all $x \in \Gamma(\omega)$, $g(x) \ge \alpha$ implying that $\inf_{x \in \Gamma(\omega)} g(x) \ge \alpha$, and hence $g_*(\omega) \in [\alpha, \infty)$ or $\omega \in g_*^{-1}([\alpha, \infty))$. Therefore

$$g_*^{-1}([\alpha, \infty)) = \left[g^{-1}([\alpha, \infty))\right]_*.$$

Since g is measurable, $g^{-1}([\alpha, \infty)) \in \mathcal{B}(\mathbb{R})$, and since Γ is strongly measurable, $\left[g^{-1}([\alpha, \infty))\right]_* \in \mathcal{A}$. The measurability of g^* follows similarly. For the converse, let $A \in \mathcal{B}(\mathbb{R})$. Then $1_A = f$ is measurable and

$$f_*(\omega) = \begin{cases} 1 & \text{if } \Gamma(\omega) \subseteq A \\ 0 & \text{otherwise} \end{cases}$$

Thus, $f^{-1}(\{1\}) = A_*$, and by hypothesis, $A_* \in \mathcal{A}$. Similarly $A^* \in \mathcal{A}$.

If we let $F_* : \mathcal{B}(\mathbb{R}) \longrightarrow [0,1]$ be defined by

$$F_*(B) = P(\{\omega : \Gamma(\omega) \subseteq B\}) = P(B_*),$$

then

$$E_*(g(X)) = \int_\Omega g_*(\omega) dP(\omega) = \int_0^\infty P(\{\omega : g_*(\omega) > t\}) dt$$
$$= \int_0^\infty P(\{g_*^{-1}(t, \infty)\}) dt = \int_0^\infty P\left[g^{-1}(t, \infty)\right]_* dt$$
$$= \int_0^\infty P(\{\omega : \Gamma(\omega) \subseteq g^{-1}(t, \infty)\}) dt$$
$$= \int_0^\infty F_*(g^{-1}(t, \infty)) dt = \int_0^\infty F_*(g > t) dt.$$

Note that Γ is assumed to be strongly measurable, so that F_* is well-defined on $\mathcal{B}(\mathbb{R})$. Similarly, letting

$$F^*(B) = P(\{\omega : \Gamma(\omega) \cap B \neq \emptyset\}) = P(B^*),$$

we have

$$E^*(g(X)) = \int_0^\infty F^*(g > t)dt.$$

In the above situation, the set-functions F^* and F_* are known (say, Γ is observable but X is not). Although F^* and F_* are not probability measures, they can be used for approximate inference procedures. Here Choquet integrals with respect to monotone set-functions represent some practical quantities of interest.

(b) **Expectation of a Function of a Random Set**. Lebesgue measure of a closed random set on $\mathbb{R}^d$ is an example of a function of a random set in the infinite case. Its expectation can be computed from knowledge of the one-point coverage function of the random set. For the finite case, say $\Theta = \{\theta_1, \theta_2, \ldots, \theta_n\}$, a random set $S : (\Omega, \mathcal{A}, P) \longrightarrow 2^\Theta$ is characterized by $F : 2^\Theta \longrightarrow [0, 1]$, where

$$F(A) = P(\omega : S(\omega) \subseteq A) = \sum_{B \subseteq A} P(\omega : S(\omega) = B) = \sum_{B \subseteq A} m(B).$$

By Moebius inverse, we have

$$m(A) = \sum_{B \subseteq A} (-1)^{|A \setminus B|} F(B).$$

It can be shown that $m : 2^\Theta \longrightarrow \mathbb{R}$ is a probability "density," i.e., $m : 2^\Theta \longrightarrow [0, 1]$ and $\sum_{A \subseteq \Theta} m(A) = 1$ if and only if $F(\emptyset) = 0$, $F(\Theta) = 1$, and F is monotone of infinite order.

Let $\phi : 2^\Theta \longrightarrow \mathbb{R}$. Then $\phi(S)$ is a random variable whose expectation is

$$E(\phi(S)) = \sum_{A \subseteq \Theta} \phi(A)m(A).$$

Consider, as above, $F = \inf_\mathcal{P} P$ and suppose F is monotone of infinite order, and $\mathcal{P} = \{P \in \mathbb{P} : F \leq P\}$, where $\mathbb{P}$ denotes the class of all probability measures on Θ. For example, let $0 < \varepsilon < 1$, $P_0 \in \mathbb{P}$, and consider

$$\mathcal{P} = \{\varepsilon P + (1 - \varepsilon)P_0 : P \in \mathbb{P}\}.$$

Then F is monotone of infinite order. Indeed, $\mathcal{A}$ is a subset of 2^Θ and $\bigcup_{A \in \mathcal{P}} A \neq \emptyset$. Then, there is a $P \in \mathbb{P}$ such that $P\left(\bigcup_{A \in \mathcal{A}} A\right) = 0$, so that

$$F\left(\bigcup_{A \in \mathcal{A}} A\right) = (1 - \varepsilon)P_0\left(\bigcup_{A \in \mathcal{A}} A\right)$$

$$\geq \sum_{\emptyset \neq T \subseteq A} (-1)^{|T|+1} P_0 \left(\bigcap_T A \right)$$

$$\geq \sum_{\emptyset \neq T \subseteq A} (-1)^{|T|+1} F \left(\bigcap_T A \right).$$

If $\cup_A A = \Theta$, then $F(\cup_A A) = 1$ and

$$\sum_{\emptyset \neq T \subseteq A} (-1)^{|T|+1} F \left(\bigcap_T A \right) \leq 1 - \varepsilon$$

unless each $A = \Theta$, in which case

$$\sum_{\emptyset \neq T \subseteq A} (-1)^{|T|+1} F \left(\bigcap_T A \right) = 1.$$

In any case, F is monotone of infinite order. Next, if $F \leq P$, then

$$
\begin{aligned}
f(\theta) &= \frac{1}{\varepsilon} [P(\{\theta\}) - F(\{\theta\})] \\
&= \frac{1}{\varepsilon} [P(\{\theta\}) - (1 - \varepsilon)P_0(\{\theta\})] \geq 0
\end{aligned}
$$

and $\sum_{\theta \in \Theta} f(\theta) = 1$. Thus

$$P = (1 - \varepsilon)P_0 + \varepsilon Q \quad \text{with} \quad Q(A) = \sum_{\theta \in A} f(\theta).$$

Hence $\mathcal{P} = \{P \in \mathbb{P} : F \leq P\}$.

Let $\mathcal{D}$ be the class of density functions on Θ such that $f \in \mathcal{D}$ if and only if $P_f \in \mathcal{P}$, where $P_f(A) = \sum_{\theta \in A} f(A)$. In the above decision procedure, based on the Choquet integral, we observe that

$$E_F(u(a, \cdot)) = E_{P_f}(u(a, \cdot)) \quad \text{for some } f \in \mathcal{D}$$

(depending not only on F but also on $u(a, \cdot)$). In other words, the Choquet integral approach leads to the selection of a density in $\mathcal{D}$ that defines the expected utility.

Now, observe that, to each $f \in \mathcal{D}$, one can find many set-functions $\phi : 2^\Theta \longrightarrow \mathbb{R}$ such that

$$E_{P_f}(u(a, \cdot)) = E(\phi(S)),$$

where S is the random set with density m. Indeed, define ϕ arbitrarily on every element of 2^Θ, except for some A for which $m(A) \neq 0$, and set

$$\phi(A) = \frac{1}{m(A)} \left[E_{P_f}(u(a, \cdot)) - \sum_{B \neq A} \phi(B)m(B) \right].$$

The point is this. Selecting ϕ and considering $E\left(\phi(S)\right) = \sum_A \phi(A)m(A)$ as expected utility is a general procedure. In practice, the choice of ϕ is guided by additional subjective information. For example, for $\rho \in [0,1]$, consider.

$$\phi_{\rho,a}(A) = \rho \max\{u(a,\theta) : \theta \in A\} + (1-\rho)\min\{u(a,\theta) : \theta \in A\}$$

and use

$$E_m\left(\phi_{\rho,a}\right) = \sum_{A \subseteq \Theta} \phi_{\rho,a}(A)m(A)$$

for decision-making. Note that $E_m\left(\phi_{\rho,a}\right) = E_{P_g}\left(u(a,\cdot)\right)$ where $g \in \mathcal{D}$ is given by

$$g(\theta) = \rho \sum_{A \in \mathcal{A}} m(A) + (1-\rho)\sum_{B \in \mathcal{B}} m(B),$$

where

$$\mathcal{A} = \{A : \theta_i \in A \subseteq \{\theta_1, \ldots, \theta_i\}\}, \quad \mathcal{B} = \{B : \theta_i \in B \subseteq \{\theta_i, \ldots, \theta_n\}\},$$

and recall that, for a given action a, the θ_i's are ordered so that

$$u(a,\theta_1) \le u(a,\theta_2) \le \cdots \le u(a,\theta_n).$$

(c) **Maximum Entropy Principle.** Consider the situation as in (b). There are other ways to select an element of $\mathcal{D}$ to form the ordinary expectation of $u(a,\cdot)$, for example by using the well-known maximum entropy principle in statistical inference. Recall that, usually, the constraints in an entropy maximization problem are given in the form of a known expectation and higher order moments. For example, with $\Theta = \{\theta_1, \theta_2, \ldots, \theta_n\}$, the canonical density on Θ which maximizes the entropy

$$H(f) = -\sum_{i=1}^n f_i \log f_i$$

subject to

$$f_i \ge 0, \quad \sum_{i=1}^n f_i = 1, \quad \text{and} \quad \sum_{i=1}^n \theta_i f_i = \alpha,$$

is given by

$$f_j = e^{-\beta\theta_j}/\Phi(\beta), \qquad j = 1, 2, \ldots, n,$$

where $\Phi(\beta) = \sum_{i=1}^n e^{-\beta\theta_i}$ and β is the unique solution of

$$d\log\Phi(\beta)/d\beta = -\alpha.$$

Now, our optimization problem is this:

$$\text{Maximize} \quad H(f) \quad \text{subject to} \quad f \in \mathcal{D}.$$

Of course, the principle of maximum entropy is sound for any kind of constrains! The problem is with the solution of the maximization problem! Note that, for $F = \inf_{\mathcal{P}} P$, $\mathcal{D} \neq \emptyset$, since the distribution functional F of a random set is convex, i.e.,

$$F(A \cup B) \geq F(A) + F(B) - F(A \cap B)$$

and

$$\mathcal{P} = \{P \in \mathbb{P} : F \leq P\} \neq \emptyset$$

(see Shapley [34]).

Here is an example. Let m be a density on 2^{Θ} with $m(\{\theta_i\}) = \alpha_i$, $i = 1, 2, \ldots, n$ and $m(\Theta) = \varepsilon = 1 - \sum_{i=1}^{n} \alpha_i$. If we write $\varepsilon = \sum_{i=1}^{n} \varepsilon_i$ with $\varepsilon_i \geq 0$, then $f \in \mathcal{D}$ if and only if it is of the form

$$f(\theta_i) = \alpha_i + \varepsilon_i, \qquad i = 1, 2, \ldots, n.$$

So the problem is to find the ε_i's so that f maximizes $H(g)$ over $g \in \mathcal{D}$. Specially, the problem is this. Determine the ε_i's which maximize

$$H(\varepsilon_1, \ldots, \varepsilon_n) = - \sum_{i=1}^{n} (\alpha_i + \varepsilon_i) \log(\alpha_i + \varepsilon_i).$$

The following observations show that nonlinear programming techniques are not needed. For details, see Nguyen and Walker [25]. There exists exactly one element $f \in \mathcal{D}$ having the largest entropy. That density is given by the following algorithm. First, rename the θ_i's so that $\alpha_1 \leq \alpha_2 \leq \cdots \leq \alpha_n$. Then

$$f(\theta_i) = \alpha_i + \varepsilon_i$$

with

$$\varepsilon_i \geq 0, \qquad \sum_{i=1}^{k} \varepsilon_i = m(\Theta),$$

and

$$\alpha_1 + \varepsilon_1 = \cdots = \alpha_k + \varepsilon_k \leq \alpha_{k+1} \leq \cdots \leq \alpha_n.$$

The construction of f is as follows. Setting

$$\delta_i = \alpha_k - \alpha_i, \qquad i = 1, 2, \ldots, k,$$

where k is the maximum index such that

$$\sum_{i=1}^{k} \delta_i \leq m(\Theta);$$

letting

$$\varepsilon_i = \delta_i + \frac{1}{k}\left(\varepsilon - \sum_{i=1}^{k} \delta_i\right), \qquad i = 1, 2, \ldots, k$$

and $\varepsilon_i = 0$ for $i > k$.

The general case is given in Meyerowitz et al. [23]. It calculates the maximum entropy density f directly from the distribution functional F as follows. Inductively define a decreasing sequence of subsets Θ_i of Θ, and numbers γ_i, as follows, quitting when Θ_i is empty:

(i) $\Theta_0 = \Theta$.

(ii) $\gamma_i = \max \{[F(A \cup \Theta_i^c) - F(\Theta_i^c)]/|A| : \emptyset \neq A \subseteq \Theta_i\}$.

(iii) A_i is the largest subset of Θ_i such that

$$F(A_i \cup \Theta_i^c) - F(\Theta_i^c) = \gamma_i |A_i|$$

(note that there is a unique such A_i).

(iv) $\Theta_{i+1} = \Theta_i \setminus A_i$.

(v)   Set $f(\theta) = \gamma_i$ for $\theta \in A_i$.

Remark 4.1. The history of the above entropy maximization problem is interesting. Starting in Nguyen and Walker [25] as a procedure for decision-making with belief functions, where only the solutions of some special cases are given, the general algorithm is presented in Meyerowitz et al. [23] (in fact, they presented two algorithms). One of their algorithms (the above one) was immediately generalized to the case of convex capacities in Jaffray [18]. Very recently, it was observed in Jaffray [19] that the general algorithm for convex capacities is the same as the one given in Dutta and Ray [7] in the context of game and welfare theory!   □

3.4.3   Uncertainty Modeling

Fuzziness is a type of uncertainty encountered in modeling linguistic information (in expert systems and control engineering) For background on fuzzy sets, see, e.g. Nguyen and Walker [26]. A formal connection with random sets was pointed out in Goodman [9]. If $f : U \longrightarrow [0,1]$ is the membership function of a fuzzy subset of U, then there exists a random set $S : (\Omega, \mathcal{A}, P) \longrightarrow 2^U$ such that

$$f(u) = P(\{\omega : u \in S(\omega)\}), \qquad \forall u \in U,$$

i.e., $f(\cdot)$ is the one-point coverage function of S. Indeed, it suffices to consider a random variable X, defined on some probability space $(\Omega, \mathcal{A}, P)$, uniformly distributed on the unit interval $[0,1]$, and define

$$S(\omega) = \{u \in U : f(u) \geq X(\omega)\}.$$

This probabilistic interpretation of fuzzy sets does not mean that fuzziness is captured by randomness! However, among other things, it suggests a very realistic way for obtaining membership functions for fuzzy sets. First of all, the specification of a membership function to a fuzzy concept can be done in many different ways. For example, by a statistical survey, or by experts (thus, subjectively). The subjectivity of experts in defining f can be understood through

$$f(u) = \mu\left(\{u \in S\}\right),$$

where S is a multi-valued mapping, say, $S : \Omega \longrightarrow 2^U$, and μ is a monotone set-function on U, not necessarily a probability measure. This is Orlowski's model [30]. In a given decision-making problem, the multi-valued mapping S is easy to specify, and the subjectivity of an expert is captured by the set-function μ. Thus, as in game theory, we are led to consider very general set-functions in uncertainty modeling and decisions. For recent works on integral representation of set-functions, see e.g., Gilboa and Schmeidler [8] and Marinacci [21].

Bibliography

[1] Aigner, M. (1979) *Combinatorial Theory*, Springer-Varlag, New York.

[2] Bell, W. C. and Hagood, J. W. (1988) Separation properties and exact Radon-Nikodym derivatives for bounded finitely additive measures, *Pacific J. Math.*, **131**, 237-248.

[3] Billingsley, P. (1695) *Ergodic Theory and Information*, J. Wiley, New York.

[4] Choquet, G. (1953/54) *Theory of capacities*, Ann. Inst. Fourier, **5**, 131–295.

[5] Debreu, G. (1967) Integration of correspondences, *Proc.* 5th *Berkeley Symp. Math. Statist. Prob.*, L. Lecam, J. Neyman, and E. L. Scott (Eds.), University of California Press, Berkeley, pp. 351-372.

[6] Durrett, R. (1996) *Probability: Theory and Examples*, (Second Edition), Duxbury Press, Belmont.

[7] Dutta, B. and Ray, D. A (1989) A concept of egalitarianism under participation constraints, *Econometrica*, **57**, 615-635.

[8] Gilboa, I. and Schmeidler, D. (1995) Canonical representation of set-functions, *Math. Oper. Res.*, **20**, 197-212.

[9] Goodman, I. R. (1982) Fuzzy sets as equivalence classes of random sets, in R. Yager (Ed.), *Fuzzy Sets and Possibility Theory*, pp. 327-343.

[10] Graf, S. (1980) A Radon-Nikodym theorem for capacities, *Journal fuer Reine und Angewandte Mathematik*, **320**, 192-214.

[11] Greco, G. H. (1981) Un teorema di Radon-Nikodym per funzioni d'insieme subadditive, *Ann. Univ. Ferrara*, soz, VII, Sc. Mat. Vol. XXVII, 13-19.

[12] Hajek, J. (1981) *Sampling from a Finite Population*, Marcel Dekker, New York.

[13] Hall, P. (1988) *Introduction to the Theory of Coverage Processes*, J. Wiley, New York.

[14] Halmos, P. (1974) *Measure Theory*, Springer-Verlag, New York.

[15] Harding, E. F. and Kendall, D. G. (Eds.), *Stochastic Geometry*, J. Wiley, New York, 1973.

[16] Huber, P. J. (1973) The use of Choquet capacities in statistics, *Bull. Internat. Statist.* **45**, 181-191.

[17] Huber, P. J. and Strassen, V. (1973) Minimax tests and Neyman-Pearson lemma for capacities, *Ann. Statist.*, **1**, 251-263.

[18] Jaffray, J. Y. (1995) On the maximum-entropy probability which is consistent with a convex capacity, *Intern. J. Uncertainty, Fuzziness and Knowledge-Based Systems*, **3**, 27-33.

[19] Jaffray, J. Y. (1996) A complement to an maximum entropy probability which is consistent with a convex capacity, Preprint.

[20] Li, X. and Yong, J. (1995) *Optimal Control Theory for Infinite Dimensional Systems*, Birkhäuser, Boston.

[21] Marinacci, M. (1996) Decomposition and representation of coalitional games, *Mathematics of Operations Research*, **21** 1000-1015.

[22] Matheron, G. (175) *Random Sets and Integral Geometry*, John Wiley, New York.

[23] Meyerowitz, A., Richman, F., and Walker, E. A. (1994) Calculating maximum entropy probability densities for belief functions, *Intern. J. Uncertainty, Fuzziness and Knowledge-Based Systems*, **2**, 377-390.

[24] Molchanov, I. S. (1993) *Limit Theorems for Unions of Random Closet Sets*, Lecture Notes in Math. No. 1561, Springer-Verlag, Berlin.

[25] Nguyen, H. T. and Walker, E. A. (1994) On decision-making using belief functions, in R. Yager et al.(eds.), *Advances in the Dempster-Shafer Theory of Evidence*, J. Wiley, New York, pp. 312-330.

[26] Nguyen, H. T. and Walker, E. A. (1996) *A First Course in Fuzzy Logic*, CRC Press, Boca Raton, Florida.

[27] Nguyen, H. T., Nguyen, N. T., and Wang, T. (1996) On capacity functionals in interval probabilities, *Intern. J. of Uncertainty, Fuzziness and Knowledge-Based Systems*, to appear.

[28] Norberg, T. (1986) Random capacities and their distributions, *Prob. Theory Relat. Fields*, **73**, 281-297.

[29] Norberg, T. (1987) Semicontinuous processes in multi-dimensional extreme-value theory, *Stoch. Process. Appl.* **25**, 27-55.

[30] Orlowski, S. A. (1994) *Calculus of Decomposable Properties, Fuzzy Sets and Decisions*, Allerton Press, New York.

[31] Robbins, H. E. (1944) On the measure of a random set, *Ann. Math. Statist.* **15**, 70-74.

[32] Rogers, C. A. (1970) *Hausdorff Dimension*, Cambridge Univ. Press, London.

[33] Rota, G. C. (1964) On the foundations of combinatorial theory I. Theory of Möbius functions, *Z. Wahrsch.*, **2**, 340-368.

[34] Shapley, L. S. (1971) Cores of convex games, *Intern. J. Game Theory*, **1**, 11-26.

[35] Shilov, G. E. and Gurevich, B. L. (1966) *Integral, Measure and Derivative: A Unified Approach*, Prentice Hall, New Jersey.

[36] Stoyan, D., Kendall, W. S., and Mecke, J. (1995) *Stochastic Geometry and Its Applications*, (Second Edition), J. Wiley, New York.

[37] Wasserman, L. A. (1987) *Some Applications of Belief Functions to Statistical Inference*, Ph.D. Thesis, University of Toronto, Toronto.

Chapter 4

Finite Random Sets

In this and the four chapters which follow, we develop the random set approach to data fusion summarized in Section 2.5 of Chapter 2. This chapter sets the stage by developing the mathematical cornerstone of this approach: the concept of a *finite random set*. The necessary mathematical preliminaries—topological spaces of finite sets, hybrid discrete-continuous spaces and hybrid integrals, etc.—are set forth in Section 4.1. Section 4.2 introduces the *set integral* and the *set derivative*. The concepts of finite random subset, absolutely continuous finite random subset, and global density of an absolutely continuous finite random subset, are explored in Section 4.3.

4.1 Mathematical Preliminaries

In this section we establish some basic mathematical concepts and results that are necessary for the sequel. In Section 4.1.1 we establish the connection between the Euclidean topologies on Euclidean spaces, and the hit-or-miss topologies on the corresponding spaces of finite subsets of Euclidean space. In Section 4.1.2 we define the concept of a *hybrid* (i.e., continuous-discrete) *space* and establish some basic properties of such spaces, including the concept of a random subset of a hybrid space. Finally, in Section 4.1.3, we briefly introduce the concept of Hausdorff distance between two compact subsets of a metric space.

4.1.1 Relationship Between the Euclidean-Space and Hit-or-Miss Topologies

In Section 2.5.4 of Chapter 2 we illustrated the concept of a global probability density using a graphically intuitive geometrical interpretation. There we represented the two-point sunsets $\{a_1, a_2\}$ of the real line $\mathbb{R}$ as points (a_1, a_2) on the plane $\mathbb{R} \times \mathbb{R} = \mathbb{R}^2$ with $a_1 < a_2$ (i.e., the half-plane). A global density was then pictorially interpreted as a conventional density on the half-plane. Obviously, this interpretation could be extended to represent the elements of any finite subset $\{a_1, ..., a_k\}$ of $\mathbb{R}$ as points $(a_1, ..., a_k)$ in $\mathbb{R}^k$. This method of interpreting finite sets as points in a suitable Euclidean space is not just a heuristic or pictorial convenience. It reflects the fact that the hit-or-miss topology on the closed subsets of a Euclidean space $\mathbb{R}^n$ is, when restricted to the finite subsets of $\mathbb{R}^n$ of fixed finite cardinality, essentially just the conventional topology on a certain subset $[\mathbb{R}^n]^k$ of the Euclidean space $(\mathbb{R}^n)^k = \mathbb{R}^{nk}$. Indeed, were this not the case–that is, if the hit-or-miss topology were inconsistent with the standard Euclidean-space topologies–then random-set techniques would be of little practical interest.

To prove the consistency of the hit-or-miss and the standard topologies, we first need some definitions:

Definition 1 *a)* $c_k(\mathbb{R}^n)$ *denotes the set of all finite subsets of* $\mathbb{R}^n$ *which contain exactly* k *elements; and* $c_{\leq k}(\mathbb{R}^n)$ *denotes the set of all finite subsets of* $\mathbb{R}^n$ *which contain no more than* k *elements.* *b)* $\chi_k : (\mathbb{R}^n)^k \to c_{\leq k}(\mathbb{R}^n)$ *is the map defined by*

$$\chi_k(\mathbf{x}_1, ..., \mathbf{x}_k) = \{\mathbf{x}_1, ..., \mathbf{x}_k\}$$

c) If $\mathbf{x} = (x_1, ..., x_n)$ *and* $\mathbf{y} = (y_1, ..., y_n)$ *are points in* $\mathbb{R}^n$ *then the lexicographic ordering, denoted by '$\prec$', is defined as follows. We have* $\mathbf{x} \prec \mathbf{y}$ *if* $x_1 < y_1$; *or if* $x_1 = y_1$ *and* $x_2 < y_2$; *or if* $x_1 = y_1$ *and* $x_2 = y_2$ *and* $x_3 < y_3$; *and so on. Note that lexicographic ordering introduces a total ordering on* $\mathbb{R}^n$. *d) Let* $[\mathbb{R}^n]^k$ *be the subset of* $(\mathbb{R}^n)^k$ *of all* $(\mathbf{x}_1, ..., \mathbf{x}_k)$ *such that* $\mathbf{x}_1 \prec ... \prec \mathbf{x}_k$. *e) Let* $\hat{\chi}_k$ *denote the restriction of the map* χ_k *to* $[\mathbb{R}^n]^k$. *Then* $\hat{\chi}_k$ *obviously sets up a one-to-one correspondence between the* k-*tuples in* $[\mathbb{R}^n]^k$ *and the elements of* $c_k(\mathbb{R}^n)$.

Note that the lexicographic ordering is not "natural" in the sense that it depends on a specific choice of vector-space basis for $\mathbb{R}^n$. In general, each choice of basis results in a different possible lexicographic ordering. Thus the subspace $[\mathbb{R}^n]^k$ also depends on the choice of basis and thus is not uniquely determined by the projection mapping χ_k.

The first thing to notice is the following:

Proposition 1 $c_{\leq k}(\mathbb{R}^n)$ *is a closed subset of the hit-or-miss topology on* $c(\mathbb{R}^n)$. *Consequently,* $c_k(\mathbb{R}^n) \in \sigma(c(\mathbb{R}^n))$. *That is,* $c_k(\mathbb{R}^n)$ *is measurable..*

Proof. Define

$$O(\geq k) = \bigcup_{G_1, ..., G_k} O_{G_1} \cap ... \cap O_{G_k}$$

where the union is taken over all lists $G_1, ..., G_k$ of k mutually disjoint open subsets of $\mathbb{R}^n$. Since $O(\geq k)$ is a union of open subsets of the hit-or-miss topology on $c(\mathbb{R}^n)$ it is itself an open set in that topology. Let $c(\geq k)$ denote the set of all closed subsets of $\mathbb{R}^n$ which have at least k elements. We will show that $O(\geq k) = c(\geq k)$. Since $c_{\leq k}(\mathbb{R}^n) = c(\geq k+1)^c$ we will then be done. On the one hand, suppose $C \in O(\geq k)$. Then there exist mutually disjoint $G_1, ..., G_k$ open subsets of $\mathbb{R}^n$ such that $C \in O_{G_1} \cap ... \cap O_{G_k}$ and hence such that

$C \cap G_1 \neq \emptyset$,   ..., $C \cap G_k \neq \emptyset$. Because the $G_1, ..., G_k$ are mutually disjoint this cannot be possible unless C has at least k elements. Thus $C \in c(\geq k)$.

Conversely, suppose that $C \in c(\geq k)$, so that C contains at least k elements. Accordingly, there exist distinct $\mathbf{x}_1, ..., \mathbf{x}_k \in \mathbb{R}^n$ such that $\mathbf{x}_1, ..., \mathbf{x}_k \in C$. Choose mutually disjoint open subsets $G_1, ..., G_k$ of $\mathbb{R}^n$ such that $\mathbf{x}_j \in G_j$ for $j = 1, ..., k$. For this choice of $G_1, ..., G_k$ we have $C \cap G_1 \neq \emptyset$,   ..., $C \cap G_k \neq \emptyset$ and hence

$$C \in O_{G_1} \cap ... \cap O_{G_k} \subseteq O(\geq k)$$

and thus $C \in O(\geq k)$ as desired.   □

The following shows that the standard subspace topology on $[\mathbb{R}^n]^k$ and the induced hit-or-miss topology on $c_k(\mathbb{R}^n)$ are closely related, in the sense that the mapping χ_k establishes an equivalence of topologies between the two spaces.

Proposition 2 *Let $\hat{\chi}_k$ be the map defined in Definition 1. Then $\hat{\chi}_k$ is a homeomorphism (equivalence of topological spaces) between $[\mathbb{R}^n]^k$ and $c_k(\mathbb{R}^n)$.*

Proof. We must show that $\hat{\chi}_k$, the restriction of χ_k to the subspace $[\mathbb{R}^n]^k$, is continuous; and that its inverse function $\hat{\chi}_k^{-1}$ is also continuous. On the one hand, let $G \in \mathbb{R}^n$ be open and define the open subset s_G of $(\mathbb{R}^n)^k$ to be

$$s_G = (G \times \mathbb{R}^n \times ... \times \mathbb{R}^n) \cup (\mathbb{R}^n \times G \times \mathbb{R}^n \times ... \times \mathbb{R}^n) \cup ... \cup (\mathbb{R}^n \times ... \times \mathbb{R}^n \times G)$$

First, notice that $\chi_k^{-1}(O_G \cap c_k(R^n)) = s_G$ where $O_G = \{C \in c(\mathbb{R}^n) | C \cap G \neq \emptyset\}$. Since the $O_G \cap c_k(\mathbb{R}^n)$ for all G forms a subbase for the induced hit-or-miss topology on $c_k(\mathbb{R}^n)$, it follows that $\chi_k^{-1}(O)$ is open in $(\mathbb{R}^n)^k$ for every open subset O of $c_k(\mathbb{R}^n)$. Thus χ_k is continuous and, consequently, so is the restriction $\hat{\chi}_k$ of χ_k to the subspace $[\mathbb{R}^n]^k$.

On the other hand, to show that $\hat{\chi}_k^{-1}$ is continuous it suffices to show that χ_k carries every open subset of $[\mathbb{R}^n]^k$ into an open subset of $c_k(\mathbb{R}^n)$. To prove this, first notice that $\chi_k(s_G) = O_G$. For on the one hand, let $\{\mathbf{x}_1, ..., \mathbf{x}_k\} \in O_G$ with $\mathbf{x}_1, ..., \mathbf{x}_k$ distinct. Then by definition, $\{\mathbf{x}_1, ..., \mathbf{x}_k\} \cap G \neq \emptyset$ and therefore $\mathbf{x}_j \in G$ for some j. Consequently, $(\mathbf{x}_1, ..., \mathbf{x}_j, ..., \mathbf{x}_k) \in s_G$ and so $\{\mathbf{x}_1, ..., \mathbf{x}_k\} \in \chi_k(s_G)$. Conversely, if $\{\mathbf{x}_1, ..., \mathbf{x}_k\} \in \chi_k(s_G)$ then

$$(\mathbf{x}_1, ..., \mathbf{x}_k) \in \mathbb{R}^n \times ... \times G \times ... \times \mathbb{R}^n$$

for G in the j'th position for some j and so $\{\mathbf{x}_1, ..., \mathbf{x}_k\} \in O_G$. Therefore, we have proved that $\chi_k(s_G) = O_G$ and hence that $\hat{\chi}_k(s_G \cap [\mathbb{R}^n]^k) = O_G$.

However, the $s_G \cap [\mathbb{R}^n]^k$ form a subbase for the subspace topology on $[\mathbb{R}^n]^k$. To prove this, it is enough to show that the sets

$$(\mathbb{R}^n \times ... \times \mathbb{R}^n \times G \times \mathbb{R}^n \times ... \times \mathbb{R}^n) \cap [\mathbb{R}^n]^k$$

can be expressed as suitable unions of finite intersections of sets of the form $s_G \cap [\mathbb{R}^n]^k$. For, because $\hat{\chi}_k$ is one-to-one and onto and because $\hat{\chi}_k(s_G \cap [\mathbb{R}^n]^k) = O_G$, it will then follow that

$$\hat{\chi}_k((\mathbb{R}^n \times ... \times \mathbb{R}^n \times G \times \mathbb{R}^n \times ... \times \mathbb{R}^n) \cap [\mathbb{R}^n]^k)$$

is a union of finite intersections of subsets of the form $s_G \cap [\mathbb{R}^n]^k$ and is therefore open. However, subsets of the form

$$(\mathbb{R}^n \times ... \times \mathbb{R}^n \times G \times \mathbb{R}^n \times ... \times \mathbb{R}^n) \cap [\mathbb{R}^n]^k$$

form an open subbase for the subspace topology on $[\mathbb{R}^n]^k$). Consequently, it will follow that $\hat{\chi}_k(H)$ is an open subset of $c_k(\mathbb{R}^n)$ for any open subset H of $[\mathbb{R}^n]^k$ and we will be finished. So, let us prove that the sets

$$(\mathbb{R}^n \times ... \times \mathbb{R}^n \times G \times \mathbb{R}^n \times ... \times \mathbb{R}^n) \cap [\mathbb{R}^n]^k$$

can be expressed as suitable unions of finite intersections of sets of the form $s_G \cap [\mathbb{R}^n]^k$. Without loss of generality we may prove the assertion for sets of the form $(G \times \mathbb{R}^n \times ... \times \mathbb{R}^n) \cap [\mathbb{R}^n]^k$. Consequently, let $\mathbf{w} \in (G \times \mathbb{R}^n \times ... \times \mathbb{R}^n) \cap [\mathbb{R}^n]^k$. Then $\mathbf{w} = (\mathbf{z}_1, ..., \mathbf{z}_k)$ where $\mathbf{z}_1 \in G$ and the list $\mathbf{z}_1 \prec ... \prec \mathbf{z}_k$ is monotone increasing with respect to lexicographic ordering on $\mathbb{R}^n$. Since the $\mathbf{z}_1, ..., \mathbf{z}_k$ are distinct it is possible to find arbitrarily small, open and mutually disjoint subsets $H_1, ..., H_k$ of $\mathbb{R}^n$ such that $\mathbf{z}_1 \in H_1, ..., \mathbf{z}_k \in H_k$. Because $H_1, ..., H_k$ are mutually disjoint it is easy to see that

$$s_{H_1} \cap ... \cap s_{H_k} = \bigcup_\pi H_{\pi 1} \times ... \times H_{\pi k}$$

where the union is taken over all permutations π on the numbers $1, ..., k$. However, note that

$$\begin{aligned} s_{H_1} \cap ... \cap s_{H_k} \cap [\mathbb{R}^n]^k &= \bigcup_\pi ((H_{\pi 1} \times ... \times H_{\pi k}) \cap [\mathbb{R}]^k) \\ &= (H_1 \times ... \times H_k) \cap [\mathbb{R}^n]^k \end{aligned}$$

For suppose that $(\mathbf{x}_1, ..., \mathbf{x}_k) \in H_{\pi 1} \times ... \times H_{\pi k}$ where π is not the identity permutation. Then since $H_1, ..., H_k$ are arbitrarily small it follows that $\mathbf{x}_j$ is arbitrarily close to $\mathbf{z}_j$ and hence that the list $\mathbf{x}_1, ..., \mathbf{x}_k$ cannot be monotone increasing in value with respect to lexicographic ordering. Consequently, $(\mathbf{x}_1, ..., \mathbf{x}_k)$ cannot be in $[\mathbb{R}^n]^k$. Therefore

$$s_{H_1} \cap ... \cap s_{H_k} \cap [\mathbb{R}^n]^k = (H_1 \times ... \times H_k) \cap [\mathbb{R}^n]^k$$

Finally, because H_1 is an arbitrarily small open neighborhood of $\mathbf{z}_1 \in G$ we can assume that $H_1 \subseteq G$ and hence that

$$\mathbf{w} \in s_{H_1} \cap ... \cap s_{H_k} \cap [\mathbb{R}^n]^k \subseteq (G \times \mathbb{R}^n \times ... \times \mathbb{R}^n) \cap [\mathbb{R}^n]^k$$

Since this holds for any $\mathbf{w} \in (G \times \mathbb{R}^n \times ... \times \mathbb{R}^n) \cap [\mathbb{R}^n]^k$ we are done.   □

An immediate consequence of Proposition 2 is that finite-set functions defined on $c_k(\mathbb{R}^n)$ can be used interchangeably with completely symmetric functions defined on $(\mathbb{R}^n)^k$. Indeed, given a completely symmetric function $f : (\mathbb{R}^n)^k \rightarrow \mathbb{R}^m$, define $f^* : c_k(\mathbb{R}^n) \rightarrow \mathbb{R}^m$ by

$$f^*(\{\mathbf{x}_1, ..., \mathbf{x}_k\}) \triangleq f(\mathbf{x}_1, ..., \mathbf{x}_k)$$

(so that $f^*\chi_k = f$). Conversely, given a set function $F : c_k(R^n) \to R^m$ define F^* by

$$F^*(\mathbf{x}_1, ..., \mathbf{x}_k) \triangleq F(\{\mathbf{x}_1, ..., \mathbf{x}_k\})$$

for all distinct $\mathbf{x}_1, ..., \mathbf{x}_k$ so that $F^* = F\chi_k$. (Note that F^* is actually undefined on a set of measure zero). Then:

Proposition 3 *The correspondences $f \to f^*$ and $F \to F^*$ set up a one-to-one correspondence between the measurable, (resp. continuous) almost everywhere defined symmetric functions on $(\mathbb{R}^n)^k$ and the measurable (resp. continuous) functions on $c_k(\mathbb{R}^n)$.*

Proof. The proof is easy and is left to the reader.   □

4.1.2   Hybrid Spaces

The *hybrid space* approach to multitarget, multisensor data fusion problems has been a relatively common one (see [1], [9], [22], [23]). A hybrid space is a Cartesian product $\mathcal{R} = \mathbb{R}^n \times W$ consisting of all pairs (z, w) with $\mathbf{z} \in \mathbb{R}^n$ and $w \in W$. The set W is finite and contains the values of discrete state or observation variables. For example, $W = U \times \{1, ..., s\}$ where U is the set of target types and $1, ..., s$ are sensor tags identifying the sensor which originated a report. Likewise, $\mathbb{R}^n$ contains the values of continuous state or observation variables. The purpose of this subsection is to summarize the basic mathematical properties (topology, measure, and integral) of hybrid spaces which will be needed in the sequel.

We begin by defining a topology on $\mathcal{R}$ and an extension of Lebesgue measure to $\mathcal{R}$.

Definition 2 *a) Let $S \subseteq \mathcal{R}$. Then for any $w \in W$ we define:*

$$S(w) \triangleq \{\mathbf{z} \in \mathbb{R}^n | (\mathbf{z}, w) \in S\}$$

Note that $S = \cup_{w \in W} S(w) \times w$ where $S(v) \times v$ and $S(w) \times w$ are disjoint if $v \neq w$. More generally let $S \subseteq \mathcal{R}^k$. Then for any $(w_1, ..., w_k) \in W^k$ define:

$$S(w_1, ..., w_k) \triangleq \{(\mathbf{z}_1, ..., \mathbf{z}_k) \in (\mathbb{R}^n)^k | ((\mathbf{z}_1, w_1), ..., (\mathbf{z}_k, w_k)) \in S\}$$

The hybrid space $\mathcal{R}$ has a topology, namely the product topology of the Euclidean topology on $\mathbb{R}^n$ and the discrete topology on W. To avoid unnecessary mathematical complication, we will treat this topology and the associated concepts of measure and integral as *definitions*.

Definition 3 *The closed (resp. open, compact) subsets of $\mathcal{R}$ are those $S \subseteq \mathcal{R}$ such that $S(w)$ are closed (resp. open, compact) for all $w \in W$. Likewise for $\mathcal{R}^k$.*

The space $\mathcal{R}$ is, like $\mathbb{R}^n$, locally compact, Hausdorff, and separable.

Definition 4 *We say that $S \subseteq \mathcal{R}$ is measurable if $S(w)$ is Lebesgue-measurable for every $w \in W$. In this case the hybrid measure (i.e., product measure) is defined by:*

$$\lambda(S) \triangleq \sum_{w \in W} \lambda(S(w))$$

Likewise, we say that $S \subseteq \mathcal{R}^k$ is measurable if $S(w_1, ..., w_k)$ is measurable for all $(w_1, ..., w_k) \in \mathcal{R}^k$ and define the hybrid measure (product measure) as

$$\lambda(S) \triangleq \sum_{(w_1, ..., w_k) \in W^k} \lambda(S(w_1, ..., w_k))$$

In the obvious way, we can also extend the concept of Lebesgue integral to the hybrid space $\mathcal{R}$.

Definition 5 *a) $f : \mathcal{R} \to \mathbb{R}^m$ is an integrable function if and only if the functions $f_w : \mathbb{R}^n \to \mathbb{R}^m$ defined by $f_w(z) \triangleq f((z, w))$ are Lebesgue-integrable for every $w \in W$. More generally, $f : \mathcal{R}^k \to \mathbb{R}^m$ is integrable if and only if the functions*

$$f_{w_1, ..., w_k}(z_1, ..., z_k) \triangleq f((z_1, w_1), ..., (z_k, w_k))$$

are Lebesgue-integrable for every $(w_1, ..., w_k) \in W^k$. b) Let $S \subseteq \mathcal{R}$ be measurable.. Then the hybrid integral of f on S is:

$$\int_S f(\xi)d\lambda(\xi) \triangleq \sum_{w \in W} \int_{S(w)} f_w(z)d\lambda(z)$$

Likewise, let $S \subseteq \mathcal{R}^k$ be measurable.. Then the integral of f on S is:

$$\int_S f(\xi_1, ..., \xi_k)d\lambda(\xi_1) \cdots d\lambda(\xi_k)$$

$$\triangleq \sum_{(w_1, ..., w_k) \in W^k} \int_{S(w_1, ..., w_k)} f_{w_1, ..., w_k}(z_1, ..., z_k)d\lambda(z_1) \cdots d\lambda(z_k)$$

Finally, we extend the concept of a probability measure to the hybrid space.

Definition 6 *A set function q defined on the measurable subsets of $\mathcal{R}$ is a probability measure if it has the form $q(S) = p(\psi \in S)$ where $\psi = (Z, y)$ is a random variable on $\mathcal{R}$. The set functions defined by $q_w(T) = q(T \times w) = p(Z \in T, y = w)$ are probability measures on $\mathbb{R}^n$ for every $w \in W$. Since S is a disjoint union $S = \cup_{w \in W} S(w) \times w$ we have: $q(S) = \sum_{w \in W} q_w(S(w))$.*

On the basis of the above definitions, it is obvious that the Radon-Nikodým theorem applies to the hybrid state space:

Proposition 4 *Let q be an (additive) measure on $\mathcal{R}$ which is absolutely continuous with respect to hybrid measure, in the sense that $q(S) = 0$ whenever $\lambda(S) = 0$. Then there is an almost everywhere unique integrable function f on $\mathcal{R}$ such that $q(S) = \int_S f(\xi)d\lambda(\xi)$ for any measurable subset $S \subseteq \mathcal{R}$.*

Proof. Each of the probability measures q_w of Definition 6 is absolutely continuous with respect to Lebesgue measure, for all $w \in W$. For suppose that $T \subseteq \mathbb{R}^n$ is such that $\lambda(T) = 0$. Then $T \times w \subseteq \mathcal{R}$ and $\lambda(T \times w) = \lambda(T) = 0$. Since q is absolutely continuous $q(T \times w) = 0$. However, by Definition 6 this means that $0 = q(S(w)) = q_w(T)$ where $S = T \times w$. Thus q_w is absolutely continuous regardless of the choice of w. By the ordinary Radon-Nikodým theorem there are almost everywhere unique f_w such that $q_w(T) = \int_T f_w(\mathbf{z}) d\lambda(\mathbf{z})$. Define the function f by $f((\mathbf{z}, w)) = f_w(\mathbf{z})$ for all $w \in W$ and $\mathbf{z} \in \mathbb{R}^n$. Then from Definition 5 it is obvious that f is integrable and that its hybrid integral is

$$\int_S f(\xi) d\lambda(\xi) = \sum_{w \in W} \int_{S(w)} f_w(\mathbf{z}) d\lambda(\mathbf{z})$$

which concludes the proof. $\quad\square$

It is clear that we can extend the hit-or-miss topology on the set of closed subsets of $\mathbb{R}^n$ to a topology on the set of closed subsets of the hybrid space $\mathcal{R}$, since $\mathcal{R}$ is locally compact, Hausdorff, and separable. This then allows us to define the concept of a random closed subset of $\mathcal{R}$ in the obvious fashion. We let $c(\mathcal{R})$ denote the class of closed subsets of $\mathcal{R}$, $c_{\leq k}(\mathcal{R})$ the set of all finite subsets of $\mathcal{R}$ which have no more than k elements, and $c_k(\mathcal{R})$ the set of all finite subsets of $\mathcal{R}$ which have exactly k elements. We also extend the lexicographic order on $\mathbb{R}^n$ (see Definition 1) to $\mathcal{R}$. Choose an arbitrary linear ordering '$\prec$' on W. Then for $\xi_1 = (\mathbf{z}_1, w_1)$ and $\xi_2 = (\mathbf{z}_2, w_2)$ in $\mathcal{R}$ define the lexicographic ordering by: $\xi_1 \prec \xi_2$ if $w_1 \prec w_2$; or if $w_1 = w_2$ and $\mathbf{z}_1 \prec \mathbf{z}_2$. Also, let $[\mathcal{R}]^k$ denote the subspace of $\mathcal{R}^k$ of all $(\xi_1, ..., \xi_k)$ such that $\xi_1 \prec ... \prec \xi_k$. Then it is clear that generalizations of Propositions 1, 2, and are true.

4.1.3   The Hausdorff Metric

In this section we will, for convenience, limit our discussion to Euclidean spaces $\mathbb{R}^n$. However, it can be made to apply to the hybrid space $\mathcal{R} = \mathbb{R}^n \times W$ as follows. Choose any metric $d(-, -)$ on the finite space W (for example, by identifying the points of W with the points of some finite integer sublattice of $\mathbb{R}^m$ for some m and using the usual Euclidean metric). The associated metric topology on W is necessarily the discrete topology since each point of W is a closed ball around itself. Extend the Euclidean metric on $\mathbb{R}^n$ to $\mathcal{R}$ by defining

$$d((\mathbf{y}, v), (\mathbf{z}, w)) \triangleq \|\mathbf{y} - \mathbf{z}\| + d(v, w)$$

The metric topology is separable, locally compact, and Hausdorff. It obviously decomposes into the disjoint union of $N = |W|$ copies of $\mathbb{R}^n$, each one of which is a disconnected component (i.e., both open and closed subset) of the topology.

Let us now restrict ourselves to $\mathcal{R} = \mathbb{R}^n$. The class $c_{\leq k}(\mathbb{R}^n)$, whose elements are the finite subsets of $\mathbb{R}^n$ containing no more than k elements, does not possess many of the characteristics that one normally takes for granted in applied mathematics. Most obviously, it is not a vector space. Fortunately, however,

$c_{\leq k}(\mathbb{R}^n)$ is (if one excludes the empty set $\emptyset$) a *metric space* since it is possible to define a distance function $d(Z, Y)$ between any two finite subsets Z and Y of $\mathbb{R}^n$, called the *Hausdorff distance* (or *Hausdorff metric* [17, p. 131], [6]). The Hausdorff distance extends the usual distance metric $d(\mathbf{z}, \mathbf{y}) = \|\mathbf{x} - \mathbf{y}\|$ on $\mathbb{R}^n$ to the space of subsets of $\mathbb{R}^n$ and, moreover, is *consistent with the hit-or-miss topology* (see [20]).

We begin with a definition. Let $\kappa(\mathbb{R}^n)$ denote the class of compact (i.e., closed and bounded) and *nonempty* subsets of $\mathbb{R}^n$. Then obviously,

$$c_{\leq k}(\mathbb{R}^n) - \{\emptyset\} \subseteq \kappa(\mathbb{R}^n) \subseteq c(\mathbb{R}^n)$$

for all $k = 1, 2, \ldots$

Definition 7 *Let* $K, L \subseteq \kappa(\mathbb{R}^n)$. *Then the Hausdorff distance between* K *and* L *is defined by:*
$$d(K, L) = \max\{d_0(K, L), \ d_0(L, K)\}$$
where
$$d_0(K, L) = \sup_{\mathbf{x} \in K} d_0(\mathbf{x}, L), \qquad d_0(\mathbf{x}, L) = \inf_{\mathbf{y} \in L} \|\mathbf{x} - \mathbf{y}\|$$

Note that since $\emptyset \notin \kappa(\mathbb{R}^n)$ by definition, then $d(\emptyset, L)$ is undefined. If we were to assign a value to $d(\emptyset, L)$ then it would have to be $d(\emptyset, L) = \infty$. The classes $O^K = \{C | C \cap K = \emptyset\}$ are open neighborhoods of $\emptyset$ for all compact $K \subseteq \mathbb{R}^n$ and their intersection is $\{\emptyset\}$. Hence if L is a fixed compact closed subset and we chose another compact $L_K \in O^K$ as K expands to fill all of $\mathbb{R}^n$, then clearly $d(L, L_K)$ increases without bound as L_K "converges" to $\emptyset$. Hence $\emptyset$ is infinitely Hausdorff-distant from every other compact subset of $\mathbb{R}^n$.

4.2   A Calculus of Set Functions

A *set function* is a function $\Phi(S)$ whose arguments S are sets. A *finite-set function* is a set function $\Phi(S)$ whose arguments S are *finite* sets. The purpose of this chapter is to develop the rudiments of an integral and differential calculus for such functions, one which *directly* extends the usual integral and differential calculus of ordinary functions. We will define the *generalized Radon-Nikodým derivative* and, more generally, the *set derivative* of a set function. Likewise, we will define the *set integral* of a finite-set function and establish the relationship between the set derivative and set integral of the belief measure of a suitably well-behaved finite random set. We also show that if a finite random set Σ is suitably well-behaved then all of its set derivatives exist and are nonnegative. We will conclude the section with a brief discussion of other approaches to integral and derivative with respect to nonadditive measures.

The chapter is organized as follows. In section 4.2.1 we set the stage by developing a difference calculus of set functions on finite universes. What follows in later subsections is a generalization of this material to the continuous (more correctly, the hybrid discrete-continuous) case. The concept of a set integral is defined in 4.2.2. The "generalized Radon-Nikodým derivative" is introduced in section 4.2.3 and is used to define the concept of the set derivative in section 4.2.4.

4.2.1   Discrete Case: Difference Calculus on Finite Sets

The material in this section provides intuitive motivation for the basic concepts underlying our integral and differential calculus of set functions. It generalizes a construction described in Grabisch, Nguyen, and Walker [12, pp. 47-48, 125-126]. We assume that we are working in a *finite* set U with $\mathcal{P}(U)$ denoting the set of subsets of U. Then:

Definition 8 *Let* $\Phi : \mathcal{P}(U) \to \mathbb{R}$ *be a set function. Then a) the first difference of* Φ *with respect to* $a \in U$ *is:*

$$(\Delta_a \Phi)(T) \triangleq \Phi(T \cup \{a\}) - \Phi(T)$$

for all $T \subseteq U$. *b) The* k'*th iterated difference of* Φ *with respect to elements* $a_1, ..., a_k \in U$ *is defined recursively by:*

$$\Delta_{a_k} \cdots \Delta_{a_1} \Phi \triangleq \Delta_{a_k}(\Delta_{a_{k-1}} \cdots \Delta_{a_1} \Phi)$$

Proposition 5 *Suppose that* $a_1, ..., a_k \in U$. *Let* $\Phi : \mathcal{P}(U) \to \mathbb{R}$ *be a set function and*

$$m_\Phi(S) = \sum_{T \subseteq S} (-1)^{|S-T|} \Phi(T)$$

its corresponding Möbius transform, for all $S \subseteq U$. *Then:*

$$(\Delta_{a_1} \cdots \Delta_{a_k} \Phi)(\emptyset) = m_\Phi(\{a_1, ..., a_k\})$$

if $a_1, ..., a_k$ *are distinct.*

Proof. Given a subset $T \subseteq U$ which is not all of U, let $\Phi_T : \mathcal{P}(U - T) \to R$ be the set function defined by $\Phi_T(R) = \Phi(R \cup T)$ for any $R \subseteq U - T$. Then we first prove that

$$(\Delta_{a_k} \cdots \Delta_{a_1} \Phi)(T \cup R) = (\Delta_{a_k} \cdots \Delta_{a_1} \Phi_T)(R)$$

whenever $a_k, ..., a_1$ are distinct elements not in $T \cup R$ for any R with $R \cap T = \emptyset$ and for any $k \geq 1$. However, this follows from the following inductive step on k,

$$
\begin{aligned}
&(\Delta_{a_k} \Delta_{a_{k-1}} \cdots \Delta_{a_1} \Phi)(T \cup R) \\
=\ &(\Delta_{a_{k-1}} \cdots \Delta_{a_1} \Phi)(T \cup R \cup \{a_k\}) - (\Delta_{a_{k-1}} \cdots \Delta_{a_1} \Phi)(T) \\
=\ &(\Delta_{a_{k-1}} \cdots \Delta_{a_1} \Phi_T)(R \cup \{a_k\}) - (\Delta_{a_{k-1}} \cdots \Delta_{a_1} \Phi_T)(R) \\
=\ &(\Delta_{a_k} \Delta_{a_{k-1}} \cdots \Delta_{a_1} \Phi_T)(R)
\end{aligned}
$$

Given this, note that

$$
\begin{aligned}
(\Delta_b \Delta_{a_k} \cdots \Delta_{a_1} \Phi)(\emptyset) &= (\Delta_{a_k} \cdots \Delta_{a_1} \Phi)(\{b\}) - (\Delta_{a_k} \cdots \Delta_{a_1} \Phi)(\emptyset) \\
&= (\Delta_{a_k} \cdots \Delta_{a_1} \Phi_{\{b\}})(\emptyset) - (\Delta_{a_k} \cdots \Delta_{a_1} \Phi|_{U-T})(\emptyset)
\end{aligned}
$$

where $\Phi|_{U-T}$ denotes the restriction of Φ to $U - T$. By induction on $|U|$ we can then conclude that

$$
\begin{aligned}
(\Delta_b \Delta_{a_k} \cdots \Delta_{a_1} \Phi)(\emptyset) &= m_{\Phi_{\{b\}}}(\{a_1, ..., a_k\}) - m_{\Phi|_{U-T}}(\{a_k, ..., a_1\}) \\
&= \sum_{R \subseteq \{a_k, ..., a_1\}} (-1)^{k-|R|} \Phi_{\{b\}}(R) \\
&\quad - \sum_{R \subseteq \{a_k, ..., a_1\}} (-1)^{k-|R|} \Phi|_{U-T}(R)
\end{aligned}
$$

On the other hand, note that

$$
\begin{aligned}
m_\Phi(\{b, a_k, ..., a_1\}) &= \sum_{S \subseteq \{b, a_k, ..., a_1\}} (-1)^{k+1-|S|} \Phi(S) \\
&= \sum_{R \subseteq \{a_k, ..., a_1\}} (-1)^{k+1-|R|} \Phi(R) \\
&\quad + \sum_{R \subseteq \{a_k, ..., a_1\}} (-1)^{k+1-|R|} \Phi(\{b\} \cup R) \\
&= -\sum_{R \subseteq \{a_k, ..., a_1\}} (-1)^{k-|R|} \Phi|_{U-T}(R) \\
&\quad + \sum_{R \subseteq \{a_k, ..., a_1\}} (-1)^{k+1-(|R|+1)} \Phi_{\{b\}}(R)
\end{aligned}
$$

which completes the proof.   □

Example 1 *The iterated difference of Φ with respect to three distinct elements $a, b, c \in U$ is easily shown to be*

$$
(\Delta_c \Delta_b \Delta_a \Phi)(T) = (\Delta_b \Delta_a \Phi)(T \cup \{c\}) - (\Delta_b \Delta_a \Phi)(T)
$$

$$
\begin{aligned}
&= (\Phi(T \cup \{a, b, c\}) - \Phi(T \cup \{b, c\}) - \Phi(T \cup \{a, c\}) + \Phi(T \cup \{c\})) \\
&\quad -(\Phi(T \cup \{a, b\}) - \Phi(T \cup \{b\}) - \Phi(T \cup \{a\}) + \Phi(T)) \\
&= \Phi(T \cup \{a, b, c\}) - \Phi(T \cup \{b, c\}) - \Phi(T \cup \{a, c\}) - \Phi(T \cup \{a, b\}) \\
&\quad + \Phi(T \cup \{c\}) + \Phi(T \cup \{b\}) + \Phi(T \cup \{a\}) - \Phi(T)
\end{aligned}
$$

Thus

$$
(\Delta_c \Delta_b \Delta_a \Phi)(\emptyset) = \sum_{T \subseteq \{a, b, c\}} (-1)^{3-|T|} \Phi(T)
$$

as claimed.

Another way of viewing Proposition 5 is as follows. Let $Z = \{a_1, ..., a_k\}$ be a subset of U with distinct elements $a_1, ..., a_k$. Then define the *set difference operator* to be:

$$(\Delta_Z \Phi)(T) \triangleq (\Delta_{a_1} \cdots \Delta_{a_k} \Phi)(T)$$

(Note that $\Delta_Z \Phi$ is well-defined by Proposition 5.) As we will shortly see, $\Delta_Z \Phi$ is the discrete counterpart of what we will call (Definition 15) the *set derivative*. By Proposition 5

$$(\Delta_Z \Phi)(\emptyset) = \sum_{T \subseteq Z} (-1)^{|Z - T|} \Phi(T)$$

Accordingly, taking the inverse Möbius transform we can write:

$$\Phi(S) = \sum_{T \subseteq S} m_\Phi(T) = \sum_{k=0}^{\infty} \sum_{\substack{T \subseteq S \\ |T| = k}} (\Delta_T \Phi)(\emptyset) = \sum_{k=0}^{\infty} \frac{1}{k!} \sum_{u_1, ..., u_k \in S} (\Delta_{\{u_1, ..., u_k\}} \Phi)(\emptyset)$$

where we are using the convention that $(\Delta_{\{u_1, ..., u_k\}} \Phi)(\emptyset) = 0$ if the $u_1, ..., u_k$ are not distinct. The double summation

$$\sum_{k=0}^{\infty} \frac{1}{k!} \sum_{u_1, ..., u_k \in S} (\Delta_{\{u_1, ..., u_k\}} \Phi)(\emptyset)$$

is the discrete counterpart of what we will shortly call (Definition 10) a *set integral*. The equation

$$\Phi(S) = \sum_{k=0}^{\infty} \frac{1}{k!} \sum_{u_1, ..., u_k \in S} (\Delta_{\{u_1, ..., u_k\}} \Phi)(\emptyset)$$

is the discrete counterpart of Proposition 20, which establishes the basic connection between the set integral and the set derivative.

4.2.2   The Set Integral

By analogy with the discussion in the previous section, we are led to make the following definition:

Definition 9 *Let Φ be a real- or vector-valued function whose domain is the finite subsets of $\mathcal{R}$. Define the functions*

$$\Phi_k(\xi_1, ..., \xi_k) = \begin{cases} \Phi(\{\xi_1, ..., \xi_k\}) & \text{if} \quad \xi_1, ..., \xi_k \text{ are distinct} \\ 0 & \text{if} \quad \text{otherwise} \end{cases}$$

and assume that (1) $\Phi_k = 0$ identically for all sufficiently large k; and (2) Φ_k are integrable (with respect to Lebesgue measure) functions for all $k \geq 0$. Then we say that Φ is an integrable finite-set function.

Note that condition 2 is equivalent to the statement that the functions defined almost everywhere by

$$z_1, ..., z_k \mapsto \Phi_{u_1,...,u_k}((z_1, u_1), ..., (z_k, u_k))$$

are Lebesgue-integrable for all $u_1, ..., u_k \in U$, for every $k \geq 1$. Also, note that as long as $n > 0$ in $\mathcal{R} = \mathbb{R}^n \times U$ then $\mathcal{R}$ is nondiscrete and $\Phi(\{\xi_1, ..., \xi_k\}) = \Phi_k(\xi_1, ..., \xi_k)$ almost everywhere. Finally, if f is an ordinary integrable function defined on $\mathcal{R}$, then f gives rise to an integrable finite-set function in the obvious way: Define $f(Z) = 0$ if Z is a finite set with $|Z| \neq 1$ and $f(\{\xi\}) = f(\xi)$ for all $\xi \in \mathcal{R}$ otherwise.

Definition 10 *Let Φ be an integrable finite-set function and let S be closed subset of $\mathcal{R}$. Then the set integral of Φ, concentrated at S, is:*

$$\int_S \Phi(Z)\delta Z \triangleq \sum_{k=0}^{\infty} \frac{1}{k!} \int_{S^k} \Phi_k(\xi_1, ..., \xi_k) d\lambda(\xi_1) \cdots d\lambda(\xi_k)$$

where $S^k = S \times ... \times S$ denotes the Cartesian product of S taken k times and where we have adopted the convention that $\int_{S^0} \Phi(\emptyset)\delta\emptyset = \Phi(\emptyset)$. Finally, suppose that $O \in \sigma(c(\mathcal{R}))$ is a measurable subset of $c(\mathcal{R})$. Then the set integral concentrated on O is defined as:

$$\int_O \Phi(Z)\delta Z \triangleq \sum_{k=0}^{\infty} \frac{1}{k!} \int_{\chi_k^{-1}(O \cap c_k(\mathcal{R}))} \Phi_k(\xi_1, ..., \xi_k) d\lambda(\xi_1) \cdots d\lambda(\xi_k)$$

where $\chi_k : \mathcal{R}^k \to c_{\leq k}(\mathcal{R})$ is the function defined in Definition 1 by $\chi_k(\xi_1, ..., \xi_k) = \{\xi_1, ..., \xi_k\}$ for all $\xi_1, ..., \xi_k \in \mathcal{R}$. Thus $\chi_k^{-1}(O \cap c_k(\mathcal{R}))$ is a measurable subset of $\mathcal{R}^k$.

If $S = \mathcal{R}$ then we will write: $\int \Phi(Z)\delta Z \triangleq \int_{\mathcal{R}} \Phi(Z)\delta Z$.

Note that the set integral of an integrable finite-set function is well-defined and always exists, and that the infinite sum is always a *finite* sum, by definition of a set-integrable function. Also note that the first definition of a set integral is a special case of the second one. For, let $S \subseteq \mathcal{R}$ be closed and let $O = (O_{S^c})^c$ where, recall, the open subset O_G of $c(\mathcal{R})$ is defined as $O_G = \{C \in c(\mathcal{R}) | C \cap G \neq \emptyset\}$ for any open $G \subseteq \mathcal{R}$. Then it is easy to show that $\chi_k^{-1}(O \cap c_k(\mathcal{R})) = S^k$ for all $k \geq 1$. Finally, since $\Phi(\{\xi_1, ..., \xi_k\}) = \Phi_k(\xi_1, ..., \xi_k)$ almost everywhere as long as $n > 0$ we may abuse notation and write

$$\int_O \Phi(Z)\delta Z = \sum_{k=0}^{\infty} \frac{1}{k!} \int_{\chi_k^{-1}(O \cap c_k(\mathcal{R}))} \Phi(\{\xi_1, ..., \xi_k\}) d\lambda(\xi_1) \cdots d\lambda(\xi_k)$$

We shall always use this convention henceforth. On the other hand, if $n = 0$ then it is easy to verify that the set integral reduces to the formula

$$\sum_{k=0}^{\infty} \frac{1}{k!} \sum_{u_1,...,u_k \in S} \Phi(\{u_1, ..., u_k\})$$

with the convention $\Phi(\{u_1, ..., u_k\}) = 0$ if $u_1, ..., u_k$ are not distinct. This is the formula for what, in section 4.2.1, we identified as the discrete counterpart of the set integral.

Example 2 *Suppose that $\Phi(Z)$ is a finite-set function which vanishes identically for all Z with $|Z| \geq 3$. Then its set integral concentrated on $S \subseteq \mathcal{R}$ has the form*

$$\int_S \Phi(Z)\delta Z = \Phi(\emptyset) + \int_S \Phi(\{\xi\})d\lambda(\xi) + \frac{1}{2}\int_{S\times S} \Phi(\{\xi_1, \xi_2\})d\lambda(\xi_1)d\lambda(\xi_2)$$

The following proposition is obvious:

Proposition 6 *Set integrals are linear:*

$$\int_O (a\Phi(Z) + b\Psi(Z))\delta Z = a\int_O \Phi(Z)\delta Z + b\int_O \Psi(Z)\delta Z$$

for all $a, b \in \mathbb{R}$, any integrable finite-set functions Φ, Ψ, and any measurable $O \in \sigma(c(\mathcal{R}))$. Moreover, they preserve inequalities:

$$\int_O \Psi(Z)\delta Z \leq \int_O \Phi(Z)\delta Z$$

if $\Psi(Z) \leq \Phi(Z)$ for all finite $Z \subseteq \mathcal{R}$. Finally, the inequality

$$\left|\int_O \Phi(Z)\delta Z\right| \leq \int_O |\Phi(Z)|\delta Z$$

holds.

Proof. Obvious.   □

In the sequel we will be interested in primarily one kind of integrable finite-set function:

Definition 11 *A global density function (or global density for short) is a non-negative, integrable finite-set function f whose total set integral is unity: $\int f(Z)\delta Z = 1$.*

Note that if f, g are global densities and $a + b = 1$ where $1 \geq a, b \geq 0$ then $af + bg$ is also a global density.

Example 3 *Given ordinary probability density functions g_1, g_2 on $\mathcal{R}$ and a constant $0 \leq q \leq 1$ define the set function f by*

$$f(\emptyset) = (1 - q)^2$$
$$f(\{\xi\}) = q(1 - q)g_1(\xi) + q(1 - q)g_2(\xi)$$
$$f(\{\xi_1, \xi_2\}) = q^2 g_1(\xi_1)g_2(\xi_2) + q^2 g_1(\xi_2)g_2(\xi_1)$$

for all $\xi, \xi_1, \xi_2 \in \mathcal{R}$, with ξ_1, ξ_2 distinct, and $f(Z) = 0$ for all Z with $|Z| \geq 3$. The reader may easily verify that f is a global density.

We first begin with a simple example of a global density: The "global" counterpart of a uniform distribution:

Definition 12 *Let T be a closed subset of $\mathcal{R}$ with finite hybrid measure $\lambda(T)$ and let $M > 0$ be an integer. Define the set function $u_{T,M}(Z)$ by*

$$u_{T,M}(Z) \triangleq \begin{cases} \frac{|Z|!}{M \lambda(T)^{|Z|}} & if \quad Z \subseteq T \text{ and } |Z| < M \\ 0 & if \quad otherwise \end{cases}$$

for all finite subsets $Z \subseteq \mathcal{R}$. (Recall that $\lambda(T) = \sum_{w \in W} T(w)$ where $T(w) \triangleq \{z \in \mathbb{R}^n | (z, w) \in T\}$.) Then we call $u_{T,M}$ a global uniform density.

In particular, let $T = D \times W$ where D is a closed bounded subset of $\mathbb{R}^n$. Let $Z = \{(z_1, w_1), ..., (z_k, w_k)\} \subseteq \mathcal{R}$. Then we write $u_{D,M} = u_{T,M}$ and note that

$$u_{D,M}(Z) = \begin{cases} \frac{|Z|!}{M N^{|Z|} \lambda(D)^{|Z|}} & if \quad z_1, ..., z_k \in D \text{ and } k < M \\ 0 & if \quad otherwise \end{cases}$$

where $N = |W|$ where $\mathcal{R} = \mathbb{R}^n \times W$.

Proposition 7 *The finite-set function $u_{T,M}$ of Definition 12 satisfies*

$$\int_S u_{T,M}(Z)\delta Z = \frac{1}{M} \sum_{k=0}^{M-1} \left(\frac{\lambda(S \cap T)}{\lambda(T)} \right)^k$$

for all closed S. (b) Consequently, it is a global density.

Proof. (a) By definition,

$$\int_S u_{T,M}(Z)\delta Z = \sum_{k=0}^{\infty} \frac{1}{k!} \int_{S^k} u_{T,M}(\{\xi_1, ..., \xi_k\}) d\lambda(\xi_1) \cdots d\lambda(\xi_k)$$

$$= \sum_{k=0}^{M-1} \frac{1}{k!} \int_{(T \cap S)^k} \frac{k!}{M \lambda(T)^k} d\lambda(\xi_1) \cdots d\lambda(\xi_k)$$

$$= \frac{1}{M} \sum_{k=0}^{M-1} \frac{\lambda(S \cap T)^k}{\lambda(T)^k}$$

(b) Thus $\int u_{T,M}(Z)\delta Z = \int_{\mathcal{R}} u_{T,M}(Z)\delta Z = \frac{1}{M} \sum_{k=0}^{M-1} \frac{\lambda(T)^k}{\lambda(T)^k} = 1$ as claimed. $\square$

4.2.3   The Generalized Radon-Nikodým Derivative

In ordinary calculus the Lebesgue integral has an "inverse" operation—the Radon-Nikodým derivative—in the sense that $\int_S \frac{dq}{d\lambda}(x) d\lambda(x) = q(S)$ and $\frac{d}{d\lambda} \int_S f(x) d\lambda(x) = f$ almost everywhere. It is natural to ask, Does the set integral have an inverse operator in the same sense? The answer is "yes."

Let us begin by quickly revisiting the concept of a Radon-Nikodým derivative. As usual, let $\lambda(S)$ denote Lebesgue measure on $\mathbb{R}^n$ for any closed Lebesgue-measurable set S of $\mathbb{R}^n$. Let q be a nonnegative measure defined on the Lebesgue-measurable subsets of $\mathbb{R}^n$ which is absolutely continuous in the sense that $q(S) = 0$ whenever $\lambda(S) = 0$. Then by the Radon-Nikodým theorem there is an almost-everywhere unique function f such that $q(S) = \int_S f(\mathbf{z})d\lambda(\mathbf{z})$ for all measurable $S \subseteq \mathbb{R}^n$ and f is called the *Radon-Nikodým derivative* of q with respect to λ, denoted by $f = \frac{dq}{d\lambda}$. Thus $\frac{dq}{d\lambda}$ is defined as an anti-integral of the Lebesgue integral.

However, there are at least three ways to define the Radon-Nikodým derivative *constructively*: via the *Lebesgue density theorem*, via *nets*, or via *Vitali systems*. One form of the Lebesgue density theorem (see Definition 1 and Theorems 5 and 6 of [26, pp. 220-222] states that

$$\lim_{\varepsilon \downarrow 0} \frac{1}{\lambda(E_{\mathbf{z};\varepsilon})} \int_{E_{\mathbf{z};\varepsilon}} f(\mathbf{y})d\lambda(\mathbf{y}) = f(\mathbf{z})$$

for almost all $\mathbf{z} \in \mathbb{R}^n$, where $E_{\mathbf{z};\varepsilon}$ denotes the *closed ball* of radius ε centered at $\mathbf{z}$. It thus follows from the Radon-Nikodým theorem that a constructive definition of the Radon-Nikodým derivative of q (with respect to Lebesgue measure) is:

$$\frac{dq}{d\lambda}(\mathbf{z}) = \lim_{\varepsilon \downarrow 0} \frac{q(E_{\mathbf{z};\varepsilon})}{\lambda(E_{\mathbf{z};\varepsilon})} = \lim_{\varepsilon \downarrow 0} \frac{1}{\lambda(E_{\mathbf{z};\varepsilon})} \int_{E_{\mathbf{z};\varepsilon}} f(\mathbf{y})d\lambda(\mathbf{y}) = f(\mathbf{z})$$

almost everywhere. An alternative approach is the use of "nets." A net is a nested, countably infinite sequence of countable partitions of $\mathbb{R}^n$ by Borel sets [26, p. 208]. That is, each partition P_j is a countable sequence $Q_{j,1}, ..., Q_{j,k}, ...$ of Borel subsets of $\mathbb{R}^n$ which are mutually disjoint and whose union is all of $\mathbb{R}^n$. This sequence of partitions is nested, in the sense that given any cell $Q_{j,k}$ of the j'th partition P_j, there is a subsequence of the partition P_{j+1} : $Q_{j+1,1}, ..., Q_{j+1,i}, ...$ which is a partition of $Q_{j,k}$. Given this, another constructive definition of the Radon-Nikodým derivative is

$$\frac{dq}{d\lambda}(\mathbf{z}) = \lim_{i \to \infty} \frac{q(E_{\mathbf{z};i})}{\lambda(E_{\mathbf{z};i})}$$

where in this case the $E_{\mathbf{z};i}$ are any sequence of sets belonging to the net which converge to the singleton set $\{\mathbf{z}\}$. A third approach involves Vitali systems [26, pp. 209-215]. Whichever approach we use we can rest assured that an existing and rigorous theory of limits exists which is rich enough to generalize the Radon-Nikodým derivative in the manner we propose. For purposes of application, however, it is simpler to use the Lebesgue density theorem version. If the limit exists then it is enough to compute it for closed balls of radius $1/i$ as $i \to \infty$. This is what we will generally assume in what follows.

Definition 13 *Let* $\Phi : c(\mathcal{R}) \to \mathbb{R}^n$ *be a set function on* $\mathcal{R}$ *and let* $\xi = (\mathbf{z}, u) \in \mathcal{R}$. *If it exists, the generalized Radon-Nikodým derivative of* Φ *at* ξ *is the set* *function defined by:*

$$\frac{\delta\Phi}{\delta\xi}(T) \triangleq \lim_{j\to\infty}\lim_{i\to\infty} \frac{\Phi((T-(F_{z;j}\times u))\cup(E_{z;i}\times u))-\Phi(T-(F_{z;j}\times u))}{\lambda(E_{z;i})}$$

where $E_{z;i}$ is a sequence of closed balls converging to $\{z\}$; and where the $F_{z;j}$ is a sequence of open balls whose closures converge to $\{z\}$. Also, we have abbreviated $T-(F_{z;j}\times u) \triangleq T\cap(F_{z;j}\times u)^c$ and $E_{z;i}\times u \triangleq E_{z;i}\times\{u\}$.

This definition is less formidable than it may at first appear. If ξ were not an element of T we could instead use the simpler definition

$$\frac{\delta\Phi}{\delta\xi}(T) \triangleq \lim_{i\to\infty} \frac{\Phi(T\cup(E_{z;i}\times u))-\Phi(T)}{\lambda(E_{z;i})}$$

If ξ is contained in T, however, this definition is no longer adequate since then the expression vanishes identically. If this occurs, we "excise out" the small piece $F_{z;j}\times u$ of T which contains ξ. Thus $T-(F_{z;j}\times u)$ does not contain ξ. If we apply the previous equation with $T-(F_{z;j}\times u)$ substituted for T then we get a set function whose argument is $T-(F_{z;j}\times u)$ rather than T. Taking the limit as $j\to\infty$ we get the equation in Definition 13.

Two special cases are worth pointing out:

Remark 1 *Note that if $|W|=1$ and thus $W=\{\cdot\}$ is a point set, then $\mathcal{R}=\mathbb{R}^n\times W\approx\mathbb{R}^n$. If q is an absolutely continuous (with respect to Lebesgue measure) measure then the conventional Radon-Nikodým derivative can be recovered as follows:*

$$\frac{\delta q}{\delta z}(\emptyset) = \lim_{i\to\infty}\frac{q(E_{z;i})-q(\emptyset)}{\lambda(E_{z;i})} = \frac{dq}{d\lambda}(z)$$

Remark 2 *Suppose that $n=0$, in which case $\mathbb{R}^0=\{\cdot\}$ and thus $\mathcal{R}=\mathbb{R}^0\times W\approx W$. In this case $E_{z;i}=\{\cdot\}=F_{z;j}$ and the generalized Radon-Nikodým derivative has the form*

$$\frac{\delta\Phi}{\delta u}(T) = \begin{cases} (\Delta_u\Phi)(T) & \text{if } u\notin T \\ (\Delta_u\Phi)(T-\{u\}) & \text{if } u\in T \end{cases}$$

where $(\Delta_u\Phi)(T)$ is the difference operator (see Definition 8). The fact that $(\Delta_u\Phi)(T)\neq 0$ when $u\in T$ does not create a problem since $u\notin\emptyset$ implies that $\frac{\delta\Phi}{\delta u}(\emptyset)=(\Delta_u\Phi)(\emptyset)$.

The following proposition shows that the generalized Radon-Nikodým derivative obeys the usual sum, product, and chain rules of differentiation:

Proposition 8 *Let Φ,Ψ be set functions, let $a,b\in\mathbb{R}$, and define $(a\Phi+b\Psi)(T)=a\Phi(T)+b\Psi(T)$ and $(\Phi\Psi)(T)=\Phi(T)\Psi(T)$ for all $T\subseteq\mathcal{R}$. Let $\xi=(z,u)\in\mathcal{R}$. Assume that the generalized Radon-Nikodým derivatives of Φ and Ψ with respect* to ξ exist at T. Then the generalized Radon-Nikodým derivatives of $a\Phi + b\Psi$ and $\Phi\Psi$ with respect to ξ also exist at T, and:

$$\frac{\delta(a\Phi + b\Psi)}{\delta\xi}(T) = a\frac{\delta\Phi}{\delta\xi}(T) + b\frac{\delta\Psi}{\delta\xi}$$

$$\frac{\delta(\Phi\Psi)}{\delta\xi}(T) = \frac{\delta\Phi}{\delta\xi}(T)\,\Psi(T) + \Phi(T)\,\frac{\delta\Psi}{\delta\xi}(T)$$

Proof. Easy and left to the reader.   □

The following three examples illustrate how generalized Radon-Nikodým derivatives are computed.

Example 4 *Let $q \in \mathbb{R}$ and let*

$$\Phi(S) = 1 - q + q\,p_f(S)$$

for all closed S, where p_f is the probability measure on $\mathbb{R}$ corresponding to the density $f(z)$. Then

$$\frac{\delta\Phi}{\delta z}(T) = \lim_{j\to\infty}\lim_{i\to\infty}\frac{\Phi((T - F_{z;j})\cup E_{z;i}) - \Phi(T - F_{z;j})}{\lambda(E_{z;i})}$$

For i sufficiently large, $E_{z;i}$ will be very small and therefore contained in $F_{z;j}$. Thus from the additivity of p_f we get

$$\frac{\delta\Phi}{\delta z}(T) = q\,\lim_{i\to\infty}\frac{p_f(E_{z;i})}{\lambda(E_{z;i})} = q\,f(z)$$

almost everywhere. Thus none of the generalized Radon-Nikodým derivatives of Φ functionally depend on the set variable T. Taking the generalized Radon-Nikodým derivative again we therefore get:

$$\frac{\delta}{\delta y}\frac{\delta\Phi}{\delta z}(T) = 0$$

Example 5 *Let*

$$\Phi(S) = (1 - q + q\,p_f(S))\,(1 - q + q\,p_g(S))$$

for all closed $S \subseteq \mathcal{R}$. By Proposition 8 we have

$$\frac{\delta\Phi}{\delta z}(T) = (1 - q + q\,p_f(S))\frac{\delta}{\delta z}(1 - q + q\,p_g(S))$$

$$+(1 - q + q\,p_g(S))\frac{\delta}{\delta z}(1 - q + q\,p_f(S))$$

$$= (1 - q + q\,p_f(S))\,q\,g(z) + (1 - q + q\,p_g(S))\,q\,f(z)$$

and

$$\frac{\delta}{\delta y}\frac{\delta\Phi}{\delta z}(T) = q^2\,f(y)\,g(z) + q^2\,g(y)\,f(z)$$

which no longer depends on the set variable T. If we take more generalized Radon-Nikodým derivatives, they vanish identically.

Example 6 *Let $f(\mathbf{x}, \mathbf{y})$ be an integrable function on $\mathbb{R}^n \times \mathbb{R}^n$ and define*

$$\Phi(S) = \int_{S \times S} f(\mathbf{x}, \mathbf{y}) d\lambda(\mathbf{x}) d\lambda(\mathbf{y})$$

for all closed $S \subseteq \mathbb{R}^n$. Then

$$
\begin{aligned}
\frac{\delta \Phi}{\delta \mathbf{z}}(S) &= \lim_{i \to \infty} \frac{1}{\lambda(E_{\mathbf{z};i})} \left(\begin{array}{c} \int_{(S \cup E_{\mathbf{z};i}) \times (S \cup E_{\mathbf{z};i})} f(\mathbf{x}, \mathbf{y}) d\lambda(\mathbf{x}) d\lambda(\mathbf{y}) \\ - \int_{S \times S} f(\mathbf{x}, \mathbf{y}) d\lambda(\mathbf{x}) d\lambda(\mathbf{y}) \end{array} \right) \\
&= \lim_{i \to \infty} \frac{1}{\lambda(E_{\mathbf{z};i})} \int_{S \times E_{\mathbf{z};i}} f(\mathbf{x}, \mathbf{y}) d\lambda(\mathbf{x}) d\lambda(\mathbf{y}) \\
&\quad + \lim_{i \to \infty} \frac{1}{\lambda(E_{\mathbf{z};i})} \int_{E_{\mathbf{z};i} \times S} f(\mathbf{x}, \mathbf{y}) d\lambda(\mathbf{x}) d\lambda(\mathbf{y}) \\
&\quad + \lim_{i \to \infty} \frac{1}{\lambda(E_{\mathbf{z};i})} \int_{E_{\mathbf{z};i} \times E_{\mathbf{z};i}} f(\mathbf{x}, \mathbf{y}) d\lambda(\mathbf{x}) d\lambda(\mathbf{y}) \\
&= \int_S f(\mathbf{x}, \mathbf{z}) d\lambda(\mathbf{x}) + \int_S f(\mathbf{z}, \mathbf{y}) d\lambda(\mathbf{y}) + 0
\end{aligned}
$$

(If we choose $E_{\mathbf{z};i}$ to be a closed ball of radius $1/i$ then the first term results by applying the Lebesgue density theorem to

$$\lim_{i \to \infty} \frac{1}{\lambda(E_{\mathbf{z};i})} \int_{S \times E_{\mathbf{z};i}} f(\mathbf{x}, \mathbf{y}) d\lambda(\mathbf{x}) d\lambda(\mathbf{y}) = \lim_{i \to \infty} \frac{1}{\lambda(E_{\mathbf{z};i})} \int_{E_{\mathbf{z};i}} g(\mathbf{y}) d\lambda(\mathbf{y}) = g(\mathbf{z})$$

where $g(\mathbf{y}) \triangleq \int_S f(\mathbf{x}, \mathbf{y}) d\lambda(\mathbf{x})$ and likewise for the second term. The third term results from the Lebesgue density theorem and the fact that

$$\int_{E_{\mathbf{z};j} \times E_{\mathbf{z};k}} f(\mathbf{x}, \mathbf{y}) d\lambda(\mathbf{x}) d\lambda(\mathbf{y}) \geq \int_{E_{\mathbf{z};i} \times E_{\mathbf{z};i}} f(\mathbf{x}, \mathbf{y}) d\lambda(\mathbf{x}) d\lambda(\mathbf{y})$$

for all $i \geq \max\{j, k\}$.) Consequently,

$$\frac{\delta}{\delta \mathbf{w}} \frac{\delta \Phi}{\delta \mathbf{z}}(S) = f(\mathbf{w}, \mathbf{z}) + f(\mathbf{z}, \mathbf{w})$$

and the higher-order derivatives vanish identically.

As these examples indicate, the generalized Radon-Nikodým derivative of a set function is again a set function and thus we can iterate the process:

Definition 14 *The iterated generalized Radon-Nikodým derivatives of order k are defined as:*

$$\frac{\delta^{k+1} \Phi}{\delta \xi_{k+1} \cdots \delta \xi_1}(T) \triangleq \frac{\delta}{\delta \xi_{k+1}} \frac{\delta^k \Phi}{\delta \xi_k \cdots \delta \xi_1}(T)$$

for any $\xi_1, \ldots, \xi_k \in \mathcal{R}$.

The following proposition establishes an explicit formula for the iterated generalized Radon-Nikodým derivatives of a set function. It also shows that, when everything is well-behaved, that the order in which one takes iterated generalized Radon-Nikodým derivatives is irrelevant. Finally, it illustrates the close relationship which exists between the iterated generalized Radon-Nikodým and the Möbius transform.

Proposition 9 *Let Φ be a set function and let $\xi_1, ..., \xi_k$ be distinct elements of $\mathcal{R}$ with $\xi_j = (z_j, u_j)$ for $j = 1, ...k$. Define the set function Φ_i by*

$$\Phi_i(\{\xi_1, ..., \xi_k\}) \triangleq \Phi((E_{z_1;i} \times u_1) \cup ... \cup (E_{z_k;i} \times u_k))$$

for all distinct $\xi_1, ..., \xi_k$, where as usual $E_{z_1;i}$ denotes a closed ball of radius $1/i$ centered at $z \in \mathbb{R}^n$. Assume that all iterated generalized Radon-Nikodým derivatives of Φ exist. If $Z \triangleq \{\xi_1, ..., \xi_k\}$ then:

$$\frac{\delta^k \Phi}{\delta \xi_1 \cdots \delta \xi_k}(\emptyset) = \lim_{i \to \infty} \frac{\sum_{Y \subseteq Z}(-1)^{|Z-Y|}\Phi_i(Y)}{\lambda(E_{z_1;i}) \cdots \lambda(E_{z_k;i})}$$

Proof. Recall (Definition 8) the difference operator $(\Delta_a \Phi)(T) = \Phi(T \cup \{a\}) - \Phi(T)$ where Φ in this case is a function of finite-set arguments. Assume without loss of generality that $\xi_1 \in \mathcal{R}$ is not in the set S. Recall that the definition of Φ_i is:

$$\Phi_i(\{\xi_1, ..., \xi_k\}) = \Phi((E_{z_1;i} \times u_1) \cup ... \cup (E_{z_k;i} \times u_k))$$

Then note that

$$\frac{\delta \Phi}{\delta \xi_1}(\emptyset) = \lim_{i \to \infty} \frac{\Phi(E_{z_1;i} \times u_1) - \Phi(\emptyset)}{\lambda(E_{z_1;i})} = \lim_{i \to \infty} \frac{(\Delta_{\xi_1} \Phi_i)(\emptyset)}{\lambda(E_{z_1;i})}$$

Also, note that the second iterated generalized Radon-Nikodým derivative is:

$$\begin{aligned}
\frac{\delta^2 \Phi}{\delta \xi_2 \delta \xi_1}(\emptyset) &= \lim_{i \to \infty} \frac{1}{\lambda(E_{z_1;i})}\left(\frac{\delta \Phi}{\delta \xi_1}(E_{z_2;i} \times u_2) - \frac{\delta \Phi}{\delta \xi_1}(\emptyset)\right) \\
&= \lim_{i \to \infty} \frac{(\Delta_{\xi_1} \Phi_i)(\{\xi_2\}) - (\Delta_{\xi_1} \Phi_i)(\emptyset)}{\lambda(E_{z_1;i})\,\lambda(E_{z_2;i})} = \lim_{i \to \infty} \frac{(\Delta_{\xi_2}\Delta_{\xi_1} \Phi_i)(\emptyset)}{\lambda(E_{z_1;i})\,\lambda(E_{z_2;i})}
\end{aligned}$$

In general we will get:

$$\frac{\delta^k \Phi}{\delta \xi_k \cdots \delta \xi_1}(\emptyset) = \lim_{i \to \infty} \frac{(\Delta_{\xi_k} \cdots \Delta_{\xi_1} \Phi_i)(\emptyset)}{\lambda(E_{z_1;i}) \cdots \lambda(E_{z_k;i})}$$

However, from Proposition 5 we know that

$$(\Delta_{\xi_k} \cdots \Delta_{\xi_1} \Phi_i)(\emptyset) = \sum_{Y \subseteq Z}(-1)^{|Z-Y|}\Phi_i(Y)$$

which then gives us the desired result. $\square$

Corollary 10 *Assume that all iterated generalized Radon-Nikodým derivatives of the set function Φ exist. Then their values at $T = \emptyset$ are independent of the order of differentiation:*

$$\frac{\delta^k \Phi}{\delta \xi_1 \cdots \delta \xi_k}(\emptyset) = \frac{\delta^k \Phi}{\delta \xi_{\pi 1} \cdots \delta \xi_{\pi k}}(\emptyset)$$

for any permutation π on the number $1, \ , k$, where $\xi_1, \ , \xi_k$ are assumed distinct.

Proof. Follows immediately from Proposition 9. $\square$

It is worth taking a moment to look at a specific case of Proposition 9:

Example 7 *Assume that $|W| = 1$ and $n = 1$, so that $\mathcal{R} = \mathbb{R}$. Note that if i is sufficiently large, all of the closed balls in Proposition 9 will be disjoint since the elements of Z are assumed distinct. Assume for convenience that $z_{1,}, z_2$ are not in S. Then by Definition 13,*

$$\frac{\delta \Phi}{\delta z_1}(S) = \lim_{i_1 \to \infty} \frac{\Phi(S \cup E_{z_1; h}) - \Phi(S)}{\lambda(E_{z_1; h})}$$

and so

$$\frac{\delta}{\delta z_2} \frac{\delta \Phi}{\delta z_1}(S) = \lim_{i_2 \to \infty} \frac{1}{\lambda(E_{z_2; i_2})} \left(\begin{array}{c} \lim_{i_1 \to \infty} \frac{\Phi(S \cup E_{z_2; i_2} \cup E_{z_1; i_1}) - \Phi(S \cup E_{z_2; i_2})}{\lambda(E_{z_1; i_1})} \\ - \lim_{i_1 \to \infty} \frac{\Phi(S \cup E_{z_1; i_1}) - \Phi(S)}{\lambda(E_{z_1; i_1})} \end{array} \right)$$

or

$$\frac{\delta^2 \Phi}{\delta z_2 \delta z_1}(S) = \lim_{i_1, i_2 \to \infty} \frac{\Phi(S \cup E_{z_2; i_2} \cup E_{z_1; i_1}) - \Phi(S \cup E_{z_2; i_2}) - \Phi(S \cup E_{z_1; i_1}) + \Phi(S)}{\lambda(E_{z_1; i_1}) \lambda(E_{z_2; i_2})}$$

Thus for $i_1 = i_2 = i$ sufficiently large we get:

$$\frac{\delta^2 \Phi}{\delta z_2 \delta z_1}(\emptyset) = \lim_{i \to \infty} \frac{\Phi_i(\{z_2, z_1\}) - \Phi_i(\{z_2\}) - \Phi_i(\{z_1\}) + \Phi(\emptyset)}{\lambda(E_{z_1; i_1}) \lambda(E_{z_2; i_2})}$$

as desired.

4.2.4   Set Derivatives

Because of Corollary 10 we can reformulate the concept of an iterated generalized Radon-Nikodým derivative as follows:

Definition 15 *Let Φ be a set function. Let $k \geq 1$ and let $Z = \{\xi_1, ..., \xi_k\} \subseteq \mathcal{R}$ be a finite subset with k distinct elements. Assume that all iterated generalized Radon-Nikodým derivatives of Φ exist. Then the set derivative of Φ is defined as:*

$$\frac{\delta \Phi}{\delta Z}(T) \triangleq \frac{\delta^k \Phi}{\delta \xi_1 \cdots \delta \xi_k}(T) \qquad (if\ Z \neq \emptyset)$$

$$\frac{\delta \Phi}{\delta \emptyset}(T) \triangleq \Phi(T)$$

Note that the set derivative is well-defined because the order of differentiation is irrelevant.

The following proposition generalizes the sum and product rules for the generalized Radon-Nikodým derivative (see Proposition 8) to the case of set derivatives:

Proposition 11 *Let Φ, Ψ be set functions and let $a, b \in \mathbb{R}$. Assume that the set derivatives of Φ and Ψ exist. Then the set derivatives of $a\Phi + b\Psi$ and $\Phi\Psi$ also exist and:*

$$\frac{\delta(a\Phi + b\Psi)}{\delta Z}(T) = a\frac{\delta\Phi}{\delta Z}(T) + b\frac{\delta\Psi}{\delta Z}(T)$$

$$\frac{\delta(\Phi\Psi)}{\delta Z}(T) = \sum_{Y \subseteq Z} \frac{\delta\Phi}{\delta Y}(T)\,\frac{\delta\Psi}{\delta(Z-Y)}(T)$$

for all $T \subseteq R$. Also, suppose that $Y, Z \subseteq \mathcal{R}$ are finite subsets with $Y \cap Z = \emptyset$. Then

$$\frac{\delta}{\delta Y}\frac{\delta\Phi}{\delta Z}(T) = \frac{\delta\Phi}{\delta(Y \cup Z)}(T)$$

for all $T \subseteq R$.

Proof. The only assertion which requires verification is the product rule. The product rule for generalized Radon-Nikodým derivatives (Proposition 8) gives us:

$$\frac{\delta(\Phi\Psi)}{\delta\xi_1} = \frac{\delta\Phi}{\delta\xi_1}\Psi + \Phi\frac{\delta\Psi}{\delta\xi_1} = \sum_{i_1=0}^{1} \frac{\delta^{i_1}\Phi}{\delta^{i_1}\xi_1}\frac{\delta^{1-i_1}\Psi}{\delta^{1-i_1}\xi_1}$$

where we have adopted the conventions $\frac{\delta^0\Phi}{\delta^0\xi} \triangleq \Phi$ and $\frac{\delta^1\Phi}{\delta^1\xi} \triangleq \frac{\delta\Phi}{\delta\xi}$. Likewise, for the second iterated generalized Radon-Nikodým derivative we can write:

$$\frac{\delta^2(\Phi\Psi)}{\delta\xi_2\delta\xi_1} = \sum_{i_1,i_2=0}^{1} \frac{\delta^{i_1+i_2}\Phi}{\delta^{i_2}\xi_2\delta^{i_1}\xi_1}\frac{\delta^{1-i_1-i_2}\Psi}{\delta^{1-i_2}\xi_2\delta^{1-i_1}\xi_1}$$

In general, the k'th iterated generalized Radon-Nikodým derivative is:

$$\frac{\delta^k(\Phi\Psi)}{\delta\xi_k\cdots\delta\xi_1} = \sum_{i_1,\ldots,i_k=0}^{1} \frac{\delta^{i_1+\ldots+i_k}\Phi}{\delta^{i_k}\xi_k\cdots\delta^{i_1}\xi_1}\frac{\delta^{1-i_1-\ldots-i_k}\Psi}{\delta^{1-i_k}\xi_k\cdots\delta^{1-i_1}\xi_1}$$

Now suppose that $Z = \{\xi_1, ..., \xi_k\}$ where the $\xi_1, ..., \xi_k$ are distinct. A little reflection allows us to see that we can rewrite the last equation as

$$\frac{\delta(\Phi\Psi)}{\delta Z}(T) = \sum_{Y \subseteq Z} \frac{\delta\Phi}{\delta Y}(T)\,\frac{\delta\Psi}{\delta(Z-Y)}(T)$$

where, for a given list $i_1, ..., i_k$ define $Y = Y(i_1, ..., i_k)$ to be the subset of all $\xi_j \in Z$ such that $i_j = 1$.   □

4.3   Basic Properties of Finite Random Subsets

We now have at our disposal the elements of a differential and integral calculus of finite-set functions for data fusion. In this section we apply this calculus to our central concept: the *global density function* of an *absolutely continuous finite random subset* of hybrid space $\mathcal{R}$.

4.3.1   Belief Measures of Finite Random Subsets

In this section we define and determine the major properties of finite random subsets:

Definition 16 *A random subset Σ of $\mathcal{R}$ is a finite random subset if there exists an integer $M \geq 1$ such that $\Sigma(\omega)$ is a finite set and $|\Sigma(\omega)| \leq M$, for all $\omega \in \Omega$. The cardinality of Σ, denoted by $card(S)$, is defined as $card(S) = \max_{\omega \in \Omega} |\Sigma(\omega)|$.*

Later we will need the following immediate consequence of Proposition 1:

Proposition 12 *Let Σ be a finite random subset of $\mathcal{R}$. Define $|\Sigma|(\omega) = |\Sigma(\omega)|$ for all $\omega \in \Omega$. Then $|\Sigma|$ is a random integer.*

Proof. Note that

$$
\begin{aligned}
|\Sigma|^{-1}(k) &= \{\omega \in \Omega \mid |\Sigma(\omega)| = k\} = \{\omega \in \Omega \mid \Sigma(\omega) \in c_k(\mathcal{R})\} \\
&= \Sigma^{-1}(c_k(\mathcal{R})) \in \sigma(\Omega)
\end{aligned}
$$

since $c_k(\mathcal{R})$ is measurable by Proposition 1 and since by definition Σ is a measurable mapping. $\square$

The following proposition provides the basic method of constructing finite random subsets (of not necessarily discrete spaces) to be used in the sequel.

Proposition 13 *Let $\mathbf{Z} : \Omega \to \mathbb{R}^n$ be a random vector on $\mathbb{R}^n$ and let $\Xi : \Omega \to \mathcal{P}(W)$ be a random subset of W (where as usual $\mathcal{R} = \mathbb{R}^n \times W$). Then $\Sigma : \Omega \to c(\mathcal{R})$ defined by*

$$
\Sigma(\omega) = \{\mathbf{Z}(\omega)\} \times \Xi
$$

for all $\omega \in \Omega$ is a finite random subset of $\mathcal{R}$.

Proof. It is enough to show that $\Sigma^{-1}(O_G) \in \sigma(\Omega)$ for any open $G \subseteq \mathcal{R}$. On the one hand,

$$
\Sigma^{-1}(O_G) = \{\omega \in \Omega \mid \Sigma(\omega) \cap G \neq \emptyset\} = \{\omega \in \Omega \mid (\{\mathbf{Z}(\omega)\} \times \Xi(\omega)) \cap G \neq \emptyset\}
$$

However, $G = \cup_{v \in W}(G(v) \times \{v\})$ where $G(v) = \{\mathbf{z} \in \mathbb{R}^n \mid (\mathbf{z}, v) \in G\}$, and so

$$
\begin{aligned}
(\{\mathbf{Z}(\omega)\} \times \Xi(\omega)) \cap G &= \bigcup_{v \in W}(\{\mathbf{Z}(\omega)\} \cap G(v)) \times (\{v\} \cap \Xi(\omega)) \\
&= \bigcup_{v \in \Xi(\omega)}(\{\mathbf{Z}(\omega)\} \cap G(v)) \times \{v\}
\end{aligned}
$$

Therefore, $(\{\mathbf{Z}(\omega)\} \times \Xi(\omega)) \cap G \neq \emptyset$ if and only if $\{\mathbf{Z}(\omega)\} \cap G(v) \neq \emptyset$ for some $v \in \Xi(\omega)$; or, equivalently, if and only if $\mathbf{Z}(\omega) \in \cup_{v \in \Xi(\omega)} G(v)$. This, in turn, holds if and only if $\Xi(\omega) = T$ and $\mathbf{Z}(\omega) \in \cup_{v \in T} G(v)$; or, alternatively, if and only if $\omega \in \Xi^{-1}(\{T\}) \cap \mathbf{Z}^{-1}(\cup_{v \in T} G(v))$ for some $T \subseteq W$. This, finally, is true if and only if $\omega \in \cup_{T \subseteq W} \Xi^{-1}(\{T\}) \cap \mathbf{Z}^{-1}(\cup_{v \in T} G(v))$. In other words we have shown that:

$$\Sigma^{-1}(O_G) = \bigcup_{T \subseteq W} \Xi^{-1}(\{T\}) \cap \mathbf{Z}^{-1}(\bigcup_{v \in T} G(v))$$

Since $\mathbf{Z}$ is a random vector in $\mathbb{R}^n$ and since Ξ is a random subset of W then $\Xi^{-1}(\{T\}) \in \sigma(\Omega)$ and $\mathbf{Z}^{-1}(\cup_{v \in T} G(v)) \in \sigma(\Omega)$ for all $T \subseteq W$. Since $\sigma(\Omega)$ is a sigma-ring it follows that $\Sigma^{-1}(O_G) \in \sigma(\Omega)$. So, Σ is a random subset of $\mathcal{R}$. $\square$

The following three results show that any finite random subset of $\mathbb{R}^n$ can be decomposed into a finite union of simpler random subsets:

Theorem 14 *Let Σ be a finite random subset of $\mathbb{R}^n$ and let $M = card(S)$. Then there exist random subsets $\Sigma_1, ..., \Sigma_M$ of $\mathbb{R}^n$ such that $card(\Sigma_j) = 1$ for $j = 1, ..., M$ and $\Sigma = \Sigma_1 \cup ... \cup \Sigma_M$.*

Proof. We prove the proposition by induction on $M = card(S)$. If $card(S) = 1$ then the assertion is trivially true. Assume that $M = card(S) \geq 2$. Embed $\mathbb{R}^n$ in $\mathbb{R}^{n+1} = \mathbb{R}^n \times \mathbb{R}$ in the obvious way and let $e = (0, 1)$ be the unit vector corresponding to the augmented dimension. Then Σ can be embedded in $\mathbb{R}^{n+1}$ as $\Sigma \times \{0\}$ (that is, $(\Sigma \times \{0\})(\omega) = (\Sigma(\omega), 0)$ for all $\omega \in \Omega$). Define the function $\Lambda : \Omega \to \mathbb{R}^{n+1}$ by

$$\Lambda(\omega) \triangleq \begin{cases} \{e\} & \text{if } \Sigma = \emptyset \\ \emptyset & \text{if } \Sigma \neq \emptyset \end{cases}$$

Then Λ is a random subset of $\mathbb{R}^{n+1}$. For, let G be a nonempty open subset of $\mathbb{R}^{n+1}$. Then

$$\Lambda^{-1}(O_G) = \{\omega \in \Omega | \Lambda(\omega) \cap G \neq \emptyset\} = \{\omega \in \Omega | \Sigma(\omega) = \emptyset, e \in G\}$$

If $e \notin G$ then this is the empty set and thus is in $\sigma(\Omega)$. Otherwise, it equals

$$\{\omega \in \Omega | \Sigma(\omega) = \emptyset\} = |\Sigma|^{-1}(0) \in \sigma(\Omega)$$

since $|\Sigma|$ is a random integer (Proposition 12). Hence $\Lambda^{-1}(O_G) \in \sigma(\Omega)$ for every open $G \subseteq \mathbb{R}^{n+1}$ and Λ is a random subset of $\mathbb{R}^{n+1}$. Accordingly, so is $\Sigma' \triangleq (\Sigma \times \{0\}) \cup \Lambda$. By construction, Σ' is a nonempty random subset. Therefore (see [2, pp. 307-308, Theorem 8.1.3]) Σ' has a selector. That is, there is a random vector $\mathbf{Z}$ on $\mathbb{R}^{n+1}$ such that $\mathbf{Z} \in \Sigma'$. By construction $\mathbf{Z}(\omega) = e$ when $\Sigma(\omega) = \emptyset$ and $\mathbf{Z}(\omega) = (z(\omega), 0)$ for some $z(\omega) \in \mathbb{R}^n$ otherwise. Also, since $\mathbf{Z}$ is a random vector on $\mathbb{R}^{n+1}$ then

$$\{\omega \in \Omega | \mathbf{Z}(\omega) \in G\} \in \sigma(\Omega)\}$$

for all open subsets $G \subseteq \mathbb{R}^{n+1}$. Now, $\{\mathbf{Z}\}$ is a random subset of $\mathbb{R}^{n+1}$ and hence so is $\Sigma'' \triangleq \Sigma' - \{\mathbf{Z}\} = \Sigma' \cap \{\mathbf{Z}\}^c$. However, Σ'' is such that $\Sigma''(\omega) = \emptyset$ if

$\Sigma(\omega) = \emptyset$ and, moreover, such that $|\Sigma''(\omega)| = |\Sigma(\omega)| - 1$ if $\Sigma(\omega) \neq \emptyset$. Therefore $card(\Sigma'') = card(\Sigma) - 1 = M - 1$.

By inductive assumption, there are random subsets $\Sigma_1'', ..., \Sigma_{M-1}''$ of $\mathbb{R}^{n+1}$ such that $\Sigma'' = \Sigma_1'' \cup ... \cup \Sigma_{M-1}''$ and such that $card(\Sigma_j'') = 1$ for all $j = 1, ..., M-1$. Then on the one hand note that by construction $\Sigma'' \subseteq \Sigma \times \{0\}$ and therefore $\Sigma_j'' \subseteq \Sigma \times \{0\}$ for all $j = 1, ..., M - 1$. Consequently $\Sigma_j'' = \Sigma_j \times \{0\}$ for all $j = 1, ..., M - 1$ where the Σ_j are random subsets of $\mathbb{R}^n$ such that $card(\Sigma_j) = 1$ for all $j = 1, ..., M - 1$. On the other hand note that we get

$$(\Sigma \times \{0\}) \cup \Lambda = \Sigma' = \Sigma'' \cup \{Z\} = (\Sigma_1 \times \{0\}) \cup ... \cup (\Sigma_{M-1} \times \{0\}) \cup \{Z\}$$

Intersecting both sides of this equation by $\{e\}^c$ and using the fact that $\Lambda \cap \{e\}^c = \emptyset$ we get

$$\Sigma \times \{0\} = (\Sigma_1 \times \{0\}) \cup ... \cup (\Sigma_{M-1} \times \{0\}) \cup (\{Z(\omega) \cap \{e\}^c)$$

However, note that $\{Z(\omega) \cap \{e\}^c = \emptyset$ when $\Sigma(\omega) = \emptyset$ and $\{Z(\omega) \cap \{e\}^c = \{(z(\omega), 0)\}$ otherwise. Define

$$\Sigma_M(\omega) \triangleq \begin{cases} \emptyset & \text{if } \Sigma(\omega) = \emptyset \\ \{z(\omega)\} & \text{if } \Sigma(\omega) \neq \emptyset \end{cases}$$

Then Σ_M is a random subset of $\mathbb{R}^n$. For, let H be an open subset of $\mathbb{R}^n$. Then

$$\begin{aligned} \Sigma_M^{-1}(O_H) &= \{\omega \in \Omega | \Sigma_M(\omega) \cap H \neq \emptyset\} = \{\omega \in \Omega | \Sigma(\omega) \neq \emptyset, \, Z(\omega) \in H \times \mathbb{R}\} \\ &= \{\omega \in \Omega | \, \Sigma(\omega) \neq \emptyset\} \cap \{\omega \in \Omega | \, Z(\omega) \in H \times \mathbb{R}\} \end{aligned}$$

The first subset in the last equation is in $\sigma(\Omega)$ as we have already noted. Since $H \times \mathbb{R}$ is open in $\mathbb{R}^{n+1}$ the second subset is in $\sigma(\Omega)$ since Z is a random vector on $\mathbb{R}^{n+1}$. Hence $\Sigma_M^{-1}(O_H) \in \sigma(\Omega)$ for all open subsets $H \subseteq \mathbb{R}^n$ and hence Σ_M is a random subset of $\mathbb{R}^n$. Hence

$$\Sigma \times \{0\} = (\Sigma_1 \times \{0\}) \cup ... \cup (\Sigma_{M-1} \times \{0\}) \cup (\Sigma_M \times \{0\})$$

where $card(\Sigma_M) = 1$. Therefore

$$\Sigma = \Sigma_1 \cup ... \cup \Sigma_{M-1} \cup \Sigma_M$$

as desired. This completes the induction and the proof. $\square$

Corollary 15 *Let Σ be a finite random subset of $\mathcal{R}$ and let $M = card(\Sigma)$. Then there exist random subsets $\Sigma_1, ..., \Sigma_m$ of $\mathcal{R}$ with $m \leq M \cdot N$ (where $N = |W|$) such that $card(\Sigma_j) = 1$ and $\Sigma = \Sigma_1 \cup ... \cup \Sigma_m$.*

Proof. As usual, $\mathcal{R} = \mathbb{R}^n \times W$. Define $\Sigma_v \triangleq \Sigma \cap (\mathbb{R}^n \times \{v\})$ for each $v \in W$. Then Σ_v is a random subset of $\mathcal{R}$ and hence also of the subspace $\mathbb{R}^n \times \{v\} \approx \mathbb{R}^n$. It is also clear that $M(v) \triangleq card(\Sigma_v) \leq M$. By Theorem 14, for each v there exist finite random subsets $\Sigma_{v;1}, ..., \Sigma_{v;M(v)}$ on $\mathbb{R}^n$ such that

$$\Sigma_v = (\Sigma_{v;1} \times \{v\}) \cup ... \cup (\Sigma_{v;M(v)} \times \{v\})$$

and $card(\Sigma_{v;j}) = 1$ for all $j = 1, ..., m$. Consequently $\Sigma = \cup_{v \in W} \Sigma_v$ is a union of no more than $M \cdot N$ random subsets each of which has no more than one element, as asserted. $\qquad\qquad\qquad\qquad\qquad\qquad\qquad\qquad\qquad\qquad\qquad\qquad\qquad\square$

Corollary 16 *(a) Let Σ be a finite random subset of R with $card(\Sigma) = 1$. Then $\Sigma = \{\mathbf{Z}\} \times \Xi$ where $\mathbf{Z}$ is some random vector on $\mathbb{R}^n$ and Ξ is a random subset of W with $card(\Xi) = 1$. (b) Consequently, Σ is a finite union of the form*

$$\Sigma = (\{\mathbf{Z}_1\} \times \Xi_1) \cup ... \cup (\{\mathbf{Z}_m\} \times \Xi_m)$$

where $\mathbf{Z}_j$ are random vectors on $\mathbb{R}^n$ and Ξ_j are random subsets of W which contain no more than one element.

Proof. (a) Let $\Sigma(\omega) = \{(\mathbf{z}(\omega), v(\omega))\}$ when $\Sigma(\omega) \neq \emptyset$. Define $\Xi(\omega) = \emptyset$ when $\Sigma(\omega) = \emptyset$ and $\Xi(\omega) = \{v(\omega)\}$ otherwise. Then Ξ is a random subset of W. To show this it is enough to show that $\Xi^{-1}(w) \in \sigma(\Omega)$ for all $w \in W$. However,

$$
\begin{aligned}
\Xi^{-1}(w) &= \{\omega \in \Omega | \Sigma(\omega) \neq \emptyset, \Sigma(\omega) \in \mathbb{R}^n \times w\} \\
&= \Sigma^{-1}(\emptyset)^c \cap \Sigma^{-1}(\mathbb{R}^n \times w) \in \sigma(\Omega)
\end{aligned}
$$

, it follows that Ξ is a random subset of W. Likewise, pick an arbitrary $e \in \mathbb{R}^n$ and define $\mathbf{Z}(\omega) = \mathbf{z}(\omega)$ when $\Sigma(\omega) \neq \emptyset$ and $\mathbf{Z}(\omega) = e$ when $\Sigma(\omega) = \emptyset$. Then $\mathbf{Z}$ is a random vector on $\mathbb{R}^n$. For, let

$$\Omega_0 = \{\omega \in \Omega | \Sigma(\omega) = \emptyset\} = \{\omega \in \Omega | \Sigma(\omega) \cap \mathcal{R} \neq \emptyset\} = \Sigma^{-1}(O_{\mathcal{R}})$$

which is in $\sigma(\Omega)$. We are to show that $\{\omega \in \Omega | \mathbf{Z}(\omega) \in G\}$ is in $\sigma(\Omega)$ for any open $G \subseteq \mathbb{R}^n$. Suppose that $e \notin G$. Then

$$\{\omega \in \Omega | \mathbf{Z}(\omega) \in G\} = \{\omega \in \Omega | \Sigma(\omega) \cap (G \times W) \neq \emptyset\} \in \sigma(\Omega)$$

On the other hand suppose that $e \in G$. Then

$$\{\omega \in \Omega | \mathbf{Z}(\omega) \in G\} = \Omega_0 \cup \{\omega \in \Omega | \Sigma(\omega) \cap (G \times W) \neq \emptyset\} \in \sigma(\Omega)$$

So we are done since $\Sigma = \{\mathbf{Z}\} \times \Xi$.

b) By Corollary 15 we can write Σ as a finite union of finite random subsets which contain no more than one element. By part a) each of these can be written in the desired form. $\qquad\qquad\qquad\qquad\qquad\qquad\qquad\qquad\qquad\qquad\qquad\qquad\qquad\square$

The following proposition shows that just as finite random subsets are decomposable into unions of simpler finite random subsets, so the belief measure of a certain type of finite random subset is decomposable into a sum of simpler belief measures.

Theorem 17 *Let Σ be a finite random subset of R and let $\beta_\Sigma(S) = p(\Sigma \subseteq S)$ be the belief measure of Σ. Then β_Σ can be written in the form* ◉

$$\beta_\Sigma(S) = a_0 + a_1 q_1(S) + a_2 q_2(S \times S) + ... + a_m q_m(S^m)$$

where $a_k \geq 0$ for all $k = 1, ..., m$ and where q_k is a probability measure on $\mathcal{R}^k$ for all $j = 1, ..., m$. (We have denoted $S^k = S \times ... \times S = $ Cartesian product of S with itself taken k times). Moreover,

$$a_0 = p(\Sigma = \emptyset), \qquad a_k = p(|\Sigma| = k), \qquad q_k(S^k) = p(\Sigma \subseteq S \mid |\Sigma| = k)$$

for all $k = 1, ..., m$.

Proof. By Corollary 16 we can write $\Sigma = \Sigma_1 \cup ... \cup \Sigma_m$ where $\Sigma_j = \{\mathbf{Z}_j\} \times \Xi_j$ where $\mathbf{Z}_j$ is a random vector on $\mathbb{R}^n$ and where Ξ_j is a random subset of W with $card(\Xi_j) = 1$ for all $j = 1, ,, .m$. We know that the Σ_j are random subsets of $\mathcal{R}$ by Proposition 13. For notational convenience define the functions X_j by $\Xi_j(\omega) = \{X_j(\omega)\}$ whenever $\Xi_j(\omega) \neq \emptyset$. Also, define $\psi_j(\omega) \triangleq (\mathbf{Z}_j(\omega), X_j(\omega))$ whenever $\Xi_j(\omega) \neq \emptyset$. We will use the following abbreviation: If X_j or ψ_j occurs in any expression in what follows, the statement $\Xi_j(\omega) \neq \emptyset$ will *be assumed to be implied.* Abbreviate $C = \{1, ..., m\}$. The joint belief measure of the $\Sigma_1, ..., \Sigma_m$ is:

$$
\begin{aligned}
\beta_{\Sigma_1, ..., \Sigma_m}(S_1, ..., S_m) &= p(\Sigma_1 \subseteq S_1, ..., \Sigma_m \subseteq S_m) \\
&= p(\Xi_1 = \emptyset \text{ or } \psi_1 \in S_1, ..., \Xi_m = \emptyset \text{ or } \psi_m \in S_m) \\
&= p\left(\bigwedge_{j \in C} (\Xi_j = \emptyset \vee \psi_j \in S_j) \right) \\
&= p\left(\bigvee_{A \subseteq C} \left[\bigwedge_{i \in A} \Xi_i = \emptyset \right] \wedge \left[\bigwedge_{k \in C-A} \psi_k \in S_k \right] \right) \\
&= \sum_{A \subseteq C} a_A q_A \left(\prod_{k \in C-A} S_k \right)
\end{aligned}
$$

where

$$
a_A \triangleq p\left(\left[\bigwedge_{j \in C-A} \Xi_j \neq \emptyset \right] \wedge \left[\bigwedge_{i \in A} \Xi_i = \emptyset \right] \right)
$$

$$
\begin{aligned}
q_A &\left(\prod_{j \in C-A} S_j \right) \\
&\triangleq p\left(\left[\bigwedge_{i \in A} \Xi_i = \emptyset \right] \wedge \left[\bigwedge_{j \in C-A} \psi_j \in S_j \right] \,\middle|\, \left[\bigwedge_{j \in C-A} \Xi_j \neq \emptyset \right] \wedge \left[\bigwedge_{i \in A} \Xi_i = \emptyset \right] \right)
\end{aligned}
$$

(Here, the notation '$\Pi_{j \in C-A} S_j$' denotes the Cartesian product of the $|C - A|$ set variables S_j which are indexed by $j \in C - A$.) Consequently we get:

$$
\beta_\Sigma(S) = \beta_{\Sigma_1, ..., \Sigma_m}(S, ..., S) = \sum_{A \subseteq C} a_A \, q_A(S^{|C-A|})
$$

Collecting terms which have the same number of set variables and normalizing the resulting terms, we get $\beta_\Sigma(S) = \sum_{k \geq 0} a_k q_k(S^k)$ for suitable $a_k \geq 0$ and suitable probability measures q_k on $\mathcal{R}^k$. Now, set $S = \mathcal{R}$ and note that

$$a_k = \sum_{A:|A|=m-k} p\left(\left[\bigwedge_{j \in C-A} \Xi_j \neq \emptyset\right] \wedge \left[\bigwedge_{i \in A} \Xi_i = \emptyset\right]\right) = p(|\Sigma| = k)$$

where $p(|\Sigma| = k)$ exists by virtue of Proposition 12. Likewise,

$$a_k q_k(S^k) = \sum_{A:|A|=m-k} p\left(\left[\bigwedge_{i \in A} \Xi_j = \emptyset\right] \wedge \left[\bigwedge_{j \in C-A} \psi_j \in S_j\right]\right) = p(|\Sigma| = k, \Sigma \subseteq S)$$

so that $q_k(S^k) = p(\Sigma \subseteq S \mid |\Sigma| = k)$ as claimed. $\qquad \square$

Remark 3 *The q's in Theorem 17 are the so-called Janossy measures of the point process N_Σ of Σ defined by $N_\Sigma(S) = |\Sigma \cap S|$ (see Section 4.3.5 below).*

4.3.2 Existence of Set Derivatives

The following proposition establishes a class of set functions (in fact, belief measures) all of whose set derivatives always exist.

Proposition 18 *(a) Let f be an integrable function on $\mathcal{R}^k$ and define*

$$\Phi(S) = \int_{S^k} f(\xi_1, ..., \xi_k) d\lambda(\xi_1) \cdots d\lambda(\xi_k)$$

for any closed $S \subseteq \mathcal{R}$. Then all generalized Radon-Nikodým derivatives

$$\frac{\delta^k \Phi}{\delta \xi_1 \cdots \delta \xi_k}(T)$$

of Φ exist for all $k \geq 0$, all $\xi_1, ..., \xi_k \in \mathcal{R}$, and all closed $T \subseteq \mathcal{R}$. Moreover, we have:

$$\frac{\delta^k \Phi}{\delta \xi_1 \cdots \delta \xi_k}(\emptyset) = \sum_\pi f(\xi_{\pi 1}, ..., \xi_{\pi k})$$

$$\frac{\delta^k \Phi}{\delta \xi_1 \cdots \delta \xi_k}(\emptyset) = 0 \qquad (if \ j \neq k)$$

where $\xi_1, ..., \xi_k$ are distinct and where the sum is taken over all permutations π on the numbers $1, ..., k$. In particular, if f is completely symmetric in its arguments then

$$\frac{\delta^k \Phi}{\delta \xi_1 \cdots \delta \xi_k}(\emptyset) = k! \ f(\xi_1, ..., \xi_k)$$

(b) Let $a_0, ..., a_M \in \mathbb{R}$ and let f_j be integrable densities on $\mathcal{R}^j$ for $j = 1, ..., M$. Then all generalized Radon-Nikodým derivatives of the set function Φ defined by

$$\Phi(S) = a_0 + \sum_{k=1}^{M} a_k \int_{S^k} f_k(\xi_1, ..., \xi_k) d\lambda(\xi_1) \cdots d\lambda(\xi_k)$$

for all $S \subseteq \mathcal{R}$ exist..

Proof. Let $\xi = (\mathbf{z}, u)$ and abbreviate $H_0 \triangleq T$ and $H_1 \triangleq E_{\mathbf{z};i} \times u$ where i is understood. Then it follows that

$$\frac{\delta \Phi}{\delta \xi}(T) = \lim_{i \to \infty} \frac{\Phi(T \cup (E_{\mathbf{z};i} \times u)) - \Phi(T)}{\lambda(E_{\mathbf{z};i})}$$

$$= \sum_{\substack{j=1 \\ }}^{k} \sum_{\substack{i_1 + ... + i_k = j \\ 0 \leq i_1, ..., i_k \leq 1}} \lim_{i \to \infty} \frac{1}{\lambda(E_{\mathbf{z};i})} \int_{H_{i_1} \times ... \times H_{i_k}} f(\zeta_1, ..., \zeta_k) d\lambda(\zeta_1) \cdots d\lambda(\zeta_k)$$

Now note that, as in Example 6, any term in this sum which contains more than one copy of $H_1 = E_{\mathbf{z};i} \times u$ vanishes. Thus we get

$$\frac{\delta \Phi}{\delta \xi}(T) = \sum_{i_1 + ... + i_k = 1} \lim_{i \to \infty} \frac{1}{\lambda(E_{\mathbf{z};i})} \int_{H_{i_1} \times ... \times H_{i_k}} f(\zeta_1, ..., \zeta_k) d\lambda(\zeta_1) \cdots d\lambda(\zeta_k)$$

By the Lebesgue density theorem this in turn becomes:

$$\frac{\delta \Phi}{\delta \xi}(T) = \sum_{i_1 = 1}^{k} \int_{T^{k-1}} f(\zeta_1, ..., \zeta_{i_1 - 1}, [\xi_1]_{i_1}, \zeta_{i_1 + 1}, ..., \zeta_k) d\lambda(\zeta_1) \cdots [d\lambda(\zeta_{i_1})] \cdots d\lambda(\zeta_k)$$

where the notation '$[\xi]_i$' indicates that the variable ζ_i has been replaced by the constant ξ and where the notation '$[d\lambda(\zeta_i)]$' indicates that the differential $d\lambda(\zeta_i)$ is not present. Similar reasoning leads to:

$$\frac{\delta^2 \Phi}{\delta \xi_2 \delta \xi_1}(T) = \sum_{0 \leq i_1 \neq i_2 \leq k} \int_{T^{k-2}} f(\zeta_1, ..., [\xi_2]_{i_2}, ..., [\xi_1]_{i_1}, ..., \zeta_k)$$

$$\cdot d\lambda(\zeta_1) \cdots [d\lambda(\zeta_{i_2})] \cdots [d\lambda(\zeta_{i_1})] \cdots d\lambda(\zeta_k)$$

and, in general, to:

$$\frac{\delta^k \Phi}{\delta \xi_1 \cdots \delta \xi_k}(T) = \sum_{i_1 \neq ... \neq i_k} f([\xi_1]_{i_1}, ..., [\xi_k]_{i_k}) = \sum_{\pi} f(\xi_{\pi 1}, ..., \xi_{\pi k})$$

as desired.   □

Virtually all of the functions defined on finite sets that we will have occasion to use in this book will be sums of simple functions of the kind defined in

Proposition 18. Suppose that Σ is a finite random set. Then recall that, from Theorem 17, we know that β_Σ can be written in the form

$$\beta_\Sigma(S) = a_0 + a_1 q_1(S) + a_2 q_2(S \times S) + ... + a_m q_m(S^m)$$

where the a_k are nonnegative constants for all $k = 0, 1, ..., m$ and the q_k are probability measures on S^k for all $k = 1, ..., m$. We define:

Definition 17 *Let Σ be a finite random set. Then Σ is absolutely continuous if the probability measures q_k of Theorem 17 are absolutely continuous (with respect to hybrid measure) for every $k = 1, ..., m$.*

As one might expect, finite random subsets need not be absolutely continuous.

Proposition 19 *Suppose that Σ is an absolutely continuous finite random subset of $\mathcal{R}$. Then all set derivatives $\frac{\delta\beta_\Sigma}{\delta Z}(\emptyset)$ exist for all finite subsets Z of $\mathcal{R}$, are nonnegative, and vanish identically for large $|Z|$.*

Proof. By the definition of an absolutely continuous finite random subset (Definition 17) and the Radon-Nikodým theorem for hybrid spaces (see Section 4.1.2) it follows that

$$q_k(S^k) = \int_{S^k} f_k(\xi_1, ..., \xi_k) d\lambda(\xi_1) \cdots d\lambda(\xi_k)$$

for some density function f_k which, without loss of generality, can be assumed to be symmetric in all of its arguments (the reader should verify this). Define $\beta_{\Sigma,k}(S) \triangleq q_k(S^k)$. By Propositions 11 and 18 and Theorem 17 we know that the set derivatives of $\beta_{\Sigma,k}$ all exist and, moreover, that

$$\frac{\delta^k \beta_{\Sigma,k}}{\delta\xi_1 \cdots \delta\xi_k}(\emptyset) = k! \, f_k(\xi_1, ..., \xi_k)$$

$$\frac{\delta^j \beta_{\Sigma,k}}{\delta\xi_1 \cdots \delta\xi_j}(\emptyset) = 0 \quad (\text{if } j \neq k)$$

for all closed $T \subseteq \mathcal{R}$. The assertion immediately follows. $\square$

In the following proposition, $c(\mathcal{R})$ denotes as usual the space of closed subsets of $\mathcal{R}$.

Proposition 20 *Let Σ be an absolutely continuous finite random subset of $\mathcal{R}$. Then*

$$p(\Sigma \in O) = \int_O \frac{\delta\beta_\Sigma}{\delta Z}(\emptyset)\delta Z$$

for any measurable subset $O \in c(\mathcal{R})$, where the integral is a set integral (Definition 10). In particular, if we set $O = (O_{S^c})^c$ where $S \subseteq \mathcal{R}$ is closed then

$$\beta_\Sigma(S) = \int_S \frac{\delta\beta_\Sigma}{\delta Z}(\emptyset)\delta Z$$

Proof. Note that

$$p(\Sigma \in O) = \sum_{k=0}^{\infty} p(|\Sigma| = k, \Sigma \in O \cap c_k(\mathcal{R}))$$

Let $\Sigma = \Sigma_1 \cup ... \cup \Sigma_m$ where $card(\Sigma_j) = 1$ for $j = 1, ..., m$ (Corollary 16). We use the notation of the proof of Theorem 17, setting $\Sigma_j = \{\psi_j\} = \{\mathbf{Z}_j\} \times \Xi_j$ whenever $\Xi_j \neq \emptyset$. Then we can write

$$p(\Sigma \in O) = \sum_{k=0}^{\infty} p(|\Sigma| = k, (\psi_{i_1}, ..., \psi_{i_k}) \in \chi_k^{-1}(O \cap c_k(\mathcal{R})))$$

where the notation $(\psi_{i_1}, ..., \psi_{i_k}) \in \chi_k^{-1}(O \cap c_k(\mathcal{R}))$ is shorthand for the following statement: "There exist distinct $i_1, ..., i_k \in \{1, ..., m\}$ such that

$$\Xi_{i_1} \neq \emptyset, ..., \Xi_{i_k} \neq \emptyset \text{ for all } i_1, ..., i_k, \text{ and } (\psi_{i_1}, ..., \psi_{i_k}) \in \chi_k^{-1}(O \cap c_k(\mathcal{R}))"$$

However, from the proof of Theorem 17 we know that

$$\sum_{k=0}^{\infty} p(|\Sigma| = k, (\psi_{i_1}, ..., \psi_{i_k}) \in S_1 \times ... \times S_k) = p(|\Sigma| = k) q_k(S_1 \times ... \times S_k)$$

where q_k is the probability measure on $\mathcal{R}^k$ defined in the proof of Theorem 17. We therefore get

$$p(\Sigma \in O) = \sum_{k=0}^{\infty} p(|\Sigma| = k) q_k(\chi_k^{-1}(O \cap c_k(\mathcal{R})))$$

However, if we define

$$\tilde{q}_k(S_1 \times ... \times S_k) \triangleq \frac{1}{k!} \sum_{\pi} q(S_{\pi 1} \times ... \times S_{\pi k})$$

(summation taken over all permutations π on $1, ..., k$) then

$$\tilde{q}_k(\chi_k^{-1}(O \cap c_k(\mathcal{R}))) = q_k(\chi_k^{-1}(O \cap c_k(\mathcal{R})))$$

and so we may assume that q_k is symmetric: $q(S_{\pi 1} \times ... \times S_{\pi k}) = q(S_1 \times ... \times S_k)$ for all π. Since Σ is absolutely continuous there are symmetric densities f_k such that

$$q(S_1 \times ... \times S_k) = \int_{S_1 \times ... \times S_k} f_k(\xi_1, ..., \xi_k) d\lambda(\xi_1) \cdots d\lambda(\xi_k)$$

Setting $S_1 = ... = S_k = S$ and proceeding as in the proof of Proposition 19 we find that

$$\frac{\delta^k \beta_\Sigma}{\delta \xi_1 \cdots \delta \xi_k}(\emptyset) = p(|\Sigma| = k) \, k! \, f_k(\xi_1, ..., \xi_k)$$

and so

$$p(\Sigma \in O) = \sum_{k=0}^{\infty} \frac{1}{k!} \int_{X_k^{-1}(O \cap c_k(\mathcal{R}))} \frac{\delta^k \beta_\Sigma}{\delta \xi_1 \cdots \delta \xi_k}(\emptyset) d\lambda(\xi_1) \cdots d\lambda(\xi_k)$$

$$= \int_O \frac{\delta \beta_\Sigma}{\delta Z}(\emptyset) \delta Z$$

as claimed.   □

Corollary 21 *Let Σ be an absolutely continuous finite random subset of $\mathcal{R}$. Then the finite-set function f_Σ defined by*

$$f_\Sigma(Z) \triangleq \frac{\delta \beta_\Sigma}{\delta Z}(\emptyset)$$

for all finite subsets $Z \subseteq \mathcal{R}$ is a global density (Definition 11). That is, it is nonnegative and $\int f_\Sigma(Z) \delta Z = 1$.

Proof. Set $S = \mathcal{R}$ in Proposition 20 and use the fact that $\beta_\Sigma(\mathcal{R}) = 1$. The fact that f_Σ is nonnegative follows from Proposition 19.   □

Example 8 *The following is an example of a belief measure whose generalized Radon-Nikodým derivatives all exist but which is not the belief measure of a finite random subset (absolutely continuous or otherwise). Define the function $f(x) \triangleq e^{-x^2}$ on the real line, let $E_{a;\varepsilon} \triangleq [a - \frac{1}{2}\varepsilon, a + \frac{1}{2}\varepsilon]$, let A be a uniformly distributed random number on $[0,1]$, and let Σ be the random subset defined by*

$$\Sigma \triangleq \{a \in \mathbb{R} \mid A \le f(a)\} = \left[-\sqrt{-\ln A}, +\sqrt{-\ln A}\right]$$

Note that $\Sigma = \{0\}$ is the only finite-set value which Σ takes. It is easy to see that the plausibility measure of Σ is $\rho_\Sigma(S) = \sup_{x \in S} f(x)$ and so its belief measure is

$$\beta(S) = 1 - \sup_{x \in S^c} f(x)$$

We begin by computing $\frac{\delta\beta}{\delta 0}(T)$ assuming that $0 \notin T$. In this case we get:

$$\frac{\delta\beta}{\delta\{0\}}(T) = \lim_{\varepsilon \downarrow 0} \frac{\beta(T \cup E_{0;\varepsilon}) - \beta(T)}{\lambda(E_{0;\varepsilon})} = \lim_{\varepsilon \downarrow 0} \frac{\sup_{x \in T^c} f(x) - \sup_{x \in (T \cup E_{0;\varepsilon})^c} f(x)}{\lambda(E_{0;\varepsilon})}$$

$$= \lim_{\varepsilon \downarrow 0} \frac{1 - \sup_{x \in (T \cup E_{0;\varepsilon})^c} f(x)}{\lambda(E_{0;\varepsilon})} = \lim_{\varepsilon \downarrow 0} \frac{1 - e^{-\varepsilon^2/4}}{\varepsilon} = -\frac{d}{dx} e^{-x^2/4}\Big|_{x=0}$$

$$= \frac{1}{2}$$

On the other hand, assume $a \notin T$ and $0 \notin T$. Then

$$\frac{\delta\beta}{\delta\{a\}}(T) = \lim_{\varepsilon \downarrow 0} \frac{1-1}{\varepsilon} = 0$$

Finally, assume $0 < a \notin T$ *and* $0 \in T$. *Then*

$$\frac{\delta\beta}{\delta\{a\}}(T) = \lim_{\varepsilon\downarrow 0}\frac{\sup_{x\in T^c}f(x) - \max\{\sup_{x\in T^c}f(x),\ \sup_{x\in E^c_{a;\varepsilon}}f(x)\}}{\lambda(E_{a;\varepsilon})}$$

$$= \lim_{\varepsilon\downarrow 0}\frac{\sup_{x\in T^c}f(x) - \max\{\sup_{x\in T^c}f(x),\ e^{-(a-\varepsilon/2)^2}\}}{\lambda(E_{a;\varepsilon})}$$

$$= \lim_{\varepsilon\downarrow 0}\frac{\sup_{x\in T^c}f(x) - \sup_{x\in T^c}f(x)}{\lambda(E_{a;\varepsilon})} = 0$$

In summary, then,

$$\frac{\delta\beta}{\delta Z}(T) = \begin{cases} \frac{1}{2} & if \quad Z = \{0\} \\ 0 & if \quad otherwise \end{cases}$$

This has an intuitive interpretation, in that $f_\Sigma(Z)$ *vanishes for all* Z *except* $Z = \{0\}$, *the only finite-set value which* Σ *takes.*

4.3.3   Global Probability Density Functions

Having proved the existence of global density functions of absolutely continuous finite random subsets, we establish their major properties and extend the concept to the *joint* global density function of a list $\Sigma_1, ..., \Sigma_m$ of i.i.d. finite random subsets.

Definition 18 *Let* Σ *be an absolutely continuous random finite subset of* $\mathcal{R}$. *Then (a) the global probability density function (global density) of* Σ *is defined by*

$$f_\Sigma(Z) \triangleq \frac{\delta\beta_\Sigma}{\delta Z}(\emptyset)$$

for all finite subsets $Z \subseteq \mathcal{R}$. *(b) The cardinality distribution of* Σ, *denoted by* $f_\Sigma(k)$, *is defined as*

$$f_\Sigma(k) \triangleq p(|\Sigma| = k) = \int_{|Z|=k} f_\Sigma(Z)\delta Z$$

$$= \frac{1}{k!}\int f_\Sigma(\{\xi_1, ..., \xi_k\})d\lambda(\xi_1)\cdots d\lambda(\xi_k)$$

for all $k = 0, 1, 2, ...$

In this notation Proposition 20 yields the following useful relationships:

$$\beta_\Sigma(S) = \int_S f_\Sigma(Z)\delta Z, \qquad 1 = \int f_\Sigma(Z)\delta Z$$

$$1 = \sum_{k=0}^{\infty} f_\Sigma(k), \qquad p(\Sigma \in O) = \int_O f_\Sigma(Z)\delta Z$$

for any closed $S \subseteq \mathcal{R}$ and any measurable subset $O \in \sigma(c(\mathcal{R}))$. Also, note that the fact that $p(|\Sigma| = k) = \int_{|Z|=k} f_\Sigma(Z)\delta Z$ follows from Theorem 17 and Proposition 20.

Just as the value $f_{\mathbf{Z}}(\mathbf{z})$ of the density function $f_{\mathbf{Z}}$ of a random vector $\mathbf{Z} \in \mathbb{R}^n$ specifies the degree to which $\mathbf{Z}$ is expected to take the value $\mathbf{Z} = \mathbf{z}$ as an instantiation, so the value $f_{\Sigma}(Z)$ of the global density specifies the likelihood with which the finite random set Σ takes the finite set Z as a specific realization.

Remark 4 *(Global densities and units of measurement). Global density functions have one peculiarity which sets them apart from conventional densities: their behavior with respect to units of measurement. Let $f_{\mathbf{Z}}$ be the density of the random vector $\mathbf{Z} \in \mathbb{R}^n$. Then for every $\mathbf{z} \in \mathbb{R}^n$ it follows that $f_{\mathbf{Z}}(\mathbf{z})$ has the units of a density (e.g., 1/meter if the coordinates of $\mathbf{z}$ are measured in meters). In like manner, let Σ be an absolutely continuous finite random subset of $\mathbb{R}^n$. Then the units of the global density $f_{\Sigma}(Z)$ vary with the cardinality of Z. For example: Suppose that the units of $\mathbb{R}^n$ are meters and that $|Z| = k$. Then the units of $f_{\Sigma}(Z)$ are $1/\text{meter}^k$.) Because of this behavior great care should be taken with respect to mathematical operations involving global densities. This is particularly true for operations involving set integrals. For example, given the conventional density $f_{\mathbf{Z}}(\mathbf{z})$ the expression $\int f_{\mathbf{Z}}(\mathbf{z})^2 d\lambda(\mathbf{z})$ is well-defined: If the units of $\mathbf{z}$ are meters, then this integral is well-defined and its units are 1/meter. However, the same is not true of the analogous set integral:*

$$\int f_{\Sigma}(Z)^2 \delta Z = f_{\Sigma}(\emptyset)^2 + \int f_{\Sigma}(\{\mathbf{z}\})^2 d\lambda(\mathbf{z}) + \frac{1}{2} \int f_{\Sigma}(\{\mathbf{z}_1, \mathbf{z}_2\})^2 d\lambda(\mathbf{z}_1) d\lambda(\mathbf{z}_2) + \dots$$

If the $\mathbf{z}$'s are measured in meters then the first term on the right is unitless, the second term has units of 1/meter, the second term has units of $1/\text{meter}^2$, and so on. Thus the integral is not well-defined because of the incongruity of the units in its individual terms. On the other hand, set integrals of the form

$$\int |f_{\Sigma}(Z) - f_{\Lambda}(Z)| \, \delta Z, \qquad \int f_{\Sigma}(Z) \, \ln\left(\frac{f_{\Sigma}(Z)}{f_{\Lambda}(Z)}\right) \delta Z$$

are well-defined.

Definition 18 can be generalized to the case of joint distributions of finite random subsets in the obvious fashion.

Definition 19 *Let $\Phi : \mathcal{P}(\mathcal{R}) \times \dots \times \mathcal{P}(\mathcal{R}) \to \mathbb{R}^m$ be a multi-variable set function on $\mathcal{R}$. Then, the partial generalized Radon-Nikodým derivative of Φ with respect to the j'th argument of Φ is:*

$$\frac{\delta \Phi}{\delta_j \xi}(S_1, \dots, S_{j-1}, T, S_{j+1}, \dots, S_m) \triangleq \frac{\delta \Phi_j}{\delta \xi}(T)$$

where the set function Φ is defined by

$$\Phi_j(T) \triangleq \Phi(S_1, \dots, S_{j-1}, T, S_{j+1}, \dots, S_m)$$

More generally, the partial set derivative of Φ with respect to its j'th argument is:

$$\frac{\delta \Phi}{\delta_j Z}(S_1, \dots, S_{j-1}, T, S_{j+1}, \dots, S_m) \triangleq \frac{\delta \Phi_j}{\delta Z}(T)$$

We then have the obvious definition:

Definition 20 *Let* $\Sigma_1, ..., \Sigma_m$ *be absolutely continuous finite random subsets of* $\mathcal{R}$. *Their joint belief measure is*

$$\beta_{\Sigma_1,...,\Sigma_m}(S_1, ..., S_m) = p(\Sigma_1 \subseteq S_1, ..., \Sigma_m \subseteq S_m)$$

The joint global density of these random finite sets is the multi-variable set function

$$f_{\Sigma_1,...,\Sigma_m}(Z_1, ..., Z_m) \triangleq \frac{\delta^m \beta_{\Sigma_1,...,\Sigma_m}}{\delta_1 Z_1 \cdots \delta_m Z_m}(\emptyset, ..., \emptyset)$$

Proposition 20 generalizes in the obvious way:

$$\beta_{\Sigma_1,...,\Sigma_m}(S_1, ..., S_m) = \int_{S_1} \cdots \int_{S_m} f_{\Sigma_1,...,\Sigma_m}(Z_1, ..., Z_m)\delta Z_1 \cdots \delta Z_m$$

In practice, however, in the sequel we will usually need to consider only the joint global densities of finite random sets which are statistically independent. The following proposition derives the obvious formula for the joint densities of such random sets:

Proposition 22 *Let* $\Sigma_1, ..., \Sigma_m$ *be statistically independent, absolutely continuous finite random subsets of* $\mathcal{R}$. *Then their joint density exists and is*

$$f_{\Sigma_1,...,\Sigma_m}(Z_1, ..., Z_m) = f_{\Sigma_1}(Z_1) \cdots f_{\Sigma_m}(Z_m)$$

for all finite $Z_1, ..., Z_m \subseteq \mathcal{R}$.

Proof. Follows immediately from the fact that if $\Sigma_1, ..., \Sigma_m$ are statistically independent then

$$\begin{aligned} \beta_{\Sigma_1,...,\Sigma_m}(S_1, ..., S_m) &= p(\Sigma_1 \subseteq S_1, ..., \Sigma_m \subseteq S_m) \\ &= p(\Sigma_1 \subseteq S_1) \cdots p(\Sigma_m \subseteq S_m) = \beta_{\Sigma_1}(S_1) \cdots \beta_{\Sigma_m}(S_m) \end{aligned}$$

which concludes the proof.   □

We conclude this section with two propositions which compute global densities for two special types of finite random subsets.

Proposition 23 *Let* Σ *be an absolutely continuous finite random subset of* $\mathcal{R}$ *and let* T *be a fixed nonempty open subset of* $\mathcal{R}$. *Then* $\Sigma \cap T$ *is also an absolutely continuous finite random subset of* $\mathcal{R}$ *and its set derivatives are given by:*

$$f_{\Sigma \cap T}(Z) = \frac{\delta \beta_\Sigma}{\delta Z}(T^c) \qquad (if \ Z \subseteq T)$$

$$f_{\Sigma \cap T}(Z) = 0 \qquad (otherwise)$$

Proof. First note that $\beta_{\Sigma \cap T}(S) = \beta_{\Sigma}(S \cup T^c)$ for all $S \subseteq \mathcal{R}$. For,

$$
\begin{aligned}
\beta_{\Sigma \cap T}(S) &= p(\Sigma \cap T \subseteq S) = p([(\Sigma \cap S) \cup (\Sigma \cap S^c)] \cap T \subseteq S) \\
&= p([\Sigma \cap S \cap T] \cup [\Sigma \cap S^c \cap T] \subseteq S) \\
&= p(\Sigma \cap S \cap T \subseteq S, \ \Sigma \cap S^c \cap T \subseteq S) = p(\Sigma \cap S^c \cap T \subseteq S) \\
&= p(\Sigma \cap S^c \cap T = \emptyset) = p(\Sigma \subseteq S \cup T^c) = \beta_{\Sigma}(S \cup T^c)
\end{aligned}
$$

as desired. It is then easily verified that:

$$
\frac{\delta^k \beta_{\Sigma \cap T}}{\delta \xi_1 \cdots \delta \xi_k}(S) = 0
$$

if any ξ_k is in T^c and, otherwise,

$$
\frac{\delta^k \beta_{\Sigma \cap T}}{\delta \xi_1 \cdots \delta \xi_k}(S) = \frac{\delta^k \beta_{\Sigma}}{\delta \xi_1 \cdots \delta \xi_k}(S \cup T^c)
$$

from which fact the proposition easily follows.   □

For the following proposition we will need a definition. Let Σ be an absolutely continuous finite random subset of $\mathcal{R}$ and let T be a fixed closed and continuously infinite subset of $\mathcal{R}$. Define the belief measure $\beta_{\Sigma}(S|T)$ by:

$$
\beta_{\Sigma}(S|T) \triangleq \frac{\beta_{\Sigma}(S \cap T)}{\beta_{\Sigma}(T)}
$$

for all closed $S \subseteq \mathcal{R}$. (This belief measure is actually a true conditional probability with respect to the hit-or-miss topology: Set $p_{\Sigma}(O) \triangleq p(\Sigma \in O)$ and note that

$$
p_{\Sigma}(O^c_{S^c}|O^c_{T^c}) = \frac{p_{\Sigma}(O^c_{S^c} \cap O^c_{T^c})}{p_{\Sigma}(O^c_{T^c})} = \frac{p(\Sigma \subseteq S, \Sigma \subseteq T)}{p(\Sigma \subseteq T)} = \frac{\beta_{\Sigma}(S \cap T)}{\beta_{\Sigma}(T)} = \beta_{\Sigma}(S|T)
$$

In the Dempster-Shafer literature, $\beta_{\Sigma}(S|T)$ is called the *geometric conditioning* of β_{Σ} with respect to the set T; see [18, pp. 125-129, 177] or [27].) Then:

Proposition 24 *All iterated generalized Radon-Nikodým derivatives of $\beta_{\Sigma}(S|T)$ exist and the corresponding global density function is given by:*

$$
\begin{aligned}
f_{\Sigma}(Z|T) &= \frac{f_{\Sigma}(Z)}{\beta_{\Sigma}(T)} \qquad \textit{(if } Z \subseteq T) \\
f_{\Sigma}(Z|T) &= 0 \qquad\qquad \textit{(otherwise)}
\end{aligned}
$$

Proof. Easy and left to the reader.   □

We now establish some important explicit formulas for global densities.

Proposition 25 *Let Σ be an absolutely continuous finite random subset of $\mathcal{R}$ such that $card(\Sigma) = 1$, i.e. Σ contains no more than one element. Then $f_\Sigma(Z) = 0$ if $|Z| \geq 2$ and*

$$f_\Sigma(\emptyset) = 1 - q_0, \qquad\qquad f_\Sigma(\{\xi\}) = q_0 f(\xi)$$

where $q_0 = p(|\Sigma| \neq \emptyset)$ and f is the probability density of the probability measure $q_1(S) = p(\Sigma \subseteq S | \Sigma \neq \emptyset)$.

Proof. By definition

$$\beta_\Sigma(S) = p(\Sigma \subseteq S) = p(\Sigma = \emptyset) + p(\Sigma \neq \emptyset, \Sigma \subseteq S) = 1 - q_0 + q_0 q_1(S)$$

Note that q_1 is a countably additive probability measure since p is countably additive and since $|\Sigma| = 1$ whenever $\Sigma \neq \emptyset$. The assertion then follows from the reasoning used in Example 4. $\qquad\qquad\square$

Proposition 26 *(a) Let $\Sigma = \Sigma_1 \cup \ldots \cup \Sigma_m$ where Σ_j are absolutely continuous and independent finite random subsets of $\mathcal{R}$ such that $card(\Sigma_j) = 1$ for all $j = 1, \ldots, m$. Assume further that the belief measures of the Σ_j are*

$$\beta_j(S) \triangleq \beta_{\Sigma_j}(S) = 1 - q_j + q_j \, q_{f_j}(S)$$

where f_j is the probability density corresponding to the probability measure q_{f_j}. Then $f_\Sigma(Z) = f_\Sigma(|Z|) = 0$ (if $|Z| > m$) and

$$f_\Sigma(\emptyset) = f_\Sigma(0) = (1 - q_1) \cdots (1 - q_m)$$

$$f_\Sigma(\{\xi_1, \ldots, \xi_k\}) = \sum_{1 \leq i_1 \neq \ldots \neq i_k \leq m} \frac{(1 - q_1) \cdots (1 - q_m)}{(1 - q_{i_1}) \cdots (1 - q_{i_k})} q_{i_1} \cdots q_{i_k} f_{i_1}(\xi_1) \cdots f_{i_k}(\xi_k)$$

$$f_\Sigma(k) = \frac{1}{k!} \sum_{1 \leq i_1 \neq \ldots \neq i_k \leq m} \frac{(1 - q_1) \cdots (1 - q_m)}{(1 - q_{i_1}) \cdots (1 - q_{i_k})} q_{i_1} \cdots q_{i_k} \ (if \ k \leq m)$$

for all distinct $\xi_1, \ldots, \xi_k \in \mathcal{R}$ and $k = 1, \ldots, m$ and where the summations are taken over all distinct $i_1, \ldots, i_k$ such that $1 \leq i_1, \ldots, i_k \leq m$. (b) Assume in particular that $q_1 = \ldots = q_m = q$. Then

$$f_\Sigma(\emptyset) = f_\Sigma(0) = (1 - q)^m$$

$$f_\Sigma(\{\xi_1, \ldots, \xi_k\}) = q^k (1 - q)^{m-k} \sum_{1 \leq i_1 \neq \ldots \neq i_k \leq m} f_{i_1}(\xi_1) \cdots f_{i_k}(\xi_k)$$

$$f_\Sigma(k) = C_{m,k} \, q^k (1 - q)^{m-k} \ (if \ k \leq m)$$

where as usual $C_{m,k} = \frac{m!}{k!(m-k)!}$. (c) Assume in addition that $f_1 = \ldots = f_m$. Then

$$f_\Sigma(\emptyset) = f_\Sigma(0) = (1 - q)^m$$

$$f_\Sigma(\{\xi_1, \ldots, \xi_k\}) = k! C_{m,k} \, q^k (1 - q)^{m-k} f(\xi_1) \cdots f(\xi_k)$$

$$f_\Sigma(k) = C_{m,k} \, q^k (1 - q)^{m-k} \qquad (if \ k \leq m)$$

Proof. (a) From Proposition 18 we know that $\beta_j(\emptyset) = 1 - q_j$ and $\frac{\delta\beta_j}{\delta\xi_i}(S) = q_j f_j(\xi_i)$ for any closed $S \subseteq \mathcal{R}$. Consequently, $\frac{\delta^2\beta_j}{\delta\zeta\delta\xi}(S) = 0$ for any $j = 1, ...m$, for any closed $S \subseteq \mathcal{R}$, and for any $\zeta, \xi \in \mathcal{R}$. From the product rule for generalized Radon-Nikodým derivatives (Proposition 8) it is easy to see that

$$\frac{\delta\beta_\Sigma}{\delta\xi_1}(S) = \sum_{i_1=1}^{m} \beta_1(S) \cdots \frac{\delta\beta_{i_1}}{\delta\xi_1}(S) \cdots \beta_m(S)$$

and therefore, more generally, that

$$\frac{\delta^k\beta_\Sigma}{\delta\xi_k \cdots \delta\xi_1}(S) = \sum_{1 \leq i_k \neq ... \neq i_1 \leq m} \beta_1(S) \cdots \frac{\delta\beta_{i_1}}{\delta\xi_1}(S) \cdots \frac{\delta\beta_{i_k}}{\delta\xi_k}(S) \cdots \beta_m(S)$$

and hence that

$$f_\Sigma(\{\xi_1, ..., \xi_k\}) = \sum_{1 \leq i_k \neq ... \neq i_1 \leq m} \beta_1(\emptyset) \cdots \frac{\delta\beta_{i_1}}{\delta\xi_1}(\emptyset) \cdots \frac{\delta\beta_{i_k}}{\delta\xi_k}(\emptyset) \cdots \beta_m(\emptyset)$$

The assertion immediately follows from the fact that $\beta_j(\emptyset) = 1 - q_j$ and $\frac{\delta\beta_j}{\delta\xi}(\emptyset) = q_j f_j(\xi)$. Thus

$$
\begin{aligned}
f_\Sigma(k) &= \frac{1}{k!} \int f_\Sigma(\{\xi_1, ..., \xi_k\}) d\lambda(\xi_1) \cdots d\lambda(\xi_k) \\
&= \frac{1}{k!} \sum_{1 \leq i_1 \neq ... \neq i_k \leq m} \frac{(1 - q_1) \cdots (1 - q_m)}{(1 - q_{i_1}) \cdots (1 - q_{i_k})} q_{i_1} \cdots q_{i_k} \\
&\qquad \cdot \int f_{i_1}(\xi_1) \cdots f_{i_k}(\xi_k) d\lambda(\xi_1) \cdots d\lambda(\xi_k) \\
&= \frac{1}{k!} \sum_{1 \leq i_1 \neq ... \neq i_k \leq m} \frac{(1 - q_1) \cdots (1 - q_m)}{(1 - q_{i_1}) \cdots (1 - q_{i_k})} q_{i_1} \cdots q_{i_k}
\end{aligned}
$$

as desired. (b) and (c) follow directly from (a). $\square$

Remark 5 *In particular let $f(\xi|\zeta)$ be a family of conventional densities on $\mathcal{R}$ parametrized by ζ and let $q(\zeta)$ be a family of constants parametrized by ζ. Let $X = \{\zeta_1, ..., \zeta_m\}$ be a set of distinct parameters and in Proposition 26 set $q_j = q(\zeta_j)$ and $f_j(\xi) = f(\xi|\zeta_j)$ for $j = 1, ..., m$. Then the formulas for f_Σ in part (a) of that proposition becomes*

$$f_\Sigma(\emptyset|\{\zeta_1, ..., \zeta_m\}) = (1 - q(\zeta_1)) \cdots (1 - q(\zeta_m))$$

$$f_\Sigma(\{\xi_1, ..., \xi_k\}|\{\zeta_1, ..., \zeta_m\})$$
$$= \sum_{1 \leq i \neq, ... \neq i_k \leq m} a_{i_1, ..., i_k}(\zeta_1, ..., \zeta_m) f(\xi_1|\zeta_{i_1}) \cdots f(\xi_k|\zeta_{i_k})$$

$$a_{i_1, ..., i_k}(\zeta_1, ..., \zeta_m) = \frac{(1 - q(\zeta_1)) \cdots (1 - q(\zeta_m))}{(1 - q(\zeta_{i_1})) \cdots (1 - q(\zeta_{i_k}))} q(\zeta_{i_1}) \cdots q(\zeta_{i_k})$$

For completeness, we define $f_{\Sigma}(\emptyset|\emptyset) \triangleq 1$ and $f_{\Sigma}(Z|\emptyset) \triangleq 0$ for $Z \neq \emptyset$, so that $\int f_{\Sigma}(Z|\emptyset)\delta Z = 1$. In other words, the global density f_{Σ} has the form $f_{\Sigma}(Z|X)$, which is to say a family of global densities parametrized by X, which we call a set parameter.

These observations lead to the following definition:

Definition 21 *The belief measure β_{Σ} of a finite random set belongs to a set-parametrized family if it functionally depends on the specific values of a finite-set parameter $X \subseteq \mathcal{R}$. In this case it is denoted $\beta_{\Sigma}(S|X)$ where X is called the set parameter of the family. The global density and the cardinality distribution both inherit the parametrization, in which case they are denoted $f_{\Sigma}(Z|X)$ and $f_{\Sigma}(k|X)$, respectively. (b) Let $f(\xi|\zeta)$ be a family of conventional densities and $0 \leq q(\zeta) \leq 1$ a family of constants, both parametrized by ζ. Then the family of global densities $f(Z|X)$ defined in Remark 5 is called the parametrized family induced by f and q.*

Remark 6 *As we shall see in Section 5.2.3 of Chapter 5, the naive perspective on set parameters reflected in Definition 21 does not take account of certain technical subtleties which are necessary to, for example, prove the statistical consistency of the "global" version of the maximum likelihood estimator.*

Example 9 *Given the belief measures*

$$\beta_{\Sigma}(S|\{\zeta\}) = 1 - q + q\, p_f(S|\zeta)$$
$$\beta_{\Sigma}(S|\{\zeta_1, \zeta_2\}) = [1 - q + q\, p_f(S|\zeta_1)]\,[1 - q + q\, p_f(S|\zeta_2)]$$

we get the following parametrized family:

$$f(\emptyset|\emptyset) = 1$$
$$f(\emptyset|\{\zeta\}) = 1 - q$$
$$f(\{\xi\}|\{\zeta\}) = q\, f(\xi|\zeta)$$
$$f(\emptyset|\{\zeta_1, \zeta_2\}) = (1 - q)^2$$
$$f(\{\xi\}|\{\zeta_1, \zeta_2\}) = q(1 - q)\, f(\xi|\zeta_1) + q(1 - q)\, f(\xi|\zeta_2)$$
$$f(\{\xi_1, \xi_2\}|\{\zeta_1, \zeta_2\}) = q^2\, f(\xi_1|\zeta_1)\, f(\xi_2|\zeta_2) + q^2\, f(\xi_2|\zeta_2)\, f(\xi_1|\zeta_2).$$

This concludes the example.

4.3.4   Global Covering Densities

The *probability hypothesis surface* (PHS) is an evidential estimation concept proposed, by Michael Stein of Oasis Research and his associates[29], [30], for use in applications such as force estimation. (Force estimation is the process of determining whether or not a grouping of platforms constitutes a tactically meaningful military formation such as a division or a brigade. See [28] for more details.) Suppose that we are looking for ground targets located on some terrain. Then the PHS for this data fusion problem is the two-dimensional surface corresponding to a (non-probability) density function $D(\mathbf{x})$ which has the following properties. First, the locations of the targets correspond to the maxima of the surface; and second, the integral $\int D(\mathbf{x})d\lambda(\mathbf{x})$ is the expected value of the number of targets.

If our domain of application is not a continuous two-dimensional terrain but rather a finite grid U of two-dimensional terrain cells, then probability hypothesis surfaces have a simple mathematical formulation in terms of one-point covering functions (see [10] or [8]). Let Γ be the randomly varying set of cells which are believed to contain a target. Then the one-point covering function $\mu_\Sigma(u) \triangleq p(u \in \Gamma)$ tells us how likely it is that any given terrain cell u contains one of the targets. Furthermore, it is easy to show that

$$\sum_{v \in U} \mu_\Gamma(v) = \sum_{v \in U} \sum_{T \subseteq U} p(v \in T, \Gamma = T) = \sum_{T \subseteq U} \left(\sum_{v \in U} \mu_T(v) \right) m_\Gamma(T)$$

$$= \sum_{T \subseteq U} |T| \, m_\Gamma(T) = \mathrm{E}[\,|\Gamma|\,]$$

where $m_\Gamma(T) = p(\Gamma = T)$, where $\mu_T(v)$ is the indicator function of the subset T, and where $\mathrm{E}[\,|\Gamma|\,]$ denotes the expected number of target-containing cells. (Note: In the fuzzy logic literature, the sum $\sum_{v \in U} \mu_\Sigma(v)$ is also called the "sigma-count" of the fuzzy membership function μ_Σ.)

This formulation of PHS's can be generalized to more general information fusion problems via the concept of a *global covering density* as follows. Because of space limitations, we will be able only to state the main results without proof. Let Σ be a finite random subset of $\mathcal{R} = \mathbb{R}^n \times W$ where $n > 0$. Then the *global covering density* of Σ, denoted by $D_\Sigma(Z)$ for any finite subset $Z \subseteq \mathcal{R}$, is defined by:

$$D_\Sigma(Z) \triangleq \int_{Y \supseteq Z} f_\Sigma(Y)\delta Y = \int f_\Sigma(Z \cup W)\delta W$$

$$= \sum_{k=0}^{\infty} \frac{1}{k!} \int f_\Sigma(Z \cup \{\xi_1, ..., \xi_k\})d\lambda(\xi_1) \cdots d\lambda(\xi_k)$$

Notice that D_Σ is well-defined. If $|Z| = 1$ then we write $D_\Sigma(\xi) \triangleq D_\Sigma(\{\xi\})$ and call D_Σ the *one-point covering density* or the *probability hypothesis density*. The following equations can then be verified to be true:

$$D_\Sigma(Z) = \frac{\delta\beta_\Sigma}{\delta Z}(\mathcal{R})$$

for all $Z \subseteq \mathcal{R}$, and

$$\mathrm{E}[\,|\Sigma \cap S|\,] = \int_S D_\Sigma(\{\xi\})d\lambda(\xi)$$

for all measurable S. (This last equation shows that $D_\Sigma(\{\xi\})$ is the density function corresponding to the mean-value measure $n_\Sigma(S) \triangleq \mathrm{E}[\,|\Sigma \cap S|\,]$ of the point process $N_\Sigma(S) \triangleq |\Sigma \cap S|$ of the finite random subset Σ.) See also [16].

4.3.5   Relationship to Other Approaches

Many other approaches to constructing an integral and/or differential calculus for set functions have been proposed. These include the Choquet integral [12, p. 42], [7], the Aumann-Debreu integral [3], [6], the Huber-Strassen derivative [15], [24, Lemma 2, p. 20], the Graf derivative and integral [13], and Goutsias' derivative [11, pp. 102, 110]. Goutsias' derivative is a special case of the generalization of the Radon-Nikodým derivative proposed in this book. The Choquet integral and our "set integral" are related as in the following proposition:

Proposition 27 *Let h be a bounded integrable real-valued function on $\mathbb{R}^n$ and let β_Σ be the belief measure of the absolutely continuous finite random subset Σ. Then*

$$\int h(z)d\beta_\Sigma(z) = \int \min_h(Z)f_\Sigma(Z)\delta Z$$

$$\int h(z)d\rho_\Sigma(z) = \int \max_h(Z)f_\Sigma(Z)\delta Z$$

where $\min_h(Z) \triangleq \min_{z \in Z} h(z)$ and where $\max_h(Z) \triangleq \max_{z \in Z} h(z)$ and $\rho_\Sigma(S) = 1 - \beta_\Sigma(S^c)$ is the plausibility measure of Σ.

Proposition 27 follows from Schmeidler's representation theorem [12] or from the fact that $\int \min_h(Z)f_\Sigma(Z)\delta Z$ is the expected value of the random number $\min_h(\Sigma)$ and the fact that $\int h(z)d\beta_\Sigma(z)$ also is its expected value [21, p. 19].

It should also be mentioned that set integrals and hybrid integrals occur regularly in the statistical theory of gases and liquids (see [14, p. 234, Equ. 37.4], [14, p. 266, Equ. 40.28]) though they are not explicitly identified in that context. This should not be surprising since there are many formal mathematical similarities between multitarget systems with multiple target types, on the one hand, and ensembles of molecules belonging to multiple molecular species.

Finally, we conclude by summarizing the basic connections between the theory of finite random subsets and point process theory. Given any finite subset $Z \subseteq \mathbb{R}^n$ define the set function $N_Z(S)$ by $N_Z(S) \triangleq |Z \cap S|$ for all measurable $S \subseteq \mathcal{R}$. Then N_Z is obviously a finite, countably additive measure. If Σ is a finite random subset of $\mathcal{R}$, define the measure-valued random variable $N_\Sigma(S)$ by $N_\Sigma(S)(\omega) \triangleq |\Sigma(\omega) \cap S|$ for all $\omega \in \Omega$. We call the random measure N_Σ the *point process* of Σ. The correspondence $Z \mapsto N_Z$ assigns a simple measure to every finite subset Z. Conversely, let θ be such a measure. Define the subset $Z_\theta \subseteq \mathcal{R}$ by $Z_\theta \triangleq \{\xi \in \mathcal{R} | \theta(\{\xi\}) > 0\}$. Then Z_θ is a finite subset. The correspondences $Z \mapsto N_Z$ and $\mu \mapsto Z_\mu$ are inverse to each other. What we have called a "global probability density" $f_\Sigma(Z)$ of a finite random set Σ is essentially the same thing as the so-called "Janossy densities" $j_{\Theta,k}$ of the corresponding point process $\Theta = N_\Sigma$, in the sense that

$$j_{\Theta,k}(z_1, ..., z_k) = f_\Sigma(\{z_1, ..., z_k\})$$

for all distinct $z_1, ..., z_k \in \mathbb{R}^n$ (see [5, pp. 122-123]).

Bibliography

[1] Alspach, D. L. (1975) A Gaussian sum Approach to the multi-target identification-tracking problem, *Automatica*, **11** , 285-296.

[2] Aubin, J. P. and Frankowska, H. (1990) *Set-Valued Analysis*, Birkhäuser.

[3] Aumann, J. (1965) Integrals of set valued functions, *Journal of Mathematical Analysis and Applications*, **12**, 1-22.

[4] Berger, J. O. (1985) *Statistical Decision Theory and Bayesian Analysis*, Second Edition, Springer-Verlag, New York.

[5] Daley, D. J. and Vere-Jones, D. (1988) *An Introduction to the Theory of Point Processes*, Springer-Verlag, New York.

[6] Debreu, G. (1967) Integration of correspondences, in L.M. Le Cam and J. Neyman (eds.), *Proceedings of the Fifth Berkeley Symposium on Mathematical Statistics and Probability*, Vol. II Part 1: Contributions to Probability Theory, June 21-July 18 1965, University of California Press.

[7] Denneberg, D. (1994) *Non-Additive Measure and Integral*, Kluwer Academic Publishers, Dordrecht.

[8] Goodman, I. R. (1982) Fuzzy sets as equivalence classes of random sets, in R. Yager (ed.), *Fuzzy Sets and Possibility Theory*, Permagon, pp. 327-343.

[9] Goodman, I. R. (1979) A general model for the multiple target correlation and tracking problem, *Proceedings of the 18th IEEE Conference on Decision and Control*, Fort Lauderdale FL, 383-388.

[10] Goodman, I. R. and Nguyen, H. T. (1985) *Uncertainty Models for Knowledge Based Systems*, North-Holland, Amsterdam.

[11] Goutsias, J. (1992) Morphological analysis of discrete random shapes, *Journal of Mathematical Imaging and Vision*, **2**, 193-215.

[12] Grabisch, M., Nguyen, H. T., and Walker, E.A. (1995) *Fundamentals of Uncertainty Calculi With Applications to Fuzzy Inference*, Kluwer Academic Publishers, Dordrecht.

[13] Graf, S. (1980) A Radon-Nikodým theorem for capacities, *Journal für Reine und Angewandte Mathematik*, **320**, 192-214.

[14] Hill, T. L. (1987) *Statistical Mechanics: Principles and Practical Applications*, Dover Publications.

[15] Huber, P. J. and Strassen, V. (1973) Minimax tests and the Neyman-Pearson lemma for capacities, *Annals of Statistics*, **1** (2), 251-263.

[16] Karr, A. F. (1991) *Point Processes and Their Statistical Inference*, Second Edition, Marcel Dekker.

[17] Kelley, J. L. (1955) *General Topology*, D. Van Nostrand Co.

[18] Kruse, R., Schwecke, E., and Heinsohn, J. (1991) *Uncertainty and Vagueness in Knowlege Based Systems: Numerical Methods*, Springer-Verlag, New York.

[19] Lyshenko, N. N. (1983) Statistics of Random Compact Sets in Euclidean Space, *Journal of Soviet Mathematics*, **21**, Plenum Publishing, 76-92.

[20] Matheron, G. (1975) *Random Sets and Integral Geometry*, J. Wiley, New York.

[21] Molchanov, I. S. (1993) *Limit Theorems for Unions of Random Closed Sets*, Springer-Verlag Lecture Notes in Mathematics No. 1561, New York.

[22] Mori, S., Chong, C.-Y., Tse, E., and Wishner, R. P. (1984) Multitarget Multisensor Tracking Problems, Part I: A General Solution and a Unified View on Bayesian Approaches (Revised Version), Technical Report TR-1048-01, Advanced Information and Decision Systems, Inc., Mountain View CA, Aug. 1984. My thanks to Dr. Mori for making this report available to me (Dr. Shozo Mori, personal communication, Feb. 28 1995).

[23] Mori, S., Chong, C.-Y., Tse, E., and Wishner, R. P. (1986) Tracking and Classifying Multiple Targets Without A Priori Identification, *IEEE Transactions on Automatic Control*, AC-31 (5), 401-409.

[24] Poor, H. V. (1982) The Rate-Distortion Function on Classes of Sources Deterined by Spectral Capacities, *IEEE Transactions on Information Theory*, **IT-28** (1), 19-26.

[25] Radstrom, H. R. (1952) An embedding theorem for spaces of convex sets, *Proceedings of the American Mathematical Society*, **3**, 165-169.

[26] Shilov, G. E and Gurevich, B. L. (1966) *Integral, Measure and Derivative: A Unified Approach*, Prentice-Hall, Englewood Cliffs, New Jersey.

[27] Smets, P. (1991) About updating, in B. D'Ambrosio, Ph. Smets, and P.P. Bonissone (eds.), *Uncertainty in Artificial Intelligence*, Morgan Kaufmann, 378-385.

[28] Stein, M. C. (1995) Recursive Bayesian fusion for force estimation, *Proc. 8th Nat'l Symp. on Sensor Fusion*, Vol. I (Unclassified), Mar. 15-17 1995, Dallas TX, 47-66.

[29] Stein, M. C., and Winter, C. L. (1993) An Additive Theory of Probabilistic Evidence Accrual, Los Alamos National Laboratories Technical Report, LA-UR-93-3336.

[30] Stein, M., and Tenney, R. R. (undated) What's the Difference Between PHS and MHT?, Oasis Research and Alphatech working paper. Also Shoenfeld, P. (undated) MHT and WEA, SAIC Corp. working paper. Our thanks to Dr. Peter Shoenfeld of SAIC Corp. for making these papers available to us.

[31] Wasserman, L. (1990) Belief functions and statistical inference, *Canadian Journal of Mathematics*, **18** (3), 183-196.

Chapter 5

Finite-Set Statistics

The purpose of this chapter is to extend the basic elements of practical statistics to multisensor, multitarget data fusion. Specifically, in section 5.1 we show that the basic concepts of elementary statistics—expectations, covariances, prior and posterior densities, etc.—have direct analogs in the multisensor, multitarget realm. In section 5.2 we show how the basic elements of parametric estimation theory—maximum likelihood estimator (MLE), maximum *a posteriori* estimator (MAPE), Bayes estimators, etc.—lead to *fully integrated Level 1 fusion algorithms*. That is, they lead to algorithms in which the numbers, I.D.s, and kinematics of multiple targets are estimated simultaneously *without* any attempt to estimate optimal report-to-track assignments. In particular, we show that two such fully integrated algorithms (analogs of the MLE and MAPE) are statistically consistent. The chapter concludes with section 5.3, in which we prove a Cramér-Rao inequality for multisensor, multitarget problems. The significance of this inequality is that it sets best-possible-theoretical-performance bounds for certain kinds of data fusion algorithms.

5.1 Basic Statistical Concepts

In this section we show that certain basic concepts of statistics—expectations, covariances, prior and posterior densities—can be directly extended to the multisensor, multitarget case.

5.1.1 Expected Values

In conventional statistics one repeatedly encounters the following situation. Suppose that $\mathbf{X}$ is a random vector on $\mathbb{R}^n$ with probability measure $p_{\mathbf{X}}$ and density $f_{\mathbf{X}}$. Let $\mathbf{F} : \mathbb{R}^n \to \mathbb{R}^m$ be a measurable vector transformation to another space. Then $\mathbf{Y} = \mathbf{F}(\mathbf{X})$ is a random vector on the space $\mathbb{R}^m$. Its probability measure is obtained in the usual manner:

$$p_{\mathbf{Y}}(S) = p(\mathbf{Y} \in S) = p(\mathbf{F}(\mathbf{X}) \in S) = p(\mathbf{X} \in \mathbf{F}^{-1}(S)) = p_{\mathbf{X}}(\mathbf{F}^{-1}(S))$$

for all measurable $S \subseteq \mathbb{R}^m$, so that $p_\mathbf{Y} = p_\mathbf{X}\mathbf{F}^{-1}$. Assume that $p_\mathbf{Y}$ has a density $f_\mathbf{Y}$ with respect to Lebesgue measure. What is the expected value of $\mathbf{Y}$? It is, of course,

$$E[\mathbf{F}(\mathbf{X})] = \int_{\mathbb{R}^m} \mathbf{y}\, f_\mathbf{Y}(\mathbf{y}) d\lambda(\mathbf{y}) = \int \mathbf{y}\, dp_\mathbf{Y}(\mathbf{y})$$

To compute the expected value of $\mathbf{Y}$, therefore, it would seem that one would have to first determine the density $f_\mathbf{Y}$ and then compute the appropriate integral. Of course, one need not do this: The expected value of $\mathbf{Y}$ can be computed using only the density $f_\mathbf{X}$ of the original random vector $\mathbf{X}$:

$$E[\mathbf{F}(\mathbf{X})] = \int_{\mathbb{R}^n} \mathbf{F}(\mathbf{x}) f_\mathbf{X}(\mathbf{x}) d\lambda(\mathbf{x}) = \int \mathbf{F}(\mathbf{x}) dp_\mathbf{X}(\mathbf{x})$$

The fact that we can do this rests on the following well-known change-of-variables formula for integrals:

Proposition 1 *Let* $\mathbf{F} : \mathbb{R}^n \to \mathbb{R}^m$ *and* $\mathbf{H} : \mathbb{R}^m \to \mathbb{R}^r$ *be measurable vector transformations. Let* $\mathbf{X}$ *be a random vector on* $\mathbb{R}^n$ *and* $p_\mathbf{X}$ *its associated probability measure. Then*

$$\int_S \mathbf{H}(\mathbf{y}) dp_\mathbf{Y}(\mathbf{y}) = \int_{\mathbf{F}^{-1}(S)} \mathbf{H}(\mathbf{F}(\mathbf{x})) dp_\mathbf{X}(\mathbf{x})$$

for all measurable $S \subseteq R^m$, *where* $p_\mathbf{Y} = p_\mathbf{X}\mathbf{F}^{-1}$ *is the probability measure of the random vector* $\mathbf{Y} = \mathbf{F}(\mathbf{X})$.

Proof. This formula is obtained from Theorem C of [10, p. 163]. Because R^r is a finite vector space we can replace $\mathbf{H}(\mathbf{y})$ by its component functions $H_i(\mathbf{y})$ for $i = 1, ..., r$. Also, because the entire space $\mathbb{R}^m$ is open and hence measurable for any m, measurable functions are also measurable transformations [10, pp. 77, 162]   □

In this section our goal is to prove analogous results for finite random sets. The major existing result of this type, *Robbin's formulas* (see [20], [9, p. 36], and also [15, p. 47]), are not much help in this regard: Let Σ be a random subset of $\mathbb{R}^n$ and let q be a sigma-finite measure on $\mathbb{R}^n$. Then $Y = q(\Sigma)$ is a random number. According to one of Robbin's formulas the expected value of Y is

$$E[q(\Sigma)] = \int \mu_\Sigma(\mathbf{z}) dq(\mathbf{z}) = \int \mu_\Sigma(\mathbf{z}) f_q(\mathbf{z}) d\lambda(\mathbf{z})$$

where $\mu_\Sigma(\mathbf{z}) \triangleq p(\mathbf{z} \in \Sigma)$ is the one-point covering function of Σ. Unfortunately, for virtually all finite random sets of interest in data fusion, Robbin's formula reduces to a triviality: If q is absolutely continuous with respect to Lebesgue measure then $q(Z) = 0$ for any finite set Z and thus $\mu_\Sigma(\mathbf{z}) = 0$ identically if Σ is a finite random subset. (For example, let $\Sigma = \{\mathbf{Z}\}$ where $\mathbf{Z}$ is a random vector on $\mathbb{R}^n$. Then $\mu_\Sigma(\mathbf{z}) = p(\mathbf{z} \in \{\mathbf{Z}\}) = p(\mathbf{z} = \mathbf{Z}) = 0$ for all $\mathbf{z}$.)

As it turns out, in finite-set statistics the correct formulas for expectations result from generalizing the classical formula of Proposition 1 to the random-set case using the global density. Specifically, let $\mathbf{T}(Z)$ be a function of a finite-set variable $Z \subseteq \mathcal{R}$ which takes its values in $\mathbb{R}^m$. Let Σ be an absolutely continuous finite random subset of $\mathcal{R}$. Then the expected value of the random vector $\mathbf{Y} = \mathbf{T}(\Sigma)$ is given by the formula (see Corollary 5):

$$E[\mathbf{T}(\Sigma)] = \int \mathbf{T}(Z) f_\Sigma(Z) \delta Z$$

Notice that we *will not* be proving a relationship of the form $E[\Sigma] = \int Z f_\Sigma(Z) \delta Z$. That is, *we do not have an integration formula for the expected value of a finite random subset itself* (as opposed to a *vector-valued transformation* of a finite random subset). The reason, of course, is that we do not know how to add and subtract finite subsets in the same way that one adds and subtracts vectors. In fact, a major unaddressed theoretical question is how to define and compute expected values of finite random subsets. We have no answer to this question other than to say that existing approaches to expectations of (continuously infinite) random sets do not seem to be applicable to the finite-set case. Various researchers have developed a concept of expectation for random *compact* closed subsets [3], [6], [2], [1] which, however, do not yield a useful concept of expected value for random finite sets. More recently, Molchanov has devised a general approach to expectations of random closed sets [16, pp. 18-27], [4]. Even in this more general framework, however, there does not seem to be a suitable definition of expectation which can be applied usefully in the finite-set case.

This section is organized as follows. We generalize a standard change-of-variables formula for integrals to the finite-set realm. We use it to derive the formula for the expected value of a vector-valued function of an absolutely continuous finite random subset. Finally, we establish finite-set analogs of some standard inequalities involving expectations: the Cauchy-Schwartz inequality and Chebychev's inequality.

Set-to-Point Transformations

In this section we introduce the concept of a transformation from finite sets to points and establish the basic properties of such transformations.

Definition 1 *A set-to-point transformation* $\mathbf{T} : c_{\leq M}(\mathcal{R}) \to \mathcal{R}$ *is measurable if* $\mathbf{T}^{-1}(G) \in \sigma(c_{\leq M}(\mathcal{R}))$, *i.e. if* $\mathbf{T}^{-1}(G)$ *is a measurable subset of the hit-or-miss topology.*

Proposition 2 *(a) Let* $\mathbf{Y} = \mathbf{T}(\Sigma)$ *where* Σ *is a finite random subset of* $\mathcal{R}$ *and* $\mathbf{T}$ *is a set-to-point transformation. (a) Then* $\mathbf{Y}$ *is a random variable on* $\mathcal{R}$. *(b) Assume that* $n > 0$ *in* $\mathcal{R} = \mathbb{R}^n \times W$ *and define the functions* $\mathbf{T}_k : \mathcal{R}^k \to \mathcal{R}$ *by*

$$\mathbf{T}_k(\xi_1, ..., \xi_k) \triangleq \begin{cases} \mathbf{T}(\{\xi_1, ..., \xi_k\}) & \text{if } \xi_1, ..., \xi_k \text{ are distinct} \\ 0 & \text{if } \text{otherwise} \end{cases}$$

for all $\xi_1, ..., \xi_k \in \mathcal{R}$ *and* $1 \leq k \leq M$, *where as usual* $\chi_k : \mathcal{R}^k \to c_{\leq M}(\mathcal{R})$ *is defined by* $\chi_k(\xi_1, ..., \xi_k) = \{\xi_1, ..., \xi_k\}$. *Then* $\mathbf{T}$ *is a measurable set-to-point* *transformation if and only if the $\mathbf{T}_k$ are measurable transformations for all $k \geq 1$. (c) Finally,*

$$\chi_k^{-1}(\mathbf{T}^{-1}(S) \cap c_k(\mathcal{R})) = \mathbf{T}_k^{-1}(S)$$

for all measurable $S \subseteq \mathcal{R}$ and all $k \geq 1$, up to a set of measure zero.

Proof. (a) Let $G \subseteq \mathcal{R}$ be an open set. Then note that

$$
\begin{aligned}
\mathbf{Y}^{-1}(G) &= \{\omega \in \Omega | \mathbf{Y}(\omega) \in G\} = \{\omega \in \Omega | \mathbf{T}(\Sigma(\omega)) \in G\} \\
&= \{\omega \in \Omega | \Sigma(\omega) \in \mathbf{T}^{-1}(G)\} = \Sigma^{-1}\mathbf{T}^{-1}(G) \in \sigma(\Omega)
\end{aligned}
$$

as claimed. (b) On the one hand, suppose that $\mathbf{T}_k$ are measurable for all $1 \leq k \leq M$. Then

$$
\begin{aligned}
\mathbf{T}^{-1}(G) &= \{Z \in c_{\leq M}(\mathcal{R}) | \mathbf{T}(Z) \in G\} = \bigcup_{k=0}^{M} \{Z \in c_k(\mathcal{R}) | \mathbf{T}(Z) \in G\} \\
&= \bigcup_{k=0}^{M} \{\{\xi_1, ..., \xi_k\} \in c_k(\mathcal{R}) | \mathbf{T}_k(\xi_1, ..., \xi_k) \in G\} \\
&= \bigcup_{k=0}^{M} \{\{\xi_1, ..., \xi_k\} \in c_k(\mathcal{R}) | (\xi_1, ..., \xi_k) \in \mathbf{T}_k^{-1}(G)\} \\
&= \bigcup_{k=0}^{M} (c_k(\mathcal{R}) \cap \chi_k(\mathbf{T}_k^{-1}(G))) = \bigcup_{k \geq 0} (\chi_k([\mathcal{R}]^k \cap \mathbf{T}_k^{-1}(G)))
\end{aligned}
$$

where $[\mathcal{R}]^k$ is the subspace of $\mathcal{R}^k$ defined in Definition 1 of Chapter 4. Since the $\mathbf{T}_k$ are measurable for all $k \geq 1$ then $\mathbf{T}_k^{-1}(G) \in \sigma(\mathcal{R}^k)$ is a measurable subset of $\mathcal{R}^k$ and so $[\mathcal{R}]^k \cap \mathbf{T}_k^{-1}(G)$ is a measurable subset of $[\mathcal{R}]^k$. However, by Proposition 2 of Chapter 4, $\chi_k : [\mathcal{R}]^k \leftrightarrow c_k(\mathcal{R})$ is a homeomorphism. Thus $\chi_k([\mathcal{R}]^k \cap \mathbf{T}_k^{-1}(G))$ is measurable in the relative hit-or-miss topology on $c_k(\mathcal{R})$ and hence so is the finite union $\bigcup_{k=0}^{M} (\chi_k([\mathcal{R}]^k \cap \mathbf{T}_k^{-1}(G))) = \mathbf{T}^{-1}(G)$.

Conversely, suppose that $\mathbf{T}$ is measurable and let G be an open subset of $\mathcal{R}^k$. If $\mathbf{0} \notin G$ then, noting that $\mathbf{T}_k(\xi_1, ..., \xi_k) \neq \mathbf{0}$ means that $\xi_1, ..., \xi_k$ must be distinct,

$$
\begin{aligned}
\mathbf{T}_k^{-1}(G) &= \{(\xi_1, ..., \xi_k) \in \mathcal{R}^k | \mathbf{T}_k(\xi_1, ..., \xi_k) \in G\} \\
&= \{(\xi_1, ..., \xi_k) \in \mathcal{R}^k | \mathbf{T}(\{\xi_1, ..., \xi_k\}) \in G, \xi_1, ..., \xi_k \text{ distinct}\} \\
&= \{(\xi_1, ..., \xi_k) \in \mathcal{R}^k | \{\xi_1, ..., \xi_k\} \in c_k(\mathcal{R}) \cap \mathbf{T}^{-1}(G)\} \\
&= \chi_k^{-1}(c_k(\mathcal{R}) \cap \mathbf{T}^{-1}(G))
\end{aligned}
$$

Since $\mathbf{T}$ is measurable then $\mathbf{T}^{-1}(G)$ is a measurable subset of $c(\mathcal{R})$ and so $\chi_k^{-1}(\mathbf{T}^{-1}(G) \cap c_k(\mathcal{R}))$ is a measurable subset of $\mathcal{R}^k$. On the other hand suppose that $\mathbf{0} \in G$ and let $S_0 \triangleq \{(\xi_1, ..., \xi_k) \in \mathcal{R}^k | \xi_1, ..., \xi_k \text{ not distinct}\}$, which is a

measurable subset of $\mathcal{R}^k$ (in fact, it is a set of hybrid measure zero since $n > 0$).
Then

$$
\begin{aligned}
\mathbf{T}_k^{-1}(G) &= \{(\xi_1, ..., \xi_k) \in \mathcal{R}^k \,|\, \mathbf{T}_k(\xi_1, ..., \xi_k) \in G\} \\
&= \{(\xi_1, ..., \xi_k) \in \mathcal{R}^k \,|\, \mathbf{T}_k(\xi_1, ..., \xi_k) \in G, \xi_1, ..., \xi_k \text{ distinct}\} \cup S_0 \\
&= \{(\xi_1, ..., \xi_k) \in \mathcal{R}^k \,|\, \mathbf{T}(\{\xi_1, ..., \xi_k\}) \in G, \xi_1, ..., \xi_k \text{ distinct}\} \cup S_0 \\
&= \{(\xi_1, ..., \xi_k) \in \mathcal{R}^k \,|\, \{\xi_1, ..., \xi_k\} \in c_k(\mathcal{R}) \cap \mathbf{T}^{-1}(G)\} \cup S_0 \\
&= \chi_k^{-1}(\mathbf{T}^{-1}(G) \cap c_k(\mathcal{R})) \cup S_0
\end{aligned}
$$

which is a measurable subset of $\mathcal{R}^k$. (c) Follows immediately from the proof of
(b).   □

Corollary 3 *Let Σ be an absolutely continuous random finite subset of $\mathcal{R}$. Define $Y(\omega) \triangleq f_\Sigma(\Sigma(\omega))$ for all $\omega \in \Omega$. Then Y is a random real number.*

Proof. By Proposition 2 the transformation $Z \mapsto f_\Sigma(Z)$ is measurable if and
only if the functions $(\xi_1, ..., \xi_k) \mapsto f_\Sigma(\{\xi_1, ..., \xi_k\})$ (for distinct $\xi_1, ..., \xi_k$) are
measurable for all $k \geq 1$. However, these functions are integrable and hence
measurable..   □

   The following give examples of set-to-point transformations.

Example 1 *Define the transformation which takes finite subsets $Z \subseteq \mathbb{R}^n$ to
integer-valued measures on $\mathbb{R}^n$ by $\mathbf{T}(Z)(S) = Z \cap S$ for all $S \subseteq \mathbb{R}^n$. Then $\mathbf{T}$
transforms the space of finite subsets into the vector space of signed measures.*

Example 2 *Let $\delta_{\mathbf{z}}(\mathbf{x})$ be the Dirac delta density on R^n concentrated at the point
$\mathbf{z}$: That is, $\delta_{\mathbf{z}}(\mathbf{x}) = 0$ if $\mathbf{x} \neq \mathbf{z}$ and $\delta_{\mathbf{x}}(\mathbf{x}) = \infty$ and $\int \delta_{\mathbf{z}}(\mathbf{x})d\lambda(\mathbf{x}) = 1$. Let $\mathcal{G}$
denote the vector space of all generalized functions on $\mathbb{R}^n$. Then for any finite
subset $\{\mathbf{z}_1, ..., \mathbf{z}_k\} \subseteq \mathbb{R}^n$ with distinct elements $\mathbf{z}_1, ..., \mathbf{z}_k$ define*

$$\mathbf{T}(\{\mathbf{z}_1, ..., \mathbf{z}_k\})(\mathbf{x}) = \delta_{\mathbf{z}_1}(\mathbf{x}) + ... + \delta_{\mathbf{z}_k}(\mathbf{x})$$

*for all $\mathbf{x} \in \mathbb{R}^n$. Then $\mathbf{T}$ is a transformation of finite sets into the vector space
of generalized functions. Clearly, $\mathbf{T}$ is a one-to-one mapping in the sense that
$\mathbf{T}(Z) = \mathbf{T}(Z')$ implies $Z = Z'$.*

Example 3 *Let $\mathbb{R}[\mathbf{s}_0, \mathbf{s}_1, ..., \mathbf{s}_n,]$ denote the set of multinomials in $n + 1$ in-
determinates $\mathbf{s}_0, \mathbf{s}_1, ..., \mathbf{s}_n$ with coefficients in the real numbers. Choose a specific orthonormal basis $\hat{\mathbf{e}}_1, .., \hat{\mathbf{e}}_n$ for $\mathbb{R}^n$ and define a linear map $L : \mathbb{R}^n \rightarrow
\mathbb{R}[\mathbf{s}_0, \mathbf{s}_1, ..., \mathbf{s}_n,]$ by*

$$a_1\hat{\mathbf{e}}_1 + ... + a_n\hat{\mathbf{e}}_n \mapsto a_1\mathbf{s}_1 + ... + a_n\mathbf{s}_n$$

Finally, define the transformation $\mathbf{T} : c_{\leq M}(\mathbb{R}) \rightarrow \mathbb{R}[\mathbf{s}_0, \mathbf{s}_1, ..., \mathbf{s}_n,]$ by

$$\mathbf{T}(\{\mathbf{z}_1, ..., \mathbf{z}_k\}) = \mathbf{s}_0^k(1 + L(\mathbf{z}_1)) \cdots (1 + L(\mathbf{z}_k))$$

*for all finite subsets $\{\mathbf{z}_1, ..., \mathbf{z}_k\} \in \mathbb{R}^n$ with distinct elements $\mathbf{z}_1, ..., \mathbf{z}_k$. Then
$\mathbf{T}$ is well-defined. Furthermore, $\mathbf{T}$ is a measurable one-to-one mapping. Note
that $\mathbf{T}$ is a generalization of the basic construction used in Kamen's symmetric
measurement equation (SME) filter [11].*

Change of Variables Formula for Set Integrals

We are now in a position to prove our change-of-variables formula for set integrals:

Proposition 4 *Let $\mathbf{T}$ be a finite-set function which takes its values in $\mathcal{R} = \mathbb{R}^n \times W$ with $n > 0$. Assume that $\mathbf{T}$ is measurable. Let $\mathbf{H} : \mathcal{R} \to \mathbb{R}^r$ be a measurable vector transformation and let Σ be an absolutely continuous finite random subset of $\mathcal{R}$. Then the following relationship exists between conventional integrals and set integrals:*

$$\int_S \mathbf{H}(\xi) f_{\mathbf{Y}}(\xi) d\lambda(\xi) = \int_{\mathbf{T}^{-1}(S)} \mathbf{H}(\mathbf{T}(Z)) f_\Sigma(Z) \delta Z$$

for all measurable $S \subseteq \mathcal{R}$, and where $f_{\mathbf{Y}}$ is the probability distribution of $\mathbf{Y} = \mathbf{T}(\Sigma)$.

Proof. By the definition of a set integral,

$$\int_{\mathbf{T}^{-1}(S)} \mathbf{H}(\mathbf{T}(Z)) f_\Sigma(Z) \delta Z$$

$$= \sum_{k=0}^{\infty} \frac{1}{k!} \int_{\chi_k^{-1}(\mathbf{T}^{-1}(S) \cap c_k(\mathcal{R}))} \mathbf{H}(\mathbf{T}(\{\xi_1, ..., \xi_k\})) f_\Sigma(\{\xi_1, ..., \xi_k\}) d\lambda(\xi_1) \cdots d\lambda(\xi_k)$$

Noting that $\chi_k^{-1}(\mathbf{T}^{-1}(S) \cap c_k(\mathcal{R})) = \mathbf{T}_k^{-1}(S)$ up to a set of measure zero (Proposition 2) the right-hand side becomes

$$\sum_{k=0}^{\infty} f_\Sigma(k) \int_{\mathbf{T}_k^{-1}(S)} \mathbf{H}(\mathbf{T}_k(\xi_1, ..., \xi_k)) \, \hat{f}_{\Sigma;k}(\xi_1, ..., \xi_k) d\lambda(\xi_1) \cdots d\lambda(\xi_k)$$

where $f_\Sigma(k) \triangleq \int_{|Z|=k} f_\Sigma(Z) \delta Z$ is the cardinality distribution of Σ (Definition 18 of Chapter 4). Also, we have defined

$$\hat{f}_{\Sigma;k}(\xi_1, ..., \xi_k) \triangleq \frac{1}{k! \, f_\Sigma(k)} f_\Sigma(\{\xi_1, ..., \xi_k\})$$

if $\xi_1, ..., \xi_k$ are distinct and $\hat{f}_{\Sigma;k}(\xi_1, ..., \xi_k) \triangleq 0$ otherwise. Note that $\hat{f}_{\Sigma;k}$ is a conventional density on the space $\mathcal{R}^k$. Thus

$$\int_{\mathbf{T}^{-1}(S)} \mathbf{H}(\mathbf{T}(Z)) f_\Sigma(Z) \delta Z = \sum_{k=0}^{\infty} f_\Sigma(k) \int_{\mathbf{T}_k^{-1}(S)} \mathbf{H}(\mathbf{T}_k(\xi_1, ..., \xi_k)) \, d\hat{q}_{\Sigma;k}(\xi_1, ..., \xi_k)$$

where $\hat{q}_{\Sigma;k}$ is the probability measure corresponding to the density $\hat{f}_{\Sigma;k}$. The $\mathbf{T}_k$ are measurable by Proposition 2 and so by the change of variables formula for conventional integrals we get:

$$\int_{\mathbf{T}^{-1}(S)} \mathbf{H}(\mathbf{T}(Z)) f_\Sigma(Z) \delta Z = \sum_{k=0}^{\infty} f_\Sigma(k) \int_S \mathbf{H}(\xi) d(\hat{q}_{\Sigma;k} \mathbf{T}_k^{-1})(\xi)$$

$$= \int_S \mathbf{H}(\xi) \, d\left(\sum_{k=0}^{\infty} f_\Sigma(k) \, \hat{q}_{\Sigma;k} \mathbf{T}_k^{-1}\right)(\xi)$$

(where we can sum under the integral sign since the summation actually involves only a finite number of terms). We will therefore be done if we show that $\sum_{k=0}^{\infty} f_{\Sigma}(k)\,\hat{q}_{\Sigma;k}\mathbf{T}_k^{-1} = p_{\mathbf{Y}}$ for all measurable $S \subseteq \mathcal{R}$. However, note that

$$p_{\mathbf{Y}}(S) = p(\mathbf{Y} \in S) = p(\mathbf{T}(\Sigma) \in S) = p(\Sigma \in \mathbf{T}^{-1}(S))$$

But by the last part of the proof of Proposition 15 of Chapter 4 we know that

$$p(\Sigma \in O) = \sum_{k=0}^{\infty} p(|\Sigma| = k)\, q_k(\chi_k^{-1}(O \cap c_k(\mathcal{R})))$$

for any measurable subset O of the hit-or-miss topology, and where $q_k = \hat{q}_{\Sigma;k}$. Setting $O = \mathbf{T}^{-1}(S)$ we get:

$$p(\Sigma \in \mathbf{T}^{-1}(S)) = \sum_{k=0}^{\infty} p(|\Sigma| = k)\, \hat{q}_{\Sigma;k}(\chi_k^{-1}(\mathbf{T}^{-1}(S) \cap c_k(\mathcal{R})))$$

$$= \sum_{k=0}^{\infty} f_{\Sigma}(k)\, \hat{q}_{\Sigma;k}(\mathbf{T}_k^{-1}(S))$$

as desired. □

Expectations of Vector Transformations of a Random Finite Set

The previous proposition leads immediately to the following result on expected values of functions of an absolutely continuous finite random set:

Corollary 5 *Let* $\mathbf{T}$ *be a set-to-point transformation and let* Σ *be an absolutely continuous finite random subset of* $\mathbb{R}^n$. *Then the expected value of the random vector* $\mathbf{Y} = \mathbf{T}(\Sigma)$ *is*

$$E[\mathbf{T}(\Sigma)] = \int \mathbf{T}(Z) f_{\Sigma}(Z)\delta Z$$

Proof. Obvious. □

In particular, note that this observation can be extended to functions $\mathbf{T}(Z_1, ..., Z_m)$ of several finite-set variables:

$$E[T(\Sigma_1, ..., \Sigma_m)] = \int T(Z_1, ..., Z_m) f_{\Sigma_1}(Z_1) \cdots f_{\Sigma m}(Z_m)\delta Z_1 \cdots \delta Z_m$$

where $\Sigma_1, ..., \Sigma_m$ are statistically independent, absolutely continuous finite random subsets of $\mathcal{R}$.

Definition 2 *If a global density is parametrized, we will use the notation*

$$E_X[\mathbf{T}(\Sigma)] \triangleq \int \mathbf{T}(Z) f_{\Sigma}(Z|X)\delta Z$$

for the parametrized expected values.

The following corollaries provide some examples of expected values.

Corollary 6 *Let Σ be an absolutely continuous finite random subset of $\mathcal{R}$. Then the mean number of elements in Σ is*

$$E[|\Sigma|] = \sum_{k=0}^{\infty} k \, f_{\Sigma}(k) = \int |Z| \, f_{\Sigma}(Z) \delta Z$$

Proof. According to Proposition 11 of Chapter 4, $|\Sigma|$ is a random number. The result then immediately follows from Proposition 4.   □

Example 4 *Given the parametrized global density*

$$
\begin{aligned}
f_{\Sigma}(\emptyset|\{\zeta\}) &= 1 - q \\
f_{\Sigma}(\{\xi\}|\{\zeta\}) &= q \, f(\xi|\zeta) \\
f_{\Sigma}(\emptyset|\{\zeta_1, \zeta_2\}) &= (1 - q)^2 \\
f_{\Sigma}(\{\xi\}|\{\zeta_1, \zeta_2\}) &= q(1 - q) \, f(\xi|\zeta_1) + q(1 - q) \, f(\xi|\zeta_2) \\
f_{\Sigma}(\{\xi_1, \xi_2\}|\{\zeta_1, \zeta_2\}) &= q^2 \, f(\xi_1|\zeta_1) f(\xi_2|\zeta_2) + q^2 \, f(\xi_2|\zeta_2) f(\xi_1|\zeta_2)
\end{aligned}
$$

It is easy to compute

$$
\begin{aligned}
E_{\{\zeta\}}[|\Sigma|] &= 0 \cdot (1 - q) + 1 \cdot q = q \\
E_{\{\zeta_1, \zeta_2\}}[|\Sigma|] &= 0 \cdot (1 - q)^2 + 1 \cdot 2q(1 - q) + 2 \cdot q^2 = 2q
\end{aligned}
$$

Inequalities Involving Expectations

We prove a random-set analog of the Schwartz inequality for expectations:

Proposition 7 *Let f, g be measurable set-to-number transformations and let Σ be an absolutely continuous finite random subset of $\mathcal{R}$. Then*

$$E[f(\Sigma)g(\Sigma)]^2 \leq E[f(\Sigma)^2] \cdot E[g(\Sigma)^2]$$

Proof. The proof is identical to that for the random-vector case. That is, let t be a real variable. From the linearity of the set integral it follows that

$$0 \leq E[(f(\Sigma) + t \, g(\Sigma))^2] = E[f(\Sigma)^2] + 2t \, E[f(\Sigma)g(\Sigma)] + t^2 E[g(\Sigma)^2]$$

independently of the value of t. Consequently, the discriminant of this quadratic polynomial must be non-positive, which concludes the proof.   □

Also, we prove an analog of the Chebychev inequality:

Proposition 8 *Let Σ be an absolutely continuous finite random subset of $\mathcal{R}$ and let $g(Z)$ be a measurable, nonnegative finite-set function. Then for any $r > 0$,*

$$p(g(\Sigma) \geq r) \leq \frac{E[g(\Sigma)]}{r}$$

Proof. The proof is essentially identical to that for the conventional case. From Corollary 5 we know that

$$
\begin{aligned}
\mathrm{E}[g(\Sigma)] &= \int g(Z) f_\Sigma(Z) \delta Z \\
&= \sum_{k=0}^{\infty} \frac{1}{k!} \int g(\{\xi_1, ..., \xi_k\}) f_\Sigma(\{\xi_1, ..., \xi_k\}) d\lambda(\xi_1) \cdots d\lambda(\xi_k)
\end{aligned}
$$

which, because g is nonnegative, gives us:

$$
\mathrm{E}[g(\Sigma)] \geq r \sum_{k=0}^{\infty} \frac{1}{k!} \int_{g(\{\xi_1, ..., \xi_k\}) \geq r} f_\Sigma(\{\xi_1, ..., \xi_k\}) d\lambda(\xi_1) \cdots d\lambda(\xi_k)
$$

where '$g(\{\xi_1, ..., \xi_k\}) \geq 0$' denotes the set of all $(\xi_1, ..., \xi_k)$ in $\mathcal{R}^k$ such that $g(\{\xi_1, ..., \xi_k\}) \geq r$. Note that this set is equal to $\chi_k^{-1}(O_r \cap c_k(\mathcal{R}))$ where, $\chi_k : \mathcal{R}^k \to c_{\leq k}(\mathcal{R})$ is the continuous map defined by $\chi_k(\xi_1, ..., \xi_k) = \{\xi_1, ..., \xi_k\}$ and where $O_r \triangleq g^{-1}([r, \infty))$ denotes the measurable subset of $c(\mathcal{R})$ consisting of all finite subsets Z with $g(Z) \geq r$. This then gives us

$$
\begin{aligned}
\mathrm{E}[g(\Sigma)] &\geq r \sum_{k=0}^{\infty} \frac{1}{k!} \int_{\chi_k^{-1}(O_r \cap c_k(\mathcal{R}))} f_\Sigma(\{\xi_1, ..., \xi_k\}) d\lambda(\xi_1) \cdots d\lambda(\xi_k) \\
&= r \int_{O_r} f_\Sigma(Z) \delta Z = r\, p(\Sigma \in O_r) = r\, p(g(\Sigma) \geq r)
\end{aligned}
$$

which establishes the result. $\qquad\square$

5.1.2 Covariances

Let $\mathbf{X}$ be a random vector on $\mathbb{R}^n$ with density $f_\mathbf{X}$ and let $\mathbf{x}_0$ be its expected value. Then in conventional statistics the *covariance* of X is defined as follows. Given two vectors $\mathbf{y}, \mathbf{z} \in \mathbb{R}^n$ the tensor product of $\mathbf{y}$ and $\mathbf{z}$ is the linear transformation $\mathbf{y} \otimes \mathbf{z} : \mathbb{R}^n \to \mathbb{R}^n$ defined by $(\mathbf{y} \otimes \mathbf{z})(\mathbf{w}) \triangleq \langle \mathbf{y}, \mathbf{w} \rangle \mathbf{z}$ for all $\mathbf{w} \in \mathbb{R}^n$, where $\langle -, - \rangle$ is the usual inner (scalar) product on $\mathbb{R}^n$ defined by $\langle (x_1, ..., x_n), (y_1, ..., y_n) \rangle = \sum_{i=1}^{n} x_i y_i$. (Note that the tensor product symbol $\otimes$ is not to be confused with our occasional usage of '$\otimes$' as the symbol for a *target*.) Now, for $\mathbf{w}$ fixed define

$$
\mathbf{F}(\mathbf{x}) \triangleq ((\mathbf{x} - \mathbf{x}_0) \otimes (\mathbf{x} - \mathbf{x}_0))(\mathbf{w})
$$

for all $\mathbf{x} \in \mathbb{R}^n$ and define the random vector $\mathbf{W}$ by $\mathbf{W}(\omega) = \mathbf{F}(\mathbf{X}(\omega))$ for all $\omega \in \Omega$. Then the expected value of $\mathbf{W}$ defines a linear transformation $cov_\mathbf{X} : \mathbb{R}^n \to \mathbb{R}^n$ by

$$
\begin{aligned}
cov_\mathbf{X}(\mathbf{w}) &= \mathrm{E}[\mathbf{F}(\mathbf{X})] = \int ((\mathbf{x} - \mathbf{x}_0) \otimes (\mathbf{x} - \mathbf{x}_0))(\mathbf{w})\, f_\mathbf{X}(\mathbf{x}) d\lambda(\mathbf{x}) \\
&= \int \langle \mathbf{x} - \mathbf{x}_0, \mathbf{w} \rangle (\mathbf{x} - \mathbf{x}_0)\, f_\mathbf{X}(\mathbf{x}) d\lambda(\mathbf{x})
\end{aligned}
$$

Thus

$$\langle \mathbf{v}, cov_{\mathbf{X}}(\mathbf{w}) \rangle = \int \langle \mathbf{x} - \mathbf{x}_0, \mathbf{w} \rangle \langle \mathbf{v}, \mathbf{x} - \mathbf{x}_0 \rangle f_{\mathbf{X}}(\mathbf{x}) d\lambda(\mathbf{x})$$

for all $\mathbf{v}, \mathbf{w} \in \mathbb{R}^n$ and, in particular,

$$\langle \mathbf{w}, cov_{\mathbf{X}}(\mathbf{w}) \rangle = \int \langle \mathbf{x} - \mathbf{x}_0, \mathbf{w} \rangle^2 f_{\mathbf{X}}(\mathbf{x}) d\lambda(\mathbf{x})$$

for all $\mathbf{w} \in \mathbb{R}^n$. If we choose an orthonormal basis $\hat{\mathbf{e}}_1, ..., \hat{\mathbf{e}}_n$ of $\mathbb{R}^n$ then $C_{i,j} = \langle \hat{\mathbf{e}}_i, cov_{\mathbf{X}}(\hat{\mathbf{e}}_j) \rangle$ for all $i, j = 1, ..., n$ is the familiar *covariance matrix* of the random vector $\mathbf{X}$.

The following proposition tells us how to compute the covariance of a random vector which is defined in terms of a set-to-point transformation:

Proposition 9 *Let $\mathbf{T}$ be a set-to-point transformation which takes its values in $\mathbb{R}^n$ and let Σ be an absolutely continuous finite random subset of $\mathbf{R}^n$. Then the covariance of the random vector $\mathbf{Y} = \mathbf{T}(\Sigma)$ is*

$$cov_{\mathbf{Y}}(\mathbf{w}) \triangleq \int \langle \mathbf{T}(Z) - \mathbf{y}_0, \mathbf{w} \rangle (\mathbf{T}(Z) - \mathbf{y}_0) f_{\Sigma}(Z) \delta Z$$

where $\mathbf{y}_0$ is the expected value of $\mathbf{Y}$.

Proof. Follows immediately from Corollary 5 and the definition of covariance. □

If a parametrized family of global densities is being used, we will write:

$$cov_{\mathbf{Y},X}(\mathbf{w}) = \int \langle \mathbf{T}(Z) - \mathbf{y}_0, \mathbf{w} \rangle (\mathbf{T}(Z) - \mathbf{y}_0) f_{\Sigma}(Z|X) \delta Z$$

Example 5 *Let $f_{\Sigma}(Z|X)$ be the parametrized global density of Example 4. Then the covariance of the random integer $|\Sigma|$ is:*

$$
\begin{aligned}
cov_{|\Sigma|,\{\mathbf{x}\}} &= (0 - q)^2(1 - q) + (1 - q)^2 q = q(1 - q) \\
cov_{|\Sigma|,\{\mathbf{x},\mathbf{y}\}} &= (0 - 2q)^2(1 - q)^2 + (1 - 2q)^2 2q(1 - q) + (2 - 2q)^2 q^2 \\
&= 2q(1 - q)
\end{aligned}
$$

5.1.3   Prior and Posterior Global Densities

It should be obvious at this point that standard Bayesian reasoning can be generalized to the finite-set case. Suppose that the global density of the absolutely continuous finite random set Σ is known to belong to a family $f_{\Sigma}(Z|X)$ of global densities parametrized by the finite set X. The parameter X itself may be assumed to be just one realization of some absolutely continuous finite random set Γ. In this case $f_{\Sigma}(Z|X)$ is denoted $f_{\Sigma|\Gamma}(Z|X)$. If $f_{\Gamma}(X)$ is the global density of Γ then it is called a *global prior distribution* or *global prior*. It describes the expected state of all information, in regard to target geodynamics, identities, and numbers.

For example, assume that X has been chosen so that its elements can be regarded as the objects in a possible "ground truth" and that X is therefore a possible ground truth. Then $f_\Gamma(X)$ describes the *a priori* belief that the configuration of targets modeled by X can occur as an actual state of reality. So, for example, $f_\Gamma(\emptyset)$ describes the *a priori* belief that there are no objects "out there" to be observed. Likewise, $f_\Gamma(\{\zeta\})$ is a measure of *a priori* belief that there is only one object and that it is in the state ζ. In general, $f_\Gamma(\{\zeta_1, ..., \zeta_t\})$ describes the likelihood to which there will be k objects in ground truth and that those objects will have the states $\zeta_1, ..., \zeta_t$.

In analogy with the conventional case we can form the *global marginal distribution*

$$f_\Sigma(Z) = \int f_{\Sigma|\Gamma}(Z|X) f_\Gamma(X) \delta X$$

and, given a "global observation" Z_1, the *global posterior distribution*

$$f_{\Gamma|\Sigma}(X|Z_1) = \frac{f_{\Sigma|\Gamma}(Z_1|X) f_\Gamma(X) \delta X}{\int f_{\Sigma|\Gamma}(Z_1|Y) f_\Gamma(Y) \delta Y}$$

The following proposition verifies that the marginal distribution is actually a global density, which in turn implies that the posterior distribution is obviously a global density:

Proposition 10 *The marginal distribution is a global density.*

Proof. We have:

$$\int f_\Sigma(Z) \delta Z = \sum_{j=0}^{\infty} \int_{|Z|=j} \left(\sum_{k=0}^{\infty} \int_{|X|=k} f_{\Sigma|\Gamma}(Z|X) f_\Gamma(X) \delta X \right) \delta Z$$

$$= \sum_{k=0}^{\infty} \int_{|X|=j} \left(\sum_{j=0}^{\infty} \int_{|Z|=k} f_{\Sigma|\Gamma}(Z|X) \delta Z \right) f_\Gamma(X) \delta X$$

$$= \sum_{k=0}^{\infty} \int_{|X|=j} f_\Gamma(X) \delta X = 1$$

where the infinite sums are actually finite. $\qquad\square$

These comments can be extended to the case of multiple independently distributed samples as follows. Suppose that $\Sigma_1, ..., \Sigma_m$ are conditionally independent, absolutely continuous finite random subsets of $\mathcal{R}$ all of which have the same global density $f_{\Sigma|\Gamma}(Z|X)$. Let $Z_1, ..., Z_m$ be finite subsets of $\mathcal{R}$. Then the joint global marginal density is

$$f_{\Sigma_1, ..., \Sigma_m}(Z_1, ..., Z_m) = \int f_{\Sigma_1, ..., \Sigma_m|\Gamma}(Z_1, ..., Z_m|Y) f_\Gamma(Y) \delta Y$$

where $f_{\Sigma_1, ..., \Sigma_m|\Gamma}(Z_1, ..., Z_m|Y) = f_{\Sigma|\Gamma}(Z_1|Y) \cdots f_{\Sigma|\Gamma}(Z_m|Y)$. Likewise, the joint global posterior is

$$f_{\Gamma|\Sigma_1, ..., \Sigma_m}(X|Z_1, ..., Z_m) = \frac{f_{\Sigma_1, ..., \Sigma_m|\Gamma}(Z_1, ..., Z_m|X) f_\Gamma(X)}{\int f_{\Sigma_1, ..., \Sigma_m|\Gamma}(Z_1, ..., Z_m|Y) f_\Gamma(Y) \delta Y}$$

We also denote the cardinality distribution of the posterior as:

$$f_{\Gamma|\Sigma}(j|Z_1, ..., Z_m) \triangleq \int_{|Y|=j} f_{\Gamma|\Sigma}(Y|Z_1, ..., Z_m)\delta Y$$

The following example provides a very simple illustration of a global posterior.

Example 6 *Consider the parametrized global densities of Example 4: Let us be given the global density ("measurement model") of Example 4:*

$$
\begin{aligned}
f_\Sigma(\emptyset|\{\zeta\}) &= 1 - q \\
f_\Sigma(\{\xi\}|\{\zeta\}) &= q\, f(\xi|\zeta) \\
f_\Sigma(\emptyset|\{\zeta_1, \zeta_2\}) &= (1-q)^2 \\
f_\Sigma(\{\xi\}|\{\zeta_1, \zeta_2\}) &= q(1-q)\, f(\xi|\zeta_1) + q(1-q)\, f(\xi|\zeta_2) \\
f_\Sigma(\{\xi_1, \xi_2\}|\{\zeta_1, \zeta_2\}) &= q^2\, f(\xi_1|\zeta_1)\, f(\xi_2|\zeta_2) + q^2\, f(\xi_2|\zeta_2)\, f(\xi_1|\zeta_2)
\end{aligned}
$$

where q is the probability of detection. Let the global prior be

$$f_\Gamma(\emptyset) = 1 - q_0, \qquad f_\Gamma(\{\zeta\}) = q_0\, \pi(\zeta)$$

for all $\zeta \in \mathcal{R}$, where $\pi(\zeta)$ denotes the conventional prior. That is, in choosing f_Γ as a global prior we are asserting that, on the basis of our previous experience, we believe that: (1) there is a probability $1 - q_0$ that there are no objects to be observed; and (2) the probability density of there being exactly one object, and that object is ζ, is $q_0\, \pi(\zeta)$. Given this, the global marginal density is given by

$$
\begin{aligned}
f_\Sigma(\emptyset) &= \int f_{\Sigma|\Gamma}(\emptyset|Y)\, f_\Gamma(Y)\delta Y \\
&= f_{\Sigma|\Gamma}(\emptyset|\emptyset) f_\Gamma(\emptyset) + \int f_{\Sigma|\Gamma}(\emptyset|\{\zeta\}) f_\Gamma(\{\zeta\}) d\lambda(\zeta) \\
&\quad + \frac{1}{2}\int f_{\Sigma|\Gamma}(\emptyset|\{\zeta_1, \zeta_2\}) f_\Gamma(\{\zeta_1, \zeta_2\}) d\lambda(\zeta_1) d\lambda(\zeta_2) \\
&= 1 - q_0 + q_0(1-q_0)\int \pi(\zeta) d\lambda(\zeta) + 0 = 1 - q_0 q
\end{aligned}
$$

and

$$
f_\Sigma(\{\xi\}) = \int f_{\Sigma|\Gamma}(\{\xi\}|Y)\, f_\Gamma(Y)\delta Y
$$

$$
f_{\Sigma|\Gamma}(\{\xi\}|\emptyset) f_\Gamma(\emptyset) + \int f_{\Sigma|\Gamma}(\{\xi\}|\{\zeta\}) f_\Gamma(\{\zeta\}) d\lambda(\zeta)
$$

$$
+ \frac{1}{2}\int f_{\Sigma|\Gamma}(\{\xi\}|\{\zeta_1, \zeta_2\}) f_\Gamma(\{\zeta_1, \zeta_2\}) d\lambda(\zeta_1) d\lambda(\zeta_2)
$$

$$
= 0 + q_0 q \int f(\xi|\zeta)\pi(\zeta) d\lambda(\zeta) + 0 = q_0 q\, m(\xi)
$$

where $m(\zeta)$ denotes the conventional marginal distribution. (The other values $f_\Sigma(Z)$ of the marginal distribution vanish.) Thus, it is easily shown that the global posterior density is:

$$f_{\Gamma|\Sigma}(\emptyset|\emptyset) = \frac{1-q_0}{1-q_0q}$$

$$f_{\Gamma|\Sigma}(\{\zeta\}|\emptyset) = \frac{q_0 - q_0q}{1-q_0q}\pi(\zeta)$$

$$f_{\Gamma|\Sigma}(\{\zeta\}|\{\xi\}) = F(\zeta|\xi)$$

where $F(\zeta|\xi)$ denotes the conventional posterior. We note that the expected number of targets given a single observation Z is, therefore,

$$E[|\Gamma|] = \frac{q_0(1-q)}{1-q_0q}$$

given $Z = \emptyset$ and $E[|\Gamma|] = 1$ given $Z = \{\xi\}$. If $q_0 = 1$ then the posterior becomes

$$f_{\Gamma|\Sigma}(\{\zeta\}|\emptyset) = \pi(\zeta), \qquad f_{\Gamma|\Sigma}(\{\zeta\}|\{\xi\}) = F(\zeta|\xi)$$

This is the result one might intuitively expect: If we observe nothing (i.e., we observe $\emptyset$) then we have obtained no additional information. Therefore, the only thing we know about ζ is the information contained in the prior distribution. If we do observe something (i.e., ξ) then ζ is constrained by the usual posterior density. However, this reasoning assumes that there is, in fact, something to observe. If $q_0 < 1$ then there is a nonzero a priori probability $1 - q_0$ that no objects are present "out there." This means that if we observe nothing then there are two possibilities: there is actually nothing to observe; or, there is an object present and the choice of which object is constrained only by the prior distribution with a suitable multiplier involving the probability of detection q.

Example 7 Consider the previous example with the global prior

$$f_\Gamma(\emptyset) = (1-q_0)^2, \qquad f_\Gamma(\{\zeta\}) = q_0(1-q_0)\left(\pi_1(\zeta) + \pi_2(\zeta)\right)$$

$$f_\Gamma(\{\zeta_1, \zeta_2\}) = q_0^2\pi_1(\zeta_1)\pi_2(\zeta_2) + q_0^2\pi_1(\zeta_2)\pi_2(\zeta_1)$$

where π_1, π_2 are two conventional density functions. Some simple but tedious algebra leads to the following formulas for the global marginal:

$$f_\Sigma(\emptyset) = (1-q_0q)^2$$
$$f_\Sigma(\{\xi\}) = q_0q(1-q_0q)\left(m_1(\xi) + m_2(\xi)\right)$$
$$f_\Gamma(\{\zeta_1, \zeta_2\}) = q_0^2q^2m_1(\xi_1)m_2(\xi_2) + q_0^2q^2m_1(\xi_2)m_2(\xi_1)$$

where $m_j(\xi) = \int f(\xi|\zeta)\pi_j(\zeta)d\lambda(\zeta)$ are the conventional marginal distributions for $j = 1, 2$. Some more algebra shows that the corresponding global posterior density is:

$$f_{\Gamma|\Sigma}(\emptyset|\emptyset) \;=\; \frac{(1-q_0)^2}{(1-q_0q)^2}$$

$$f_{\Gamma|\Sigma}(\{\zeta\}|\emptyset) \;=\; \frac{(1-q)(1-q_0)q_0}{(1-q_0q)^2}(\pi_1(\zeta)+\pi_2(\zeta))$$

$$f_{\Gamma|\Sigma}(\{\zeta_1,\zeta_2\}|\emptyset) \;=\; \frac{(1-q)^2 q_0^2}{(1-q_0q)^2}(\pi_1(\zeta_1)\pi_2(\zeta_2)+\pi_1(\zeta_2)\pi_2(\zeta_1))$$

$$f_{\Gamma|\Sigma}(\{\zeta\}|\{\xi\}) \;=\; \frac{1-q_0}{1-q_0q}\, f(\xi|\zeta)\,\frac{\pi_1(\zeta)+\pi_2(\zeta)}{m_1(\xi)+m_2(\xi)}$$

$$f_{\Gamma|\Sigma}(\{\zeta_1,\zeta_2\}|\{\xi\}) \;=\; \frac{(1-q)q_0}{1-q_0q}\cdot(f(\xi|\zeta_1)+f(\xi|\zeta_2))$$
$$\cdot\frac{\pi_1(\zeta_1)\pi_2(\zeta_2)+\pi_1(\zeta_2)\pi_2(\zeta_1)}{m_1(\xi)+m_2(\xi)}$$

$$f_{\Gamma|\Sigma}(\{\zeta_1,\zeta_2\}|\{\xi_1,\xi_2\}) \;=\; (f(\xi_1|\zeta_1)f(\xi_2|\zeta_2)+f(\xi_2|\zeta_1)f(\xi_1|\zeta_2))$$
$$\cdot\frac{\pi_1(\zeta_1)\pi_2(\zeta_2)+\pi_1(\zeta_2)\pi_2(\zeta_1)}{m_1(\xi_1)m_2(\xi_2)+m_1(\xi_2)m_2(\xi_1)}$$

The expected number of targets, given a single global observation, is therefore given by:

$$E[|\Gamma|] = \frac{q_0(1-q)}{1-q_0q}\left(1+\frac{q_0(1-q)}{1-q_0q}\right)$$

$$E[|\Gamma|] = 1 + \frac{q_0(1-q)}{1-q_0q}, \qquad E[|\Gamma|] = 2$$

for $Z = \emptyset$, $Z = \{\xi\}$, and $Z = \{\xi_1,\xi_2\}$, respectively.

5.2   Global Parametric Estimation

Let us be given a simple single-sensor, single-target "localizer": that is, an algorithm which accepts independent observations $z_1, ..., z_m$ of corrupted data as input and tries to construct a more accurate location estimate of the target location. The noise statistics of the sensor are modeled by a random vector $\mathbf{Z}$. Let us also assume that the density $f_{\mathbf{Z}}$ of $\mathbf{Z}$ has the parametrized form $f_{\mathbf{Z}}(\mathbf{z}|\mathbf{x})$. If $\mathbf{Z}_1, ..., \mathbf{Z}_m$ are statistically independent random vectors all of which have $f_{\mathbf{Z}}(\mathbf{z}|\mathbf{x})$ as their densities, then an "estimator" of the parameter x is a random vector $\mathbf{X} = \mathbf{J}(\mathbf{Z}_1, ..., \mathbf{Z}_m)$ of the random measurements which is, in some explicitly specified sense, more "closely" clustered around $\mathbf{x}$ than the random vector $\mathbf{Z}$ itself.

More generally, one may attempt not to estimate $\mathbf{x}$ directly, but rather some transformation (some statistic) $\mathbf{T}(\mathbf{x})$ of $\mathbf{x}$. For example, $\mathbf{J}(z_1, ..., z_m) = \frac{1}{m}(z_1 + ... + z_m)$ is an estimator of $\mathbf{T}(\mathbf{x}) = \mathbf{E}_{\mathbf{x}}[\mathbf{Z}]$, the expected value of $\mathbf{Z}$.

The most familiar and generally well-behaved estimator is the *maximum likelihood estimator (MLE)*. Let $z_1, ..., z_m$ be a fixed set of independent samples.

Given the likelihood function

$$L(\mathbf{x}|\mathbf{z}_1, ..., \mathbf{z}_m) = f_{\mathbf{Z}_1,...,\mathbf{Z}_m}(\mathbf{z}_1, ..., \mathbf{z}_m|\mathbf{x}) = f_{\mathbf{Z}}(\mathbf{z}_1|\mathbf{x}) \cdots f_{\mathbf{Z}}(\mathbf{z}_m|\mathbf{x})$$

the maximum likelihood estimator $\mathbf{J}_{MLE}(\mathbf{z}_1, ..., \mathbf{z}_m)$ of the parameter $\mathbf{x}$, if it is defined, is that (presumably unique) value of $\mathbf{x}$ which maximizes $L(\mathbf{x}|\mathbf{z}_1, ..., \mathbf{z}_m)$ over all values $\mathbf{x} \in \mathbb{R}^n$. Another familiar example is the *maximum a posteriori (MAP)* estimator, also known as the "generalized MLE." Here, we assume that the vector parameter $\mathbf{x}$ is itself a random quantity $\mathbf{X}$ governed by an *a priori* probability distribution $f_{\mathbf{X}}(\mathbf{x})$, so that the density governing $\mathbf{Z}$ has the form $f_{\mathbf{Z}|\mathbf{X}}(\mathbf{z}_1|\mathbf{x})$. The MAP estimate of $\mathbf{x}$ is that value of $\mathbf{x}$ which maximizes the posterior density

$$f_{\mathbf{X}|\mathbf{Z}_1,...,\mathbf{Z}_m}(\mathbf{x}|\mathbf{z}_1, ..., \mathbf{z}_m) = \frac{f_{\mathbf{Z}|\mathbf{X}}(\mathbf{z}_1|\mathbf{x}) \cdots f_{\mathbf{Z}|\mathbf{X}}(\mathbf{z}_m|\mathbf{x}) \, f_{\mathbf{X}}(\mathbf{x})}{f_{\mathbf{Z}_1,...,\mathbf{Z}_m}(\mathbf{z}_1, ..., \mathbf{z}_m)}$$

In this chapter we directly generalize the elements of parametric estimation to the case when the random quantities of interest are *randomly varying finite sets*.

5.2.1   Global Estimators of Vector-Valued Functions

Definition 3 *(a) Let $\Sigma_1, ..., \Sigma_m$ be independent finite random subsets all of which have a common global density function $f_{\Sigma}(Z|X)$, where X is a set parameter. Then a global estimator of the set parameter X is a finite random set of the form $\Lambda = J(\Sigma_1, ..., \Sigma_m)$. More generally, let $\mathbf{T}(X)$ be a measurable vector-valued transformation of X. Then a global estimator of $\mathbf{T}(X)$ is a random vector of the form $\mathbf{Y} = \mathbf{J}(\Sigma_1, ..., \Sigma_m)$. (b) A global estimator $\mathbf{J}$ of $\mathbf{T}(X)$ is unbiased if*

$$E_X[\mathbf{J}(\Sigma_1, ..., \Sigma_m)] = \mathbf{T}(X)$$

where, as defined in Section 5.1.1,

$$E_X[\mathbf{J}(\Sigma_1, ..., \Sigma_m)] \triangleq \int \mathbf{J}(Z_1, ..., Z_m) f_{\Sigma}(Z_1|X) \cdots f_{\Sigma}(Z_m|X) \delta Z_1 \cdots \delta Z_m$$

Example 8 *Define $J(Z_1, ..., Z_m) \triangleq \frac{1}{m}(|Z_1| + ... + |Z_m|)$. Then it is easily seen that*

$$E_X[J(\Sigma_1, ..., \Sigma_m)] = E_X[|\Sigma|]$$

That is, J is an unbiased estimator of $T(X) = E_X[|\Sigma|]$, the expected number of elements in Σ.

The following generalizes the concept of a Bayesian estimator [17, pp. 87, 300], [21, pp. 54-58]:

Definition 4 *Let Σ be an absolutely continuous finite random subset of $\mathcal{R}$ with global density $f_{\Sigma}(Z|X)$, let $\Sigma_1, ..., \Sigma_m$ be i.i.d. finite random subsets which have $f_{\Sigma}(Z|X)$ as their common global density, and let $f_{\Gamma}(X)$ be the global density* *of the set parameter* X. *Then: (a) Let* $C(Y, Z)$ *be a nonnegative-valued cost function of two finite-set variables* Y, Z *and let* J *be an estimator of the set parameter* X. *The global risk associated with the estimator* J *is the expected cost* $E_X[C(X, J(\Sigma_1, ..., \Sigma_m))]$, *i.e.:*

$$R_J(X) \triangleq \int C(X, J(Z_1, ..., Z_m)) f_{\Sigma|\Gamma}(Z_1|X) \cdots f_{\Sigma|\Gamma}(Z_m|X) \delta Z_1 \cdots \delta Z_m$$

The global Bayes risk is the expected value $R(J) = E[R_J(\Gamma)]$ *of the risk:*

$$R(J) \triangleq \int \int C(X, J(Z_1, ..., Z_m))$$
$$\cdot f_{\Sigma|\Gamma}(Z_1|X) \cdots f_{\Sigma|\Gamma}(Z_m|X) f_\Gamma(X) \delta Z_1 \cdots \delta Z_m \delta X$$

A global Bayes estimator is any estimator J *which minimizes the global Bayes risk. (b) Likewise, suppose that* $C(\mathbf{x}, \mathbf{y})$ *is a nonnegative cost function defined on* $\mathbb{R}^r$. *Then the global risk associated with the estimator* $\mathbf{J}$ *of* $\mathbf{T}(X)$ *is*

$$R_J(\mathbf{x}) \triangleq E_\mathbf{x}[C(\mathbf{T}(X), \mathbf{J}(\Sigma_1, ..., \Sigma_m))]$$

and the global Bayes risk is $R(\mathbf{J}) \triangleq E[R_J(\Gamma)]$. *A global Bayes estimator for* $\mathbf{T}(X)$ *is one which minimizes* $R(\mathbf{J})$.

5.2.2   The Global ML and MAP Estimators

We can now define the "global" analogs of the ML and MAP estimators:

Definition 5 *Let* Σ *be an absolutely continuous finite random subset of* $\mathcal{R}$ *and let* $\Sigma_1, ..., \Sigma_m$ *be independent, absolutely continuous finite random subsets all of which have* $f_\Sigma(Z|X)$ *as their global densities. Let* $Z_1, ..., Z_m$ *be a sequence of independent global observations. Then the global maximum likelihood estimator (global MLE) is*

$$J_{GML}(Z_1, ..., Z_m) \triangleq \arg \sup_{X \subseteq \mathcal{R}} L(X|Z_1, ..., Z_m)$$

Here, $L(X|Z_1, ..., Z_m) \triangleq f_\Sigma(Z_1|X) \cdots f_\Sigma(Z_m|X)$ is the "global" likelihood function and "$X_0 = \arg \max L(X)$" means that $L(X_0)$ is the supremal value of $L(X)$ among all finite $X \subseteq \mathcal{R}$. Notice that for fixed $Z_1, ..., Z_m$, the units of $L(X|Z_1, ..., Z_m)$ are always the same regardless of the value of X. Thus $J_{GML}(\Sigma_1, ..., \Sigma_m)$ is well-defined. For example, if the units of $\mathcal{R}$ are meters then $L(X|Z_1, ..., Z_m)$ has units of $meter^{-e}$ where $e = |Z_1| + ... + |Z_m|$.

The GMLE can be extended in the usual way to define estimators of *vector-valued functions* of the set parameter. Suppose that $\mathbf{F}(X)$ is the vector-valued function whose value we are to estimate. Define the generalized likelihood function as

$$L^*(\mathbf{y}|Z_1, ..., Z_m) \triangleq \sup_{\mathbf{F}(X)=\mathbf{y}} f_\Sigma(Z_1|X) \cdots f_\Sigma(Z_m|X)$$

Then the GMLE is determined by finding that $\mathbf{y}$ which maximizes L^*. The GMLE obviously satisfies the usual MLE invariance principle, in that the GML estimate of $\mathbf{y} = \mathbf{F}(X)$ is just $\hat{\mathbf{y}} = \mathbf{F}(\hat{X})$ where $\hat{X}$ is the GML estimate of X.

Example 9 *Consider the simple parametrized global density*

$$f_{\Sigma|\Gamma}(\emptyset|\emptyset) = 1, \qquad f_{\Sigma|\Gamma}(\emptyset|\{x\}) = 1 - q, \qquad f_{\Sigma|\Gamma}(\{z\}|\{x\}) = q \, f(z|x).$$

Suppose further that the following global observations are collected: $\Sigma = \{a_1\}$, $\Sigma = \emptyset$, *and* $\Sigma = \{a_2\}$. *The corresponding likelihood function is given by* $L(\emptyset|\{a_1\}, \emptyset, \{a_2\}) = 0$ *and by*

$$
\begin{aligned}
L(\{x\}|\{a_1\}, \emptyset, \{a_2\}) &= f_{\Sigma|\Gamma}(\{a_1\}|\{x\}) \, f_{\Sigma|\Gamma}(\emptyset|\{x\}) \, f_{\Sigma|\Gamma}(\{a_2\}|\{x\}) \\
&= q^2(1 - q) \, f(a_1|x) \, f(a_2|x),
\end{aligned}
$$

where $L(x|a_1, a_2)$ *is the conventional likelihood function corresponding to the reports* a_1, a_2. *Suppose further that the parametrized family is extended to include*

$$
\begin{aligned}
f_{\Sigma|\Gamma}(\emptyset|\{x_1, x_2\}) &= (1 - q)^2 \\
f_{\Sigma|\Gamma}(\{z\}|\{x_1, x_2\}) &= q(1 - q) \, f(z|x_1) + q(1 - q) \, f(z|x_2) \\
f_{\Sigma|\Gamma}(\{z_1, z_2\}|\{x_1, x_2\}) &= q^2 \, f(z_1|x_1) f(z_2|x_2) + q^2 \, f(z_2|x_1) f(z_1|x_2)
\end{aligned}
$$

Suppose that two reports $\Sigma = \{a\}$ *and* $\Sigma = \{b_1, b_2\}$ *are collected. Then the global likelihood function is given by*

$$L(\emptyset|\{a\}, \{b_1, b_2\}) = 0 = L(x|\{a\}, \{b_1, b_2\})$$

$$
\begin{aligned}
L(\{x_1, x_2\}|\{a\}, \{b_1, b_2\}) &= q^3(1 - q) \, (f(a|x_1) + f(a|x_2)) \\
&\quad \cdot [f(b_1|x_1) f(b_2|x_2) + f(b_2|x_1) f(b_1|x_2)].
\end{aligned}
$$

The global ML estimate consists of those x_1, x_2 *which maximize* $L(\{x_1, x_2\}|\{a\}, \{b_1, b_2\})$. $\qquad\square$

We introduce two generalizations of the MAP estimator, as follows:

Definition 6 *(a) The global MAP estimator of the first kind (GMAPE-I) is*

$$J_{GMAP-I}(Z_1, ..., Z_m) \triangleq \arg \max_{X \subseteq \mathcal{R}} f_{\Gamma|\Sigma_1, ..., \Sigma_m}(X|Z_1, ..., Z_m)$$

(b) Let $K_0 \subseteq \mathcal{R}$ *be a fixed compact subset whose measure* $\lambda(K_0)$ *is very small. Then the global MAP estimator of the second kind (GMAPE-II), with respect to* K_0, *is*

$$J_{GMAP-II}(Z_1, ..., Z_m) \triangleq \arg \sup_{X \subseteq \mathcal{R}} f_{\Gamma|\Sigma_1, ..., \Sigma_m}(X|Z_1, ..., Z_m) \lambda(K_0)^{|X|}$$

Here, "arg max" is shorthand for the following procedure. First, one computes the cardinality distribution (Definition 18 of Chapter 4)

$$f_{\Gamma|\Sigma_1, ..., \Sigma_m}(k|Z_1, ..., Z_m) \triangleq \int_{|X|=k} f_{\Gamma|\Sigma_1, ..., \Sigma_m}(X|Z_1, ..., Z_m) \delta X$$

of the posterior distribution. Then one determines that value $k = j_{MAP}$ which maximizes the cardinality distribution. Finally, one maximizes the posterior $f_{\Gamma|\Sigma_1,...,\Sigma_m}(X|Z_1,...,Z_m)$ subject to the constraint that $|X| = j_{MAP}$.

Both of the global MAP estimators just defined are Bayes estimators. To show this, however, let us first define the following cost functions. Let $Z = \{\xi_1,...,\xi_r\}$ and $Y = \{\zeta_1,...,\zeta_s\}$ with $|Z| = r$ and $|Y| = s$ and let $\xi_i = (\mathbf{x}_i, w_i)$ for all $l = 1,...,r$ and $\zeta_j = (\mathbf{y}_j, v_j)$ for all $j = 1,...,s$. In analogy with van Trees [21, p. 55-58] define the "infinitesimal notch" cost functions C_1 and C as follows:

$$C_0(Y, Z) = \begin{cases} 0 & if \quad |Y| = |Z| \\ 1 & if \quad |Y| \neq |Z| \end{cases}$$

$$C_1(Y, Z) = \begin{cases} 0 & if \quad \begin{cases} r = s \text{ and} \\ (w_1,...,w_r) = (v_1,...,v_r) \text{ and} \\ (\mathbf{x}_1,...,\mathbf{x}_r), (\mathbf{y}_{\pi 1},...,\mathbf{y}_{\pi r}) \in K \text{ for some } \pi \end{cases} \\ 1 & if \quad otherwise \end{cases}$$

and

$$C(Y, Z) = C_0(Y, Z) + C_1(Y, Z)$$

where K is some closed ball in $(\mathbb{R}^n)^r$ with $\lambda(K) = \lambda(K_0)^r$ and where π denotes a permutation on the numbers $1,...,r$. Then:

Proposition 11 *GMAP-I and GMAP-II are the Bayes estimators corresponding to the cost functions C_0 and C, respectively.*

Proof. We deal with the case for GMAP-II. The case for GMAP-I is similar and will be left to the reader. We follow the proof of van Trees [21, pp. 56-57]. We are to minimize the risk

$$R(J) = \int \int C(X, J(Z_1,...,Z_m)) f_{\Sigma_1,...,\Sigma_m|\Gamma}(Z_1,...,Z_m|X)\delta Z_1 \cdots \delta Z_m \delta X$$

$$= \int \int C(X, J(Z_1,...,Z_m)) f_{\Gamma|\Sigma_1,...,\Sigma_m}(X|Z_1,...,Z_m)$$
$$\cdot f_{\Sigma_1,...,\Sigma_m}(Z_1,...,Z_m)\delta Z_1 \cdots \delta Z_m \delta X$$

$$= \int \left(\int C(X, J(Z_1,...,Z_m)) f_{\Gamma|\Sigma_1,...,\Sigma_m}(X|Z_1,...,Z_m)\delta X \right)$$
$$\cdot f_{\Sigma_1,...,\Sigma_m}(Z_1,...,Z_m)\delta Z_1 \cdots \delta Z_m$$

The risk will be minimized if the innermost set integral is minimized for all possible choices of $Z_1,...,Z_m$. However, by the definition of a set integral and of the cost function C this is just:

$$\sum_{s=0}^{\infty} \frac{1}{s!} \int C(\{\zeta_1,...,\zeta_s\}, J(Z_1,...,Z_m)) f_{\Gamma|\Sigma_1,...,\Sigma_m}(\{\zeta_1,...,\zeta_s\}|Z_1,...,Z_m)$$
$$\cdot d\lambda(\zeta_1) \cdots d\lambda(\zeta_s)$$

$$= 2 \sum_{s \neq |J|}^{\infty} \frac{1}{s!} \int f_{\Gamma|\Sigma_1,\ldots,\Sigma_m}(\{\zeta_1,\ldots,\zeta_s\}|Z_1,\ldots,Z_m)d\lambda(\zeta_1)\cdots d\lambda(\zeta_s)$$

$$+ \frac{1}{|J|!} \int_{G(|J|)^c} f_{\Gamma|\Sigma_1,\ldots,\Sigma_m}(\{\zeta_1,\ldots,\zeta_{|J|}\}|Z_1,\ldots,Z_m)d\lambda(\zeta_1)\cdots d\lambda(\zeta_{|J|})$$

where $G(|J|)$ is some very small closed ball surrounding the point $J = J(Z_1,\ldots,Z_m)$ in $\mathcal{R}^{|J|}$. Now on the one hand,

$$\sum_{s \neq |J|}^{\infty} \frac{1}{s!} \int f_{\Gamma|\Sigma_1,\ldots,\Sigma_m}(\{\zeta_1,\ldots,\zeta_s\}|Z_1,\ldots,Z_m)d\lambda(\zeta_1)\cdots d\lambda(\zeta_s)$$

$$= 1 - f_{\Gamma|\Sigma_1,\ldots,\Sigma_m}(|J|)$$

where $f_{\Gamma|\Sigma_1,\ldots,\Sigma_m}(|J|)$ denotes the cardinality distribution of Γ (Definition 18 of Chapter 4) evaluated at $s = |J|$. On the other hand,

$$1 = \sum_{s \neq |J|}^{\infty} \frac{1}{s!} \int f_{\Gamma|\Sigma_1,\ldots,\Sigma_m}(\{\zeta_1,\ldots,\zeta_s\}|Z_1,\ldots,Z_m)d\lambda(\zeta_1)\cdots d\lambda(\zeta_s)$$

$$+ \frac{1}{|J|!} \int_{G(|J|)^c} f_{\Gamma|\Sigma_1,\ldots,\Sigma_m}(\{\zeta_1,\ldots,\zeta_{|J|}\}|Z_1,\ldots,Z_m)d\lambda(\zeta_1)\cdots d\lambda(\zeta_{|J|})$$

$$+ \int_{G(|J|)} f_{\Gamma|\Sigma_1,\ldots,\Sigma_m}(\{\zeta_1,\ldots,\zeta_{|J|}\}|Z_1,\ldots,Z_m)d\lambda(\zeta_1)\cdots d\lambda(\zeta_{|J|})$$

where the factorial divisor $|J|!$ does not occur in front of the final integral since the closed ball $G(|J|)$ is so small that it surrounds a vector whose components are only one possible permutation of the elements $\zeta_1,\ldots,\zeta_{|J|}$. Therefore it follows that

$$\int C(X, J(Z_1,\ldots,Z_m))f_{\Gamma|\Sigma_1,\ldots,\Sigma_m}(X|Z_1,\ldots,Z_m)\delta X$$

$$\approx 2 - f_{\Gamma|\Sigma_1,\ldots,\Sigma_m}(|J|) - f_{\Gamma|\Sigma_1,\ldots,\Sigma_m}(\{\zeta_1,\ldots,\zeta_{|J|}\}|Z_1,\ldots,Z_m)\lambda(G(|J|))$$

by the Lebesgue density theorem, where $\lambda(G(|J|))$ denotes the Lebesgue measure of $G(|J|)$. Now, note that in the last approximate equality,

$$f_{\Gamma|\Sigma_1,\ldots,\Sigma_m}(|J|) >> f_{\Gamma|\Sigma_1,\ldots,\Sigma_m}(\{\zeta_1,\ldots,\zeta_{|J|}\}|Z_1,\ldots,Z_m)\lambda(G(|J|))$$

where '$>>$' denotes "much greater than". Consequently the integral

$$\int C(X, J(Z_1,\ldots,Z_m))f_{\Gamma|\Sigma_1,\ldots,\Sigma_m}(X|Z_1,\ldots,Z_m)\delta X$$

will be minimized only if $f_{\Gamma|\Sigma_1,\ldots,\Sigma_m}(|J|)$ is the maximum of the values $f_{\Gamma|\Sigma_1,\ldots,\Sigma_m}(k)$ of the cardinality distribution. Given that fact, it will be minimized only if $f_{\Gamma|\Sigma_1,\ldots,\Sigma_m}(X|Z_1,\ldots,Z_m)$ is maximized over all finite subsets $X \subseteq \mathcal{R}$ with $|X| = |J|$ for any given list $Z_1,\ldots,Z_m$ of global samples. $\square$

Example 10 *Consider the global posterior density computed in Example 6:*

$$f_{\Gamma|\Sigma}(\emptyset|\emptyset) = \frac{1 - q_0}{1 - q_0 q}$$

$$f_{\Gamma|\Sigma}(\{\mathbf{x}\}|\emptyset) = \frac{q_0 - q_0 q}{1 - q_0 q}\pi(\mathbf{x})$$

$$f_{\Gamma|\Sigma}(\{\mathbf{x}\}|\{\mathbf{z}\}) = F(\mathbf{x}|\mathbf{z})$$

Suppose that we have exactly one observation, $Z = \{\mathbf{a}\}$. Then $J_{MAP-I}(Z) = \arg\sup_{\mathbf{x}} F(\mathbf{x}|\mathbf{z})$ which is just the conventional single-observation MAP estimate. On the other hand suppose that $Z = \emptyset$ (i.e., we did not see anything). Then $J_{MAP-I}(Z) = \arg\max_{X \subseteq \mathbb{R}^n} f_{\Gamma|\Sigma}(X|\emptyset)$. Since the cardinality distribution of the global posterior is

$$f_{\Gamma|\Sigma}(0|\emptyset) = \frac{1 - q_0}{1 - q_0 q}, \qquad f_{\Gamma|\Sigma}(1|\emptyset) = \frac{q_0(1 - q)}{1 - q_0 q}, \qquad f_{\Gamma|\Sigma}(1|\{\mathbf{z}\}) = 1$$

it follows that $J_{MAP-I}(Z) = \emptyset$ if $f_{\Gamma|\Sigma}(0|\emptyset) > f_{\Gamma|\Sigma}(1|\emptyset)$ and hence if $q > 2 - \frac{1}{q_0}$ which, note, will always be the case if $q_0 < \frac{1}{2}$. Otherwise, if $q_0 > \frac{1}{2}$ and $q < 2 - \frac{1}{q_0}$ then $J_{MAP-I}(Z) = \arg\sup_{\mathbf{x} \in \mathbb{R}^n} \pi(\mathbf{x})$. □

5.2.3 "Set Parameters" and the Statistical Consistency of the Global ML and MAP Estimators

Set Parameters or "Multiset" Parameters?

In our discussions of global densities we have casually referred to a "set parameter" as an obvious generalization of the concept of a parameter to the case of finite set-valued random variates. In actuality this naive perspective glosses over important subtleties. The most important of these is the fact that a critical statistical property—the consistency of the maximum likelihood estimator (see Theorem 14)—*depends on the topological good behavior of "set parameters."* Specifically, we need to know two things:

Property 1: The space of "set parameters" is a complete metric space which satisfies the following property: every bounded closed subset is compact; and

Property 2: The parametrized global density $f_\Sigma(Z|X)$ is continuous in the parameter-variable X with respect to the metric topology.

(Recall that the *diameter* of a subset S of a metric space is $diam(S) = \sup_{x,y \in S} d(x,y)$ and that S is *bounded* if $diam(S) < \infty$.)

Unfortunately it turns out that, if we take the naive perspective regarding set parameters, global densities $f_{\Sigma|\Gamma}(Z|X)$ are *discontinuous functions (with respect to the hit-or-miss topology) in both the observation variable Z and the parameter variable X.* That is, if $X_i \in c_{\leq M}(\mathbb{R}^n)$ is a sequence of finite subsets of $\mathbb{R}^n$ which converges to X_0, i.e., $d(X_i, X_0) \to 0$ as $i \to \infty$ where $d(-, -)$ denotes the Hausdorff metric (see Section 4.1.3), then it will often be the case that $\lim_{i \to \infty} f_{\Sigma|\Gamma}(Z|X_i) \neq f_{\Sigma|\Gamma}(Z|X_0)$ and likewise for the observation variable Z. This fact is illustrated in the following example:

Example 11 *Consider the parametrized family of global densities of Example 4:*

$$f_\Sigma(\emptyset|\{\mathbf{x}\}) = 1 - q$$
$$f_\Sigma(\{\mathbf{z}\}|\{\mathbf{x}\}) = q\, f(\mathbf{z}|\mathbf{x})$$
$$f_\Sigma(\emptyset|\{\mathbf{x}_1,\mathbf{x}_2\}) = (1-q)^2$$
$$f_\Sigma(\{\mathbf{z}\}|\{\mathbf{x}_1,\mathbf{x}_2\}) = q(1-q)\, f(\mathbf{z}|\mathbf{x}_1) + q(1-q)\, f(\mathbf{z}|\mathbf{x}_2)$$
$$f_\Sigma(\{\mathbf{z}_1,\mathbf{z}_2\}|\{\mathbf{x}_1,\mathbf{x}_2\}) = q^2\, f(\mathbf{z}_1|\mathbf{x}_1)\, f(\mathbf{z}_2|\mathbf{x}_2) + q^2\, f(\mathbf{z}_2|\mathbf{x}_2)\, f(\mathbf{z}_1|\mathbf{x}_2)$$

where now we assume that $f(\mathbf{z}|\mathbf{x})$ is continuous in $\mathbf{x}$. Let a_i be a sequence of real numbers which converges to zero and let $\mathbf{y}$ be a nonzero vector in $\mathbb{R}^n$. Then, with respect to the Hausdorff metric, $X_i = \{a_i\mathbf{y}, \mathbf{0}\}$ converges to $X_0 = \{\mathbf{0}\}$ as $i \to \infty$:

$$d(\{a_i\mathbf{y}, \mathbf{0}\}, \{\mathbf{0}\}) = |a_i| \cdot \|\mathbf{y}\| \to 0$$

Nevertheless, $f_\Sigma(\{\mathbf{z}\}|\{\mathbf{0}\}) = q\, f(\mathbf{z}|\mathbf{0})$ but

$$f_\Sigma(\{\mathbf{z}\}|\{a_i\mathbf{y}, \mathbf{0}\}) \to 2q(1-q)\, f(\mathbf{z}|\mathbf{x})$$

The reason for this behavior is the fact that, when we estimate the state of a continuous system using a global density $f_{\Sigma|\Gamma}(Z|X)$, it may happen that the "set parameter" X may actually consist of a list of the form $X = \{\mathbf{x}_1, ..., \mathbf{x}_t\}$ where $\mathbf{x}_i = \mathbf{x}_j$ for some $i \neq j$. For example, there may be t targets two of which happen to have identical geokinematic states.

The following example demonstrates that such discontinuous behavior arises because we are assuming a multitarget system in which target states are incompletely specified: i.e., *the unique identifying tag of any target is being ignored.*

Example 12 *Consider the parametrized family*

$$f_\Sigma(\emptyset|\{\zeta\}) = 1 - q$$
$$f_\Sigma(\{\xi\}|\{\zeta\}) = q\, f(\xi|\zeta)$$
$$f_\Sigma(\emptyset|\{\zeta_1,\zeta_2\}) = (1-q)^2$$
$$f_\Sigma(\{\xi\}|\{\zeta_1,\zeta_2\}) = q(1-q)\, f(\xi|\zeta_1) + q(1-q)\, f(\xi|\zeta_2)$$
$$f_\Sigma(\{\xi_1,\xi_2\}|\{\zeta_1,\zeta_2\}) = q^2\, f(\xi_1|\zeta_1)\, f(\xi_2|\zeta_2) + q^2\, f(\xi_2|\zeta_2)\, f(\xi_1|\zeta_2)$$

where ξ, ζ have the form $\xi = (\mathbf{z}, v), \zeta = (\mathbf{x}, w)$, where v, w are identifying target tags—that is, it is acknowledged that any given target has a unique identity which differentiates it from all other targets. In this case we are forced to reformulate Example 11 as follows: $X_i = \{(a_i\mathbf{y}, v), (\mathbf{0}, w)\}$ and $X_0 = \{(\mathbf{0}, u)\}$. In this case

$$d(X_i, X_0)^2 = d(\{(a_i\mathbf{y}, v), (\mathbf{0}, w)\}, \{(\mathbf{0}, u)\})^2$$
$$= \max\{d((a_i\mathbf{y}, v), (\mathbf{0}, u))^2, \ d((\mathbf{0}, w), (\mathbf{0}, u))^2\}$$
$$= \max\{a_i^2 \|\mathbf{y}\|^2 + d(v, u)^2, \ d(w, u)^2\}.$$

Thus $d(X_i, X_0) \to 0$ is impossible unless $v = w = u$. However, $v = w$ means that the parameter set $X_i = \{(a_i\mathbf{y}, v), (\mathbf{0}, v)\}$ is describing a situation in which *the target with identifying tag v is occupying two different kinematic states simultaneously—a physically meaningless possibility. As long as $v \neq w$, limits $X_i \rightarrow X_0$ such that $\lim_{i \to \infty} |X_i| \neq |X_0|$ are not possible. Thus discontinuities $\lim_{i \to \infty} f_{\Sigma|\Gamma}(Z|X_i) \neq f_{\Sigma|\Gamma}(Z|X_0)$ of the kind in Example 11 cannot occur.*

Accordingly, as long as target states are completely specified (in the sense that each target platform has a unique identifying tag which specifies its unique identity) then discontinuities in the set parameter will not occur. However, completely specifying the state of targets is often inconvenient in situations in which sensors can supply only positional and/or velocity information but not target-I.D. information. In this case common practice—especially in the multitarget tracking literature—is to ignore the (implicit) tag part of the state and regard the full state of a target as though it had geodynamic state variables only.

If states are not uniquely specified—for example, if targets are describable only up to gross target class—then discontinuities *will* occur. The technical reason is that $c_k(\mathbb{R}^n)$ (the set of all finite subsets of $\mathbb{R}^n$ with exactly k elements) is not a *complete* metric space with respect to the Hausdorff metric: It does not contain all of its limit points. The sequence X_i in Example 11 converges to $\{0, 0\}$ which, being a set, is the same thing as the singleton set $\{0\}$. But $\{0\}$ is not in $c_2(\mathbb{R}^n)$. For $c_k(\mathbb{R}^n)$ to contain its limit points for each $k \geq 1$ we would in effect have to generalize the concept of a finite set $\{x_1, ..., x_t\}$ to what, in the literature (e.g., [19, p. 57]), is called a *multiset*: that is, a unordered list $/x_1, ..., x_t/$ which allows *repetition of elements, e.g.*:

$$/x, y/ = /y, x/, \qquad /x, x/ \neq /x/$$

For completeness we need, therefore, to take account of the fact that, when states are incompletely specified, a different topological/metrical structure will be needed on the parameter space.

We consider the two possible cases which must be taken into account when defining spaces of set parameters:

(1) states are completely specified (unique identifying target-tags)

(2) state are incompletely specified (no unique identifying target-tags)

In the first case, the parameter space will turn out to be a space of set parameters, whereas in the second case it will turn out to be a space of *multiset* parameters.

Case I: Completely Specified States

We assume that target states are completely specified and construct a well-behaved space of set parameters satisfying Property 1 as follows. Let K be a closed, compact domain of $\mathbb{R}^n$ (e.g., the geokinematic region of interest) and let $\mathcal{R}_K \triangleq K \times W$. Let $c(\mathcal{R}_K)$ be the space of closed subsets of $\mathcal{R}_K$, endowed with the hit-or-miss topology. Then $c(\mathcal{R}_K)$ is compact [15, p. 13, Prop. 1-4-1] and the empty set is an isolated point in $c(\mathcal{R}_K)$ [15, p. 13, Cor. 1]. Thus $c(\mathcal{R}_K) - \{\emptyset\}$ is also compact. So is $c_{\leq M}(\mathcal{R}_K) - \{\emptyset\}$ since $c_{\leq M}(\mathcal{R}_K)$ is closed in $c(\mathcal{R}_K)$ and hence $c_{\leq M}(\mathcal{R}_K) - \{\emptyset\} = (c(\mathcal{R}_K) - \{\emptyset\}) \cap c_{\leq M}(\mathcal{R}_K)$ is a closed subset of a compact space and hence also is compact. Finally, the topology on $c(\mathcal{R}_K) - \{\emptyset\}$ is equivalent to the metric topology induced by the Hausdorff metric (see [15, p. 15, Prop. 1-4-4]). The same will therefore be true for $c_{\leq M}(\mathcal{R}_K) - \{\emptyset\}$. Consequently we get a metric space $c_{\leq M}(\mathcal{R}_K) - \{\emptyset\}$ of set parameters which trivially satisfies the property that every closed bounded subset is compact.

However, this still does not get us out of the woods since we need Property 2 as well as Property 1. We thus specify that multitarget states of the general form

$$\{\zeta_1, ..., (\mathbf{x}_i, v), ..., (\mathbf{x}_j, v), ..., \zeta_t\} \in c_{\leq M}(\mathcal{R}_K) - \{\emptyset\}$$

for $\mathbf{x}_i \neq \mathbf{x}_j$ are *not legitimate states of the system* since, otherwise, it would be possible to have the same target (with identifying tag v) occupying two different geokinematic states $\mathbf{x}_i$ and $\mathbf{x}_j$ simultaneously. Thus $c_{\leq M}(\mathcal{R}_K) - \{\emptyset\}$ is not the metric space we are looking for.

Let $\pi : \mathcal{R}_K \to W$ be the projection operator defined by $\pi(\mathbf{x}, v) = v$ for all $\mathbf{x} \in K$ and $v \in W$. Extend π to a map $\pi : c(\mathcal{R}_K) \to \mathcal{P}(W)$ on closed subsets by defining $\pi(C) \triangleq \{\pi(\zeta) | \zeta \in C\}$ for all closed subsets $C \subseteq \mathcal{R}_K$.

The physically unrealizable states of the multitarget system consist of all finite subsets X such that $|\pi(X)| < |X|$ and so the physically realizable states are those X such that $\pi(X) \in \mathcal{P}_k(W)$ whenever $X \in c_k(\mathcal{R}_K)$, where $\mathcal{P}_k(W)$ denotes the set of all k-element subsets of W. In other words the class of all physically realizable set parameters is the subspace

$$\check{c}_{\leq M}(\mathcal{R}_K) \triangleq \bigcup_{k=0}^{M} \left(c_{\leq k}(\mathcal{R}_K) \cap \pi^{-1}(\mathcal{P}_k(W)) \right)$$

of all finite subsets $X \subseteq \mathcal{R}_K$ such that $|\pi(X)| = |X|$.

The following shows that $\check{c}_{\leq M}(\mathcal{R}_K)$ is a closed subset of the hit-or-miss topology and that, consequently, it contains all of its limit points.

Proposition 12 *(a) For any $S \in \mathcal{P}(W)$ the set $\pi^{-1}(S) \subseteq \mathcal{R}_K$ is closed in the hit-or-miss topology. (b) The subspace*

$$\check{c}_{\leq M}(\mathcal{R}_K) \triangleq \bigcup_{k=0}^{M} \left(c_{\leq k}(\mathcal{R}_K) \cap \pi^{-1}(\mathcal{P}_k(W)) \right)$$

is closed in the hit-or-miss topology.

Proof. (a) Since $\mathcal{P}(W)$ is endowed with the discrete topology (every subset of $\mathcal{P}(W)$ is both open and closed), it is enough to prove that $\pi^{-1}(\{S\})$ is closed in the relative hit-or-miss topology on $c_{\leq M}(\mathcal{R}_K)$ for any subset $S \subseteq W$. For, if $O \subseteq \mathcal{P}_k(W)$ then $\pi^{-1}(O) = \pi^{-1}(\bigcup_{S \in O} \{S\}) = \bigcup_{S \in O} \pi^{-1}(\{S\})$. Since $\mathcal{P}(W)$ is a finite set so is O and so the union is a finite union of closed subsets and therefore is also closed. Thus all we need prove is that the set of all $X \in$

$c_{\leq M}(\mathcal{R}_K)$ such that $\pi(X) = S$ is closed in the relative hit-or-miss topology on $c_{\leq M}(\mathcal{R})$. However,

$$\pi^{-1}(\{S\}) = \bigcap_{w \in W-S} O^c_{K \times w}$$

where, as always, for any open subset G of $\mathcal{R}$ the open subset O_G of the hit-or-miss topology on $c(\mathcal{R}_K)$ is $O_G \triangleq \{C \in c(\mathcal{R}_K)| \ C \cap G \neq \emptyset\}$.(Note that $K \times w$ is both open and closed in $\mathcal{R}_K$ for any $w \in W$.) For, suppose that there is a $C \in c(\mathcal{R}_K)$ such that $C \in \pi^{-1}(\{S\})$, so that $\pi(C) = S$. This simply means that $\pi(\zeta) \in S$ for all $\zeta \in C$, which means that $(K \times w) \cap C = \emptyset$ for all $w \in W-S$, which means that $C \in O^c_{K \times w}$ for all $w \in W-S$ and so $C \in \bigcap_{w \in W-S} O^c_{K \times w}$. Thus $\pi^{-1}(\{S\}) \subseteq \bigcap_{w \in W-S} O^c_{K \times w}$. Conversely, suppose that $C \in \bigcap_{w \in W-S} O^c_{K \times w}$. Then $(K \times w) \cap C = \emptyset$ for all $w \in W-S$, which means that C must be such that $\pi(\zeta) \in S$ for all $\zeta \in C$. Thus $\pi(C) = S$ and we are done. (b) Each space $c_{\leq k}(\mathcal{R})$ is closed in the hit-or-miss topology (Proposition 1 of Chapter 4) and $\pi^{-1}(\mathcal{P}_k(W))$ is closed for each $k \geq 1$ by part (a). Therefore $\check{c}_{\leq M}(\mathcal{R}_K)$ is closed in the hit-or-miss topology since it is a finite union of closed subsets of that topology.   □

Therefore we have:

Proposition 13 *The space $\check{c}_{\leq M}(\mathcal{R}_K) - \{\emptyset\}$ of set parameters is a complete metric space (with respect to the Hausdorff metric) which has Property 1: Every bounded closed subset of $\check{c}_{\leq M}(\mathcal{R}_K) - \{\emptyset\}$ is compact.*

Proof. Note that

$$\check{c}_{\leq M}(\mathcal{R}_K) - \{\emptyset\} = (c_{\leq M}(\mathcal{R}_K) - \{\emptyset\}) \cap \check{c}_{\leq M}(\mathcal{R}_K)$$

is a closed subset of the compact space $c_{\leq M}(\mathcal{R}_K) - \{\emptyset\}$ and therefore is itself compact. Completeness follows from the fact that $c_{\leq M}(\mathcal{R}_K)$ is closed in $c(\mathcal{R}_K)$. □

Finally, we note that we can extend $\check{c}_{\leq M}(\mathcal{R}_K) - \{\emptyset\}$, endowed with the Hausdorff metric $d(-,-)$, to the space $\check{c}_{\leq M}(\mathcal{R}_K)$ with a new metric $\check{D}$ as follows. Define $\check{D}(X,Y) \triangleq d(X,Y)$ if $X, Y \in \check{c}_{\leq M}(\mathcal{R}_K) - \{\emptyset\}$ and also

$$\check{D}(X,\emptyset) \triangleq \check{D}(\emptyset,X) = diam(\mathcal{R}_K), \qquad \check{D}(\emptyset,\emptyset) \triangleq 0$$

for all finite $X \in \check{c}_{\leq M}(\mathcal{R}_K) - \{\emptyset\}$. It is easy to show that we still have a metric and that the metric topology induced by $\check{D}$ is the same as the hit-or-miss topology. (We call $\check{D}$ the "modified" Hausdorff metric.) Also, $\check{c}_{\leq M}(\mathcal{R}_K)$ is a complete metric space which has the property that every closed bounded subset is compact (since $\check{c}_{\leq M}(\mathcal{R}_K)$ is also compact).

Case II: Incompletely Specified States

We will only sketch the main features of what is required. We will introduce a new metric D on $c_k(\mathcal{R})$. With respect to this metric, $c_k(\mathcal{R})$ is not complete since it does not contain all of its limit points. We will therefore extend both

$c_k(\mathcal{R})$ and D to their completions $\hat{c}_k(\mathcal{R})$ and $\hat{D}$. This space contains, in effect, all of the multisets which are the limit points of countable sequences of finite subsets. We will then demonstrate that $\hat{c}_k(\mathcal{R})$ satisfies the following property: Every closed subset of $\hat{c}_k(\mathcal{R})$ which is bounded with respect to the metric $\hat{D}$ is also compact.

What we need to do is obvious. Endow the finite subset W with an arbitrary metric $d(-,-)$ and define a metric on $\mathcal{R} = \mathbb{R}^n \times W$ by

$$d((\mathbf{x},v),(\mathbf{y},w))^2 \triangleq \|\mathbf{x} - \mathbf{y}\|^2 + d(v,w)^2$$

and a metric on $\mathcal{R}^j$ by

$$d_j((\zeta_1,...,\zeta_j),(\eta_1,...,\eta_j))^2 \triangleq d(\zeta_1,\eta_1)^2 + ... + d(\zeta_j,\eta_j)^2$$

Define $\mathcal{R}^{\leq k}$ to be the disjoint union (i.e., direct sum of topological spaces)

$$\mathcal{R}^{(\leq k)} \triangleq \mathcal{R}^0 \cup \mathcal{R}^1 \cup \mathcal{R}^2 \cup ... \cup \mathcal{R}^k$$

where $\mathcal{R}^0 \triangleq \{*\}$ consists of a single point, and define the metric D on $\mathcal{R}^{\leq k}$ by

$$D(\zeta^{(i)}, \eta^{(j)}) \triangleq \left\{ \begin{array}{ll} d_i(\zeta^{(i)}, \eta^{(j)}) & \text{if } i = j \\ |i - j| & \text{if } i \neq j \end{array} \right.$$

where $\zeta^{(i)}, \eta^{(i)} \in \mathcal{R}^i$. By construction the subsets $\mathcal{R}^k$ are both open and closed with respect to the metric topology induced on $\mathcal{R}^{\leq k}$ by D, and the restriction of D to $\mathcal{R}^k$ for any given k yields usual hybrid-space topology. From Proposition 2 of Chapter 4 we recall that the mappings $\chi_j : \mathcal{R}^j \rightarrow c_{\leq j}(\mathcal{R})$ defined by $\chi_j(\zeta_1,...,\zeta_j) = \{\zeta_1,...,\zeta_k\}$ induce homeomorphisms (equivalences of topologies) $\hat{\chi}_j : [\mathcal{R}]^j \leftrightarrow c_{\leq j}(\mathcal{R})$ between the relative hybrid topology on $[\mathcal{R}]^j$ and the hit-or-miss topology on $c_j(\mathcal{R})$. So, define a new metric on $\mathcal{R}^{\leq k}$ by

$$D(X,Y) \triangleq D(\hat{\chi}_{|X|}^{-1}(X),\ \hat{\chi}_{|Y|}^{-1}(Y))$$

for all $X,Y \in c_{\leq k}(\mathcal{R})$. The subsets $c_j(\mathcal{R})$ are both open and closed in the induced topology. Complete $c_{\leq k}(\mathcal{R})$ and D to $\hat{c}_{\leq k}(\mathcal{R})$ and $\hat{D}$. It is clear that $\hat{c}_k(\mathcal{R})$ are complete under $\hat{D}$ as well, and that $\hat{c}_k(\mathcal{R})$ can be identified with the finite sub-multisets of $\mathcal{R}$.

It is also clear that $\hat{c}_{\leq k}(\mathcal{R})$ has the property that any subset of $\hat{c}_k(\mathcal{R})$ which is closed and bounded also is compact (since Euclidean spaces have this property).

Finally, let $f(Z|X)$ be a parametrized family of global densities where X is now assumed to be an element of $\hat{c}_k(\mathcal{R})$. Then the function $X \mapsto f(Z|X)$ is continuous if and only if the functions defined almost everywhere by $(\zeta_1,...,\zeta_j) \mapsto f(Z|\{\zeta_1,...,\zeta_j\})$ are continuous with respect to the hybrid topology.

Consistency of the Global ML and MAP Estimators

In this section we show that, under certain assumptions, the global ML estimator (Definition 5) and the global MAP-II estimator (Definition 6-b) converge almost surely to the true value X_0 of the set parameter X which is being estimated. That is, suppose that the global maximum likelihood estimate

$$X_m = J_{GML}(Z_1, ..., Z_m) = \arg \sup_{X \subseteq \mathcal{R}} f_{\Sigma}(Z_1|X) \cdots f_{\Sigma}(Z_m|X)$$

exists. Then $p(\lim_{m \to \infty} X_m = X_0) = 1$. Likewise, suppose that

$$Y_m = J_{GMAP-I}(Z_1, ..., Z_m) = \arg \sup_{X \subseteq \mathcal{R}} f_{\Sigma}(X|Z_1, ..., Z_m) \lambda(K_0)^{|X|}$$

exists. Then $p(\lim_{m \to \infty} Y_m = X_0) = 1$. That is, we want to show that these two global estimators are "statistically consistent." Here, as outlined in section 5.2.3, limits are being taken with respect to one or other of the following two metrics:

(1) the modified Hausdorff metric $\check{D}(-, -)$ on the space $\check{c}_{\leq M}(\mathcal{R}_D)$ of set parameters; or

(2) the "multiset" metric $\hat{D}(-, -)$ on the space $\hat{c}_{\leq M}(\mathcal{R})$ of multiset parameters.

Abusing notation, we will use the generic notation "$D(-, -)$" to denote whichever metric is being assumed. We will also use the symbol $\mathcal{R}$ to denote either $\mathcal{R} = \mathbb{R}^n \times W$ or $\mathcal{R}_K = K \times W$.

We show that Wald's classic proof [22] of the consistency of the ML estimator can be directly generalized to cover the global ML and MAP-II estimators. This is possible because Wald's proof applies to estimators of parameters in *any* parameter space which satisfies the following properties:

(1) It is a complete metric space; and

(2) Any bounded closed subset of the parameter space is compact.

In what follows the value $X = X_0$ denotes the true value of the set parameter X. For any real-valued function $g(\Sigma)$ of the absolutely continuous finite random subset Σ of $\mathcal{R}$ the notation $E_0[g(\Sigma)] \triangleq \int g(Z) f_{\Sigma}(Z|X_0) \delta Z$ denotes the expected value of $g(\Sigma)$ with respect to X_0. Let Y_0 be a fixed finite subset of $\mathcal{R}$. For any $\rho, r > 0$ we denote

$$f_{\Sigma}(X|X; \rho) \triangleq \sup_{d(Y,X) \leq \rho} f_{\Sigma}(Z|Y), \qquad \varphi_{\Sigma}(Z|r) \triangleq \sup_{d(Y,Y_0) > r} f_{\Sigma}(Z|Y)$$

$$f_{\Sigma}^*(Z|X; \rho) \triangleq \max\{1, f_{\Sigma}(Z|X; \rho)\}, \qquad \varphi_{\Sigma}^*(Z|r) \triangleq \max\{1, \varphi_{\Sigma}(Z|r)\}$$

We then make the following assumptions (to make matters less complex Wald's assumptions 3 and 5 have been altered so that we do not have to worry about exceptional sets of parameters):

ASSUMPTION 1: The global density $f_{\Sigma}(Z|X)$ is set-integrable as a function of Z for every X.

ASSUMPTION 2: For sufficiently small ρ and sufficiently large r, the following expected values are finite:

$$E_0[\ln f_{\Sigma}^*(\Sigma|X; \rho)] < \infty, \qquad E_0[\ln \varphi_{\Sigma}^*(\Sigma|r)] < \infty$$

ASSUMPTION 3: Suppose that X_j is a sequence of finite subsets such that $\lim_{i \to \infty} X_i = X$. Then $\lim_{i \to \infty} f_\Sigma(Z|X_i) = f_\Sigma(Z|X)$ for all Z.

ASSUMPTION 4: If $X_1 \neq X_0$ then $\beta_\Sigma(S|X_1) \neq \beta_\Sigma(S|X_0)$ for at least one $S \subseteq \mathcal{R}$, where β_Σ denotes the belief measure of f_Σ.

ASSUMPTION 5: Let Y_0 be some fixed subset of $\mathcal{R}$. Suppose that X_i is a sequence such that $\lim_{i \to \infty} D(X_i, Y_0) = \infty$. Then $\lim_{i \to \infty} f_\Sigma(Z|X_i) = 0$ for any Z.

ASSUMPTION 6: The following is true: $E_0[|\ln f_\Sigma(\Sigma|X_0)|] < \infty$

ASSUMPTION 7: The particular subclass of global parameters X under consideration is a closed subset of $\check{c}_{\leq M}(\mathcal{R}_D)$ with respect to the modified Hausdorff metric or of $\hat{c}_{\leq M}(\mathcal{R})$ with respect to the multiset metric.

Immediate consequences of these assumptions are that:

(1) $f_\Sigma(Z|X; \rho)$ are measurable functions of Z for every X, ρ
(2) $f_\Sigma^*(Z|X, \rho)$ and $\varphi_\Sigma^*(Z|r)$ are measurable functions of Z for any X, ρ, r.
We can then prove:

Theorem 14 *(a) Suppose that the above assumptions are satisfied. Assume that the global maximum likelihood estimate exists. Then the estimator is consistent. That is, the global ML estimator converges almost surely to the true parameter value. (b) The same is true of the global MAP-II estimator.*

Proof. Following Wald, we will first prove three lemmas and two theorems. The assertion will then follow immediately from the proof of the second theorem.

Lemma A: For any $X \subseteq X_0$ we have: $E_0[\ln f_\Sigma(\Sigma|X)] < E_0[\ln f_\Sigma(\Sigma|X_0)]$.

Proof. First, note that $f_\Sigma(Z|X) \leq f_\Sigma^*(Z|X; \rho)$ for all Z, X, ρ and hence that, by Assumption 2, the expectations $E_0[\ln f_\Sigma(\Sigma|X)]$ and $E_0[\ln f_\Sigma(\Sigma|X_0)]$ are finite. Also, $E_0[|\ln f_\Sigma(\Sigma|X_0)|]$ is finite by Assumption 6. Now, if we know that $E_0[\ln f_\Sigma(\Sigma|X)] = -\infty$ then the lemma holds trivially. Therefore we may assume that $-\infty < E_0[\ln f_\Sigma(\Sigma|X)] < \infty$ and hence that $E_0[|\ln f_\Sigma(\Sigma|X)|] < \infty$. Define $u(Z) \triangleq \ln f_\Sigma(Z|X) - \ln f_\Sigma(Z|X_0)$ so that $E[|u|] = E_0[|u(\Sigma)|] < \infty$. Then as in Wald's proof we apply the inequality $E[u] < \ln E[e^u]$, which holds for any random number u such that $E[u] < \infty$ and which, with probability one, is not a constant. However, by definition of u we get:

$$E_0[e^{u(\Sigma)}] = E_0 \left[\frac{f_\Sigma(\Sigma|X)}{f_\Sigma(\Sigma|X_0)} \right] = \int f_\Sigma(Z|X) \delta X = 1$$

and so $E_0[u(\Sigma)] < \ln(1) = 0$. This finishes the proof of Lemma A. □

Lemma B: The following is true:

$$\lim_{\rho \to 0} E_0[\ln f_\Sigma(\Sigma|X; \rho)] = E_0[\ln f_\Sigma(\Sigma|X)]$$

for any X.

Proof. We introduce the notation

$$f_\Sigma^*(Z|X) \triangleq \max\{1, f_\Sigma(Z|X)\}, \qquad f_\Sigma^{**}(Z|X) \triangleq \min\{1, f_\Sigma(Z|X)\}$$
$$f_\Sigma^{**}(Z|X;\rho) \triangleq \min\{1, f_\Sigma(Z|X;\rho)\}$$

from which it follows that

$$f_\Sigma(Z|X) = f_\Sigma^{**}(Z|X) + f_\Sigma^*(Z|X) - 1$$
$$f_\Sigma(Z|X;\rho) = f_\Sigma^{**}(Z|X;\rho) + f_\Sigma^*(Z|X;\rho) - 1$$

If we prove the assertion for the functions f_Σ^* and f_Σ^{**} separately then we will be done. First note that $\lim_{\rho \to 0} \ln f_\Sigma^*(Z|X;\rho) = f_\Sigma^*(Z|X)$. For on the one hand

$$f_\Sigma^*(Z|X;\rho) = \max\{1, \sup_{d(Y,X) \le \rho} f_\Sigma(Z|Y)\}$$

by definition. Pick ρ_i with $\rho_i \to 0$ and X_i such that $d(X, X_i) < \rho_i$. Then

$$\lim_{\rho \to 0} f_\Sigma^*(Z|X;\rho) = \lim_{i \to \infty} f_\Sigma^*(Z|X;\rho_i) = \max\{1, \lim_{i \to \infty} f_\Sigma(Z|X_i)\}$$

By Assumption 3 the quantity on the rightmost side becomes

$$\max\{1, \lim_{i \to \infty} f_\Sigma(Z|X_i)\} = \max\{1, f_\Sigma(Z|X)\} = f_\Sigma^*(Z|X)$$

Now note that, by construction, $\ln f_\Sigma^*(Z|X;\rho)$ is a nonnegative, increasing function in ρ. Therefore

$$\lim_{\rho \to 0} E_0[\ln f_\Sigma^*(\Sigma|X;\rho)] = E_0[\lim_{\rho \to 0} \ln f_\Sigma^*(\Sigma|X;\rho)] = E_0[f_\Sigma(\Sigma|X)]$$

as desired.

In the case of the function f_Σ^{**} we begin by noting that $|\ln f_\Sigma^{**}(\Sigma|X;\rho)| \le |\ln f_\Sigma^{**}(\Sigma|X)|$. For by construction $\ln f_\Sigma^{**}(\Sigma|X;\rho) \le 0$ and $\ln f_\Sigma^{**}(\Sigma|X) \le 0$. Therefore, from $\ln f_\Sigma^{**}(\Sigma|X;\rho) \ge \ln f_\Sigma^{**}(\Sigma|X)$ we get $-\ln f_\Sigma^{**}(\Sigma|X;\rho) \le -\ln f_\Sigma^{**}(\Sigma|X)$ as desired. Just as before we also have: $\lim_{\rho \to 0} \ln f_\Sigma^{**}(\Sigma|X;\rho) = \ln f_\Sigma^{**}(\Sigma|X)$.

This done, there are two cases to consider: either $E_0[f_\Sigma^{**}(\Sigma|X)] = -\infty$ or $E_0[f_\Sigma^{**}(\Sigma|X)]$ is finite. The former case follows from $-\ln f_\Sigma^{**}(\Sigma|X;\rho) \le -\ln f_\Sigma^{**}(\Sigma|X)$ and in the latter case we also know that $-E_0[\ln f_\Sigma^{**}(\Sigma|X;\rho)]$ is finite. Again by construction, $f_\Sigma^{**}(\Sigma|X;\rho)$ is an increasing function in ρ and so as before we get

$$\lim_{\rho \to 0} E_0[\ln f_\Sigma^{**}(\Sigma|X;\rho)] = E_0[\lim_{\rho \to 0} \ln f_\Sigma^{**}(\Sigma|X;\rho)] = E_0[\ln f_\Sigma^{**}(\Sigma|X)]$$

as desired. This concludes the proof of Lemma B. $\square$

Lemma C: The following is true: $\lim_{r \to \infty} E_0[\ln \varphi_\Sigma(\Sigma|r)] = -\infty$.

Proof. First, recall that for a fixed finite $Y_0 \subseteq \mathcal{R}$, $\varphi_\Sigma(\Sigma|r) = \sup_{d(Y,Y_0)>r} f_\Sigma(Z|Y)$. Thus note that, by construction, $\ln \varphi_\Sigma(\Sigma|r)$ is a nonnegative, decreasing function in r. Also, $\ln \varphi_\Sigma(\Sigma|r) - \ln \varphi_\Sigma^*(\Sigma|r) \leq 0$. But

$$\ln \varphi_\Sigma(\Sigma|r) = (\ln \varphi_\Sigma(\Sigma|r) - \ln \varphi_\Sigma^*(\Sigma|r)) + \ln \varphi_\Sigma^*(\Sigma|r)$$

and so

$$E_0[\ln \varphi_\Sigma(\Sigma|r)] = E_0[(\ln \varphi_\Sigma(\Sigma|r) - \ln \varphi_\Sigma^*(\Sigma|r))] + E_0[\ln \varphi_\Sigma^*(\Sigma|r)]$$

By Assumption 5, if X_i is a sequence such that $\lim_{r\to\infty} D(Y_0, X_i) = \infty$ then $\lim_{i\to\infty} f_\Sigma(Z|X_i) = 0$ for any Z. So, choose $r_i \to \infty$ and choose X_i such that $d(Y_0, X_i) > r_i$. Then

$$\lim_{r\to\infty} \ln \varphi_\Sigma(\Sigma|r) = \lim_{r\to\infty} \ln \sup_{d(Y,Y_0)>r} f_\Sigma(Z|Y) = \lim_{i\to\infty} \ln f_\Sigma(Z|X_i) = -\infty$$

Since $\varphi_\Sigma^*(\Sigma|r)$ is decreasing in r it follows that

$$\lim_{r\to\infty} E_0[\ln \varphi_\Sigma^*(\Sigma|r)] = E_0[\lim_{r\to\infty} \ln \varphi_\Sigma^*(\Sigma|r)] = -\infty$$

and thus $\lim_{r\to\infty} E_0[\ln \varphi_\Sigma(\Sigma|r)] = -\infty$. This completes the proof of Lemma C.
$\square$

Theorem A: *Let P be any closed subset of the space of set parameters not containing X_0. Then*

$$p\left(\lim_{m\to\infty} \frac{\sup_{X\in P} f_\Sigma(\Sigma_1|X)\cdots f_\Sigma(\Sigma_m|X)}{f_\Sigma(\Sigma_1|X_0)\cdots f_\Sigma(\Sigma_m|X_0)} = 0\right) = 1$$

Proof. From Lemma C we know that $E_0[\ln \varphi_\Sigma(\Sigma|r)]$ decreases without bound. Thus there exists an $r_0 > 0$ sufficiently large that

$$E_0[\ln \varphi_\Sigma(\Sigma|r_0)] < E_0[\ln f_\Sigma(\Sigma|X_0)]$$

Let B_0 be the closed ball consisting of all X with $D(X, Y_0) \leq r_0$ and let $P_0 = P \cap B_0$. Since every bounded closed subset is compact then in particular P_0 is compact. Next, from Lemma A we know that for any $X \in P_0$

$$E_0[\ln f_\Sigma(\Sigma|X)] < E_0[\ln f_\Sigma(\Sigma|X_0)]$$

and from Lemma B we know that

$$\lim_{\rho\to 0} E_0[\ln f_\Sigma(\Sigma|X;\rho)] = E_0[\ln f_\Sigma(\Sigma|X)]$$

Hence, for each $X \in P_0$ there exists a $\rho_X > 0$ which is sufficiently small that

$$E_0[\ln f_\Sigma(\Sigma|X;\rho_X)] < E_0[\ln f_\Sigma(\Sigma|X_0)]$$

holds. Now, the open balls $B_{\rho X}$ for all $X \in P_0$ form an open cover of P_0. Since P_0 is compact, by definition there is an open subcover

$$B_{\rho X_1}(X_1) \cup ... \cup B_{\rho X_h}(X_h) \supseteq P_0$$

Abbreviate $\rho_i = \rho_{X_i}$ and $B_i = B_{\rho_i}(X_i)$ for $i = 1, ..., h$. Note that $P \subseteq B_1 \cup ... \cup D_h \cup B_0^c$. Then, using the general inequalities

$$\sup_{X \in S \cup T} f(X) = \max\{\sup_{X \in S} f(X), \sup_{X \in T} g(X)\}, \qquad \max\{x, y\} \le x + y$$

$$\sup_{X \in S} f(X)g(X) \le \left(\sup_{X \in S} f(X) \right) \left(\sup_{X \in S} g(S) \right)$$

it is easy to show that

$$\sup_{X \in P} f_\Sigma(Z_1|X) \cdots f_\Sigma(Z_m|X) \ \le \ \sum_{i=1}^{h} f_\Sigma(Z_1|X_i; \rho_i) \cdots f_\Sigma(Z_m|X_i; \rho_i)$$
$$+ \varphi_\Sigma(Z_1|r_0) \cdots \varphi_\Sigma(Z_m|r_0)$$

Accordingly, to prove Theorem A it will be enough to prove that

$$p \left(\lim_{m \to \infty} \frac{f_\Sigma(\Sigma_1|X_i; \rho_i) \cdots f_\Sigma(\Sigma_m|X_i; \rho_i)}{f_\Sigma(\Sigma_1|X_0) \cdots f_\Sigma(\Sigma_m|X_0)} = 0 \right) = 1$$

for all $i = 1, ..., h$ and that

$$p \left(\lim_{m \to \infty} \frac{\varphi_\Sigma(\Sigma_1|r_0) \cdots \varphi_\Sigma(\Sigma_m|r_0)}{f_\Sigma(\Sigma_1|X_0) \cdots f_\Sigma(\Sigma_m|X_0)} = 0 \right) = 1$$

However, fix i and define the random numbers $Q_{\alpha;i} = \ln f_\Sigma(Z_\alpha|X_i; \rho_i) - \ln f(Z_\alpha|X_0)$ for all $i = 1, ..., m$. For fixed i the $Q_{\alpha;i}$ are independent, identically distributed random numbers with expected value

$$E[Q_{\alpha;i}] = E_0[\ln f_\Sigma(\Sigma|X_i; \rho)] - E_0[\ln f_\Sigma(\Sigma|X_0)] = -\mu_i$$

The numbers μ_i are nonnegative by construction. Thus by the strong law of large numbers, for each $i = 1, ..., h$ we have:

$$1 \ = \ p \left(\lim_{m \to \infty} \frac{1}{m} \sum_{\alpha=1}^{m} Q_{\alpha;i} = -\mu_i \right) = p \left(\lim_{m \to \infty} \sum_{\alpha=1}^{m} Q_{\alpha;i} = -\infty \right)$$

$$= \ p \left(\lim_{m \to \infty} \prod_{\alpha=1}^{m} e^{Q_{\alpha;i}} = -\infty \right)$$

$$= \ p \left(\lim_{m \to \infty} \frac{f_\Sigma(\Sigma_1|X_i; \rho_i) \cdots f_\Sigma(\Sigma_m|X_i; \rho_i)}{f_\Sigma(\Sigma_1|X_0) \cdots f_\Sigma(\Sigma_m|X_0)} = 0 \right)$$

as desired. The other result follows similarly. This concludes the proof of Theorem A. $\qquad \qquad \Box$

Theorem B: *Let $X_m(Z_1, .., Z_m)$ be any function of the global observations $Z_1, ..., Z_m$ such that*

$$\frac{f_\Sigma(Z_1|X_m) \cdots f_\Sigma(Z_m|X_m)}{f_\Sigma(Z_1|X_0) \cdots f_\Sigma(Z_m|X_0)} \geq c > 0$$

for all m and for all $Z_1, ..., Z_m$. Then $p(\lim_{m \to \infty} X_m = X_0) = 1$.

Proof. Suppose that there is a limit point of the sequence X_m such that $d(X_m, X_0) > \varepsilon$ implies

$$\sup_{d(X_m, X_0) \geq \varepsilon} f_\Sigma(Z_1|X) \cdots f_\Sigma(Z_m|X) \geq f_\Sigma(Z_1|X_\infty) \cdots f_\Sigma(Z_m|X_\infty)$$

for infinitely many m. By assumption this implies that

$$\frac{\sup_{d(X_m, X_0) \geq \varepsilon} f_\Sigma(Z_1|X) \cdots f_\Sigma(Z_m|X)}{f_\Sigma(Z_1|X_0) \cdots f_\Sigma(Z_m|X_0)} \geq c > 0$$

for infinitely many m. By Theorem A this is an event with probability zero. Consequently, the event that all limit points X_∞ of $\{X_m\}$ satisfy $d(X_\infty, X_0) \leq \varepsilon$ has probability one, for arbitrary $\varepsilon > 0$. Thus $d(X_\infty, X_0) = 0$ with probability one for all limit points X_∞, from which the theorem follows. This concludes the proof of Theorem B. $\quad\square$

Therefore:

Proof of Theorem 14 If the global MLE $X_m = X_m(Z_1, ..., Z_m)$ exists then

$$f_\Sigma(Z_1|X_m) \cdots f_\Sigma(Z_m|X_m) \geq f_\Sigma(Z_1|X_0) \cdots f_\Sigma(Z_m|X_0)$$

by definition. Thus the assertion follows from Theorem B with $c = 1$. $\quad\square$

5.3   Information Theory and Information Fusion

5.3.1   Global Information Theory

In this section we define the random-set analog of the Kullback-Leibler cross-entropy metric and show that it satisfies analogs of the standard discrimination-minimization theorems (e.g. [13, Chapter 3]).

Definition 7 *Let f, g be global densities with the property that $\int_S g(Z)\delta Z = 0$ implies $\int_S f(Z)\delta Z = 0$ for every closed $S \subseteq \mathcal{R}$. Then we say that f is absolutely continuous with respect to g.*

Note that this is the same as saying that, for every $k \geq 1$, the measure corresponding to the function $(\xi_1, ..., \xi_k) \mapsto f(\{\xi_1, ..., \xi_k\})$ in k distinct variables is absolutely continuous with respect to the measure corresponding to the function $(\xi_1, ..., \xi_k) \mapsto g(\{\xi_1, ..., \xi_k\})$ in k distinct variables.

Definition 8 *Let f, g be integrable global densities such that f is absolutely continuous with respect to g. Then the global Kullback-Leibler cross-entropy, or global discrimination, is:*

$$I(f;g) \triangleq \int f(Z) \ln \left(\frac{f(Z)}{g(Z)} \right) \delta Z$$

where as usual we use the convention $x \ln x = 0$ if $x = 0$. If $f(Z) = f(Z|X)$ and $g(Z) = f(Z|Y)$ are parametrized densities then we abbreviate:

$$I_f(X;Y) \triangleq \int f(Z|X) \ln \left(\frac{f(Z|X)}{g(Z|Y)} \right) \delta Z$$

Example 13 *Let $D' \subseteq D$ be two closed bounded domains in $\mathbb{R}^n$ and let $M \leq M'$ be two positive integers. Recall (Definition 12 of Chapter 4) that the "uniform" global density $u_{D,M}$ is defined as*

$$u_{D,M}(Z) \triangleq \frac{|Z|!}{M \, N^{|Z|} \, \lambda(D)^{|Z|}}$$

whenever $Z = \{(\mathbf{z}_1, w_1), ..., (\mathbf{z}_k, w_k)\}$ is such that $\mathbf{z}_1, ..., \mathbf{z}_k \in D$ and $|Z| < M$; and where $u_{D,M}(Z) \triangleq 0$ otherwise. Then note that $u_{D',M'}$ is absolutely continuous with respect to $u_{D,M}$. The reader may verify that

$$I(u_{D',M'}; u_{D,M}) = \ln \left(\frac{M}{M'} \right) + \frac{1}{2}(M' - 1) \ln \left(\frac{\lambda(D)}{\lambda(D')} \right)$$

Note that the discrimination breaks down into two terms that can be regarded as "detection" (number of elements) and "geopositional" components of the cross-entropy.   □

We now show that the global discrimination satisfies the usual properties associated with the Kullback-Leibler cross-entropy.

Proposition 15 *Let f, g be integrable global densities such that f is absolutely continuous with respect to g. Then $I(f;g) \geq 0$ and $I(f;g) = 0$ if and only if $f = g$ almost everywhere.*

Proof. We show that one of the standard proofs [12, p. 156] carries over. Define the real-valued function h of a set variable X by $g(X) = f(X)(1+h(X))$ for all finite subsets X such that $f(X) \neq 0$, and define $h(X) = 1$ otherwise. Also, define the function of a real variable $r(x) = x - \ln x$. Then it follows that $1 + h(X) > 0$ for all X and that

$$1 = \int g(Z)\delta Z = \int f(Z)(1 + h(Z))\delta Z = 1 + \int f(Z)h(Z)\delta Z$$

and hence $\int f(Z)h(Z)\delta Z = 0$. Therefore

$$I(f;g) = \int f(Z)\ln(1 + h(Z))\delta Z = \int f(Z)(h(Z) - \ln(1 + h(Z)))\delta Z$$

$$= \int f(Z)r(h(Z))\delta Z$$

The function $r(x)$ is concave, nonnegative, and $r(x) = 0$ if and only if $x = 0$. Consequently $I(f; g) \geq 0$. If $I(f; g) = 0$ then $r(h(Z)) = 0$ almost everywhere and we are done.   □

Corollary 16 *Let g be a global density and h a nonnegative-valued but unitless function of a finite-set variable such that $\int h(Z)g(Z) = 1$. Let*

$$\varepsilon_g(h) \triangleq \int h(Z)\ln(h(Z))g(Z)\delta Z$$

denote the "differential entropy" of h with respect to g. Then $\varepsilon_g(h) \geq 0$ with equality if and only if $h(Z) = 1$ identically.

Proof. Set $h'(Z) = h(Z)g(Z)$ and use the fact that $\varepsilon_g(h) = I(h'; g)$.   □

We now turn to the problem of determining minimal-discrimination global densities—that is, global densities $f = g^*$ which minimize $I(f : g)$ subject to some constraint. First, we need the following generalization of a standard result from the calculus of variations:

Lemma 17 *Let $F(x, Z)$ be a real-valued function of a nonnegative real variable x and a finite-set variable Z which is infinitely differentiable in x. Let $g(Z)$ be a global density, let $f(Z)$ be a non-negative integrable finite-set function, and let $D \subseteq \mathbb{R}^n$ be a bounded closed domain. If f minimizes or maximizes the value of the set integral*

$$J(f) = \int_D F(f(Z), Z)g(Z)\delta Z$$

then f must satisfy the equation

$$\frac{dF}{dx}(f(Z), Z) = 0$$

for almost all $Z \subseteq D$. Moreover, if

$$\frac{d^2 F}{dx^2}(f(Z), Z) \geq 0 \qquad resp. \qquad \frac{d^2 F}{dx^2}(f(Z), Z) \leq 0$$

identically then f minimizes (resp. maximizes) $J(f)$.

Proof. We follow the classic but somewhat informal proof in [7, pp. 13-14]. Let κ be a very small real number of arbitrary sign and let $s(Z)$ be a bounded integrable set function defined on the domain D. Then for each Z we can write a Taylor's series expansion

$$F(f(Z) + \kappa\, s(Z),\, Z) \;=\; F(f(Z), Z) + \kappa\, s(Z)\, \frac{dF}{dx}(f(Z), Z)$$

$$+ \frac{1}{2}\kappa^2\, s(Z)^2\, \frac{d^2 F}{dx^2}(f(Z), Z) + \dots$$

Now perturb the value of $J(f)$ by allowing f to vary to the nearby set function $f + \kappa s$. The variation ΔJ in $J(f)$ due to this perturbation is

$$
\begin{aligned}
\Delta J &= J(f + \kappa s) - J(f) \\
&= \kappa \int_D s(Z) \frac{dF}{dx}(f(Z), Z) g(Z) \delta Z + \kappa^2 \int_D s(Z)^2 \frac{d^2 F}{dx^2}(f(Z), Z) g(Z) \delta Z + \dots
\end{aligned}
$$

Since κ is arbitrarily small the first term on the left-hand side of this last equation will dominate the value of ΔJ. Depending on the sign of κ the value of ΔJ will be positive or negative. Assume that f minimizes $J(f)$. Then $J(f + \kappa s) \geq J(f)$ and hence $\Delta J \geq 0$ for all s. This is possible only if the set integral

$$
\int_D s(Z) \frac{dF}{dx}(f(Z), Z) g(Z) \delta Z
$$

vanishes for all s and hence only if $\frac{dF}{dx}(f(Z), Z) = 0$ almost everywhere. Likewise, if f maximizes $J(f)$ then $J(f + \kappa s) \leq J(f)$ and $\Delta J \leq 0$ for all s. Once again this is possible only if the same set integral as before vanishes. In either case the value of ΔJ will be dominated by the term

$$
\kappa^2 \int s(Z)^2 \frac{d^2 F}{dx^2}(f(Z), Z) g(Z) \delta Z
$$

If $\frac{d^2 F}{dx^2}(f(Z), Z) \geq 0$ identically then ΔJ will always be positive regardless of the value of s and so f must minimize $J(f)$. Likewise, if $\frac{d^2 F}{dx^2}(f(Z), Z) \leq 0$ identically then ΔJ will always be negative and f must maximize $J(f)$.   □

Proposition 18 *Suppose that* $\mathbf{T}(Z)$ *is an integrable vector-valued set function of a finite-set variable and that* f, g *are global densities such that* f *is absolutely continuous with respect to* g. *Then any global density* $f = g^*$ *which minimizes the global discrimination* $I(f; g)$, *subject to the constraint that the vector moment* $\int f(Z)\mathbf{T}(Z)\delta Z$ *of* f *is a known constant vector* θ, *must have the form*

$$
g^*(Z|\tau) = \frac{g(Z)}{M_g(\tau)} e^{\langle \tau, \mathbf{T}(Z) \rangle}
$$

for some choice of $\tau = (\tau_1, ..., \tau_m)$, *where*

$$
M_g(\tau) \triangleq \int g(Y) e^{\langle \tau, \mathbf{T}(Y) \rangle} \delta Y
$$

is the normalization constant. Furthermore, the minimal value of $I(f; g)$ *is*

$$
I(g^*; g) = \frac{\partial \ln M_g}{\partial \tau}(\tau) - \ln M_g(\tau) = \langle \tau, \theta \rangle - \ln M_g(\tau)
$$

Proof. We show that the standard proof [12, pp. 59-60] can be extended to our case. Let $\theta = (\theta_1, ..., \theta_m)$ and $\mathbf{T}(Z) = (T_1(Z), ..., T_m(Z))$. Then the vector constraint equation becomes m scalar constraint equations $\theta_j = \int f(Z) T_j(Z) \delta Z$.

Writing the Lagrangian multiplier method in vector notation, we are to minimize the quantity

$$I(f; g; \tau_0, \tau) = I(f; g) - (\tau_0 + 1)\left(\int f(Z)\delta Z - 1\right) - \langle \tau, \int \mathbf{T}(Z)g(Z)\delta Z - \theta\rangle$$

Let $h(Z) = f(Z)/g(Z)$. Then the previous equation becomes:

$$
\begin{aligned}
I(f; g; \tau_0, \tau) &= \int h(Z)\ln h(Z) - (\tau_0 + 1)(h(Z) - 1)g(Z)\delta Z \\
&\quad - \int \langle \tau, \mathbf{T}(Z)h(Z) - \theta\rangle g(Z)\delta Z \\
&= \int F(h(Z), Z)g(Z)\delta Z
\end{aligned}
$$

where $F(x, Z) = x \ln x - (\tau_0 + 1)(x - 1) - \langle \tau, x\mathbf{T}(Z) - \theta\rangle$. Taking the first and second derivatives of F we get:

$$
\begin{aligned}
\frac{dF}{dx}(x, Z) &= \ln x - \tau_0 - \langle \tau, \mathbf{T}(Z)\rangle \\
\frac{d^2 F}{dx^2}(x, Z) &= \frac{1}{x} > 0
\end{aligned}
$$

identically. Thus any solution of the equation $0 = \ln h(Z) - \tau_0 - \langle \tau, \mathbf{T}(Z)\rangle$ must minimize $I(f; g; \tau_0, \tau)$. Solving for h we find that

$$h(Z) = g^*(Z|\tau) = e^{\tau_0} g(Z)\, e^{\langle \tau, \mathbf{T}(Z)\rangle}$$

Since g^* is a global density by assumption we get the desired form for g^*. Moreover,

$$I(g^*; g) = -\ln M_g(\tau) + \int g^*(Z|\tau)\langle \tau, \mathbf{T}(Z)\rangle \delta Z = -\ln M_g(\tau) + \langle \tau, \tau\rangle$$

On the other hand,

$$
\begin{aligned}
\frac{\partial \ln M_g}{\partial \tau}(\tau) &= \frac{1}{M_g}\frac{\partial}{\partial \tau}\int g(Y)e^{\langle \tau, \mathbf{T}(Y)\rangle}\delta Y = \frac{1}{M_g}\int g(Y)\frac{\partial}{\partial \tau}e^{\langle \tau, \mathbf{T}(Y)\rangle}\delta Y \\
&= \langle \tau, \int g^*(Y)\mathbf{T}(Y)\delta Y\rangle = \langle \tau, \theta\rangle
\end{aligned}
$$

as desired.   □

5.3.2   Global Best-Performance Inequalities

In this section we will prove an analog of the Cramér-Rao inequality for estimators of a *vector-valued function* of a set parameter. In what follows, we will simplify notation by considering only the *single-observation* case. That is, normally we would consider the estimator $\mathbf{T}(Z_1, ..., Z_m)$ with several global observations $Z_1, ..., Z_m$. Instead we will assume without loss of generality that $m = 1$.

Let $\mathbf{Z} \in \mathbb{R}^n$ be an random vector with density $f_{\mathbf{Z}}(z|\mathbf{x})$ parametrized by $\mathbf{x} \in \mathbb{R}^m$. Let $\mathbf{T} : \mathbb{R}^n \to \mathbb{R}^m$ be an estimator of $\mathbf{x}$ and $\mathbf{X} = \mathbf{T}(\mathbf{Z})$. Recall that the *covariance* of $\mathbf{X}$ is (Proposition 9) the linear transformation $cov_{\mathbf{T},\mathbf{x}} = cov_{\mathbf{X},\mathbf{x}} : \mathbb{R}^m \to \mathbb{R}^m$ defined by

$$
\begin{aligned}
cov_{\mathbf{T},\mathbf{x}}(\mathbf{y}) &= E_{\mathbf{x}}\left[\langle \mathbf{X} - \mathbf{x}_0, \mathbf{y} \rangle (\mathbf{X} - \mathbf{x}_0)\right] \\
&= \int \langle \mathbf{T}(z) - \mathbf{x}_0, \mathbf{y} \rangle (\mathbf{T}(z) - \mathbf{x}_0) f_{\mathbf{Z}}(z|\mathbf{x}) d\lambda(z)
\end{aligned}
$$

where $\mathbf{x}_0 = \int \mathbf{T}(z) f_{\mathbf{Z}}(z|\mathbf{x}) d\lambda(z)$ is the expected value of $\mathbf{X}$. Finally, define the linear transformation $\mathbf{L}_{\mathbf{x}} : \mathbb{R}^n \to \mathbb{R}^n$ by

$$
\begin{aligned}
\langle \mathbf{v}, \mathbf{L}_{\mathbf{x}}(\mathbf{w}) \rangle &= E_{\mathbf{x}}\left[\left(\frac{\partial \ln f}{\partial \mathbf{v}}\right)(\mathbf{Z}|\mathbf{x}) \left(\frac{\partial \ln f}{\partial \mathbf{w}}\right)(\mathbf{Z}|\mathbf{x})\right] \\
&= \int \left(\frac{\partial \ln f}{\partial \mathbf{v}}\right)(z|\mathbf{x}) \left(\frac{\partial \ln f}{\partial \mathbf{w}}\right)(z|\mathbf{x}) f_{\mathbf{Z}}(z|\mathbf{x}) d\lambda(z)
\end{aligned}
$$

for all $\mathbf{v}, \mathbf{w} \in \mathbb{R}^n$. Suppose that the estimator $\mathbf{T}$ is unbiased in the sense that $E_{\mathbf{x}}[\mathbf{T}(\mathbf{Z})] = \int \mathbf{T}(z) f_{\mathbf{Z}}(z|\mathbf{x}) d\lambda(z) = \mathbf{T}(\mathbf{x})$ for all $\mathbf{x} \in \mathbb{R}^m$. The *Cramér-Rao inequality* states that, in this case, the covariance of $\mathbf{X} = \mathbf{T}(\mathbf{Z})$ obeys the inequality

$$
\langle \mathbf{w}, cov_{\mathbf{T},\mathbf{x}}(\mathbf{w}) \rangle \geq \langle \mathbf{w}, \mathbf{L}_{\mathbf{x}}^{-1}(\mathbf{w}) \rangle
$$

for all $\mathbf{w} \in \mathbb{R}^m$. A generalization of this inequality to biased estimators also exists and is

$$
\langle \mathbf{w}, cov_{\mathbf{T},\mathbf{x}}(\mathbf{w}) \rangle \cdot \langle \mathbf{v}, \mathbf{L}_{\mathbf{x}}(\mathbf{v}) \rangle \geq \langle \mathbf{w}, \frac{\partial}{\partial \mathbf{v}} E_{\mathbf{x}}[\mathbf{T}(\mathbf{Z})] \rangle^2
$$

for all $\mathbf{v}, \mathbf{w} \in \mathbb{R}^n$. The Cramér-Rao inequality for unbiased estimators is easily seen to be a special case of this generalized Cramér-Rao inequality. For if $\mathbf{T}$ is unbiased then $E_{\mathbf{x}}[\mathbf{T}(\mathbf{Z})] = \mathbf{x}$ and so $\frac{\partial}{\partial \mathbf{v}} E_{\mathbf{x}}[\mathbf{T}(\mathbf{Z})] = \frac{\partial}{\partial \mathbf{v}} \mathbf{x} = \mathbf{v}$. Thus

$$
\langle \mathbf{w}, cov_{\mathbf{T},\mathbf{x}}(\mathbf{w}) \rangle \cdot \langle \mathbf{v}, \mathbf{L}_{\mathbf{x}}(\mathbf{v}) \rangle \geq \langle \mathbf{w}, \mathbf{v} \rangle^2
$$

and so setting $\mathbf{v} = \mathbf{L}^{-1}(\mathbf{w})$ this becomes

$$
\langle \mathbf{w}, cov_{\mathbf{T},\mathbf{x}}(\mathbf{w}) \rangle \cdot \langle \mathbf{L}_{\mathbf{x}}^{-1}(\mathbf{w}), \mathbf{w} \rangle \geq \langle \mathbf{w}, \mathbf{L}_{\mathbf{x}}^{-1}(\mathbf{w}) \rangle^2
$$

which yields the inequality for the unbiased estimator.

Our goal is to extend the Cramér-Rao inequality for biased estimators to a Cramér-Rao inequality for *vector-valued estimators of a set parameter*. In order to do this we must first slightly extend the standard Newtonian differential calculus as follows:

Definition 9 *Let $f(X)$ be a scalar-valued function of a finite-set variable. The directional derivative of f at $\zeta = (\mathbf{x}, v) \in \mathcal{R}$ in the direction of $\mathbf{y}$ is*

$$\frac{\partial f}{\partial_\zeta \mathbf{y}}(X) \triangleq \lim_{\varepsilon \to 0} \frac{f((X - \{(\mathbf{x}, v)\}) \cup \{(\mathbf{x} + \varepsilon \mathbf{y}, v)\}) - f(X)}{\varepsilon}$$

if $\zeta \in X$ and

$$\frac{\partial f}{\partial_\zeta \mathbf{y}}(X) \triangleq 0$$

otherwise (provided, of course, that the limit exists).

This is obviously just the usual directional derivative in different notation. For, let $X = \{(\mathbf{x}, v), \zeta_2, ..., \zeta_k\}$ and $|X| = k$. Then

$$\frac{\partial f}{\partial_\zeta \mathbf{y}}(X) = \lim_{\varepsilon \to 0} \frac{f(\{(\mathbf{x} + \varepsilon \mathbf{y}, v), \zeta_2, ..., \zeta_k\}) - f(\{(\mathbf{x}, v), \zeta_2, ..., \zeta_k\})}{\varepsilon}$$

$$= \frac{\partial}{\partial \mathbf{y}} f(\{(\mathbf{x}, v), \zeta_2, ..., \zeta_k\})$$

where the partial derivative is taken with respect to the vector variable $\mathbf{x}$.
Now, we define:

Definition 10 *The discrimination rate of the global density f evaluated at the finite subset X, if it exists, is:*

$$I_{f, X, \zeta}(\mathbf{v}) \triangleq 2 \lim_{\eta \to 0} \frac{I_f((X - \{\zeta\}) \cup \{(\mathbf{x} + \eta \mathbf{v}, v)\}; X)}{\eta^2}$$

for all $\mathbf{v} \in \mathbb{R}^n$ and all $\zeta = (\mathbf{x}, v) \in \mathcal{R}$.

We begin by proving a generalization of an information inequality proved in [5, Theorem 8.1.1, p. 301].

Proposition 19 *Let $\mathbf{T}(Z) \in \mathbf{R}^m$ be a vector-valued estimator of the set parameter X of the parametrized family $f_\Sigma(Z|X)$ of global densities, and let $cov_{\mathbf{T}, X}$ denote the covariance of the random vector $\mathbf{Y} = \mathbf{T}(\Sigma)$ when Σ is drawn from the population governed by $f_\Sigma(Z|X)$. Define*

$$\Delta_X(Y) \triangleq E_Y \left[\frac{f(Z|Y) - f(Z|X)}{f(Z|X)} \right] = \int f(Z|Y) \left[\frac{f(Z|Y)}{f(Z|X)} - 1 \right] \delta Z$$

Then $\Delta_X(Y) \geq 0$ for all set parameters X, Y and

$$\langle \mathbf{w}, cov_{\mathbf{T}, X}(\mathbf{w}) \rangle \cdot \Delta_X(Y) \geq \langle \mathbf{w}, E_Y[\mathbf{T}(\Sigma)] - E_X[\mathbf{T}(\Sigma)] \rangle^2$$

for all $\mathbf{w} \in \mathbb{R}^m$ and all X, Y.

Proof. The Schwartz inequality (Proposition 7) gives us

$$E_X[\langle \mathbf{w}, \mathbf{T}(\Sigma) - \mathbf{x}_0 \rangle^2] \cdot E_X\left[\left(1 - \frac{f(\Sigma|Y)}{f(\Sigma|X)}\right)^2\right]$$

$$\geq E_X\left[\langle \mathbf{w}, \mathbf{T}(\Sigma) - \mathbf{x}_0 \rangle \cdot \left(1 - \frac{f(\Sigma|Y)}{f(\Sigma|X)}\right)\right]^2$$

However, the right-hand side of the inequality can be written as:

$$\left[\int f(Z|X)\langle \mathbf{w}, \mathbf{T}(Z) - \mathbf{x}_0 \rangle \cdot \left(1 - \frac{f(Z|Y)}{f(Z|X)}\right) \delta Z\right]^2$$

$$= \left[\int (\langle \mathbf{w}, \mathbf{T}(Z) \rangle - \langle \mathbf{w}, \mathbf{x}_0 \rangle)(f(Z|X) - f(Z|Y)) \delta Z\right]^2$$

$$= \langle \mathbf{w}, E_Y[\mathbf{T}(\Sigma)] - E_X[\mathbf{T}(\Sigma)] \rangle^2$$

On the other hand, note that

$$0 \leq E_X\left[\left(1 - \frac{f(\Sigma|Y)}{f(\Sigma|X)}\right)^2\right] = \int f(Z|X)\left(1 - \frac{f(Z|Y)}{f(Z|X)}\right)^2 \delta Z$$

$$= \int \left(f(Z|X) - 2f(Z|Y) + \frac{f(Z|Y)^2}{f(Z|X)}\right) \delta Z = \int \left(-1 + \frac{f(Z|Y)^2}{f(Z|X)}\right) \delta Z$$

$$= \int f(Z|Y)\left(-1 + \frac{f(Z|Y)^2}{f(Z|X)}\right) \delta Z = E_Y\left[\frac{f(Z|Y)}{f(Z|X)} - 1\right] = \Delta_X(Y)$$

as desired.   □

Suppose now that, in the inequality just proved, the finite set Y becomes arbitrarily close to the finite set X. Then what form does the inequality take in the limit?

Corollary 20 *Suppose that the parametrized family $f(Z|X)$ of global densities is uniformly continuous in the sense that*

$$l(\mathbf{y}, \eta|X) \triangleq \sup_{Z \subseteq \mathcal{R}} \left|\frac{f(Z|(X - \{\zeta\}) \cup \{(\mathbf{x} + \eta\mathbf{y}, v)\}) - f(Z|X)}{f(Z|X)}\right|$$

is finite and converges to zero as $\eta \to 0$, for all $\mathbf{y} \in \mathbb{R}^m$. Assume furthermore that the directional derivatives $\frac{\partial}{\partial_\zeta \mathbf{y}} f(Z|X)$ (taken with respect to the set variable X) exists for all $\zeta \in \mathcal{R}$ and $\mathbf{y} \in \mathbb{R}^m$. Then $I_{f,X,\zeta}(\mathbf{v}) = \langle \mathbf{v}, \mathbf{L}_{X,\zeta}(\mathbf{v}) \rangle$ for all $\mathbf{v}$ and

$$\langle \mathbf{w}, cov_{\mathbf{T},X}(\mathbf{w}) \rangle \cdot \langle \mathbf{v}, \mathbf{L}_{X,\zeta}(\mathbf{v}) \rangle \geq \langle \mathbf{w}, \frac{\partial}{\partial_\zeta \mathbf{v}} E_X[\mathbf{T}(\Sigma)] \rangle^2$$

where $\mathbf{L}_{X,\zeta} : \mathbb{R}^m \to \mathbb{R}^m$ is the linear transformation defined by

$$\langle \mathbf{v}, \mathbf{L}_{X,\zeta}(\mathbf{w}) \rangle = E_X\left[\left(\frac{\partial \ln f}{\partial_\zeta \mathbf{v}}\right)(\Sigma|X)\left(\frac{\partial \ln f}{\partial_\zeta \mathbf{w}}\right)(\Sigma|X)\right]$$

for all $\mathbf{v}, \mathbf{w} \in \mathbb{R}^m$ *and where*

$$cov_{\mathbf{T},X}(\mathbf{y}) = E_X\left[\langle \mathbf{T}(\Sigma) - \mathbf{x}_0, \mathbf{y} \rangle (\mathbf{T}(\Sigma) - \mathbf{x}_0)\right]$$

is the covariance of $\mathbf{T}(\Sigma)$. *Furthermore,*

$$\langle \mathbf{v}, \mathbf{L}_{X,\varsigma}(\mathbf{v}) \rangle = E_X\left[\left(\frac{\partial \ln f}{\partial_\varsigma \mathbf{v}}(\Sigma|X)\right)^2\right]$$

for all $\mathbf{v} \in \mathbb{R}^m$.

Proof. We adapt the proof of [5, Theorem 8.1.3, p. 302]. In Proposition 19 set

$$Y = (X - \{\varsigma\}) \cup \{(\mathbf{x} + \eta\mathbf{v}, v)\} = X_{\eta,\varsigma,\mathbf{v}}$$

and take the limits of both sides as $\eta \to 0$ where $X_{\eta,\varsigma,\mathbf{v}}$ is an abbreviation. Then by definition of $I_{f,X,\varsigma}(\mathbf{v})$ we have

$$\langle \mathbf{w}, cov_{\mathbf{T},X}(\mathbf{w}) \rangle \cdot I_{f,X,\varsigma}(\mathbf{v}) \geq \langle \mathbf{w}, \frac{\partial}{\partial_\varsigma \mathbf{v}} E_X[\mathbf{T}(\Sigma)]\rangle^2$$

assuming that the limits exist. Therefore it will be enough to prove that $I_{f,X,\varsigma}(\mathbf{v}) = \langle \mathbf{v}, \mathbf{L}_{X,\varsigma}(\mathbf{v}) \rangle$ for all $\mathbf{v} \in \mathbb{R}^m$. Begin by expanding the function $g(a) = a \ln a$ in a Taylor's series about 1:

$$a \ln a = (a-1) + \frac{1}{2}(a-1)^2 + \frac{1}{6}(a-1)^3 + \dots$$

Substitute

$$a = \frac{f(Z|X_{\eta,\varsigma,\mathbf{v}})}{f(Z|X)}$$

into the power series, multiply both sides by $f(Z|X)$ and take the set integral of both sides with respect to Z. Then we have:

$$I_f(X_{\eta,\varsigma,\mathbf{v}}; X) = \int (f(Z|X_{\eta,\varsigma,\mathbf{v}}) - f(Z|X))\, \delta Z$$

$$+ \int \left(\frac{f(Z|X_{\eta,\varsigma,\mathbf{v}})}{f(Z|X)} - 1\right)^2 f(Z|X)\delta Z + \text{higher order terms}$$

The first term vanishes identically and so

$$I_f(X_{\eta,\varsigma,\mathbf{v}}; X) = \int \left(\frac{f(Z|X_{\eta,\varsigma,\mathbf{v}})}{f(Z|X)} - 1\right)^2 f(Z|X)\delta Z + \text{higher order terms}$$

Now, each of the higher-order terms will be a constant times an integral of the form

$$\int \left(\frac{f(Z|X_{\eta,\varsigma,\mathbf{v}})}{f(Z|X)} - 1\right)^{i+2} f(Z|X)\delta Z$$

for $i \geq 1$. By the assumption of uniform continuity,

$$\left| \int \left(\frac{f(Z|X_{\eta,\zeta,\mathbf{v}})}{f(Z|X)} - 1 \right)^{i+2} f(Z|X) \right| \leq \sup_{Z \subseteq \mathcal{R}} \left| \frac{f(Z|X_{\eta,\zeta,\mathbf{v}})}{f(Z|X)} - 1 \right|^{i+2}$$

and so

$$\lim_{\eta \to 0} \frac{2}{\eta^2} \left| \int \left(\frac{f(Z|X_{\eta,\zeta,\mathbf{v}})}{f(Z|X)} - 1 \right)^{i+2} f(Z|X) \right|$$

$$\leq \lim_{\eta \to 0} \sup_{Z} \left| \frac{f(Z|X_{\eta,\zeta,\mathbf{v}})}{f(Z|X)} - 1 \right|^{i} \cdot \lim_{\eta \to 0} \sup_{Z} \left| \frac{f(Z|X_{\eta,\zeta,\mathbf{v}}) - f(Z|X)}{\eta \, f(Z|X)} \right|^2 = 0$$

where the third limit exists because of the differentiability assumption. Consequently,

$$\begin{aligned}
I_{f,X,\zeta}(\mathbf{v}) &= \lim_{\eta \to 0} \frac{\Delta_X(X_{\eta,\zeta,\mathbf{v}})}{\eta^2} = \lim_{\eta \to 0} \frac{1}{\eta^2} \int \left(\frac{f(Z|X_{\eta,\zeta,\mathbf{v}})}{f(Z|X)} - 1 \right)^2 f(Z|X) \delta Z \\
&= \int \frac{1}{f(Z|X)} \left(\lim_{\eta \to 0} \frac{f(Z|X_{\eta,\zeta,\mathbf{v}}) - f(Z|X)}{\eta} \right)^2 \delta Z \\
&= \int \frac{1}{f(Z|X)} \left(\frac{\partial f}{\partial_\zeta \mathbf{v}}(Z|X) \right)^2 \delta Z = E_X \left[\left(\frac{\partial \ln f}{\partial_\zeta \mathbf{v}}(\Sigma|X) \right)^2 \right] \\
&= \langle \mathbf{v}, \mathbf{L}_{X,\zeta}(\mathbf{v}) \rangle
\end{aligned}$$

as desired.   □

Finally, we show that the discrimination rate can be written in a more familiar-looking form:

Proposition 21 *Suppose that the parametrized family $f(Z|X)$ is uniformly continuous in the sense of Corollary 20 and that second-order directional derivatives of f with respect to the set parameter X exist. Suppose further that the following regularity conditions are satisfied:*

$$0 = \frac{\partial}{\partial_\zeta \mathbf{v}} \int f(Z|X) \delta Z = \int \frac{\partial f}{\partial_\zeta \mathbf{v}}(Z|X) \delta Z$$

$$0 = \frac{\partial^2}{\partial_\zeta \mathbf{v}^2} \int f(Z|X) \delta Z = \int \frac{\partial^2 f}{\partial_\zeta \mathbf{v}^2}(Z|X) \delta Z$$

for all $\zeta \in \mathcal{R}$ and $\mathbf{v} \in \mathbb{R}^m$. Then the discrimination rate can also be written as:

$$I_{f,X,\zeta}(\mathbf{v}) = -E_X \left[\frac{\partial^2 \ln f}{\partial_\zeta \mathbf{v}^2}(\Sigma|X) \right] = \frac{\partial^2}{\partial_\zeta \mathbf{v}^2} I_f(Y;X) \Big|_{Y=X}$$

where in the second equation the second-order directional derivative is taken with respect to the set variable Y.

Proof. Note that

$$\frac{\partial^2 \ln f}{\partial_\zeta \mathbf{v}^2}(Z|X) \;=\; \frac{\partial}{\partial_\zeta \mathbf{v}} \left(\frac{1}{f(Z|X)} \frac{\partial f}{\partial_\zeta \mathbf{v}}(Z|X) \right)$$

$$= \; -\frac{1}{f(Z|X)^2} \left(\frac{\partial f}{\partial_\zeta \mathbf{v}}(Z|X) \right)^2 + \frac{1}{f(Z|X)} \frac{\partial^2 f}{\partial_\zeta \mathbf{v}^2}(Z|X)$$

and so

$$E_X \left[\frac{\partial^2 \ln f}{\partial_\zeta \mathbf{v}^2}(\Sigma|X) \right] \;=\; -\int \frac{1}{f(Z|X)} \left(\frac{\partial f}{\partial_\zeta \mathbf{v}}(Z|X) \right)^2 \delta Z + \int \frac{\partial^2 f}{\partial_\zeta \mathbf{v}^2}(Z|X)\delta Z$$

$$= \; -I_{f,X,\zeta}(\mathbf{v}) + \frac{\partial^2}{\partial_\zeta \mathbf{v}^2} \int f(Z|X)\delta Z = -I_{f,X,\zeta}(\mathbf{v})$$

Next, taking the directional derivative of both sides of

$$I_f(Y;X) = \int f(Z|Y) \, \ln\left(\frac{f(Z|Y)}{f(Z|X)} \right) \delta Z$$

yields

$$\frac{\partial}{\partial_\zeta \mathbf{v}} I_f(Y;X) \;=\; \frac{\partial}{\partial_\zeta \mathbf{v}} \int f(Z|Y) \, \ln\left(\frac{f(Z|Y)}{f(Z|X)} \right) \delta Z$$

$$= \; \int \left(\frac{\partial f}{\partial_\zeta \mathbf{v}}(Z|Y) \, \ln\left(\frac{f(Z|Y)}{f(Z|X)} \right) + \frac{\partial f}{\partial_\zeta \mathbf{v}}(Z|Y) \right) \delta Z$$

and so

$$\frac{\partial^2}{\partial_\zeta \mathbf{w} \partial_\zeta \mathbf{v}} I_f(Y;X) \;=\; \int \frac{\partial^2 f}{\partial_\zeta \mathbf{w} \partial_\zeta \mathbf{v}}(Z|Y) \, \ln\left(\frac{f(Z|Y)}{f(Z|X)} \right) \delta Z$$

$$+ \int \frac{1}{f(Z|X)} \left(\frac{\partial f}{\partial_\zeta \mathbf{w}}(Z|Y) \right) \left(\frac{\partial f}{\partial_\zeta \mathbf{v}}(Z|Y) \right) \delta Z$$

$$+ \int \frac{\partial^2 f}{\partial_\zeta \mathbf{w} \partial_\zeta \mathbf{v}}(Z|Y)\delta Z$$

Therefore, setting $Y = X$ and $\mathbf{w} = \mathbf{v}$ yields:

$$\left.\frac{\partial^2}{\partial_\zeta \mathbf{v}^2} I_f(Y;X)\right|_{Y=X} \;=\; \int \frac{1}{f(Z|X)} \left(\frac{\partial f}{\partial_\zeta \mathbf{v}}(Z|X) \right)^2 \delta Z + \int \frac{\partial^2 f}{\partial_\zeta \mathbf{v}^2}(Z|X)\delta Z$$

$$= \; \int \frac{1}{f(Z|X)} \left(\frac{\partial f}{\partial_\zeta \mathbf{v}}(Z|X) \right)^2 \delta Z = I_{f,X,\zeta}(\mathbf{v})$$

This concludes the proof.   □

Bibliography

[1] Artstein, Z. and Burns, J.A. (1975) Integration of compact set-valued functions, *Pacific Journal of Mathematics*, **58** (2), 297-307.

[2] Artstein, Z. and Vitale, R. A. (1975) A strong law of large numbers for random compact sets, *Annals of Probability*, **3**, 879-882.

[3] Aumann, J. (1965) Integrals of set valued functions, *Journal of Mathematical Analysis and Applications*, **12**, 1-22.

[4] Baddeley, A. J., and Molchanov, I. S. (1995) Averaging of random sets based on their distance functions, Centrum voor Wiskunde en Informatica Technical Report BS-R9528, Amsterdam, Oct. 1995.

[5] Blahut, R. E. (1987) *Principles and Practice of Information Theory*, Addison-Wesley, Chapters 7 and 8.

[6] Debreu, G. (1967) Integration of correspondences, in L.M. Le Cam and J. Neyman (eds.), *Proceedings of the Fifth Berkeley Symposium on Mathematical Statistics and Probability, Vol. II Part 1: Contributions to Probability Theory*, June 21-July 18 1965, University of California Press.

[7] Forsyth, A.R. (1960) *Calculus of Variations*, Dover Publications.

[8] Freĭman, G.A. (1973) *Foundations of a Structural Theory of Set Addition*, American Mathematical Society.

[9] Grabisch, M., Nguyen, H. T., and Walker, E. A. (1995) *Fundamentals of Uncertainty Calculi With Applications to Fuzzy Inference*, Kluwer Academic Publishers, Dordrecht.

[10] Halmos, P. R. (1974) *Measure Theory*, Springer-Verlag, New York.

[11] Kamen, E. W., and Sastry, C. R. (1993) Multiple target tracking using products of position measurements, *IEEE Transactions on Aerospace and Electronics Systems*, **29** (2), 476-493.

[12] Kapur, J. N. and Kesavan, H. K. (1992) *Entropy Optimization Principles with Applications*, Academic Press, New York.

[13] Kullback, S. (1968) *Information Theory and Statistics*, Dover Publications.

[14] Lang, S. (1971) *Algebra*, Addison-Wesley, Massachusetts.

[15] Mathéron, G. (1975) *Random Sets and Integral Geometry*, J. Wiley, New York.

[16] Molchanov, I. S. (1993) *Limit Theorems for Unions of Random Closed Sets*, Springer-Verlag Lecture Notes in Mathematics No. 1561, New York.

[17] Nguyen, H. T., and Rogers, G. S. (1989) *Fundamentals of Mathematical Statistics, Vol. II: Statistical Inference*, Springer-Verlag, New York.

[18] Radstrom, H. (1952) An embedding theorem for spaces of convex sets, *Proceedings of the American Mathematical Society*, **3**, 165-169.

[19] Reingold, E. M. and Nievergelt, J. (1977) *Combinatorial Algorithms: Theory and Practice*, Prentice-Hall, Englewood Cliffs, New Jersey.

[20] Robbins, H. E. (1943) On the measure of a random set, *Annals of Mathematical Statistics*, **14**, 70-74.

[21] van Trees, H. L. (1968) *Detection, Estimation, and Modulation Theory, Part I: Detection, Estimation, and Linear Modulation Theory*, J. Wiley, New York.

[22] Wald, A. (1949) Note on the consistency of the maximum likelihood estimator, *The Annals of Mathematical Statistics*, **20**, 595-601.

Chapter 6

Fusion of Unambiguous Observations

In this chapter we describe the basic elements of a unified approach to data fusion, assuming that observations are *unambiguous* (the approach will be generalized to include ambiguous observations in the next chapter). Section 6.1 briefly describes the basic problem under consideration. Section 6.2 describes how random set techniques can be used to derive measurement models for entire sensor suites, and prior knowledge in data fusion is discussed in Section 6.3. Section 6.4 describes "global" motion models and "global" discrete-time recursive Bayesian nonlinear filtering, and how they can be used to fuse data generated by dynamic targets. The problem of determining global densities for the *output* of a data fusion algorithm is considered in Section 6.5, which specific application to multihypothesis-type algorithms. The chapter concludes with three detailed, simple examples in Section 6.6.

6.1 The Central-Fusion Problem

Understood mathematically, the basic problem in any information fusion problem is that of determining an estimate of what "state" an information fusion system—that is, a system of sensors, targets, and noise models—is in. We therefore must first know what the term "state" in information fusion means. We are to locate, identify, and determine the number of targets existing in, say, three-dimensional space. We assume that these targets belong to a fixed and finite set V of possible known targets (or, alternatively, target classes/types). At its simplest, an individual point-target is completely characterized by a 10-tuple of the form $(x, y, z, u, v, w, a, b, c, \Delta)$ where

$$x, y, z \;=\; \text{components of position}$$
$$u, v, w \;=\; \text{components of velocity}$$
$$a, b, c \;=\; \text{components of acceleration}$$

$$\Delta \;=\; \text{unique descriptor of target I.D.}$$

A state is then just a *finite set* of these 10-tuples.

Thus defined, states are obviously highly simplified. For example attribute variables used in target classification can be, and often are, continuous rather than discrete. The components of a state need not be expressed in terms of Cartesian coordinates. There may be additional geodynamic state variables (e.g. aspect angle; yaw, pitch, and roll). There also may be additional discrete state variables (e.g. threat state, threat intent, etc.).

Accordingly, we generalize as follows:

Definition 1 *An information fusion state is a finite, possibly empty, subset of the set $\mathcal{R} \triangleq \mathbb{R}^m \times V$.*

Measurements can be—and usually are—even more complex than states. Besides geopositional variables (range, azimuth, and elevation, range-rate, azimuth-rate, and elevation-rate, as measured by a radar) continuous observation variables can include things such as signal intensity (e.g. doppler, as measured by a moving-target-indicator radar), signal "attributes" (e.g. RF center frequency, as detected by an electronic countermeasures receiver), signal "signatures" (e.g. radar signal intensity expressed as a function of a range interval), and so on. At their most general, signal variables can include complex acoustic phenomena, chemical-species attributes, and so on.

We will make a distinction between two kinds of observation: *data* and *evidence*. The basic distinction is that data is *Bayesian* in form (i.e., datums are mutually exclusive and exhaustive) whereas evidence is *non-Bayesian* (i.e., possibly non-exclusive and non-exhaustive). Data ("unambiguous" observations) will be treated in this chapter whereas evidence ("ambiguous" observations) will be treated in the next. For simplicity we will assume that single-sensor observations have the same general form as states: $\mathcal{R} = \mathbb{R}^n \times U$. In addition, however, we usually also know which sensor originated a report: A report is usually accompanied by a "sensor tag" which identifies which sensor originated the observation. Therefore:

Definition 2 *An information fusion observation is an element of $\mathcal{R}_\odot \triangleq \mathbb{R}^n \times U \times U_\odot$ where $U_\odot \triangleq \{1, ..., s\}$ denotes the set of sensor tags. We will write $W \triangleq U \times U_\odot$.*

In principle, Definition 2 can be used to model most spaces of unambiguous observations. For example, suppose that observations consist of "signatures" which are (randomly varying) functions $s(x)$ of a continuous variable x. Then we can discretize the range of the variable x by partitioning it into "bins" centered around fixed values $x = i \Delta x$ for $i = 0, 1, ..., M$. By this means, the random function s can be modeled as the random vector $(s(0), s(\Delta x), ..., s(M \Delta x))$.

NOTE: For the sake of convenience we will sometimes abuse notation by using the symbol '$\mathcal{R}$' to represent both $\mathcal{R} = \mathbb{R}^n \times U$ and $\mathcal{R} = \mathbb{R}^m \times V$.

It should also be noted that the techniques described here can, in principle, be extended to *distributed fusion* problems by employing the techniques of, say, [2].

6.2 Random Set Measurement Models

In what follows we will show how to construct random set observation models for a variety of multisensor and multitarget scenarios. The basic procedure which is followed in all cases is as follows:

(1) From a knowledge of the sensors in the sensor suite, construct the typical (random) observation-set Σ collected by the sensor suite;

(2) Determine the belief measure $\beta_\Sigma(S) = p(\Sigma \subseteq S)$ of the random observation-set; and

(3) Construct the global density $f_\Sigma(Z)$ of the belief measure $\beta_\Sigma(S)$ using the set derivative.

For the sake of clarity, we will usually assume that observations are individually generated by individual targets, though it should be clear that more general assumptions concerning data generation can be accommodated by our approach. We also note that measurement models for "extended targets" (i.e., targets whose reflecting surfaces generate observations consisting of small sets of point images) can in principle be included in the purview of our approach.

In Sections 6.2.1 and 6.2.2 we will describe how to construct measurement models for multisensor, mulitarget systems. In Section 6.2.3 we show that these measurement models are consistent with conventional (i.e., point-variate rather than set-variate) probability theory.

6.2.1 Single-Sensor, Single-Target Measurement Models

Single-Sensor, Single-Target Case Without Clutter

Assume initially that a single sensor $\odot$ observes a single target $\otimes$ and has possibly non-unity probability of detection, p_D, but zero false alarm rate, $p_{FA} = 0$. The observation-set Σ will contain either exactly one element or it will be the empty set. Assume that the internal self-noise of the sensor is modeled by a random variable $\psi = (\mathbf{Z}, z)$ on $\mathcal{R} = \mathbb{R}^n \times V$. Define $\Sigma = \{\mathbf{Z}\} \times \Xi$ where Ξ is a random subset of V such that $p(\Xi = \emptyset) = 1 - p_D$ and $\Xi = \{z\}$ if $\Xi \neq \emptyset$.

Remark 1 *Such a subset Ξ can always be constructed. For, construct a random subset Θ of V such that $p(\Theta = \emptyset) = 1 - p_D$ and $p(\Theta = V) = p_D$. Let $\Xi = \Theta \cap \{z\}$. Then $p(\Xi = \emptyset) = p(\Theta = \emptyset) = 1 - p_D$ and $p(\Xi = V) = p_D$.*

Now by Proposition 12 of Chapter 4, $\{\mathbf{Z}\} \times \Xi$ is a random subset of $\mathcal{R}$ and the belief measure of Σ is

$$\begin{aligned} \beta_\Sigma(S) &= p(\{\mathbf{Z}\} \times \Xi \subseteq S) = p(\Xi = \emptyset) + p(\Xi \neq \emptyset, (\mathbf{Z}, z) \in S) \\ &= 1 - p_D + p_D \, q(S) \end{aligned}$$

where $q(S) \triangleq p((\mathbf{Z}, z) \in S | \Xi \neq \emptyset)$ is a probability measure on $\mathcal{R}$ (see [14, p. 30] for a similar analysis). We may assume that β_Σ has the form

$$\beta_\Sigma(S|\zeta) = 1 - p_D(\zeta) + p_D(\zeta)\, q_f(S|\zeta) = 1 - p_D(\zeta) + p_D(\zeta) \int_S f(\xi|\zeta)d\lambda(\xi)$$

where $\zeta \subseteq \mathcal{R}$ is a parameter and f is the density of q_f. Since $p_D(\zeta)$ is parametrized it can (in general) vary with such quantities as: range from sensor, deviation angle from sensor beam; aspect angle of target with respect to sensor; and pitch, roll, and yaw. Note that the global density of $\beta_\Sigma(S|\zeta)$ is:

$$f_\Sigma(\emptyset|\zeta) = 1 - p_D(\zeta), \qquad f_\Sigma(\{\xi\}|\zeta) = p_D(\zeta)\, f(\xi|\zeta)$$

Remark 2 *The belief measure $\beta_\Sigma(S|\zeta)$ has a simple probabilistic interpretation. Since $1 - p_D$ is the probability of missed detection, the quantity $\beta_\Sigma(\emptyset|\zeta) = 1 - p_D(\zeta)$ is the total probability that the sensor will return the "empty observation"— i.e., no observation at all. On the other hand $p_D(\zeta)\, q(S|\zeta)$ is the probability that the sensor will return a nonempty observation and, moreover, return one which is contained in the set S. Hence*

$$\beta_\Sigma(S|\zeta) = 1 - p_D(\zeta) + p_D(\zeta) \int_S f(\xi|\zeta)d\lambda(\xi)$$

is the probability that the sensor will return any observation (empty or otherwise) contained in S.

Single-Sensor, Single-Target Case With Clutter

Let us now take clutter and false alarms into account. The typical observation set Σ will have the general form

$$\Sigma = \text{"signal"} \cup \text{"clutter"} = \Sigma' \cup \Lambda$$

Here Σ' is the random subset of reports due to the target. It contains either one report or is the empty set. The subset Λ of observations, on the other hand, models the clutter process. We assume that clutter observations are generated by "clutter objects" which can be modeled in the same way as the targets of actual interest, but whose noise statistics may differ from those of actual target objects.

For a clutter object, a zero probability of detection means that the clutter object is never observed by the sensor. Likewise, a unity probability of detection for a clutter object means that the clutter object always generates a spurious observation. Thus for a clutter object, the probability of detection is p_{FA} where p_{FA} is the probability of false alarm. Hence the belief measure of the random set of observations generated by a clutter object is

$$\beta_c(S) = 1 - p_{FA} + p_{FA}\, q_c(S) = 1 - p_{FA} + p_{FA} \int c(\xi)d\lambda(\xi)$$

for some probability measure q_c with density c. (This model describes point clutter which is state-independent. Clearly, it is also possible to model state-dependent point clutter by allowing β_c to be functionally dependent on the parameter ζ.)

It follows that the typical sensor observation-set Σ will have the form $\Sigma = \Sigma' \cup \Lambda$ where Σ' models the target $\otimes$ and where $\Lambda = \Lambda_1 \cup ... \cup \Lambda_M$, where each Λ_j models an individual clutter source. At one extreme each Λ_j could have its own probability measure $q_{c;j}$ and its own probability of false alarm $p_{FA;j}$. At the other extreme, the belief measures of the Λ_j could be identical.

With suitable independence assumptions we can derive relatively simple, explicit formulas for β_Σ and f_Σ. First, if Σ', Λ are independent then

$$\beta_\Sigma(S|\zeta) = p(\Sigma' \subseteq S, \Lambda \subseteq S) = p(\Sigma' \subseteq S)\, p(\Lambda \subseteq S) = \beta_{\Sigma'}(S|\zeta)\, \beta_\Lambda(S)$$

From the product rule for set derivatives and the definition of a global density, we get:

$$f_\Sigma(Z|\zeta) = f_{\Sigma' \cup \Lambda}(Z|\zeta) = \sum_{Y \subseteq Z} f_{\Sigma'}(Y|\zeta)\, f_\Lambda(Z - Y)$$

for all $Z \subseteq \mathcal{R}$. Assume furthermore that the $\Lambda_1, ..., \Lambda_M$ are independent and identically distributed. Then the belief measure of Σ is

$$\beta_\Sigma(S|\zeta) = p(\Sigma' \subseteq S, \Lambda_1 \subseteq S, ..., \Lambda_M \subseteq S) = \beta_{\Sigma'}(S|\zeta)\, \beta_c(S)^M$$

Using the fact that

$$\beta_c(S) = 1 - p_{FA} + p_{FA}\, q_c(S)$$
$$\beta_{\Sigma'}(S|\zeta) = 1 - p_D(\zeta) + p_D(\zeta)\, q_f(S|\zeta)$$

we know (Proposition 20 of Chapter 4) that the global density of the clutter process is

$$f_\Lambda(\{\xi_1, ..., \xi_k\}) = k!\, C_{M,k}\, p_{FA}^k\, (1 - p_{FA})^{M-k}\, c(\xi_1) \cdots c(\xi_k)$$

and that the corresponding cardinality distribution (Definition 18 of Chapter 4)

$$f_\Lambda(k) = p(|\Lambda| = k) = C_{M,k}\, p_{FA}^k\, (1 - p_{FA})^{M-k}$$

for $0 \le k \le M$ is binomial. If M is very large and p_{FA} is very small then $\Lambda = \Lambda_1 \cup ... \cup \Lambda_M$ approximates a Poisson process.

Modeling Bearings-Only Sensors and Other Sensors With Extended Fields of View

So far we have illustrated how one might model sensors which, like many surveillance and tracking radars, are relatively ideal in the sense that they collect detections which are individual points in space. Many sensors—typical examples are passive sonars or electronic warfare receivers—collect very incomplete observations which are not easily localizable as distinct points. For example, passive sonar systems often will provide only an indication of the line-of-sight to target (the bearing) and perhaps the rate of change of the bearing line (the bearing rate). They may provide no range-to-target information at all, or at best very limited range information (e.g., localization only to within some geographically extended "convergence zone"; see below). As another example, real-world long-range surveillance radars do not collect precise point detections with Gaussian uncertainties. Rather, they only constrain the probable detection of a target to inclusion within a three-dimensional "smile"-shaped region whose shape is described by an interval of range (the linear width of a range cell) and intervals of azimuth and elevation (the angular widths of azimuth and elevation cells). All such sensors have "detection zones" or "fields of view" within which targets can be detected, and outside of which they cannot.

Such sensors can be modeled in the random set framework by exploiting the state-dependence of the probability of detection $p_D(\zeta)$. As a simple but specific example, we construct a crude observation model of a two-dimensional passive-sonar sensor located at the origin of a polar coordinate system. The observation space consists of measured bearing-angles in $[-\frac{1}{2}\pi, \frac{1}{2}\pi] \subseteq \mathbb{R}$. Let $\zeta = (r, \alpha, v)$ denote target state, where r denotes range to target, α denotes bearing angle to target, v denotes target type. Then a model for the belief measure of the sensor will have the form

$$\beta(S|r, \alpha, v) = 1 - p_D(r)p_D(\alpha)p_D(v) + p_D(r)p_D(\alpha)p_D(v) \int_S g(\theta|\alpha)d\lambda(\theta)$$

for all $S \subseteq [-\frac{1}{2}\pi, \frac{1}{2}\pi]$ and where g is a density describing the sensor noise involved in measuring the angle α. Also, $p_D(r)$ and $p_D(\alpha)$ describe detectability as functions of range and bearing angle, respectively, and $p_D(v)$ characterizes the detectability of individual target types. For example, choose

$$p_D(r) \triangleq \begin{cases} \frac{1}{2}e^{-r/R_1}\left(1 + \cos\left(\frac{\pi r}{R_0}\right)\right) & \text{if} \quad 0 \leq r \leq \frac{5R_0}{2} \\ 0 & \text{if} \quad \text{otherwise} \end{cases}$$

$$p_D(\alpha) \triangleq e^{-K|\sin\alpha|} \qquad (\text{all } \alpha \in [-\frac{\pi}{2}, \frac{\pi}{2}])$$

This is a crude model of a single lobe of a passive sonar sensor with three "convergence zones" symmetric about the x-axis. The unitless constant $K > 0$ determines the degree of attenuation in detectability as a function of angular separation from $\alpha = 0$ radians. The range constant R_1 determines the degree of detection attenuation in range and the range constant R_0 governs the placement of the detection peaks and nulls of the convergence zones. The peak of the first zone is at the origin, the peak of the second zone is at $r = R_0$, and the peak of the third zone is at $r = 2R_0$. Detection nulls occur at the odd multiples of $\frac{1}{2}R_0$.

The global measurement-model density is, obviously,

$$f(\emptyset|r, \alpha, v) = 1 - p_D(r)p_D(\alpha)p_D(v)$$
$$f(\{\theta\}|r, \alpha, v) = p_D(r)p_D(\alpha)p_D(v)\, g(\theta|\alpha)$$

6.2.2   Multisensor, Multitarget Measurement Models

In passing from the single-sensor to the multi-sensor case, a new consideration must be taken into account: We must know which sensor originated which observation—that is, we must know the statistical model corresponding to that particular sensor. Thus an observation must have the general form $\xi = (z, w, j)$ where $j \in U_\odot = \{1, ..., s\}$ is the unique identifying tag of the sensor which originated observation (z, w). In general, the random observation supplied by the j'th sensor will have the signal-plus-clutter form

$$\Sigma'_j \times j = (\Sigma_j \cup \Lambda_{j;1} \cup ... \cup \Lambda_{j;M(j)}) \times j$$

described in the previous section (where as usual we are abusing notation by writing $S \times j \triangleq \times \{j\}$). In general the random observation supplied by the entire suite of sensors will have the form

$$\Sigma = (\Sigma'_1 \times 1) \cup ... \cup (\Sigma'_s \times s)$$

Note that for any $S \subseteq \mathcal{R}_\odot = \mathbb{R}^n \times U \times U_\odot$ the belief measure of $\Sigma'_j \times j$ is

$$
\begin{aligned}
\beta_{\Sigma'_j \times j}(S) &= p(\Sigma'_j \times j \subseteq S) = p\left(\Sigma'_j \times j \subseteq \bigcup_{i=1}^{s} S(i) \times i\right) \\
&= p(\Sigma'_j \times j \subseteq S(j) \times j) = p(\Sigma'_j \subseteq S(j)) = \beta_{\Sigma'_j}(S(j))
\end{aligned}
$$

where $S(j) \triangleq \{(z, w) \in \mathcal{R} \mid (z, w, j) \in S\}$. Suppose in particular that $S = T \times i$ for some $T \subseteq \mathcal{R} = \mathbb{R}^n \times U$. Then

$$
\beta_{\Sigma'_j \times j}(T \times i) = \begin{cases} \beta_{\Sigma'_j}(T) & \text{if } j = i \\ \beta_{\Sigma'_j}(\emptyset) & \text{if } j \neq i \end{cases}
$$

Thus the belief measure of the general observation for the entire sensor suite will be

$$
\begin{aligned}
\beta_\Sigma(S) &= p(\Sigma'_1 \times 1 \subseteq S, ..., \Sigma'_s \times s \subseteq S) \\
&= p(\Sigma'_1 \subseteq S(1), ..., \Sigma'_s \subseteq S(s))
\end{aligned}
$$

If $\Sigma'_1, ..., \Sigma'_s$ are independent this becomes $\beta_\Sigma(S) = \beta_{\Sigma'_1}(S(1)) \cdots \beta_{\Sigma'_s}(S(s))$. This last equation leads us to the following result, which tells us how to compute the global density of a sensor suite consisting of two sensors which supply independent observations. As expected, it allows us to write the global density of a multisensor scenario to the product of global densities of the separate single-sensor scenarios.

Proposition 1 *Let* $\Sigma = (\Sigma'_1 \times 1) \cup (\Sigma'_2 \times 2)$ *where* Σ'_1, Σ'_2 *are independent random subsets of* $\mathcal{R} = \mathbb{R}^n \times U$. *Then:*

$$f_\Sigma(Z_1 \times 1 \cup Z_2 \times 2) = f_{\Sigma'_1}(Z_1)\, f_{\Sigma'_2}(Z_2)$$

for all finite subsets $Z_1, Z_2 \subseteq \mathcal{R}$.

Proof. Under these assumptions $\beta_\Sigma(S) = \beta_{\Sigma_1'}(S(1))\,\beta_{\Sigma_2'}(S(2))$ and so for $i = 1, 2$ and $\xi \in \mathcal{R}$ we get

$$\frac{\delta\beta_\Sigma}{\delta(\xi \times i)}(S) = \beta_{\Sigma_1'}(S(1))\,\frac{\delta}{\delta(\xi \times i)}\beta_{\Sigma_2'}(S(2)) + \beta_{\Sigma_2'}(S(2))\,\frac{\delta}{\delta(\xi \times i)}\beta_{\Sigma_1'}(S(1))$$

However, note that if $\xi = (z, u)$ and $S = T \cup (E_{z;l} \times u \times i)$ with $\xi \notin T$ then

$$S(j) = \left\{ \begin{array}{ll} T(j) \cup (E_{z;l} \times u) & \text{if}\quad j = i \\ T(j) & \text{if}\quad j \neq i \end{array} \right.$$

and so if $j \neq i$ then

$$\frac{\delta}{\delta(\xi \times i)}\beta_{\Sigma_j'}(S(j)) = \lim_{l\to\infty}\frac{\beta_{\Sigma_j'}(T(j)) - \beta_{\Sigma_j'}(T(j))}{\lambda(E_{z;l})} = 0$$

whereas if $j = i$ then

$$\frac{\delta}{\delta(\xi \times i)}\beta_{\Sigma_i'}(S(i)) = \lim_{l\to\infty}\frac{\beta_{\Sigma_i'}(T(i) \cup (E_{z;l} \times u)) - \beta_{\Sigma_i'}(T(i))}{\lambda(E_{z;l})} = \frac{\delta\beta_{\Sigma_i'}}{\delta\xi}(T(i))$$

Thus

$$\frac{\delta\beta_\Sigma}{\delta(\xi \times 1)}(S) = \beta_{\Sigma_2'}(S(2))\,\frac{\delta\beta_{\Sigma_1'}}{\delta\xi}(S(1))$$

$$\frac{\delta\beta_\Sigma}{\delta(\xi \times 2)}(S) = \beta_{\Sigma_1'}(S(1))\,\frac{\delta\beta_{\Sigma_2'}}{\delta\xi}(S(2))$$

The proposition immediately follows from these two equations by using the definition

$$f_\Sigma(Z_1 \times 1 \cup Z_2 \times 2) \triangleq \frac{\delta}{\delta(\xi_1 \times 1)}\cdots\frac{\delta}{\delta(\xi_k \times 1)}\frac{\delta}{\delta(\eta_1 \times 2)}\cdots\frac{\delta}{\delta(\eta_l \times 2)}\beta_\Sigma$$

for $Z_1 = \{\xi_1, ..., \xi_k\}$ and $Z_2 = \{\eta_1, ..., \eta_l\}$ such that $|Z_1| = k$ and $|Z_2| = l$. $\square$

No Missed Detections, No False Alarms

For purposes of illustration let us derive the belief-measure model and global density of the multisensor problem assuming that there are no missed detections or false alarms. Given the presence of t targets the global observation collected by the j'th sensor will have the form $\Sigma_j' = \Sigma_j = \{\psi_{j;1}, ..., \psi_{j;t}\}$ where $\psi_{j;1} = (\mathbf{Z}_{j;1}, A_{j;1}), ..., \psi_{j;t} = (\mathbf{Z}_{j;t}, A_{j;t})$ for random vectors $\mathbf{Z}_{j;i}$ on $\mathbb{R}^n$ and random variables $A_{j;i}$ on U. The total global observation of the sensor suite is the random set

$$\Sigma = \{(\psi_{1;1}, 1), ..., (\psi_{1;t}, 1), ..., (\psi_{s;1}, s), ..., (\psi_{s;t}, s)\}$$

which always contains exactly $s \cdot t$ elements. The belief measure is

$$\beta_\Sigma(S|\{\zeta_1, ..., \zeta_t\})$$

$$= p((\psi_{1;1}, 1) \in S, ..., (\psi_{1;t}, 1) \in S, ..., (\psi_{s;1}, s) \in S, ..., (\psi_{s;t}, s) \in S)$$
$$= p(\psi_{1;1} \in S(1), ..., \psi_{1;t} \in S(1), ..., \psi_{s;1} \in S(s), ..., \psi_{s;t} \in S(s))$$

and, if $\psi_{1;1}, ..., \psi_{1;t}, ..., \psi_{s;1}, ..., \psi_{s;t}$ are independent with respective probability measures

$$q_1(S|\zeta_1), ..., q_1(S|\zeta_t), ..., q_s(S|\zeta_1), ..., q_s(S|\zeta_t)$$

then

$$\beta_\Sigma(S|\{\zeta_1, ..., \zeta_t\}) = q_1(S|\zeta_1) \cdots q_1(S|\zeta_t) \cdots q_s(S|\zeta_1) \cdots q_s(S|\zeta_t)$$

In light of Proposition 1 we know that the corresponding global density is

$$f_\Sigma(Z_1 \times 1 \cup ... \cup Z_s \times s|X) = f_1(Z_1|X) \cdots f_s(Z_s|X)$$

where $f_j(-|X)$ is the global density corresponding to the belief measure

$$\beta_j(S|\{\zeta_1, ..., \zeta_t\}) = q_j(S|\zeta_1) \cdots q_j(S|\zeta_t)$$

for all $j = 1, ..., s$. From Proposition 20 of Chapter 4 we know that this is

$$f_j(\{\xi_1, ..., \xi_t\}|\{\zeta_1, ..., \zeta_t\}) = \sum_\pi f_j(\xi_{\pi 1}|\zeta_1) \cdots f_j(\xi_{\pi t}|\zeta_t)$$

where $f_j(\xi|\zeta)$ is the parametrized (single-target) observation-noise density for the j'th sensor. Accordingly, the global density for the entire multisensor problem is

$$f_\Sigma(\{\xi_{1;1}, ..., \xi_{1;t}\} \times 1 \cup ... \cup \{\xi_{s;1}, ..., \xi_{s;t}\} \times s|\{\zeta_1, ..., \zeta_t\})$$
$$= \left(\sum_\pi f_1(\xi_{1;\pi 1}|\zeta_1) \cdots f_1(\xi_{1;\pi t}|\zeta_t) \right) \cdots \left(\sum_\pi f_s(\xi_{s;\pi 1}|\zeta_1) \cdots f_s(\xi_{s;\pi t}|\zeta_t) \right)$$

Missed Detections, No False Alarms/Clutter

Suppose now that we generalize by allowing missed detections, with $p_{D,j}(\zeta)$ denoting the probability of detection for the j'th sensor as a function of individual-target state ζ. In this case the observation collected by the j'th sensor from the i'th target, which we denote by $\Sigma_{j;i}$, is either empty or consists of a single element. The belief measure has the form

$$\beta_j(S|\zeta_i) = \beta_{\Sigma_{j;i}}(S) = 1 - p_{D;i}(\zeta_i) + p_{D;j}(\zeta_i) \, q_j(S|\zeta_i)$$

where $q_j(S|\zeta_i) = p(\Sigma_{j;i} \subseteq S|\Sigma_{j;i} \neq \emptyset)$ is a probability measure. Accordingly, the global observation collected by the entire sensor suite is

$$\Sigma = (\Sigma_{1;1} \times 1) \cup ... \cup (\Sigma_{1;t} \times 1) \cup ... \cup (\Sigma_{s;1} \times s) \cup ... \cup (\Sigma_{s;t} \times s)$$

Thus Σ can contain anywhere from $s \cdot t$ elements to no elements at all. If we assume that $\Sigma_{1;1}, ..., \Sigma_{1;t}, ..., \Sigma_{s;1}, ..., \Sigma_{s;t}$ are independent then the belief measure of the global observation has the form

$$\beta_\Sigma(S|\{\zeta_1, ..., \zeta_t\}) = \beta_1(S|\zeta_1) \cdots \beta_1(S|\zeta_t) \cdots \beta_s(S|\zeta_1) \cdots \beta_s(S|\zeta_t)$$

Once again we know that the corresponding global density is

$$f_\Sigma(Z_1 \times 1 \cup ... \cup Z_s \times s | X) = f_1(Z_1|X) \cdots f_s(Z_s|X)$$

where $f_j(-|X)$ is the global density corresponding to the belief measure

$$\beta_j(S|\{\zeta_1, ..., \zeta_t\}) = \beta_j(S|\zeta_1) \cdots \beta_j(S|\zeta_t)$$

for all $j = 1, ..., s$. From Proposition 20 of Chapter 4 we know that

$$
\begin{aligned}
f_\Sigma(\emptyset|\{\zeta_1, ..., \zeta_t\}) &= (1 - q(\zeta_1)) \cdots (1 - q(\zeta_t)) \\
f_\Sigma(\{\xi_1, ..., \xi_k\}|\{\zeta_1, ..., \zeta_t\}) &= \sum_{1 \leq i_1 \neq ... \neq i_t \leq t} a_{i_1,...,i_k}(\zeta_1, ..., \zeta_t) \\
&\quad \cdot f(\xi_1|\zeta_{i_1}) \cdots f(\xi_k|\zeta_{i_k})
\end{aligned}
$$

where the summation is taken over all distinct $i_1, ..., i_t$ such that $1 \leq i_1, ..., i_t \leq t$ and where

$$a_{i_1,...,i_k}(\zeta_1, ..., \zeta_t) \triangleq \frac{(1 - q(\zeta_1)) \cdots (1 - q(\zeta_m))}{(1 - q(\zeta_{i_1})) \cdots (1 - q(\zeta_{i_k}))} q(\zeta_{i_1}) \cdots q(\zeta_{i_k})$$

This concludes our discussion.

6.2.3 Conventional Interpretation of Global Measurement Models

As developed so far, "global densities" are abstract concepts formally derived from finite random sets by iteration of a differentiation operator. If the random set approach to information fusion is to be useful, however, it should be possible to provide an interpretation of a global measurement density $f_\Sigma(Z|X)$ in terms of conventional probability theory. This is the purpose of the present section, in which we intend to establish the following two interpretations of a global measurement density:

Random set interpretation: $f_\Sigma(Z|X)$ is the likelihood that Σ takes the global observation Z as a value, given the set of parameters X.

Conventional interpretation: $f_\Sigma(Z|X)$ is the total probability density of association between the observations in Z and the parameters in X.

To motivate what follows, let us begin by assuming that t targets are being observed by a single sensor with density $f(\mathbf{z}|\mathbf{x})$ which supplies independent observations and which has no missed detections and no false alarms. Let $\mathbf{x}_1, ..., \mathbf{x}_t \in \mathbb{R}^m$ be the state (parameter) vectors of the t targets. Let $\mathbf{z}_1, ..., \mathbf{z}_t \in \mathbb{R}^m$ be the observations collected (with probability one) by the sensor at any given instant. In general there are $t!$ possible ways—the possible report-to-track associations—to account for the observed production of observations $\mathbf{z}_1, ..., \mathbf{z}_t$ by targets $\mathbf{x}_1, ..., \mathbf{x}_t$. Each association will have the form

$$\mathbf{z}_{\pi 1} \leftrightarrow \mathbf{x}_1 \qquad , ..., \qquad \mathbf{z}_{\pi t} \leftrightarrow \mathbf{x}_t$$

where π is some permutation on the numbers $1, ..., t$. The probability density which describes how likely it is that observation z_i will arise given state x_j is $f(z_i|x_j)$. Thus given an association π, the probability density which describes how likely it is that $x_1, ..., x_t$ generated $z_{\pi 1}, ..., z_{\pi t}$ is

$$f(z_1, ..., z_t|x_1, ..., x_t, \pi) = f(z_{\pi 1}|x_1) \cdots f(z_{\pi t}|x_t)$$

where, note, the association π is treated as an additional undetermined state-parameter of the multitarget system. Let $p(\pi|x_1, ..., x_t)$ be the *a priori* probability that the association π will occur given the configuration of states $x_1, ..., x_t$. Then the total probability density that observations $z_1, ..., z_t$ will be seen (independent of association) given the configuration $x_1, ..., x_t$ is

$$f_{ass}(z_1, ..., z_t|x_1, ..., x_t) = \sum_{\pi} f(z_{\pi 1}|x_1) \cdots f(z_{\pi t}|x_t) \, p(\pi|x_1, ..., x_t)$$

As argued in Chapter 2, *the correct report-to-track association is an unobservable of the multitarget system.* Given that this is true, $p(\pi|x_1, ..., x_t)$ cannot be determined and so let us choose, as a least informative prior on the set of associations, $p(\pi|x_1, ..., x_t) = 1/t$ for all π and all $x_1, ..., x_t \in \mathbb{R}^m$. In this case the total probability density of observing the observation-set $\{z_1, ..., z_t\}$ given the parameter-set $\{x_1, ..., x_t\}$ is

$$f_{ass}(z_1, ..., z_t|x_1, ..., x_t) = \frac{1}{t!} \sum_{\pi} f(z_{\pi 1}|x_1) \cdots f(z_{\pi t}|x_t)$$

Now, on the other hand, we already know from Section 6.2.2 that the global measurement density for this problem is just

$$f_{\Sigma}(\{z_1, ..., z_t\}|\{x_1, ..., x_t\}) = \sum_{\pi} f(z_{\pi 1}|x_1) \cdots f(z_{\pi t}|x_t)$$

for distinct $\{z_1, ..., z_t\}$ and distinct $\{x_1, ..., x_t\}$. That is,

$$f_{\Sigma}(\{z_1, ..., z_t\}|\{x_1, ..., x_t\}) = t! \, f_{ass}(z_1, ..., z_t|x_1, ..., x_t)$$

In other words, the global density $f_{\Sigma}(Z|X)$ for this single-sensor problem can be interpreted as the total probability density of association between the reports $z_1, ..., z_t$ and the states $x_1, ..., x_t$. Our goal in this section is to show that this last formula holds under much more general single-sensor circumstances.

Conventional Formulation of the Problem

Specifically, let us continue to assume that a single sensor with parametrized (single-target) noise density $f(\xi|\zeta)$ interrogates a group of t targets, but now we assume additionally that there is a constant but possibly non-unity probability of detection p_D and a nonzero false alarm probability p_{FA}.

To model missed detections we introduce the "null observation," denoted by θ. If the sensor interrogates a target and no observation is returned then we will say that it returns the null observation. The probability that a target with state x_j gives rise to a non-null observation *and* that this observation is z_i is, therefore, $g(\xi_i|\zeta_j) = p_D\ f(\xi_i|\zeta_j)$.

To model false alarms we assume that they are generated by "virtual targets," all of which have noise statistics of the form $c(\xi) = c(\xi|\varepsilon)$ where ε is a dummy (i.e. constant) parameter whose value will eventually be ignored. For a virtual target a zero probability of detection means that it never generates a false alarm; whereas a unity probability of detection means that it always generates a false alarm. Thus as in our analysis of Section 6.2.1 the probability of detection for a virtual target is actually the probability of false alarm p_{FA}. Hence the probability density that a virtual target with state ε_j will not generate a false alarm is $1 - p_{FA}$; whereas the probability density that a virtual target will generate a false alarm *and* that this false alarm is ξ_i is $p_{FA}\ c(\xi_i|\varepsilon_j)$. Finally, to complete our model, we must assume one additional arbitrary constant: the maximum possible number M of false alarms.

Assume that there are n missed detections and e false alarms. Then there are $t = k - e + n$ actually-existing targets and $N = t + M = k - e + n + M$ targets which are either actual or virtual. Likewise there are $r = k - e$ actual observations and $N = r + n + M$ observations which are actual, null, or false. Denote $\zeta_{t+j} = \varepsilon_j$ for $j = 1, ...M$ and write

$$
\begin{aligned}
h(\theta|\zeta_i) &= 1 - p_D   &&(\text{all } i = 1, ..., t)\\
h(\theta|\zeta_j) &= 1 - p_{FA} &&(\text{all } j = t+1, ..., N)\\
h(\xi|\zeta_i) &= p_D\ f(\xi|\zeta_i) &&(\text{all } i = 1, ..., t)\\
h(\xi|\zeta_j) &= p_{FA}\ c(\xi|\zeta_j) &&(\text{all } j = t+1, ..., N)
\end{aligned}
$$

The analysis just completed for the $p_D = 1, p_{FA} = 0$ case now directly carries over, provided that we note that the number of occurrences of the null observation (associated now with both actual and virtual targets) is $d + M - e = N - k$. Likewise the number of occurrences of non-null observations (i.e., actual and false observations) is k. Accordingly the probability density of observing N observations (actual, null, or false) given N targets (actual and virtual) is

$$
h(\eta_1, ..., \eta_N|\zeta_1, ..., \zeta_N) = \frac{1}{N!}\sum_\sigma \prod_{i=1}^{N} h(\eta_i|\zeta_{\sigma i})
$$

where the summation is taken over all permutations σ on the numbers $1, ..., N$.

Now let us assume that $\eta_1, ..., \eta_N$ consists of $k = r + e$ actual or false observations $\xi_1, ..., \xi_k$ and $N - k$ null observations. Then we can write:

$$
h(\theta, ..., \xi_1, ..., \theta, ..., \xi_k, ..., \theta|\zeta_1, ..., \zeta_N)
$$
$$
= \frac{1}{N!}\sum_\sigma \left(\prod_{i=1}^{k} h(\xi_i|\zeta_{\sigma i})\right)\left(\prod_{i=k+1}^{N} h(\theta|\zeta_{\sigma i})\right)
$$

This value is independent of the particular combination of null observations, of which are $C_{N,k}$. So we can (while also suppressing the dummy parameters in the left-hand side of the equation) write

$$f_{ass}(\xi_1, ..., \xi_k | \zeta_1, ..., \zeta_t) = \frac{C_{N,k}}{N!} \sum_\sigma \left(\prod_{i=1}^k h(\xi_i | \zeta_{\sigma i}) \right) \left(\prod_{i=k+1}^N h(\theta | \zeta_{\sigma i}) \right)$$

where we have set $f_{ass} = h$.

Random Set Formulation of the Problem

Now, we reformulate the problem in purely random set terms. The randomly varying observation set has the form

$$\Sigma' = \Sigma \cup \Lambda = \Sigma_1 \cup ... \cup \Sigma_t \cup \Lambda_1 \cup ... \cup \Lambda_M$$

where Σ_i for $i = 1, ..., t$ denotes the observation collected from the i'th target and where Λ_l for $l = 1, ..., M$ denotes the (false) observation collected from the l'th virtual target. Also, $\Sigma = \Sigma_1 \cup ... \cup \Sigma_t$ and $\Lambda = \Lambda_1 \cup .... \cup \Lambda_M$. The belief measure for all of the Λ_l is $\beta_c(S) = 1 - p_{FA} + p_{FA} \, p_c(S)$ where p_c is the probability measure associated with the density c. The belief measure for the observation-set Σ_i is $\beta(S | \mathbf{x}_i) = 1 - p_D + p_D \, p_f(S | \mathbf{x}_i)$. Because of independence,

$$\beta_{\Sigma'}(S | \{\mathbf{x}_1, ..., \mathbf{x}_t\}) = \beta(S | \mathbf{x}_1) \cdots \beta(S | \mathbf{x}_t) \, \beta_c(S)^M$$

The global density $f_{\Sigma'}(Z | X)$ for our problem is computed from this belief measure.

Equivalence of Conventional and Random Set Formulations

Proposition 2 *The following is true:*

$$f_{\Sigma'}(\{\xi_1, ..., \xi_k\} | \{\zeta_1, ..., \zeta_t\}) = k! \, f_{ass}(\xi_1, ..., \xi_k | \zeta_1, ..., \zeta_t)$$

for all distinct $\xi_1, ..., \xi_k$ and all distinct $\zeta_1, ..., \zeta_t$ and where

$$f_{\Sigma'}(\{\xi_1, ..., \xi_k\} | \{\zeta_1, ..., \zeta_t\}) = \sum_{T \subseteq Z} \frac{p_D^{|T|}(1 - p_D)^{|X| - |T|} p_{FA}^{|T|}(1 - p_{FA})^{M - |T|} M!}{(M - |Z| + |T|)!}$$

$$\cdot \left(\sum_o \prod_{\xi \in T} f(\xi_1 | o(\xi)) \right) \cdot \left(\prod_{\xi \in T} c(\xi) \right)$$

and

$$f_{ass}(\xi_1, ..., \xi_k | \zeta_1, ..., \zeta_t) = \frac{C_{N,k}}{N!} \sum_\sigma \left(\prod_{i=1}^k h(\xi_i | \zeta_{\sigma i}) \right) \left(\prod_{i=k+1}^N h(\theta | \zeta_{\sigma i}) \right)$$

as defined earlier in the section.

Proof. For the sake of space we will abbreviate the proof, leaving some details to the reader. From $\beta_{\Sigma'}(S|X) = \beta_{\Sigma}(S|X)\beta_{\Lambda}(S)$ the product rule for set derivatives (Proposition 10 of Chapter 4) yields:

$$\frac{\delta\beta_{\Sigma'}}{\delta Z}(\emptyset) = \sum_{T \subseteq Z} \frac{\delta\beta_{\Sigma}}{\delta T}(\emptyset) \, \frac{\delta\beta_{\Lambda}}{\delta(Z-T)}(\emptyset)$$

Since $\beta_{\Lambda}(R) = \beta(R)^M$ it follows from Proposition 20 of Chapter 4 that, for $T = \{\xi_1, ..., \xi_j\}$ with $|T| = j$:

$$\frac{\delta\beta_{\Lambda}}{\delta T}(\emptyset) = \frac{M!}{(M-|T|)!} \, p_{FA}^{|T|}(1 - p_{FA})^{M-|T|} \prod_{\xi \in T} c(\xi)$$

Of course, if $|T| > M$ then this quantity is zero. Likewise, since $\beta_{\Sigma}(R) = \beta(R|\zeta_1)\cdots\beta(R|\zeta_t)$ we know from Proposition 20 of Chapter 4 that

$$\frac{\delta\beta_{\Sigma}}{\delta T}(\emptyset) = p_D^{|T|}(1 - p_D)^{|X|-|T|} \sum_{o} \prod_{\xi \in T} f(\xi_1|o(\xi))$$

where the sum is taken over all one-to-one functions $o : T \to X$ (provided, of course, that $j = |T| \leq |X| = t$; if $|T| > t$ then the right hand side of the equation is zero). Now in the first equation of this proof, we see that in this sum any term for which $|T| > t$ or $|Z - T| > M$ will vanish identically. As a consequence the entire sum will vanish identically if $k > M + t$. Substituting the formulas just derived into this equation we get:

$$\frac{\delta\beta_{\Sigma'}}{\delta Z}(\emptyset) = \sum_{T \subseteq Z} \left(p_D^{|T|}(1 - p_D)^{|X|-|T|} \sum_{o} \prod_{\xi \in T} f(\xi_1|o(\xi)) \right)$$
$$\cdot \left(\frac{M!}{(M-|Z|+|T|)!} \, p_{FA}^{|T|}(1 - p_{FA})^{M-|T|} \prod_{\xi \in T} c(\xi) \right)$$

where any term vanishes if $|T| > |X| = t$ or $|Z| - |T| > M$. Accordingly:

$$\frac{\delta\beta_{\Sigma'}}{\delta Z}(\emptyset) = \sum_{T \subseteq Z} \frac{p_D^{|T|}(1 - p_D)^{|X|-|T|} p_{FA}^{|T|}(1 - p_{FA})^{M-|T|} M!}{(M-|Z|+|T|)!}$$
$$\cdot \left(\sum_{o} \prod_{\xi \in T} f(\xi_1|o(\xi)) \right) \cdot \left(\prod_{\xi \in T} c(\xi) \right)$$

which completes the derivation of the global density for Σ'.

We must now show that the Bayesian formula

$$h(\eta_1, ..., \eta_N|\zeta_1, ..., \zeta_N) = \frac{1}{N!} \sum_{\sigma} \prod_{i=1}^{N} h(\eta_i|\zeta_{\sigma i})$$

can be transformed into a formula of similar form.

Let o be any one-to-one function from the set $\{\eta_1, ..., \eta_N\}$ to the set $\{\zeta_1, ..., \zeta_N\}$. Ordering the η's so that the first k are the $\xi_1, ..., \xi_k$ we can rewrite the last equation as

$$h(\xi_1, ..., \xi_k | \zeta_1, ..., \zeta_t) = \frac{C_{N,k}}{N!} \sum_o \left(\prod_{i=1}^{k} h(\xi_i | o(\zeta_i)) \right) \left(\prod_{i=k+1}^{N} h(\theta | o(\eta_i)) \right)$$

Next, let $T(o)$ denote the subset of $Z = \{\xi_1, ..., \xi_k\}$ such that $o(T) \subseteq X = \{\zeta_1, ..., \zeta_t\}$. That is, if $T(o)$ is nonempty then it consists of all $\xi \in Z$ such that $o(\xi)$ is an *actual*, not a virtual, target. For example, if $M = 0$ then $T(o) = Z$ for all o (since $N = M + t$). If $M = 1$ then either $T(o) = Z$ (but only if $|Z| \le |X|$) or $T(o)$ is a subset of Z with $|Z| - 1$ elements. In general, $\max\{0, |Z| - M\} \le |T(o)| \le \min\{|Z|, |X|\}$. Accordingly $o(\xi)$ is a virtual object for all $\xi \in Z - T(o)$. This leaves $t - |T(o)| = |X| - |T(o)|$ actual targets unaccounted for, and they are images of the null observations $\eta_{k+1}, ..., \eta_N$ under o. Likewise there are $M - |Z - T(o)| = M - |Z| + |T(o)|$ virtual targets unaccounted for, and they also are images of the $\eta_{k+1}, ..., \eta_N$ under o. Accordingly

$$h(\xi_1, ..., \xi_k | \zeta_1, ..., \zeta_t) = \frac{C_{N,|Z|}}{N!} \sum_o \left(\prod_{\xi \in T(o)} h(\xi | o(\xi)) \right) \left(\prod_{\xi \in Z - T(o)} h(\xi | o(\xi)) \right)$$
$$\cdot (1 - p_D)^{|X| - |T(o)|} (1 - p_{FA})^{M - |Z| + |T(o)|}$$

where we adopt the convention that terms with $|T| > |X|$ vanish. Thus we can also write

$$h(\xi_1, ..., \xi_k | \zeta_1, ..., \zeta_t) =$$

$$\frac{C_{N,|Z|}}{N!} \sum_o \left(\prod_{\xi \in T(o)} f(\xi | o(\xi)) \right) \left(\prod_{\xi \in Z - T(o)} c(\xi) \right)$$
$$\cdot p_D^{|T(o)|} p_{FA}^{|Z| - |T(o)|}$$
$$\cdot (1 - p_D)^{|X| - |T(o)|} (1 - p_{FA})^{M - |Z| + |T(o)|}$$

Now, notice that if o and o' are such that $T(o) = T(o')$ and such that $o(\xi) = o'(\xi)$ for every $\xi \in T(o)$ then the two terms in the sum which arise from o and o' are identical. Moreover, the relationship $T(o) = T(o')$ and $o(\xi) = o'(\xi)$ for every $\xi \in T(o)$ defines an equivalence relation $o \equiv o'$ on the set of all functions $o : \{\eta_1, ..., \eta_N\} \to \{\zeta_1, ..., \zeta_N\}$. Thus the last equation can be rewritten in the form

$$h(\xi_1, ..., \xi_k | \zeta_1, ..., \zeta_t) = \frac{C_{N,|Z|}}{N!} \sum_\kappa |\kappa| \left(\prod_{\xi \in T(\kappa)} f(\xi | o_\kappa(\xi)) \right) \left(\prod_{\xi \in Z - T(\kappa)} c(\xi) \right)$$
$$\cdot p_D^{|T(\kappa)|} p_{FA}^{|Z| - |T(\kappa)|}$$
$$\cdot (1 - p_D)^{|X| - |T(\kappa)|} (1 - p_{FA})^{M - |Z| + |T(\kappa)|}$$

where the sum is now taken over all equivalence classes κ of functions o, where $|\kappa|$ denotes the number of functions o in the equivalence class κ and where $T(\kappa)$ and $o_\kappa(\xi)$ are self-explanatory.

The reader may verify that the set of equivalence classes κ is in one-to-one correspondence with pairs of the form (T, o_T) where T is any subset of Z such that $|Z| - M \leq |T| \leq \min\{|Z|, |X|\}$ and where $o_T : T \to X$ is a one-to-one function. Accordingly we can now rewrite

$$
h(\xi_1, ..., \xi_k | \zeta_1, ..., \zeta_t) = \frac{C_{N,|Z|}}{N!} \sum_{T \subseteq Z} \sum_{o_T : T \to X}
$$
$$
\cdot |(T, o_T)| \left(\prod_{\xi \in T} f(\xi | o_T(\xi)) \right) \left(\prod_{\xi \in Z - T} c(\xi) \right)
$$
$$
\cdot p_D^{|T|} \, p_{FA}^{|Z|-|T|} \, (1 - p_D)^{|X|-|T|} (1 - p_{FA})^{M-|Z|+|T|}
$$

where $|(T, o_T)|$ is defined by the equation $|\kappa| = |(T, o_T)|$ and where the first summation is taken over all T such that $|Z| - M \leq |T| \leq \min\{|Z|, |X|\}$.

We now adopt the following convention: Terms involving those subsets T for which $|Z - T| \geq M$ (or for which $|T| > |Z|$ or $|T| > |X|$) vanish identically. Then

$$
h(\xi_1, ..., \xi_k | \zeta_1, ..., \zeta_t) = \frac{C_{N,|Z|}}{N!} \sum_{T \subseteq Z} \sum_{o_T : T \to X}
$$
$$
\cdot |(T, o_T)| \left(\prod_{\xi \in T} f(\xi | o_T(\xi)) \right) \left(\prod_{\xi \in Z - T} c(\xi) \right)
$$
$$
\cdot p_D^{|T|} \, p_{FA}^{|Z|-|T|} \, (1 - p_D)^{|X|-|T|} (1 - p_{FA})^{M-|Z|+|T|}
$$

where now the first summation is allowed to range over *all* subsets of Z.

The reader may verify that the value of $|(T, o_T)|$ is

$$
|(T, o_T)| = C_{M,|Z|-|T|}(|Z| - |T|)!(N - |Z|)! = \frac{M! \, (N - |Z|)!}{(M - |Z| + |T|)!}
$$

We therefore get

$$
h(\xi_1, ..., \xi_k | \zeta_1, ..., \zeta_t) = \frac{C_{N,|Z|}}{N!} \sum_{T \subseteq Z} \sum_{o_T : T \to X} \left(\prod_{\xi \in T} f(\xi | o_T(\xi)) \right) \left(\prod_{\xi \in Z - T} c(\xi) \right)
$$
$$
\cdot \frac{p_D^{|T|} \, p_{FA}^{|Z|-|T|} \, (1 - p_D)^{|X|-|T|} (1 - p_{FA})^{M-|Z|+|T|} \, M!}{|Z|! \, (M - |Z| + |T|)!}
$$

This equation agrees with the random set result except for the factor of $|Z|! = k!$. Thus we get the desired relationship. $\qquad\square$

Remark 3 *EAMLE, the multitarget extended Kalman filter [5], [6], [7] described in Section 2.2.2 of Chapter 2, is closely related to the global MLE (Definition 5 of Chapter 5). Assume that: (1) there is a single Gaussian kinematics-reporting sensor; (2) the number t of targets is known a priori; (3) observations are independent; (4) clutter is spacially dense and uniform with Poisson arrivals; and (5) the global measurement density is linearized using a Gaussian approximation. With these assumptions the single-scan case of the EAMLE filter can be shown to be a special case of the global ML estimator. Specifically, given the formula for $f_{\Sigma'}(Z|X)$ in Proposition 2 assume in addition that the number M of clutter objects is very large but that the probability of false alarm p_{FA} is very small. Assume also that all clutter objects are independently, identically, and uniformly distributed within a given region of interest R which has volume L. In this case $c(\mathbf{z}) = L^{-1}$ for all $\mathbf{z} \in R$ and $c(\mathbf{z}) = 0$ otherwise. Then we can approximate the binomial distribution $C_{M,k} p_{FA}^{k} (1 - p_{FA})^{M-k}$ by the Poisson distribution $Q^{k} e^{-Q}/k!$ where $Q \cong M p_{FA}$. The density of Proposition 2 becomes*

$$f_{ass}(Z|X) \cong \frac{e^{-Q}}{k!} \sum_{T \subseteq Z} p_{D}^{|T|} (1 - p_{D})^{|X|-|Z|} \rho^{|Z-T|} \left(\sum_{o} \prod_{\mathbf{z} \in T} f(\mathbf{z}|o(\mathbf{z})) \right)$$

where $Z = \{\mathbf{z}_{1}, ..., \mathbf{z}_{k}\}$, $X = \{\mathbf{x}_{1}, ..., \mathbf{x}_{t}\}$ and where $\rho = Q/L \cong M p_{FA} L$ is the clutter density. If we interpret the set variable T as the set of possible missed detections then $|T| = n$ and $|Z - T| = k - n = e = no.$ of false alarms. Substituting these values the above formula for f_{ass} becomes (except for slight differences in notation) the same as that for the single-scan case of EAMLE [5, equ's 6-8].

6.3 Modeling Prior Knowledge

In data fusion prior knowledge is of three major types: *geodynamic, identity-type*, and *detection-type*.

Geodynamic prior knowledge reflects prior knowledge in regard to such state variables as position, velocity, acceleration, and so on. It reflects the fact that the locations of targets are constrained, or at least influenced, by the existence of such features as regions of land or water, forests, roads, mountains and hills, and so on. Mathematically, pure geopositional/geodynamic prior knowledge can be modelled by an (ordinary) density function $f(\mathbf{x})$. When the value $f(\mathbf{x})$ is very small then targets are very unlikely to be found which have state $\mathbf{x}$. When it is large, targets are much more likely to be found which have that state.

Identity-type prior knowledge reflects the fact that, in any given theater of operations, some target types are more rarely observed than others. Prior knowledge in this case can be modeled by a probability distribution $p(u)$ on the set U of possible target identities/types.

Detection-type prior knowledge describes the degree to which we expect, on an a priori basis, certain numbers of targets to occur rather than others. Mathematically, it is modelled as a probability distribution $p(j)$ on the nonnegative integers j which vanishes beyond a certain point: $n(j) = 0$ for all sufficiently large j. If $p(j)$ is small then groups of j targets are unlikely to be observed. If $p(j)$ is large then it is much more likely that that number of targets will be observed.

In practice, these three types of prior knowledge are not easily separated from one another. For example, it is very unlikely that a cargo plane would be observed executing a jet fighter-like turn rate showing that, in general, geodynamic and identity-type prior knowledge will be correlated. Mathematically, prior knowledge involving both geodynamic and identity-type prior knowledge can be modelled as a probability density of the form $f(\zeta) = f(\mathbf{x}, u)$. That is, $f(\mathbf{x}, u)$ is a nonnegative function such that

$$\int f(\zeta)d\lambda(\zeta) \triangleq \sum_{u \in U} \int f(\mathbf{x}, u)d\lambda(\mathbf{x}) = 1.$$

Prior knowledge of this sort—the kind that can be expressed in terms of conventional prior densities—is, of course, a standard feature of probabilistic inference. Still other kinds of prior knowledge, however, require the concept of a *global prior* in the sense considered in Section 5.1.3. A typical instance is prior knowledge in regard to *groups* of targets rather than individual targets. For example, an aircraft carrier will virtually never be detected in isolation from a group of other, supporting platforms. All such group-based information can be expressed in terms of global priors of the form $f_\Gamma(X)$ where $X = \{(\mathbf{x}_1, v_1), ..., (\mathbf{x}_t, v_t)\}$ and $|X| = t$ and where $f_\Gamma(X)$ takes greater or smaller values depending on how likely it is that (for example) targets of type $v_{i_1}, ..., v_{i_s}$ are likely to be found jointly with kinematic states $\mathbf{x}_{i_1}, ..., \mathbf{x}_{i_s}$.

Example 1 *Let $f(\zeta|\varepsilon) = f(\mathbf{x}, v|\mathbf{e}, e) = N_C(\mathbf{x} - \mathbf{e}) h(v|e)$ be a prior distribution on state space with parameter $\varepsilon = (\mathbf{e}, e)$ in state space. Then*

$$f(\{\zeta_1, ..., \zeta_t\}) = \sum_\pi f(\zeta_{\pi 1}|\varepsilon_1) \cdots f(\zeta_{\pi_t}|\varepsilon_t)$$

for distinct $\zeta_1, ..., \zeta_t$ and fixed $\varepsilon_1, ..., \varepsilon_t$ defines a global prior distribution in which, intuitively speaking, the "expected value" of the prior consists of exactly t targets with (hybrid) states $\varepsilon_1, ..., \varepsilon_t$. More generally, let $q_0, q_1, ..., q_t$ be a probability distribution on the integers such that $q_i = 0$ if $i < 0$ or $i > t$. Then

$$f(\{\zeta_1, ..., \zeta_d\}) = \frac{q_d}{C_{t,d}} \sum_{1 \le i_1 \ne ... \ne i_d \le t} f(\zeta_{i_1}|\varepsilon_1) \cdots f(\zeta_{i_d}|\varepsilon_t)$$

for distinct $\zeta_1, ..., \zeta_d$ defines a global prior in which the initial number of targets is not presumed to be known but, rather, is governed by the probability distribution $q_i = \int_{|X|=i} f(X)\delta X$ for all $i = 0, 1, ..., t$.

6.4   Random Set Motion Models

Up until now we have been looking only at the statistics of *static* problems, that is non-time-varying (motionless targets) problems. However, it should be clear by now that it is possible to extend the conventional statistics of dynamic, i.e. time-varying (moving targets) problems to finite-set statistics as well. Specifically, in this section we show how to extend the familiar *Bayes-Markov nonlinear filtering* approach to multisensor, multitarget problems.

6.4.1 Bayesian Recursive Nonlinear Filtering

First we begin with a brief summary of conventional single-sensor, single-target, discrete-time Bayes-Markov filtering theory. In ordinary Kalman filtering, one begins by assuming two models. First, a linear, discrete-time *sensor measurement equation* $\mathbf{z}_\alpha = A\mathbf{x}_\alpha + B\mathbf{v}_\alpha$ which models the uncertainty due to noisy and/or incomplete measurements $\mathbf{z}_\alpha$ of the state-parameter vector $\mathbf{x}_\alpha$ of the target at time-instant $t = \alpha \cdot \Delta t$. Second, a linear, recursive, discrete-time *system motion equation* $\mathbf{x}_{\alpha+1} = C_\alpha\mathbf{x}_\alpha + D_\alpha\mathbf{w}_\alpha$ which models the uncertainty in our knowledge of the trajectory of the target between measurement time-instants $t = \alpha \cdot \Delta t$ and $t = (\alpha + 1) \cdot \Delta t$. Here, A, B, C_α, D_α are matrices and $\mathbf{v}_\alpha, \mathbf{w}_\alpha$ are statistically independent Gaussian white noise processes ($\mathbf{v}_\alpha$ is the "sensor noise" and $\mathbf{w}_\alpha$ is the "system noise"). Given this linear model, the time-evolving state $\mathbf{x}_\alpha$ of the target can be estimated from a time-sequence of measurements using the so-called Kalman time-update and information-update equations.

The approach just described can be generalized (see [9], [1]) to take account of *nonlinear and non-Gaussian* measurement models $\mathbf{z}_\alpha = g(\mathbf{x}_\alpha, \mathbf{w}_\alpha)$ and motion models $\mathbf{x}_{\alpha+1} = h_\alpha(\mathbf{x}_\alpha, \mathbf{v}_\alpha)$ where g, h_α are nonlinear functions of $\mathbf{x}_\alpha, \mathbf{v}_\alpha, \mathbf{w}_\alpha$ and where $\mathbf{v}_\alpha, \mathbf{w}_\alpha$ are (not necessarily Gaussian) noise processes. These vector-level models can, in turn, be expressed in terms of density-level models of the form $f(\mathbf{z}_\alpha|\mathbf{x}_\alpha), f_{\alpha+1|\alpha}(\mathbf{x}_{\alpha+1}|\mathbf{x}_\alpha)$, and $f_0(\mathbf{x}_0)$ for $\alpha = 0, 1, \dots$ and where $\mathbf{x}_0$, the *initial state* of the target, must be specified on an *a priori* basis. The first of these three densities is called the *measurement density*, the second is the *Markov state-transition density*, and the third is the *initial-state density*. The state-transition density must satisfy the *Chapman-Kolmogorov equation*

$$f_{\alpha+1|\alpha-1}(\mathbf{x}_{\alpha+1}|\mathbf{x}_{\alpha-1}) = \int f_{\alpha+1|\alpha}(\mathbf{x}_{\alpha+1}|\mathbf{x}_\alpha) f_{\alpha|\alpha-1}(\mathbf{x}_\alpha|\mathbf{x}_{\alpha-1}) d\lambda(\mathbf{x}_\alpha)$$

where $f_{\alpha+1|\alpha-1}(\mathbf{x}_{\alpha+1}|\mathbf{x}_{\alpha-1})$ describes the state-transition from time-step $\alpha - 1$ to time-step $\alpha + 1$.

For example, suppose that the noise processes $\mathbf{v}_\alpha, \mathbf{w}_\alpha$ have densities $f(\mathbf{x})$ and $f_\alpha(\mathbf{x})$, respectively, and that the models have additive noise: $\mathbf{z}_\alpha = g(\mathbf{x}_\alpha) + \mathbf{v}_\alpha$ and $\mathbf{x}_{\alpha+1} = h_\alpha(\mathbf{x}_\alpha) + \mathbf{w}_\alpha$. Then the corresponding measurement and state-transition densities are $f(\mathbf{z}_\alpha|\mathbf{x}_\alpha) = f(\mathbf{z}_\alpha - g(\mathbf{x}_\alpha))$ and $f_{\alpha+1|\alpha}(\mathbf{x}_{\alpha+1}|\mathbf{x}_\alpha) = f_\alpha(\mathbf{x}_{\alpha+1} - h_\alpha(\mathbf{x}_\alpha))$. Suppose now that a time-sequence of measurements $\mathbf{z}_1, \dots, \mathbf{z}_\alpha$, $\mathbf{z}_{\alpha+1}$ has been collected by the sensor. Denote the cumulative measurement sets as $Z^\alpha \triangleq \{\mathbf{z}_1, \dots, \mathbf{z}_\alpha\}$ for any $\alpha \geq 1$. At any given time-step α our knowledge of the state $\mathbf{x}_\alpha$ of the target, based on the collected measurements Z^α, is completely specified by the posterior distribution $f_{\alpha|\alpha}(\mathbf{x}_\alpha|Z^\alpha)$. If we wish to estimate the value of $\mathbf{x}_\alpha$ we can do so by finding the expected value of f_α, the maximizing value, etc. Given an additional measurement $\mathbf{z}_{\alpha+1}$, it is possible to recursively deduce $f_{\alpha+1}(\mathbf{x}_{\alpha+1}|Z^{\alpha+1})$ from $f_\alpha(\mathbf{x}_\alpha|Z^\alpha)$ as follows. First, note that from the total probability theorem,

$$f_{\alpha+1}(\mathbf{x}_{\alpha+1}|Z^\alpha) = \int f_{\alpha+1|\alpha}(\mathbf{x}_{\alpha+1}|\mathbf{x}_\alpha, Z^\alpha)\, f_{\alpha|\alpha}(\mathbf{x}_\alpha|Z^\alpha) d\lambda(\mathbf{x}_\alpha)$$

Suppose that, for any α, the dependence of the updated state $\mathbf{x}_{\alpha+1}$ on past data Z^α is expressed only through dependence upon the current state $\mathbf{x}_\alpha$. Then $f_{\alpha+1|\alpha}(\mathbf{x}_{\alpha+1}|\mathbf{x}_\alpha, Z^\alpha) = f_{\alpha+1|\alpha}(\mathbf{x}_{\alpha+1}|\mathbf{x}_\alpha)$ and we get the so-called *prediction integral*

$$f_{\alpha+1|\alpha}(\mathbf{x}_{\alpha+1}|Z^\alpha) = \int f_{\alpha+1|\alpha}(\mathbf{x}_{\alpha+1}|\mathbf{x}_\alpha) f_{\alpha|\alpha}(\mathbf{x}_\alpha|Z^\alpha) d\lambda(\mathbf{x}_\alpha)$$

which updates $f_\alpha(\mathbf{x}_\alpha|Z^\alpha)$ to the current time-instant α. The posterior distribution at time-instant $\alpha+1$ can then be computed from Bayes' rule:

$$f_{\alpha+1|\alpha+1}(\mathbf{x}_{\alpha+1}|Z^{\alpha+1}) = \frac{f_{\alpha+1}(\mathbf{z}_{\alpha+1}|\mathbf{x}_{\alpha+1}, Z^\alpha) f_{\alpha+1|\alpha}(\mathbf{x}_{\alpha+1}|Z^\alpha)}{\int f_{\alpha+1}(\mathbf{z}_{\alpha+1}|\mathbf{y}_{\alpha+1}, Z^\alpha) f_{\alpha+1|\alpha}(\mathbf{y}_{\alpha+1}|Z^\alpha) d\lambda(\mathbf{y}_{\alpha+1})}$$

If we further suppose that, for any α, the generation of current data $\mathbf{z}_\alpha$ depends only the current state $\mathbf{x}_\alpha$ and not on past data $Z^{\alpha-1}$, then

$$f_{\alpha+1}(\mathbf{z}_{\alpha+1}|\mathbf{x}_{\alpha+1}, Z^\alpha) = f_{\alpha+1}(\mathbf{z}_{\alpha+1}|\mathbf{x}_{\alpha+1})$$

Furthermore, assume that the functional form of data generation is the same at all time instants: $f_{\alpha+1}(\mathbf{z}_{\alpha+1}|\mathbf{x}_{\alpha+1}) = f(\mathbf{z}_{\alpha+1}|\mathbf{x}_{\alpha+1})$. Then

$$f_{\alpha+1|\alpha+1}(\mathbf{x}_{\alpha+1}|Z^{\alpha+1}) = \frac{f(\mathbf{z}_{\alpha+1}|\mathbf{x}_{\alpha+1}) f_{\alpha+1|\alpha}(\mathbf{x}_{\alpha+1}|Z^\alpha)}{\int f(\mathbf{z}_{\alpha+1}|\mathbf{y}_{\alpha+1}) f_{\alpha+1|\alpha}(\mathbf{y}_{\alpha+1}|Z^\alpha) d\lambda(\mathbf{y}_{\alpha+1})}$$

This last equation, together with the prediction integral

$$f_{\alpha+1|\alpha}(\mathbf{x}_{\alpha+1}|Z^\alpha) = \int f_{\alpha+1|\alpha}(\mathbf{x}_{\alpha+1}|\mathbf{x}_\alpha) f_{\alpha|\alpha}(\mathbf{x}_\alpha|Z^\alpha) d\lambda(\mathbf{x}_\alpha)$$

are known as the *Bayesian discrete-time recursive nonlinear filtering equations*. When all densities are Gaussian and linear these last two equations reduce to the familiar Kalman filter information-update and time-update equations, respectively [9]. (A more general approach to discrete-time Bayesian recursive nonlinear filtering can be found in [4].)

6.4.2   Global Nonlinear Filtering

The previous remarks can be generalized in the obvious fashion. That is, we assume that we have collected a sequence $Z^{(a)} = \{Z_1, ..., Z_a\}$ of global observations and that, at any given time-instant $\alpha+1$ we wish to update the global posterior $f_{\alpha|\alpha}(X_\alpha|Z^{(\alpha)})$ to a new global posterior $f_{\alpha|\alpha}(X_\alpha|Z^{(\alpha)})$. Then under analogous assumptions to those of the last section we get:

$$f_{\alpha+1|\alpha+1}(X_{\alpha+1}|Z^{(\alpha+1)}) = \frac{f(Z_{\alpha+1}|X_{\alpha+1})\, f_{\alpha+1|\alpha}(X_{\alpha+1}|Z^{(\alpha)})}{\int f(Z_{\alpha+1}|Y_{\alpha+1})\, f_{\alpha+1|\alpha}(Y_{\alpha+1}|Z^{(\alpha)})\delta Y_{\alpha+1}}$$

$$f_{\alpha+1|\alpha}(X_{\alpha+1}|Z^{(\alpha)}) = \int f_{\alpha+1|\alpha}(X_{\alpha+1}|X_\alpha)f_{\alpha|\alpha}(X_\alpha|Z^{(\alpha)})\delta X_\alpha$$

Note that the global state-transition density $f_{\alpha+1|\alpha}(X_{\alpha+1}|X_\alpha)$—what we will call a "global motion model" for the global target—may have $|X_{\alpha+1}| \neq |X_\alpha|$, thus expressing the fact that the number of targets may change with time due to "death" and "birth" processes.

6.4.3 Constructing Global Motion Models

In this section we demonstrate that it is possible to extend conventional single-sensor, single-target motion models to multitarget motion models—what we will call "canonical global motion models"—in such a way that there is complete consistency between the canonical global models and the conventional models from which they are generated. In what follows we assume that we are given conventional Markov state transition densities $f_{\alpha+1|\alpha}(\zeta_{\alpha+1}|\zeta_\alpha)$. They thereby satisfy the usual Chapman-Kolmogorov equations [8, p. 8]:

$$\int f_{\alpha+1|\alpha}(\zeta_{\alpha+1}|\zeta_\alpha)\, f_{\alpha|\alpha-1}(\zeta_\alpha|\zeta_{\alpha-1})d\lambda(\zeta_\alpha) = f_{\alpha+1|\alpha-1}(\zeta_{\alpha+1}|\zeta_{\alpha-1})$$

for all $\zeta_{\alpha+1}, \zeta_\alpha$ in state space. We also assume discrete state-transition probabilities $q_{\alpha+1|\alpha}(j_{\alpha+1}|j_\alpha)$ satisfying the following property: $q_{\alpha+1|\alpha}(j_{\alpha+1}|j_\alpha) = 0$ if $j_{\alpha+1} > j_\alpha$ or if $j_{\alpha+1} < 0$. They satisfy the discrete Chapman-Kolmogorov equations

$$\sum_{j_\alpha=j_{\alpha+1}}^{j_{\alpha-1}} q_{\alpha+1|\alpha}(j_{\alpha+1}|j_\alpha)\, q_{\alpha|\alpha-1}(j_\alpha|j_{\alpha-1}) = q_{\alpha+1|\alpha-1}(j_{\alpha+1}|j_{\alpha-1})$$

for all $j_{\alpha+1}, j_{\alpha-1} = 0, 1, \dots$.

Remark 4 *Note that $f_{\alpha+1|\alpha}(\zeta_{\alpha+1}|\zeta_\alpha)$ has enough structure to model the motion of any target type being considered. If $\zeta_{\alpha+1} = (\mathbf{x}_{\alpha+1}, v_{\alpha+1})$ and $\zeta_\alpha = (\mathbf{x}_\alpha, v_\alpha)$ then suppose that*

$$f_{\alpha+1|\alpha}(\mathbf{x}_{\alpha+1}, v_{\alpha+1}|\mathbf{x}_\alpha, v_\alpha) = \begin{cases} f_{\alpha+1|\alpha}(\mathbf{x}_{\alpha+1}|\mathbf{x}_\alpha; v_\alpha) & if \quad v_{\alpha+1} = v_\alpha \\ 0 & if \quad v_{\alpha+1} \neq v_\alpha \end{cases}$$

where $f_{\alpha+1|\alpha}(\mathbf{x}_{\alpha+1}|\mathbf{x}_\alpha; v_\alpha)$ denotes the motion model for a particular target type v_α. Then motion models for all target types can be taken into consideration.

We begin with a definition:

Definition 3 *Let $g(\eta|\zeta)$ be a parametrized density where η, ζ are both drawn from parameter space. Let $q(j|i)$ be a discrete parametrized distribution on integers $0 \leq i, j$ such that $q(j|i) = 0$ if $j > i$ and if $j < 0$. Define the parametrized global density $\breve{g}(X|Y; q)$ by*

$$\breve{g}(\emptyset|\{\zeta_1, ..., \zeta_l\}; q) \triangleq q(0|l)$$

$$\breve{g}(\{\eta_1, ..., \eta_k\}|\{\zeta_1, ..., \zeta_l\}; q) \triangleq q(k|l) \, C_{l,k}^{-1} \sum_{1 \leq i_1 \neq ... \neq i_k \leq l} g(\eta_1|\zeta_{i_1}) \cdots g(\eta_k|\zeta_{i_k})$$

where as usual the summation is taken over all distinct $i_1, ..., i_k$ such that $1 \leq i_1, ..., i_k \leq l$, and where $\zeta_1, ..., \zeta_l$ are distinct and $\eta_1, ..., \eta_k$ are distinct.

Note that the cardinality distribution of $\breve{g}$ is

$$\breve{g}(k|Y; q) \triangleq \int_{|X|=k} \breve{g}(X|Y)\delta X = q(k| |Y|)$$

which is therefore dependent upon Y only via its cardinality $|Y|$.

The following proposition establishes the existence of a broad class of global Markov state-transition densities:

Proposition 3 *Let $f_{\alpha+1|\alpha}(\zeta_{\alpha+1}|\zeta_\alpha)$ be a Markov state-transition density on parameter space and $q_{\alpha+1|\alpha}(j_{\alpha+1}|j_\alpha)$ discrete state-transition probabilities such that $q_{\alpha+1|\alpha}(j_{\alpha+1}|j_\alpha) = 0$ if $j_{\alpha+1} > j_\alpha$ or if $j_{\alpha+1} < 0$. Then the following "global" Chapman-Kolmogorov equation*

$$\int \breve{f}_{\alpha+1|\alpha}(X_{\alpha+1}|X_\alpha; q_{\alpha+1|\alpha}) \, \breve{f}_{\alpha|\alpha-1}(X_\alpha|X_{\alpha-1}; q_{\alpha|\alpha-1}) \, \delta X_\alpha$$

$$= \breve{f}_{\alpha+1|\alpha-1}(X_{\alpha+1}|X_{\alpha-1}; q_{\alpha+1|\alpha-1})$$

is true for all finite subsets $X_{\alpha+1}, X_{\alpha-1}$ of state space, where $\breve{f}_{\alpha+1|\alpha}, \breve{f}_{\alpha|\alpha-1}$ and $\breve{f}_{\alpha+1|\alpha-1}$ are as in Definition 3.

Proof. By definition,

$$\int \breve{f}_{\alpha+1|\alpha}(\{\eta_1, ..., \eta_a\}|X_\alpha; q_{\alpha+1|\alpha}) \, \breve{f}_{\alpha|\alpha-1}(X_\alpha|\{\zeta_1, ..., \zeta_c\}; q_{\alpha|\alpha-1}) \, \delta X_\alpha$$

$$= \sum_{b=0}^{\infty} \frac{1}{b!} \int \breve{f}_{\alpha+1|\alpha}(\{\eta_1, ..., \eta_a\}|\{\gamma_1, ..., \gamma_b\}; q_{\alpha+1|\alpha})$$

$$\cdot \breve{f}_{\alpha|\alpha-1}(\{\gamma_1, ..., \gamma_b\}|\{\zeta_1, ..., \zeta_c\}; q_{\alpha|\alpha-1}) \, d\lambda(\gamma_1) \cdots d\lambda(\gamma_b)$$

$$= \sum_{b=a}^{c} \frac{1}{b!} \int \left(q_{\alpha+1|\alpha}(a|b) \, C_{b,a}^{-1} \sum_{1 \leq i_1 \neq ... < \neq i_a \leq b} f_{\alpha+1|\alpha}(\eta_1|\gamma_{i_1}) \cdots f_{\alpha+1|\alpha}(\eta_a|\gamma_{i_a}) \right)$$

$$\cdot \left(q_{\alpha|\alpha-1}(b|c) \, C_{c,b}^{-1} \sum_{1 \leq j_1 \neq ... \neq j_b \leq c} f_{\alpha|\alpha-1}(\gamma_1|\zeta_{j_1}) \cdots f_{\alpha|\alpha-1}(\gamma_b|\zeta_{j_b}) \right)$$

$$\cdot d\lambda(\gamma_1) \cdots d\lambda(\gamma_b)$$

$$= \sum_{b=a}^{c} \frac{1}{b!} q_{\alpha+1|\alpha}(a|b) q_{\alpha|\alpha-1}(b|c) \, C_{b,a}^{-1} C_{c,b}^{-1} \sum_{1 \leq i_1 \neq \ldots \neq i_a \leq b} \sum_{1 \leq j_1 \neq \ldots \neq j_b \leq c}$$

$$\cdot \int f_{\alpha+1|\alpha}(\eta_1|\gamma_{i_1}) \cdots f_{\alpha+1|\alpha}(\eta_a|\gamma_{i_a}) \, f_{\alpha|\alpha-1}(\gamma_1|\zeta_{j_1}) \cdots f_{\alpha|\alpha-1}(\gamma_b|\zeta_{j_b})$$

$$\cdot d\lambda(\gamma_1) \cdots d\lambda(\gamma_b)$$

Suppose that we knew that

$$\sum_{1 \leq i_1 \neq \ldots \neq i_a \leq b} \sum_{1 \leq j_1 \neq \ldots \neq j_b \leq c} \int f_{\alpha+1|\alpha}(\eta_1|\gamma_{i_1}) \cdots f_{\alpha+1|\alpha}(\eta_a|\gamma_{i_a})$$

$$\cdot f_{\alpha|\alpha-1}(\gamma_1|\zeta_{j_1}) \cdots f_{\alpha|\alpha-1}(\gamma_b|\zeta_{j_b}) \, d\lambda(\gamma_1) \cdots d\lambda(\gamma_b)$$

$$= b! C_{c-a,b-a} \sum_{1 \leq i_1 \neq \ldots \neq i_a \leq c} f_{\alpha+1|\alpha-1}(\eta_1|\zeta_{i_1}) \cdots f_{\alpha+1|\alpha-1}(\eta_a|\zeta_{i_a})$$

Then we would have

$$\int \check{f}_{\alpha+1|\alpha}(\{\eta_1, ..., \eta_a\}|X_\alpha; q_{\alpha+1|\alpha}) \, \check{f}_{\alpha|\alpha-1}(X_\alpha|\{\zeta_1, ..., \zeta_c\}; q_{\alpha|\alpha-1}) \, \delta X_\alpha$$

$$= \left(\sum_{b=a}^{c} q_{\alpha+1|\alpha}(a|b) q_{\alpha|\alpha-1}(b|c) \, C_{b,a}^{-1} C_{c,b}^{-1} C_{c-a,b-a} \right)$$

$$\cdot \sum_{1 \leq i_1 \neq \ldots \neq i_a \leq c} f_{\alpha+1|\alpha-1}(\eta_1|\zeta_{i_1}) \cdots f_{\alpha+1|\alpha-1}(\eta_a|\zeta_{i_a})$$

$$= C_{c,a}^{-1} \left(\sum_{b=a}^{c} q_{\alpha+1|\alpha}(a|b) q_{\alpha|\alpha-1}(b|c) \right)$$

$$\cdot \sum_{1 \leq i_1 \neq \ldots \neq i_a \leq c} f_{\alpha+1|\alpha-1}(\eta_1|\zeta_{i_1}) \cdots f_{\alpha+1|\alpha-1}(\eta_a|\zeta_{i_a})$$

$$= C_{c,a}^{-1} q_{\alpha+1|\alpha-1}(a|c) \sum_{1 \leq i_1 \neq \ldots \neq i_a \leq c} f_{\alpha+1|\alpha-1}(\eta_1|\zeta_{i_1}) \cdots f_{\alpha+1|\alpha-1}(\eta_a|\zeta_{i_a})$$

$$= \check{f}_{\alpha+1|\alpha-1}(\{\eta_1, ..., \eta_a\}|\{\zeta_1, ..., \zeta_c\}; q_{\alpha+1|\alpha-1})$$

as claimed.

Now, let π be a permutation π of the numbers $1, ..., b$ such that $\pi 1 = i_1, ..., \pi a = i_a$ and such that $\pi(a+1), ..., \pi b$ are chosen arbitrarily from $\{1, ..., b\} - \{i_1, ..., i_a\}$. Then we can write

$$\int f_{\alpha+1|\alpha}(\eta_1|\gamma_{i_1}) \cdots f_{\alpha+1|\alpha}(\eta_a|\gamma_{i_a}) \, f_{\alpha|\alpha-1}(\gamma_1|\zeta_{j_1}) \cdots f_{\alpha|\alpha-1}(\gamma_b|\zeta_{j_b})$$

$$\cdot d\lambda(\gamma_1) \cdots d\lambda(\gamma_b)$$

$$= \int f_{\alpha+1|\alpha}(\eta_1|\gamma_{\pi 1}) \cdots f_{\alpha+1|\alpha}(\eta_a|\gamma_{\pi a}) \, f_{\alpha|\alpha-1}(\gamma_{\pi 1}|\zeta_{j_{\pi 1}}) \cdots f_{\alpha|\alpha-1}(\gamma_{\pi b}|\zeta_{j_{\pi b}})$$

$$\cdot d\lambda(\gamma_{\pi 1}) \cdots d\lambda(\gamma_{\pi a}) d\lambda(\gamma_{\pi(a+1)}) \cdots d\lambda(\gamma_{\pi b})$$

$$= \left(\frac{\int f_{\alpha+1|\alpha}(\eta_1|\gamma_{\pi 1})f_{\alpha|\alpha-1}(\gamma_{\pi 1}|\zeta_{j_{\pi 1}})\cdots f_{\alpha+1|\alpha}(\eta_a|\gamma_{\pi a})f_{\alpha|\alpha-1}(\gamma_{\pi a}|\zeta_{j_{\pi a}})}{\cdot\, d\lambda(\gamma_{\pi 1})\cdots d\lambda(\gamma_{\pi a})} \right)$$

$$\cdot\left(\int f_{\alpha|\alpha-1}(\gamma_{\pi(a+1)}|\zeta_{j_{\pi(a+1)}})\cdots f_{\alpha|\alpha-1}(\gamma_{\pi b}|\zeta_{j_{\pi b}})\, d\lambda(\gamma_{\pi(a+1)})\cdots d\lambda(\gamma_{\pi b}) \right)$$

Because of the Chapman-Kolmogorov equation for the conventional state-transition density $f_{\alpha+1|\alpha}$ this becomes

$$\int f_{\alpha+1|\alpha}(\eta_1|\gamma_{i_1})\cdots f_{\alpha+1|\alpha}(\eta_a|\gamma_{i_a})$$

$$\cdot f_{\alpha|\alpha-1}(\gamma_1|\zeta_{j_1})\cdots f_{\alpha|\alpha-1}(\gamma_b|\zeta_{j_b})\, d\lambda(\gamma_1)\cdots d\lambda(\gamma_b)$$

$$= f_{\alpha+1|\alpha}(\eta_1|\zeta_{j_{\pi 1}})\cdots f_{\alpha+1|\alpha}(\eta_a|\zeta_{j_{\pi a}})$$

Thus

$$\sum_{1\le i_1\ne...\ne i_a\le b}\ \sum_{1\le j_1\ne...\ne j_b\le c} \int f_{\alpha+1|\alpha}(\eta_1|\gamma_{i_1})\cdots f_{\alpha+1|\alpha}(\eta_a|\gamma_{i_a})$$

$$\cdot f_{\alpha|\alpha-1}(\gamma_1|\zeta_{j_1})\cdots f_{\alpha|\alpha-1}(\gamma_b|\zeta_{j_b})\, d\lambda(\gamma_1)\cdots d\lambda(\gamma_b)$$

$$= \sum_{\pi}\ \sum_{1\le j_1\ne...\ne j_b\le c} f_{\alpha+1|\alpha-1}(\eta_1|\zeta_{j_{\pi 1}})\cdots f_{\alpha+1|\alpha-1}(\eta_a|\zeta_{j_{\pi a}})$$

$$= \sum_{\pi}\ \sum_{1\le e_{\pi-1_1}\ne...\ne e_{\pi-1_a}\ne j_{a+1}\ne...\ne j_b\le c} f_{\alpha+1|\alpha-1}(\eta_1|\zeta_{e_1})\cdots f_{\alpha+1|\alpha-1}(\eta_a|\zeta_{e_a})$$

$$= \sum_{\pi}\ \sum_{1\le e_1\ne...\ne e_a\ne j_{a+1}\ne...\ne j_b\le c} f_{\alpha+1|\alpha-1}(\eta_1|\zeta_{e_1})\cdots f_{\alpha+1|\alpha-1}(\eta_a|\zeta_{e_a})$$

$$= K_{a,b,c}\ \sum_{1\le d_1\ne...\ne d_a\le c} f_{\alpha+1|\alpha-1}(\eta_1|\zeta_{d_1})\cdots f_{\alpha+1|\alpha-1}(\eta_a|\zeta_{d_a})$$

for some integer constant $K_{a,b,c}$ and where we have set $e_1 = j_{\pi 1}, ..., e_a = j_{\pi a}$. Integrate both sides of the equation

$$\sum_{1\le i_1\ne...\ne i_a\le b}\ \sum_{1\le j_1\ne...\ne j_b\le c} \int f_{\alpha+1|\alpha}(\eta_1|\gamma_{i_1})\cdots f_{\alpha+1|\alpha}(\eta_a|\gamma_{i_a})$$

$$\cdot f_{\alpha|\alpha-1}(\gamma_1|\zeta_{j_1})\cdots f_{\alpha|\alpha-1}(\gamma_b|\zeta_{j_b})\, d\lambda(\gamma_1)\cdots d\lambda(\gamma_b)$$

$$= K_{a,b,c}\ \sum_{1\le d_1\ne...\ne d_a\le c} f_{\alpha+1|\alpha-1}(\eta_1|\zeta_{d_1})\cdots f_{\alpha+1|\alpha-1}(\eta_a|\zeta_{d_a})$$

with respect to $\eta_1, ..., \eta_a$ Then we get $(a!C_{b,a})(b!C_{c,b}) = K_{a,b,c}(a!C_{c,a})$ from which follows $K_{a,b,c} = b!C_{c-a,b-a}$. This completes the proof. $\square$

Proposition 3 tells us that a conventional motion model $f_{\alpha+1|\alpha}(\zeta_{\alpha+1}|\zeta_\alpha)$ can be used to generate a multitarget motion model $\check{f}_{\alpha+1|\alpha}(X_{\alpha+1}|X_\alpha; q_{\alpha+1|\alpha-1})$ in such a way that the global model is completely consistent (as a Markov state-transition model) with the conventional model from which it was generated. We are therefore justified in making the following definition:

Definition 4 *Let $f_{\alpha+1|\alpha}(\zeta_{\alpha+1}|\zeta_\alpha)$ be a Markov state-transition density on parameter space and $q_{\alpha+1|\alpha}(j_{\alpha+1}|j_\alpha)$ discrete state-transition probabilities such that $q_{\alpha+1|\alpha}(j_{\alpha+1}|j_\alpha) = 0$ if $j_{\alpha+1} > j_\alpha$ or if $j_{\alpha+1} < 0$. Then the global densities $\check{f}_{\alpha+1|\alpha}(X_{\alpha+1}|X_\alpha; q_{\alpha+1|\alpha})$ defined in Proposition 3 are called the canonical global state-transition densities generated by $f_{\alpha+1|\alpha}(\zeta_{\alpha+1}|\zeta_\alpha)$ and $q_{\alpha+1|\alpha}(j_{\alpha+1}|j_\alpha)$.*

Notice that the cardinality distribution of $\check{f}_{\alpha+1|\alpha}(X_{\alpha+1}|X_\alpha; q_{\alpha+1|\alpha})$ is

$$\check{f}_{\alpha+1|\alpha}(k_{\alpha+1}|X_\alpha; q_{\alpha+1|\alpha}) = q_{\alpha+1|\alpha}(k_{\alpha+1}| |X_\alpha|)$$

This means in particular that in a global transition $X_\alpha \to X_{\alpha+1}$ the probability that $X_{\alpha+1}$ will have a given number of targets depends only on the number of targets in the previous global state X_α. This number is *completely independent of the previous states of the individual targets.*

The canonical global transition densities $\check{f}_{\alpha+1|\alpha}(X_{\alpha+1}|X_\alpha; q_{\alpha+1|\alpha-1})$ have one obvious limitation: The global transition $X_\alpha \to X_{\alpha+1}$ has the property that $|X_{\alpha+1}| \leq |X_\alpha|$ for any time-instant α—that is, the number of targets may decrease but cannot increase. Consequently, death processes but not birth processes can be modeled by these state transitions. In particular, the reader may verify that a simple choice for the discrete distribution $q_{\alpha+1|\alpha}$ is $q_{\alpha+1|\alpha}(i|j) = C_{j,i}\, q_0^i (1 - q_0)^{j-i}$ if $i \leq j$ and $q_{\alpha+1|\alpha}(i|j) = 0$ if $i > j$.

6.4.4 Closure Properties of Canonical Global Densities

Proposition 3 leads to the following *closure property* for canonical global densities:

Corollary 4 *Suppose that $f_{\alpha|\alpha}(\zeta_\alpha)$ is a density on state space and that $q_{\alpha|\alpha}(k_\alpha)$ is a discrete distribution with $q_{\alpha|\alpha}(k_\alpha) = 0$ if $k_\alpha > M$ for some $M > 0$ or if $k_\alpha < 0$. Also suppose that $f_{\alpha+1|\alpha}(\zeta_{\alpha+1}|\zeta_\alpha)$ and $q_{\alpha+1|\alpha}(k_{\alpha+1}|k_\alpha)$ are state-transition densities with $q_{\alpha+1|\alpha}(k_{\alpha+1}|k_\alpha) = 0$ if $k_{\alpha+1} > k_\alpha$ or if $k_{\alpha+1} < 0$. Let*

$$f_{\alpha+1|\alpha}(\zeta_{\alpha+1}) \triangleq \int f_{\alpha+1|\alpha}(\zeta_{\alpha+1}|\zeta_\alpha)\, f_{\alpha|\alpha}(\zeta_\alpha)\, d\lambda(\zeta_\alpha)$$

$$q^*_{\alpha+1|\alpha}(k_{\alpha+1}) \triangleq \sum_{k_\alpha = k_{\alpha+1}}^{M} q_{\alpha+1|\alpha}(k_{\alpha+1}|k_\alpha)\, q_{\alpha|\alpha}(k_\alpha)$$

be the corresponding time-extrapolated distributions. Then

$$\check{f}_{\alpha+1|\alpha}(X_{\alpha+1}; q^*_{\alpha+1|\alpha}) = \int \check{f}_{\alpha+1|\alpha}(X_{\alpha+1}|X_\alpha; q_{\alpha+1|\alpha})\, \check{f}_{\alpha|\alpha}(X_\alpha; q_{\alpha|\alpha})\, \delta X_\alpha$$

for all finite subsets $X_{\alpha+1}$ of state space.

Proof. An immediate consequence of Proposition 3. □

The significance of Corollary 4 is that global posterior densities which have a certain form are closed under the global prediction integral, provided that the corresponding global motion model is a canonical model. That is, if we know that a posterior at time-instant α is a canonical global density $\check{f}_{\alpha|\alpha}(X_\alpha; q_{\alpha|\alpha})$ then the time-extrapolation of $\check{f}_{\alpha|\alpha}(X_\alpha; q_{\alpha|\alpha})$ is also a canonical global density.

It is useful to consider the implications of Corollary 4 under Gaussian assumptions. Given conventional multivariate Gaussian distributions $N_A(\mathbf{x} - \mathbf{a})$ and $N_B(\mathbf{x} - \mathbf{b})$ we know that

$$\int N_A(\mathbf{x} - \mathbf{a})\, N_B(\mathbf{x} - \mathbf{b})\, d\lambda(\mathbf{x}) = N_{A+B}(\mathbf{a} - \mathbf{b})$$

If, therefore, $f_{\alpha|\alpha}(\mathbf{x}_\alpha|Z_\alpha) = N_Q(\mathbf{x}_\alpha - \hat{\mathbf{x}}_\alpha)$ and $f_{\alpha+1|\alpha}(\mathbf{x}_{\alpha+1}|\mathbf{x}_\alpha) = N_P(\mathbf{x}_{\alpha+1} - \Phi\mathbf{x}_\alpha)$ where Q, P are covariance matrices and Φ is a state transition matrix then

$$\begin{aligned} f_{\alpha+1|\alpha}(\mathbf{x}_{\alpha+1}|Z_\alpha) &= \int f_{\alpha+1|\alpha}(\mathbf{x}_{\alpha+1}|\mathbf{x}_\alpha)\, f_{\alpha|\alpha}(\mathbf{x}_\alpha|Z_\alpha)\, d\lambda(\mathbf{x}_\alpha) \\ &= N_{P+\Phi Q\Phi^T}(\mathbf{x}_{\alpha+1} - \Phi\hat{\mathbf{x}}_\alpha) \end{aligned}$$

Define the "para-Gaussian" density $\check{N}_C(X|X_0; q)$ by $\check{N}_C(\emptyset|\{\mathbf{x}_1, ..., \mathbf{x}_l\}; q) \triangleq q(0|l)$ and by

$$\check{N}_C(\{\mathbf{y}_1, ..., \mathbf{y}_k\}|\{\mathbf{x}_1, ..., \mathbf{x}_l\}; q)$$
$$\triangleq q(k|l)\, C_{l,k}^{-1} \sum_{1 \le i_1 \ne ... \ne i_k \le l} N_C(\mathbf{y}_1 - \mathbf{x}_{i_1}) \cdots N_C(\mathbf{y}_k - \mathbf{x}_{i_k})$$

where q satisfies $q(k|l) = 0$ if $k > l$ or if $k < 0$. Then $\check{N}_C(X|X_0; q)$ is completely specified by the finite list of parameters C, q, and $X_0 = \{\mathbf{x}_1, ..., \mathbf{x}_l\}$. It follows that

$$\int \check{N}_A(Y|X; q)\, \check{N}_B(X|X_0; q_0)\, \delta X = \check{N}_{A+B}(Y|X_0; q^*)$$

where $q^*(j|l) \triangleq \sum_{k=j}^{l} q(j|k)\, q_0(k|l)$. In particular, let $\Phi\{\mathbf{x}_1, ..., \mathbf{x}_l\} = \{\Phi\mathbf{x}_1, ..., \Phi\mathbf{x}_l\}$ for all distinct $\mathbf{x}_1, ..., \mathbf{x}_l$. Then

$$\check{N}_{P+\Phi Q\Phi^T}(Y|\Phi X_0; q^*) = \int \check{N}_P(Y|\Phi X; q)\, \check{N}_Q(X|X_0; q_0)\, \delta X$$

6.5   Random Set Outputs of Estimators

Let $\Sigma_1, ..., \Sigma_m$ be i.i.d. finite random subsets of observation space and suppose that $\Gamma = T(\Sigma_1, ..., \Sigma_m)$ is the output of a global estimator, such as the global ML or global MAP estimators discussed in Section 5.2.2 of Chapter 5. Then Γ is a finite random subset of the parameter space. One can ask the following questions: Does Γ have a global density? If so, what is it? We will not be able to answer these questions definitively here. However, in this section we will show

that the random set output of one very common type of data fusion algorithm—
multihypothesis-type estimators—do have explicitly computable global densities
(see Proposition 6).

We first begin with a definition of the type of instantaneous output most
commonly produced by multihypothesis-type algorithms.

Definition 5 *(a) A track is a random variable ψ on the state space $\mathcal{R}$ with
density $f_\psi(\zeta)$. (b) A track table is a finite list $\Upsilon = \{\psi_1, ..., \psi_d\}$ of statistically
independent tracks. (c) A hypothesis is a subset $H \subseteq \Upsilon$ of the track table.
(d) A multihypothesis is a finite list of pairs $L = \{(H_1, q_1), ..., (H_c, q_c)\}$ where
$H_1, ..., H_c$ are hypotheses and where $q_1, ..., q_c$ are nonnegative numbers which
sum to 1. If $(H, q) \in L$ then q is called the hypothesis probability of the hypothesis
H and we also write: $q = p_{hyp}(H)$. (Note: We will abuse notation and write
$H \in L$ if $(H, q) \in L$ for some $0 \le q \le 1$.) (e) Given a multihypothesis L
and a track $\psi \in \Upsilon$, the track probability of ψ, denoted $p_L(\psi)$, is the cumulative
probability of all hypotheses that contain ψ: $p_L(\psi) = \sum_{L \ni H \ni \psi} p_{hyp}(H)$ where
the summation is taken over all $H \in L$ such that $\psi \in H$.*

Remark 5 *More typically, the uncertainties in the tracks of a multihypothesis-
type algorithm are assumed to be Gaussian. In this case a track consists of a
pair $\hat{\mathbf{x}}, C$ where $\hat{\mathbf{x}}$ is the track estimate and C is the error-covariance matrix of
$\hat{\mathbf{x}}$.*

The output of a typical multihypothesis-type data fusion algorithm is a multihypothesis. Yet we also know that the output is a randomly varying finite set
of parameter space. How can we construct this finite random subset? The
following discussion suggests what should be done. Consider the simple multihypothesis $L = \{(H, q), (H', q')\}$ where $H = \{\psi\}$, $H' = \{\psi, \psi'\}$ and $q + q' = 1$.
This multihypothesis is asserting that there are two possible interpretations of
ground truth: The hypothesis H postulates the existence of a single target; and
the hypothesis H' asserts the existence of *two* targets, one of which is the same
target postulated by H. The uncertainty in our knowledge of the states of these
two targets is expressed by the fact that their states are specified only to the
degree of certainty imposed by their corresponding densities $f_\psi(\zeta)$ and $f_{\psi'}(\zeta)$.
One way of thinking about L, therefore, is that it is actually a random set of
the form $\Sigma = \{\psi, \psi'\}$ where ψ is a random vector which *always occurs* and ψ'
is a random vector which *occasionally is null*. The probability that ψ occurs is
just $p_L(\psi) = p_{hyp}(H) + p_{hyp}(H') = q + q' = 1$ and, likewise, the probability that
ψ' occurs is $p_L(\psi') = p_{hyp}(H') = q'$. Since random variables cannot actually
"vanish," however, it is clear that the multihypothesis L is actually a means
of representing a random set of the form $\Sigma = \{\psi\} \cup \Sigma_2$ where Σ_2 is a random
subset such that $p(\Sigma_2 = \emptyset) = 1 - q'$ and such that $\Sigma_2 = \{\psi'\}$ when $\Sigma_2 \ne \emptyset$.

This leads to the following result:

Proposition 5 *Suppose that L is a multihypothesis whose tracks are drawn
from a track table $\Upsilon = \{\psi_1, ..., \psi_d\}$. Then there exist statistically independent*

random subsets $\Sigma_1, ..., \Sigma_d$ *such that*

$$p(\Sigma_i \neq \emptyset) = p_L(\psi_i)$$
$$\Sigma_i = \{\psi_i\} \qquad (\text{if } \Sigma_i \neq \emptyset)$$
$$\beta_{\Sigma_i}(S) = 1 - p_L(\psi_i) + p_L(\psi_i)\, p_{\psi_i}(S)$$

for all $i = 1, .., d;$ *and such that*

$$p(\Sigma_1 \subseteq S_1, ..., \Sigma_d \subseteq S_d | \Sigma_1 \neq \emptyset, ..., \Sigma_d \neq \emptyset) = p_{\psi_1}(S_1) \cdots p_{\psi_d}(S_d)$$

Proof. Construct independent random subsets $\Theta_1, ..., \Theta_d$ which satisfy the following properties:

$$p(\Theta_i = \emptyset) = 1 - p_L(\psi_i), \qquad p(\Theta_i = \mathcal{R}) = p_L(\psi_i)$$

$$p(\psi_1 \in S_1, ..., \psi_d \in S_d, \Theta_1 = T_1, ..., \Theta_d = T_d)$$
$$= p(\psi_1 \in S_1) \cdots p(\psi_d \in S_d)\, p(\Theta_1 = T_1) \cdots p(\Theta_d = T_d)$$

for all $i = 1, ..., d$. Define $\Sigma_i \triangleq \{\psi_i\} \cap \Theta_i$ for all $i = 1, ..., d$. Then note that $\Sigma_i = \{\psi_i\}$ if $\Sigma_i \neq \emptyset$ and that $p(\Sigma_i \neq \emptyset) = p(\Theta_i \neq \emptyset) = p_L(\psi_i)$ and

$$\beta_{\Sigma_i}(S) = p(\Sigma_i \subseteq S) = p(\Sigma_i = \emptyset) + p(\Sigma_i \neq \emptyset)\, p(\psi_i \in S | \Theta_i = \mathcal{R})$$
$$= 1 - p_L(\psi_i) + p_L(\psi_i)\, p_{\psi_i}(S)$$

Finally, note that

$$p(\Sigma_1 \subseteq S_1, ..., \Sigma_d \subseteq S_d | \Sigma_1 \neq \emptyset, ..., \Sigma_d \neq \emptyset)$$
$$= p(\{\psi_1\} \subseteq S_1, ..., \{\psi_d\} \subseteq S_d | \Theta_1 = \mathcal{R}, ..., \Theta_d = \mathcal{R})$$
$$= p(\psi_1 \in S_1) \cdots p(\psi_d \in S_d) = p_{\psi_1}(S_1) \cdots p_{\psi_d}(S_d)$$

as claimed.   □

Note that the previous proposition could be extended to include the possibility that the tracks $\psi_1, ..., \psi_d$ are not statistically independent.

The reasoning which underlies the following definition is self-evident:

Definition 6 *Suppose that L is a multihypothesis drawn from a track table $\Upsilon = \{\psi_1, ..., \psi_d\}$ and that $\Sigma_1, ..., \Sigma_d$ are the random subsets of state space constructed in Proposition 5. Then the random subset $\Sigma_L = \Sigma_1 \cup ... \cup \Sigma_d$ is called the random output subset corresponding to the multihypothesis L.*

The following proposition tells us that the global density of Σ_L exists and also tells us what it actually is:

Proposition 6 *Let L be a multihypothesis drawn from a track table $\Upsilon = \{\psi_1, ..., \psi_d\}$ and let Σ_L its corresponding random subset of parameter space. Abbreviate $q_i \triangleq p_L(\psi_i)$ for all $i = 1, ..., d$. Then the global density of Σ_L exists and is:*

$$f_{\Sigma_L}(\emptyset) = (1 - q_1)\cdots(1 - q_d)$$

$$f_{\Sigma_L}(\{\zeta_1, ..., \zeta_k\}) = \sum_{1 \le i_1 \ne ... \ne i_k \le d} \frac{(1 - q_1)\cdots(1 - q_m)}{(1 - q_{i_1})\cdots(1 - q_{i_k})} q_{i_1}\cdots q_{i_k}$$

$$\cdot f_{\psi_{i_1}}(\zeta_1)\cdots f_{\psi_{i_k}}(\zeta_k)$$

where the summation is taken over all distinct $i_1, ..., i_k$ such that $1 \le i_1, ..., i_k$

$\le d$.

Proof. The belief measure of Σ_L is

$$\beta_{\Sigma_L}(S) = p(\Sigma_L \subseteq S) = p(\Sigma_1 \subseteq S, ..., \Sigma_d \subseteq S) = p(\Sigma_1 \subseteq S)\cdots p(\Sigma_d \subseteq S)$$
$$= (1 - p_L(\psi_1) + p_L(\psi_1)\, p_{\psi_1}(S))\cdots(1 - p_L(\psi_d) + p_L(\psi_d)\, p_{\psi_d}(S))$$

Suppose that there are m integers i such that $p_L(\psi_i) \ne 0$. Let $q_i = p_L(\psi_i)$ for all $i = 1, ..., d$. Then by Proposition 20 of Chapter 4 we know that $f_{\Sigma_L}(Z) = f_{\Sigma_L}(|Z|) = 0$ (if $|Z| > m$) and

$$f_{\Sigma_L}(\emptyset) = (1 - q_1)\cdots(1 - q_d)$$

$$f_{\Sigma_L}(\{\zeta_1, ..., \zeta_k\}) = \sum_{1 \le i_1 \ne ... \ne i_k \le d} \frac{(1 - q_1)\cdots(1 - q_m)}{(1 - q_{i_1})\cdots(1 - q_{i_k})} q_{i_1}\cdots q_{i_k}$$

$$\cdot f_{\psi_{i_1}}(\zeta_1)\cdots f_{\psi_{i_k}}(\zeta_k)$$

for all distinct $\zeta_1, ..., \zeta_k \in \mathcal{R}$ and $k = 1, ..., m$. $\qquad\square$

6.6 Simple Examples

The purpose of this section is to illustrate the concepts of this chapter using simple one- or two-dimensional examples. Section 6.6.1 is devoted to two examples involving two stationary or two moving targets in one dimension. In section 6.6.2 we illustrate the global GML estimator using multitarget two-dimensional examples.

6.6.1 Two Targets in One Dimension

In this section we examine two simple one-dimensional examples. In the first subsection, we examine the geometrical behavior of the global measurement density for a two-target problem. The second subsection illustrates the application of global nonlinear filtering to a simple dynamic two-target problem.

Random Set Measurement Models

We compute a detailed examination of a very simple global measurement-model density based on the results of Proposition 20 of Chapter 4. We assume the existence of two targets of differing type existing in one dimension. They are observed by a single Gaussian sensor with no missed detections and no false alarms. This sensor cannot distinguish between the targets directly. Rather, the targets are distinguishable only very indirectly, in that observations generated by one target are more uncertain than observations generated by the other.

Translated into mathematical notation, these assumptions become: (1) dimension of Euclidean space, $n = \dim \mathbb{R}^n = 1$; (2) cardinality of discrete space, $N = |V| = 2$ with $V = \{1, 2\}$; (3) thus $\mathcal{R} = \mathbb{R} \times \{1, 2\}$; (4) number of parameters, $m = |X| = 2$; and (5) the sensor noise density is Gaussian:

$$f(z|(x,i)) = \frac{1}{\sqrt{2\pi}\sigma_i} \, e^{-\frac{1}{2}(z-x)^2/\sigma_i^2} \qquad (i = 1, 2)$$

With these assumptions the only nonvanishing member of the parametrized family of global densities is

$$f_\Sigma(\{z_1, z_2\}|\{\zeta_1, \zeta_2\}) = f(z_1|\zeta_1)\,f(z_2|\zeta_2) + f(z_2|\zeta_1)\,f(z_1|\zeta_2)$$

where $\zeta_1 = (x_1, j)$ and $\zeta_2 = (x_2, k)$ with $x_1, x_2 \in \mathbb{R}$ and $j, k \in \{1, 2\}$.

We first determine the extrema of the global density f_Σ. Taking the partial derivatives of f_Σ with respect to z_1 and z_2 and setting them to zero, we get:

$$0 = -\frac{z_1 - x_1}{\sigma_j^2} f(z_1|(x_1, j))\,f(z_2|(x_2, k)) - \frac{z_1 - x_2}{\sigma_k^2} f(z_2|(x_1, j))\,f(z_1|(x_2, k))$$

$$0 = -\frac{z_2 - x_2}{\sigma_k^2} f(z_1|(x_1, j))\,f(z_2|(x_2, k)) - \frac{z_2 - x_1}{\sigma_j^2} f(z_2|(x_1, j))\,f(z_1|(x_2, k))$$

for $j = 1, 2$. The quantities $f(z_1|\zeta_1)$, $f(z_2|\zeta_2)$, $f(z_2|\zeta_1)$, $f(z_1|\zeta_2)$ can be eliminated from these two equations to yield:

$$(z_1 - x_1)(z_2 - x_1)\sigma_k^4 = (z_1 - x_2)(z_2 - x_2)\sigma_j^4$$

For any fixed j, k and x_1, x_2 this equation defines a hyperbola in the $z_1 \times z_2$ plane. The points $(z_1, z_2) = (x_1, x_2)$ and $(z_1, z_2) = (x_2, x_1)$ are solutions to this equation but not to the original system of equations. On the other hand the covariance-weighted average

$$z_1 = z_2 = \sigma_{j,k}^2 \left(\frac{x_1}{\sigma_j^2} + \frac{x_2}{\sigma_k^2} \right) \quad \text{where} \quad \frac{1}{\sigma_{j,k}^2} = \frac{1}{\sigma_j^2} + \frac{1}{\sigma_k^2}$$

is a true solution. It corresponds either to a saddle point or to an absolute maximum. Abbreviate this solution as $z_1 = z_2 = P$.

When does this solution determine a maximal point of f_Σ or, equivalently, of $\ln f_\Sigma$? The components of the second-derivative matrix of $\ln f_\Sigma$ at (P, P) are easily determined to be

$$\frac{\partial^2 \ln f_\Sigma}{\partial x_1^2}(P,P) = \frac{1}{2}\left(-\frac{1}{\sigma_{j,k}^2} + 2\frac{(x_1-x_2)^2}{(\sigma_j^2+\sigma_k^2)^2}\right) = \frac{\partial^2 \ln f_\Sigma}{\partial x_2^2}(P,P)$$

$$\frac{\partial^2 \ln f_\Sigma}{\partial x_1 \partial x_2}(P,P) = \frac{(x_1-x_2)^2}{(\sigma_j^2+\sigma_k^2)^2}$$

Consequently the determinant of the second-derivative matrix of $\ln f_\Sigma$ is

$$\frac{1}{4\sigma_{j,k}^2}\left(\frac{1}{\sigma_{j,k}^2} - 4\frac{(x_1-x_2)^2}{(\sigma_j^2+\sigma_k^2)^2}\right)$$

By the second derivative test (P,P) will be a saddle point if and only if $|x_1 - x_2| > \Delta x_{j,k}$ where

$$\Delta x_{j,k} \triangleq \frac{(\sigma_j^2+\sigma_k^2)^{3/2}}{2\sigma_j\sigma_k}$$

and a local maximum if and only if $|x_1 - x_2| < \Delta x_{j,k}$. Thus the most likely value of the finite random subset Σ is some two-point extremum $\{a, b\}$ with $a \neq b$ if $|x_1 - x_2| > \Delta x_{j,k}$.

Finally, let us compute covariance matrix $C_\Sigma(P,P)$ of the Gaussian approximation of f_Σ at the single-point solution (P,P). By definition, this computed by taking the Taylor's series expansion of $\ln f_\Sigma$ at the point (P,P), in which case $C_\Sigma(P,P)^{-1}$ is -1 times the second-derivative matrix of $\ln f_\Sigma$ evaluated at $(z_1, z_2) = (P,P)$:

$$C_\Sigma(P,P)^{-1} = \begin{pmatrix} \frac{1}{2\sigma_{j,k}^2} - \frac{(x_1-x_2)^2}{(\sigma_j^2+\sigma_k^2)^2} & -\frac{(x_1-x_2)^2}{(\sigma_j^2+\sigma_k^2)^2} \\ -\frac{(x_1-x_2)^2}{(\sigma_j^2+\sigma_k^2)^2} & \frac{1}{2\sigma_{j,k}^2} - \frac{(x_1-x_2)^2}{(\sigma_j^2+\sigma_k^2)^2} \end{pmatrix}$$

To get a more intuitive feel for the meaning of this approximate "system-level" measurement-model covariance matrix we compute the corresponding one-sigma error ellipse when (P,P) is a maximum. This ellipse has one axis oriented along the line $x_1 = x_2$. If b_0 denotes the half-length of the axis along this line, a_0 denotes the half-length of the other axis, and if A is the area of the ellipse then:

$$b_0^2 = \frac{\frac{1}{2}(\sigma_j^2+\sigma_k^2)^2}{\Delta x_{j,k}^2 - |x_1-x_2|^2}, \qquad a_0^2 = 2\sigma_{j,k}^2, \qquad A^2 = \frac{\pi\sigma_j^2\sigma_k^2(\sigma_j^2+\sigma_k^2)}{\Delta x_{j,k}^2 - |x_1-x_2|^2}$$

As the separation distance $|x_1 - x_2|$ between the parameters approaches the critical distance $\Delta x_{j,k}$ the area of the ellipse tends to infinity. That is, the ambiguity near the critical distance is very large since it is increasingly unclear as to whether there are one or two parameters.

Remark 6 *Given a two-dimensional error-covariance matrix*

$$Q = \begin{pmatrix} A & B \\ B & C \end{pmatrix}$$

then the semi-minor axis a_0 and the semi-major axis b_0 are given by $a_0^2 - 2D/E$ and $b_0^2 = 1/2E$, where $D = AC - B^2$ and $E = A + B + \sqrt{(A - B)^2 + 4C^2}$. Also, the area of the ellipse is $A = \pi a_0 b_0 = \pi\sqrt{D}$. The angle of inclination τ of the b_0-axis of the ellipse (measured counterclockwise from the horizontal axis) is given by $\tau = \frac{1}{2}\{\pi - \arctan(2C/(A - B))\}$.

The graph of the global density $f(z_1, z_2) \triangleq f_\Sigma(\{z_1, z_2\}|\{(x_1, j), (x_2, k)\}$ is exhibited in the following six figures, assuming that $x_2 = 5.5$, that $j = 1, k = 2$, that $\sigma_1 = 1.0, \sigma_2 = 2.0$, and that x_1 takes one of the six values $x_1 = 1.5, 2, 0, 2.5, 3.0, 3.5$ or 4.0.

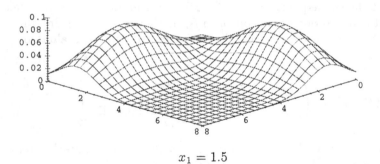

$$x_1 = 1.5$$

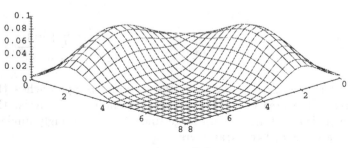

$$x_1 = 2.0$$

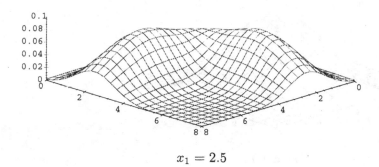

$$x_1 = 2.5$$

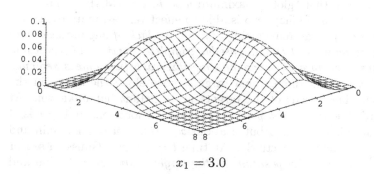

$$x_1 = 3.0$$

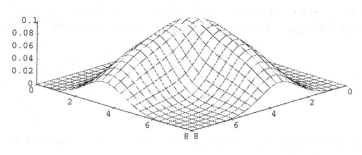

$$x_1 = 3.5$$

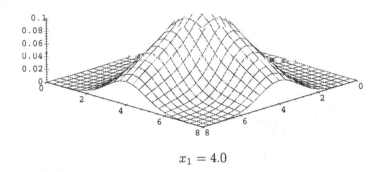

$$x_1 = 4.0$$

Global Nonlinear Filtering

We illustrate random set multitarget filtering by applying it to a simple one-dimensional, two-target model problem originally posed by O.E. Drummond [3]. The example illustrates that "global maximum *a posteriori* (MAP) estimation" as defined in Definition 5 of Chapter 5 is able to effectively estimate the states of multiple targets despite extreme ambiguity due to: *lack of any measurement information describing target I.D*; and *possible complete "mixing" of (confusion between) identities of the targets during the time-interval between measurements.*

Specifically, let two one-dimensional targets move along the x-axis in the positive direction. The speeds of the targets are constant but unknown. At time $t = 0$ min it is known on an *a priori* basis that the locations and speeds of the two targets are Gaussian-distributed about $x = 0$ km, $u = 4$ km/min and $x = 2$ km, $u = 2$ km/min, respectively. At time $t = 1$ min a Gaussian sensor with variance σ^2 measures the *positions of the targets only*: $z_1 = 4.8$ km and $z_2 = 5.6$ km.

We then ask the following question: *What are the positions and speeds of the two targets at time $t = 1$ min.?* We will answer this question by computing the global maximum *a posteriori* (GMAP) estimate of the state. (Note that if the problem were entirely deterministic then the answer would be $(x, u) = (4.0, 4.0), (4.0, 2.0).$)

Let the position variance for both targets at time $t = 0$ be $\sigma_{x,0}^2$ and let the velocity variance be $\sigma_{u,0}^2$. Then the prior distributions describing the two targets are

$$f_0((x, u)^T) = N_E((x, u)^T - (0, 4)^T), \qquad f_0((x, u)^T) = N_E((x, u)^T - (2, 2)^T)$$

$$E = \begin{pmatrix} \sigma_{x,0}^2 & 0 \\ 0 & \sigma_{u,0}^2 \end{pmatrix}$$

where $N_E(x)$ denotes the multivariate normal distribution with covariance matrix E.

The measurement density is $f(z|(y,v)^T) = N_R(z - H(y,v)^T)$ where $R = (\sigma^2)$ and $H = (1,0)$. We assume that the between-measurements dynamics of both targets are governed by the Markov state-transition density

$$f_{1|0}((y,v)^T|(x,u)^T) = N_Q((y,v)^T - \Phi(x,u)^T)$$

$$\Phi = \begin{pmatrix} 1 & \Delta t \\ 0 & 1 \end{pmatrix}, \qquad Q = \begin{pmatrix} \sigma_x^2 & 0 \\ 0 & \sigma_u^2 \end{pmatrix}$$

Finally, we attach identifying labels '1' and '2' to the targets with initial states $(0,4)$ and $(2,2)$, respectively. (These labels can be either actual target types or arbitrarily assigned tags. See below.). Thus the complete initial states of the targets are $(0.0, 4.0, 1)$ and $(2.0, 2.0, 2)$, respectively.

We then pose the following question: What are the positions and speeds of the two targets at time $t = 1$ min?

All of this can be summarized as follows. The prior density describing the initial states of the two targets is

$$f_0(x, u, a|e, s, n) = N_{\sigma_{x,0}^2}(x - e)\, N_{\sigma_{u,0}^2}(u - s)\, q_0(a|n)$$

where $(e, s, n) = (0, 4, 1)$ or $(e, s, n) = (2, 2, 2)$ and where $q_0(a|n)$ is a discrete prior distribution on the labels $a = 1, 2$ with parameter $n = 1, 2$. Likewise, the state transition density for both targets is

$$f_{1|0}(y, v, b|x, u, a) = N_Q((y,v)^T - \Phi(x,u)^T)\, q_{1|0}(b|a)$$

and the measurement density is $f(z|y, v, b) = N_{\sigma^2}(z - y)$.

In the random set formalism the states of a group of targets are thought of as a single system (a "global state"). So, let $Y = \{\eta_1, \eta_2\}, X = \{\zeta_1, \zeta_2\}$ denote the global states of a two-target system at times $t = 1$ and $t = 0$, respectively, where $\eta_i = (y_i, v_i, b_i)$ and $\zeta_i = (x_i, u_i, a_i)$ for $i = 1, 2$. Also, let $Z = \{4.8, 5.6\}$ denote the "global measurement" collected by the sensor at $t = 1$. From the random-set calculus developed in the previous chapters we know that the global measurement-model density must have the form

$$f(\{z_1, z_2\}|\{\eta_1, \eta_2\}) = f(z_1|\eta_1)f(z_2|\eta_2) + f(z_2|\eta_1)f(z_1|\eta_2)$$

or

$$\begin{aligned} f(\{z_1, z_2\}|\{(y_1, v_1, b_1), (y_2, v_2, b_2)\}) &= N_{\sigma^2}(z_1 - y_1)N_{\sigma^2}(z_2 - y_2) \\ &\quad + N_{\sigma^2}(z_2 - y_1)N_{\sigma^2}(z_1 - y_2) \end{aligned}$$

We will assume a global state-transition model of the form

$$f_{1|0}(\{\eta_1, \eta_2\}|\{\zeta_1, \zeta_2\}) = f_{1|0}(\eta_1|\zeta_1)f_{1|0}(\eta_2|\zeta_2) + f_{1|0}(\eta_2|\zeta_1)f_{1|0}(\eta_1|\zeta_2)$$

or, in full notation,

$$\begin{aligned} &f_{1|0}(\{(y_1, v_1, b_1), (y_2, v_2, b_2)\}|\{(x_1, u_1, a_1), (x_2, u_2, a_2)\}) \\ =\ & N_Q((y_1, v_1)^T - \Phi(x_1, u_1)^T)\, N_Q((y_2, v_2)^T - \Phi(x_2, u_2)^T)\, q_{1|0}(b_1|a_1)\, q_{1|0}(b_2|a_2) \\ &+ N_Q((y_2, v_2)^T - \Phi(x_1, u_1)^T)\, N_Q((y_1, v_1)^T - \Phi(x_2, u_2)^T)\, q_{1|0}(b_2|a_1)\, q_{1|0}(b_1|a_2) \end{aligned}$$

Among other things, this motion model assumes that the number of targets remains the same (always two). In addition, to model the maximal possible uncertainty we will assume that the tag-transition distribution $q_{1|0}$ is $q_{1|0}(a|b) = \frac{1}{2}$ for all $a, b = 1, 2$. That is, we will assume that the state transition

$$(x, u, 1) \rightarrow (y, v, 1), \qquad (x, u, 2) \rightarrow (y, v, 2)$$

and the state transition

$$(x, u, 1) \rightarrow (y, v, 2), \qquad (x, u, 2) \rightarrow (y, v, 1)$$

are *equally likely*. We are therefore modeling a situation in which our ability to sort out the proper tags of the two targets is completely lost due to "mixing" of (i.e., confusion between) tags, between measurements. Thus the motion model becomes

$$f_{1|0}(\{(y_1, v_1), (y_2, v_2)\}|\{(x_1, u_1), (x_2, u_2)\})$$
$$= \frac{1}{4} N_Q((y_1, v_1)^T - \Phi(x_1, u_1)^T) \, N_Q((y_2, v_2)^T - \Phi(x_2, u_2)^T)$$
$$+ \frac{1}{4} N_Q((y_2, v_2)^T - \Phi(x_1, u_1)^T) \, N_Q((y_1, v_1)^T - \Phi(x_2, u_2)^T)$$

Remark 7 *Though this motion model is somewhat unusual it has at least one precedent in the literature. In [15, equ. 3.5, p. 2661] Xie and Evans apply a discrete-state version of this motion model to the tracking of multiple frequency lines without computing optimal report-track associations. They describe a "mixed track" (a "global state" in our terminology) as a "nonordered array" or "symbol" in which the target states must be considered as an unit because I.D. tags cannot be attached to individual tracks. They then use the motion model $p_{\{i,j\}}(\{k, l\}) = p_{il} \cdot p_{jk} + p_{ik} \cdot p_{jl}$ where p_{il} denotes the transition probability of an individual target moving from state a_i at time K to state a_l at time $K + 1$.*

Finally we choose the global prior density

$$f_0(\{\zeta_1, \zeta_2\}|\{\varepsilon_1, \varepsilon_2\}) = f_0(\zeta_1|\varepsilon_1) f_0(\zeta_2|\varepsilon_2) + f_0(\zeta_2|\varepsilon_1) f_0(\zeta_1|\varepsilon_2)$$

$$= N_{\sigma_{x,0}^2}(x_1) \, N_{\sigma_{x,0}^2}(x_2 - 2) \, N_{\sigma_{u,0}^2}(u_1 - 4) \, N_{\sigma_{u,0}^2}(u_2 - 2) \, q_0(a_1|1) \, q_0(a_2|2)$$
$$+ N_{\sigma_{x,0}^2}(x_2) \, N_{\sigma_{x,0}^2}(x_1 - 2) \, N_{\sigma_{u,0}^2}(u_2 - 4) \, N_{\sigma_{u,0}^2}(u_1 - 2) \, q_0(a_2|1) \, q_0(a_1|2)$$

where $\varepsilon_1 = (0, 4, 1)$ and $\varepsilon_2 = (2, 2, 2)$.

We now turn to the solution of the problem. First, a little bit of algebra shows that the prediction integral is

$$f_{1|0}(\{\eta_1, \eta_2\}|\{\varepsilon_1, \varepsilon_2\}) = \int f_{1|0}(\{\eta_1, \eta_2\}|Y) \, f_0(Y|\{\varepsilon_1, \varepsilon_2\}) \delta Y$$

$$= \frac{1}{2} \int f_{1|0}(\{\eta_1, \eta_2\}|\{\zeta_1, \zeta_2\}) \, f_0(\{\zeta_1, \zeta_2\}|\{\varepsilon_1, \varepsilon_2\}) d\lambda(\zeta_1) d\lambda(\zeta_2)$$

$$= \frac{1}{2} \sum_{a_1,a_2=1,2} \int f_{1|0}(\{\eta_1,\eta_2\}|\{(x_1,u_1,a_1),(x_2,u_2,a_2)\})$$

$$\cdot f_0(\{(x_1,u_1,a_1),(x_2,u_2,a_2)\}|\{\varepsilon_1,\varepsilon_2\})dx_1du_1dx_2du_2$$

$$= \frac{1}{2} \int f_{1|0}(\{\eta_1,\eta_2\}|\{(x_1,u_1),(x_2,u_2)\})$$

$$\cdot \left(\sum_{a_1,a_2=1,2} f_0(\{(x_1,u_1,a_1),(x_2,u_2,a_2)\}|\{\varepsilon_1,\varepsilon_2\}) \right) dx_1du_1dx_2du_2$$

where the last equation is true since $f_{1|0}(X|Y)$ is independent of the parameters a_1, a_2. However, also note that the summation in parentheses is

$$= N_{\sigma^2_{x,0}}(x_1)\, N_{\sigma^2_{x,0}}(x_2-2)\, N_{\sigma^2_{u,0}}(u_1-4)\, N_{\sigma^2_{u,0}}(u_2-2)$$

$$+ N_{\sigma^2_{x,0}}(x_2)\, N_{\sigma^2_{x,0}}(x_1-2)\, N_{\sigma^2_{u,0}}(u_2-4)\, N_{\sigma^2_{u,0}}(u_1-2)$$

Then, a little more algebra using the standard matrix identity

$$N_A(\mathbf{x}-\mathbf{a})\, N_B(\mathbf{x}-\mathbf{b}) = N_C(\mathbf{x}-\mathbf{c})\, N_{A+B}(\mathbf{a}-\mathbf{b})$$

(where $C^{-1} = A^{-1} + B^{-1}$ and $C^{-1}\mathbf{c} = A^{-1}\mathbf{a} + B^{-1}\mathbf{b}$) shows that the time-prediction of the global prior is just

$$f_{1|0}(\{(y_1,v_1),(y_2,v_2)\}|\{(0,4),(2,2)\})$$

$$= N_D((y_1,v_1)^T - \Phi(0,4)^T)\, N_D((y_2,v_2)^T - \Phi(2,4)^T)$$

$$+ N_D((y_2,v_2)^T - \Phi(0,4)^T)\, N_D((y_1,v_1)^T - \Phi(2,4)^T)$$

where $D = Q + \Phi E \Phi^T$. (In other words, the result of applying the "global" motion model is consistent with what one would get if one predicted the priors $\varepsilon_1, \varepsilon_2$ separately, with the same increase in uncertainty represented by the transition $E \rightarrow Q + \Phi E \Phi^T$. Compare this result to the theoretical prediction of Section 6.4.4.)

For what follows we do not need to know the normalizing constant in the non-linear filtering equation and so we will not compute it here. Since the number of targets is known *a priori*, the maximum *a posteriori* (MAP) estimate of the target positions is just the value of the global state variable $Y = \{(y_1,v_1),(y_2,v_2)\}$ which maximizes the global posterior $f_{1|1}(Y|Z) \propto f(Z|Y)\, f_{1|0}(Y|E)$ where $E = \{4.8, 5.6\}$. That is, we are to maximize the product

$$f_{1|1}(\{(y_1,v_1),(y_2,v_2)\}|\{4.8,5.6\})$$

$$= \{N_{\sigma^2}(4.8-y_1)N_{\sigma^2}(5.6-y_2) + N_{\sigma^2}(5.6-y_1)N_{\sigma^2}(4.8-y_2)\}$$

$$\cdot \{N_D((y_1,v_1)^T - \Phi(0,4)^T)\, N_D((y_2,v_2)^T - \Phi(2,4)^T)$$

$$+ N_D((y_2,v_2)^T - \Phi(0,4)^T)\, N_D((y_1,v_1)^T - \Phi(2,4)^T)\}$$

with respect to the variables y_1, y_2, v_1, v_2.

We do this for a number of cases:

Case I (Basic problem): The following assumptions are made:

variances: $\sigma^2 = \sigma_x^2 = \sigma_{x,0}^2 = 1.0$ and $\sigma_u^2 = \sigma_{u,0}^2 = 1.0$.

initial state:$\{(0,4), (2,2)\}$

observation:$\{4.8, 5.6\}$

Then the global MAP estimate is: $\{(4.90, 3.96), (4.90, 2.64)\}$.

Case II (Small variances): The following assumptions are made:

variances: $\sigma^2 = \sigma_x^2 = \sigma_{x,0}^2 = .001$ and $\sigma_u^2 = \sigma_{u,0}^2 = .001$.

initial state:$\{(0,4), (2,2)\}$

observation:$\{4.8, 5.6\}$

GMAP estimate: $\{(4.00, 4.00), (4.00, 2.00)\}$

Case III (Different initial velocity):

variances: $\sigma^2 = \sigma_x^2 = \sigma_{x,0}^2 = 1.0$ and $\sigma_u^2 = \sigma_{u,0}^2 = 1.0$.

initial state:$\{(0,4.1), (2,2)\}$

observation:$\{4.8, 5.6\}$

GMAP estimate: $\{(4.92, 4.10), (4.90, 2.57)\}$

Case IV (Different observations):

variances: $\sigma^2 = \sigma_x^2 = \sigma_{x,0}^2 = 1.0$ and $\sigma_u^2 = \sigma_{u,0}^2 = 1.0$.

initial state:$\{(0,4), (2,2)\}$

observation:$\{4.3, 3.8\}$

GMAP estimate: $\{(4.04, 3.67), (4.04, 2.35)\}$

Case V (Same different observations, smaller velocity transition variance):

variances: $\sigma^2 = \sigma_x^2 = \sigma_{x,0}^2 = 1.0$ but $\sigma_u^2 = \sigma_{u,0}^2 = 0.5$.

initial state:$\{(0,4), (2,2)\}$

observation:$\{4.3, 3.8\}$

GMAP estimate: $\{(4.04, 3.93), (4.04, 2.10)\}$

In this example we have posed a problem in which the ability to distinguish between two targets during the between-measurements interval is severely limited. Despite the fact that both target I.D. and velocity are unobservables, the global MAP estimation technique is capable of estimating the "system level" kinematic state of the entire two-target system taken as a whole. Note, however, that it is not possible to determine which estimate corresponds to which original target. That is, is the target at $t = 1$ with state $(4.04, 3.93)$ the same as the target at time $t = 0$ with state (0.4), or the one with state $(2,2)$?

6.6.2   Multiple Targets in Two Dimensions

In this section we use a very simple example in two dimensions to illustrate how the global ML estimator (Definition 5 of Chapter 5) can be used to simultaneously estimate the number and the positions of targets in the presence of point clutter (see [10], [11], [12], [13] for more a more extensive set of examples). The problem to be solved is as follows:

1) Observations are collected by a single position-reporting Gaussian sensor with noise density

$$f(a, b|x, y) = \frac{1}{2\pi\sigma^2}\, e^{-\frac{1}{2}(a-x)^2/\sigma^2 - \frac{1}{2}(b-y)^2/\sigma^2}$$

where $\sigma^2 = 4$.

2) The sensor has constant probability of detection $q \neq 1$ and constant probability of false alarm $Q \neq 0$.

3) For any given observation, no more than one report can be a clutter report.

4) Clutter objects are Gaussian-distributed in the area of interest according to the density

$$c(a, b) = \frac{1}{2\pi\sigma_c^2} \, e^{-\frac{1}{2}(a-e_1)^2/\sigma_c^2 - \frac{1}{2}(b-e_2)^2/\sigma_c^2}$$

where $(e_1, e_2) = (8, 8)$ and $\sigma_c^2 = 100$. Thus in the area of interest, clutter objects are approximately uniformly distributed.

5) The sensor interrogates the region of interest at three nearly simultaneous instants and collects the following reports:

Response to query 1: $(3.0, 4.0)$
Response to query 2: $(4.0, 5.0)$, $(8.0, 9.0)$
Response to query 3: $(3.5, 2.9)$, $(6.9, 7.0)$

We ask:

a) What *one-*, *two-*, or *three*-target configurations, respectively, best accounts for these observations?

b) Which of these possibilities (one, two or three targets) best account for the data?

The solution proceeds as follows. The belief measure of the clutter model is just

$$\beta_c(S) = 1 - Q + Q \int_S c(a, b) \, da \, db$$

Likewise, the belief measure describing the sensor under single-target conditions is

$$\beta_f(S|x, y) = 1 - q + q \int_S f(a, b|x, y) \, da \, db$$

Accordingly, the complete sensor model for the one-, two-, and three-target situations is the parametrized belief measure:

$$\begin{aligned}
\beta(S|\emptyset) &= \beta_c(S) \\
\beta(S|\{(x_1, y_1)\}) &= \beta_f(S|x_1, y_1) \, \beta_c(S) \\
\beta(S|\{(x_1, y_1), (x_2, y_2)\}) &= \beta_f(S|x_1, y_1) \, \beta_f(S|x_2, y_2) \, \beta_c(S) \\
\beta(S|\{(x_1, y_1), (x_2, y_2), (x_3, y_3)\}) &= \beta_f(S|x_1, y_1) \, \beta_f(S|x_2, y_2) \, \beta_f(S|x_3, y_3) \, \beta_c(S)
\end{aligned}$$

The corresponding parametrized global density is described as follows (note that since in our problem there are no more than two observations, we do not need to determine the formulas for the global density $f(Z|X)$ for $|Z| > 2$.) We abbreviate $\mathbf{x}_j \triangleq (x_j, y_j)$ and $\mathbf{a}_j \triangleq (a_j, b_j)$ for $j = 1, 2, 3, 4.$:

    I. *Zero-target case*

$$\begin{aligned}
f(\emptyset|\emptyset) &= 1 \\
f(\{\mathbf{a}_1\}|\emptyset) &= c(\mathbf{a}_1)
\end{aligned}$$

II. *One-target case.*

$$f(\emptyset|\{\mathbf{x}_1\}) = (1-q)(1-Q)$$
$$f(\{\mathbf{a}_1\}|\{\mathbf{x}_1\}) = q(1-Q)\,f(\mathbf{a}_1|\mathbf{x}_1) + (1-q)Q\,c(\mathbf{a}_1)$$
$$f(\{\mathbf{a}_1,\mathbf{a}_2\}|\{\mathbf{x}_1\}) = qQ\,f(\mathbf{a}_1|\mathbf{x}_1)\,c(\mathbf{a}_2) + qQ\,f(\mathbf{a}_2|\mathbf{x}_1)\,c(\mathbf{a}_1)$$

III. *Two-target case*

$$f(\emptyset|\{\mathbf{x}_1,\mathbf{x}_2\}) = (1-q)^2(1-Q)$$
$$\begin{aligned}
f(\{\mathbf{a}_1\}|\{\mathbf{x}_1,\mathbf{x}_2\}) = {}& q(1-q)(1-Q)\,f(\mathbf{a}_1|\mathbf{x}_1)\\
&+ q(1-q)(1-Q)\,f(\mathbf{a}_1|\mathbf{x}_2)\\
&+ (1-q)^2 Q\,c(\mathbf{a}_1)
\end{aligned}$$
$$\begin{aligned}
f(\{\mathbf{a}_1,\mathbf{a}_2\}|\{\mathbf{x}_1,\mathbf{x}_2\}) = {}& q^2(1-Q)\,f(\mathbf{a}_1|\mathbf{x}_1)\,f(\mathbf{a}_2|\mathbf{x}_2)\\
&+ q^2(1-Q)\,f(\mathbf{a}_2|\mathbf{x}_1)\,f(\mathbf{a}_1|\mathbf{x}_2)\\
&+ q(1-q)Q\,f(\mathbf{a}_1|\mathbf{x}_1)\,c(\mathbf{a}_2)\\
&+ q(1-q)Q\,f(\mathbf{a}_1|x_2)\,c(\mathbf{a}_2)\\
&+ q(1-q)Q\,f(\mathbf{a}_2|\mathbf{x}_1)\,c(\mathbf{a}_1)\\
&+ q(1-q)Q\,f(\mathbf{a}_2|x_2)\,c(\mathbf{a}_1)
\end{aligned}$$

IV. *Three-target case*

$$f(\emptyset|\{\mathbf{x}_1,\mathbf{x}_2,\mathbf{x}_3\}) = (1-q)^3(1-Q)$$
$$\begin{aligned}
f(\{\mathbf{a}_1\}|\{\mathbf{x}_1,\mathbf{x}_2,\mathbf{x}_3\}) = {}& q(1-q)^2(1-Q)\,f(\mathbf{a}_1|\mathbf{x}_1)\\
&+ q(1-q)^2(1-Q)\,f(\mathbf{a}_1|\mathbf{x}_2) + q(1-q)^2(1-Q)\,f(\mathbf{a}_1|\mathbf{x}_3) + (1-q)^3 Q c(\mathbf{a}_1)
\end{aligned}$$
$$\begin{aligned}
f(\{\mathbf{a}_1,\mathbf{a}_2\}|\{\mathbf{x}_1,\mathbf{x}_2,\mathbf{x}_3\}) = {}& q^2(1-q)(1-Q)\,f(\mathbf{a}_1|\mathbf{x}_1)f(\mathbf{a}_2|\mathbf{x}_2)\\
&+ q^2(1-q)(1-Q)\,f(\mathbf{a}_2|\mathbf{x}_1)f(\mathbf{a}_1|\mathbf{x}_2)\\
&+ q^2(1-q)(1-Q)\,f(\mathbf{a}_2|\mathbf{x}_1)\,f(\mathbf{a}_1|\mathbf{x}_3)\\
&+ q^2(1-q)(1-Q)\,f(\mathbf{a}_1|\mathbf{x}_1)\,f(\mathbf{a}_2|\mathbf{x}_3)\\
&+ q^2(1-q)(1-Q)\,f(\mathbf{a}_1|\mathbf{x}_2)\,f(\mathbf{a}_2|\mathbf{x}_3)\\
&+ q^2(1-q)(1-Q)\,f(\mathbf{a}_2|\mathbf{x}_2)\,f(\mathbf{a}_1|\mathbf{x}_3)\\
&+ q(1-q)^2 Q\,f(\mathbf{a}_1|\mathbf{x}_1)\,c(\mathbf{a}_2)\\
&+ q(1-q)^2 Q\,f(\mathbf{a}_1|\mathbf{x}_2)\,c(\mathbf{a}_2)\\
&+ q(1-q)^2 Q\,f(\mathbf{a}_1|\mathbf{x}_3)\,c(\mathbf{a}_2)\\
&+ q(1-q)^2 Q\,f(\mathbf{a}_2|\mathbf{x}_1)\,c(\mathbf{a}_1)\\
&+ q(1-q)^2 Q\,f(\mathbf{a}_2|\mathbf{x}_2)\,c(\mathbf{a}_1)\\
&+ q(1-q)^2 Q\,f(\mathbf{a}_2|\mathbf{x}_3)\,c(\mathbf{a}_1)
\end{aligned}$$

We now form the global likelihood function. We have three global observations $Z_1 = \{(3.0, 4.0)\}$, $Z_2 = \{(4.0, 5.0), (8.0, 9.0)\}$, and $Z_3 = \{(3.5, 2.9), (6.9, 7.0)\}$. The global likelihood function thus has the form

$$L(\{\mathbf{x}_1\}|Z_1, Z_2, Z_3) = f(Z_1|\{\mathbf{x}_1\})\,f(Z_3|\{\mathbf{x}_1\})\,f(Z_3|\{\mathbf{x}_1\})$$
$$L(\{\mathbf{x}_1, \mathbf{x}_2\}|Z_1, Z_2, Z_3) = f(Z_1|\{\mathbf{x}_1, \mathbf{x}_2\})\,f(Z_3|\{\mathbf{x}_1, \mathbf{x}_2\})\,f(Z_3|\{\mathbf{x}_1, \mathbf{x}_2\})$$

$$L(\{\mathbf{x}_1, \mathbf{x}_2, \mathbf{x}_3\} | Z_1, Z_2, Z_3)$$
$$= f(Z_1|\{\mathbf{x}_1, \mathbf{x}_2, \mathbf{x}_3\}) \, f(Z_2|\{\mathbf{x}_1, \mathbf{x}_2, \mathbf{x}_3\}) \, f(Z_3|\{\mathbf{x}_1, \mathbf{x}_2, \mathbf{x}_3\})$$

To solve our estimation problem we must maximize each of these three functions. Let us do so assuming that $q = .92, Q = .80$. We get:

L	X
$.219 \times 10^{-11}$	$\{(3.596, \ 4.079)\}$
2.360×10^{-11}	$\{(3.504, \ 3.968), \ (7.418, \ 7.970)\}$
1.034×10^{-13}	$\{(3.613, \ 4.069), \ (7.459, \ 8.031), \ (3.614, \ 4.069)\}$

These results tell us that the most likely source of the reports is two targets located at $X = \{(3.5, \ 4.0), \ (7.4, \ 8.0)\}$. This solution is about eleven times as likely as the next most likely case, namely the single-target possibility. Note that even the least likely three-target possibility strongly resembles the two-target case, with one target located at $(7.4, 8.0)$ and two targets co-located at $(3.6, 4.1)$.

The situation changes drastically if we now assume a very low probability of false alarm: $q = .92, Q = .05$. Then:

L	X
4.540×10^{-15}	$\{(6.528, \ 2.604)\}$
5.704×10^{-14}	$\{(6.0208, \ 2.983), \ (11.963, \ 5.512)\}$
1.700×10^{-12}	$\{(4.015, \ 3.993), \ (15.001, \ 9.996), \ (9.494, \ 1.001)\}$

The three-target configuration is now overwhelmingly the most likely possibility.

Bibliography

[1] Bar-Shalom, Y., and Fortmann, T.E. (1988) *Tracking and Data Association*, Academic Press, New York.

[2] Chong, C.-Y., Mori, S., and Chang, K.-C. (1990) Distributed multitarget multisensor tracking, in Y. Bar-Shalom (ed.), *Multitarget-Multisensor Tracking: Advanced Applications*, Artech House, Chapter 8.

[3] Drummond, O. E. (1975) Multiple-Object Estimation, Ph.D. Dissertation, University of Southern California, University Microfilms No. 75-26, 954.

[4] Elliott, R. J., Aggoun, L., and Moore, J. B. (1995) *Hidden Markov Models*, Springer-Verlag, New York.

[5] Kastella, K. (1995) Event averaged maximum likelihood estimation and mean-Field theory in multi-target tracking, *IEEE Transactions on Automatic Control*, **40** (6), 1070-1074.

[6] Kastella, K. (1993) A maximum likelihood estimator for report-to-track association, *Proceedings of the 1993 SPIE Conference on OE/Aerospace and Remote Sensing*, SPIE Proceedings **1954**, 386-393.

[7] Kastella, K., and Lutes, C. (1993) Coherent maximum likelihood estimation and mean-field theory in multi-target tracking, *Proceedings of the Sixth Joint Service Data Fusion Symposium*, Vol. I (Part 2), Johns Hopkins Applied Physics Laboratory, Laurel, Maryland, June 14-18 1993, 971-982.

[8] Larson, H. J. and Shubert, B. O. (1979) *Probabilistic Models in Engineering Sciences, Vol. II: Random noise, signals, and dynamic systems*, J. Wiley, New York.

[9] Ho, Y. C., and Lee, R. C. K. (1964) A Bayesian Approach to problems in stochastic estimation and control, *IEEE Transactions on Automatic Control*, **AC-9**, 333-339.

[10] Mahler, R. (1994) Global Maximum Likelihood Estimation: A Numerical Example, Lockheed Martin Technical Report, Feb. 7 1994.

[11] Mahler, R. (1994) Integrated Detection in Clutter: Numerical Example, Lockheed Martin Technical Report, Mar. 3 1994.

[12] Mahler, R. (1994) Integrated Localization and Classification Using Global MLE: Numerical Example, Lockheed Martin Technical Report, Feb. 3 1994.

[13] Mahler, R. (1994) Integrated Random-Set Estimation and Detection: Numerical Example, Lockheed Martin Technical Report, Feb. 8 1994.

[14] Mori, S., Chong, C.-Y., Tse, E., and Wishner, R. P. (1984) Multitarget Multisensor Tracking Problems, Part 1: A General Solution and a Unified View on Bayesian Approaches (Revised Version), Advanced Information and Decision Systems Technical Report TR-1048-01, Mountain View CA, August 1984. My thanks to Dr. Mori for making this report available to me (Dr. Shozo Mori, personal communication, Feb. 28 1995).

[15] Xie, X., and Evans, R. J. (1991) Multiple target tracking and multiple frequency line tracking using hidden markov models, *IEEE Trans. on Signal Processing*, **39** (12), 2659-2676.

Chapter 7

Fusion of Ambiguous Observations

In the previous chapter we showed how the theory of finite random sets can be used to reformulate multisensor, multitarget estimation problems as single-sensor, single-target problems. We then showed how this fact allowed us to directly generalize conventional and very well understood "Statistics 101" techniques to multisensor, multitarget estimation problems. It nevertheless could be argued that the random set formulation is just a fancy "bookkeeping" scheme which introduces complication while adding little value over, for example, more familiar vector-oriented approaches. In this chapter we turn to two problems which are difficult to attack using any approach other than random set theory:

(1) Incorporating *ambiguous evidence* (e.g. natural-language reports, rules) into multisensor, multitarget estimation in a mathematically defensible manner; and

(2) Incorporating various *expert systems methodologies*—e.g., fuzzy logic or rule-based inference—into multisensor, multitarget estimation in a rigorous fashion.

In the previous chapter it was always assumed that the observations collected by individual sensors were "unambiguous" or, more precisely, *Bayesian*. That is, it was always assumed that observations are *mutually exclusive* and *exhaustive* in the sense that a comprehensive list of all of the observations that can possibly occur has been compiled beforehand, and that each such observation is distinct and easily distinguished from any other observation. The totality of all unambiguous observations is called the *observation space* or *measurement space*. In previous chapters we identified it with a generic hybrid space of the form $\mathcal{R}_\odot = \mathbb{R}^n \times U \times U_\odot$ where $\mathbb{R}^n$ is the space of continuous observations, U is a space of discrete observations, and $U_\odot = \{1, ..., s\}$ is the set of sensor identifiers ("sensor tags").

As argued earlier in the book, many observations encountered in real-world applications are *not* unambiguous in the sense just described. It is often not

possible to exhaustively denumerate all possible observations beforehand even
if we knew what they should be. Not all of the observations that can occur may
be known to us ahead of time. The possible observations may be too variable,
or their numbers too vast, to be compiled with a reasonable amount of effort.
The nature of the observations themselves may be so conceptually murky as to
make it difficult to clearly demarcate one observation from another or even to
mathematically model individual observations.

In addition, measurement spaces are normally used in conjunction with
measurement models—that is, some means of expressing our understanding of
which unambiguous observations will be observed and how frequently, given the
presence of a particular target. The most typical measurement model for pre-
cise observations is a collection of density values $f_j(z, u|x, v)$ where each value
expresses the likelihood that the j'th sensor will observe measurement (z, u),
given the parametric representation (x, v) of a particular target in state space
$\mathcal{R} = \mathbb{R}^m \times V$.

In like manner, it is not enough merely to mathematically model ambiguous
observations: One must also find mathematical models of the mechanisms by
which these kinds of observations are generated by targets. That is the purpose
of the current chapter. We will show how ill-characterized evidence, as well as
many expert-systems methodologies often regarded as heuristic, can be system-
atically incorporated into the random set multisource, multitarget estimation
approach developed in earlier chapters.

The chapter is divided into two major sections. In Section 7.1 we introduce
the basic concepts by first restricting ourselves to the *finite-universe* case. There
we introduce our three basic concepts: strong-consistency, weak-consistency,
and data-dependent measurement models for ambiguous observations. We then
extend these concepts to the hybrid continuous-discrete case in Section 7.2.
The result is a *unified* approach to dynamically fusing both ambiguous and
unambiguous observations, based on an extension of standard Bayes-Markov
nonlinear filtering theory.

7.1   Overview of the Approach: The Finite Uni-
verse Case

In order to provide a relatively simple overview of the proposed approach we
begin by first examining the *finite, single-sensor* case: If $n = 0$ in $\mathcal{R} = \mathbb{R}^n \times U$ we
may assume that both state and observation spaces have the general form $\mathcal{R} = U$
where as usual U is a finite set (of, say, target types or of discrete observations).
First, however, briefly recall the discussion of Section 2.3 of Chapter 2, in which
we summarized random set models of various kinds of evidence in finite spaces:

1) *Random set models of imprecise evidence:*

$$m_\Sigma(S) \triangleq p(\Sigma = S)$$

2) *Canonical random set models of fuzzy evidence:*

$$\Sigma_A(f) \triangleq \{u \in U \mid A \le f(u)\}$$

where A is a uniformly distributed random variable on the unit interval $[0,1]$ and f is a fuzzy membership function. Also,

$$\Sigma_A(f) \cap \Sigma_A(g) = \Sigma_A(f \wedge g)$$
$$\Sigma_A(f) \cup \Sigma_A(g) = \Sigma_A(f \vee g)$$

3) *Random set models of contingent evidence:*

$$\Sigma_\Phi(S|X) \triangleq (S \cap X) \cup (X^c \cap \Phi)$$

where Φ is a uniformly distributed random subset of U and $(S|X)$ is a conditional event. Also,

$$\Sigma_\Phi(S|X) \cap \Sigma_\Phi(T|Y) = \Sigma_\Phi((S|X) \wedge (T|Y))$$
$$\Sigma_\Phi(S|X) \cup \Sigma_\Phi(T|Y) = \Sigma_\Phi((S|X) \vee (T|Y))$$

4) *Random set models of partial-probabilistic evidence:*

$$p(\Sigma = S) = \sum_{T \subseteq S} (-1)^{|S-T|} p_*(T)$$

where $p_*(T) = \inf_q q(T)$ and the infimum is taken over all probability measures q on U which are consistent with the constraint on probabilities.

5) *General canonical random set models:* Finally, it is worth considering a general means of constructing canonical random set models, making use of a mathematical construction due to Y. Li [3]. Li's purpose was to create a probabilistic foundation for fuzzy logic, but it serves equally well as a general methodology for constructing examples of random subsets of finite universes. Let $I = [0,1]$ denote the unit interval and define $U^* \triangleq U \times I$. Let $\pi_U : U^* \to U$ be the projection mapping defined by $\pi_U(u,a) \triangleq u$ for all $u \in U$ be measurably and let $a \in I$. The universe U is assumed to have an *a priori* probability measure p, I has the usual Lebesgue measure (which in this case is a probability measure), and the set $U \times I$ is given the product probability measure. Let $W \subseteq U^*$ and for any $u \in U$ define

$$p(W|u) \triangleq \frac{p(W \cap (u \times I))}{p(u)}$$

Now, let A be a random number uniformly distributed on I. Then the "canonical" random subset of U generated by W, denoted by $\Sigma_A(W)$, is defined by: $\Sigma_A(W)(\omega) = \pi_U(W \cap (U \times A(\omega)))$ for all $\omega \in \Omega$. The following relationships are easily established:

$$\Sigma_A(V \cap W) = \Sigma_A(V) \cap \Sigma_A(W), \quad \Sigma_A(V \cup W) = \Sigma_A(V) \cup \Sigma_A(W)$$
$$\Sigma_A(W^c) = \Sigma_A(W)^c, \quad \Sigma_A(\emptyset) = \emptyset, \quad \Sigma_A(U^*) = U$$

Also, it is easy to show that $\mu_{\Sigma_A(W)}(u) = p(u \in \Sigma_A(W)) = p(W|u)$.

This generalizes the canonical random set representation of a fuzzy set. For, let $f : U \rightarrow I$ be a fuzzy membership function on U. Define $W_f \triangleq \{(u, a) \in U^*| a \leq f(u)\}$. Then it is easily verified that $p(W_f|u) = f(u)$, that

$$W_{f \wedge g} = W_f \cap W_g, \qquad W_{f \vee g} = W_f \cup W_g$$

(where '$\wedge$' and '$\vee$' denote the usual min/max conjunction and disjunction operators of the standard Zadeh fuzzy logic) and that $\Sigma_A(W_f) = \{u \in U| A(\omega) \leq f(u)\} = \Sigma_A(f)$.

7.1.1   Evidence as a Constraint on Data

If we are to successfully incorporate ambiguous evidence into multisensor, multitarget estimation we need to describe the process by which such observations are generated by targets. Conventional linear measurement models have the form $z = Cx + v$ where z is an observation, x is the state variable, and v is some noise process. If C is the identity matrix or, more generally, a nonsingular square matrix then (except for random variation) the state can be completely characterized via observations. In general, however, C may be singular and the measurement process results in loss of information about a state. In this case some state variables are not completely accessible.

The observation z assumed in such models is *precise* in the sense that it is clearly distinguishable (mutually exclusive) from alternative observations, even though uncertainty is present in the sense that z is a random quantity. We will call such observations *precise observations* or, more succinctly, *data*.

Imprecise Observations

Other observations are *imprecise* in the sense that they are not necessarily mutually exclusive. Suppose, for example, that a near-sighted observer examines a distant object which is known to belong to the set of objects

$$U \triangleq \{\text{red ball, green ball, red cube, green cube, blue cube}\}$$

Suppose that the actual object is a red ball and that the observer is allowed to choose only from the list U. Since he cannot see the object perfectly sometimes he may choose "red ball," occasionally "red cube," and perhaps (rarely) "green ball." In any case, his observation is precise even though uncertainty is involved. It has the general form of a random variable expressed as a function $z = z(o, n)$ of both o (the observation) and n (some noise process). This function could, for example, be described by a measurement-model probability distribution such as:

$$f(\text{red ball}|\text{red ball}) = .7$$
$$f(\text{red cube}|\text{red ball}) = .2$$
$$f(\text{green ball}|\text{red ball}) = .1$$

More generally, suppose that the observer is permitted to choose from a more general set of observations of the form "red," "green," "blue," "round," "square," "uncertain." These "imprecise" observations correspond to certain subsets of U :

$$RED \quad \leftrightarrow \quad \{\text{red ball, red cube}\}$$
$$GREEN \quad \leftrightarrow \quad \{\text{green ball, green cube}\}$$
$$BLUE \quad \leftrightarrow \quad \{\text{blue cube}\}$$
$$ROUND \quad \leftrightarrow \quad \{\text{red ball, green ball}\}$$
$$SQUARE \quad \leftrightarrow \quad \{\text{red cube, green cube, blue cube}\}$$
$$UNCERTAIN \quad \leftrightarrow \quad U$$

Any given imprecise observation implies only a *constraint* on what the identity—
that is, the state—of the object can be. What is implicitly being stated is that, for example, 'object $\in RED$' or 'object $\in ROUND$'. Thus "object $\in RED$" means that the object can be a red ball or a red cube.

Vague Observations

The observation may be more ambiguous than merely "imprecise": it may be *vague*. For example, suppose that it is believed that the observation most strongly supports the contention that the object is red, that there is less support for the contention that it is, in fact, a red ball, and a small possibility that either of these two hypotheses are completely mistaken. That is:

$$\begin{array}{llll} \text{object} & \in & RED\_BALL & \text{(somewhat likely)} \\ \text{object} & \in & RED & \text{(most likely)} \\ \text{object} & \in & UNCERTAIN & \text{(least likely)} \end{array}$$

Evidence of this sort, which consists of a nested sequence

$$RED\_BALL \subseteq RED \subseteq UNCERTAIN$$

is commonly referred to as *vague* or, more commonly, as *fuzzy evidence* [1]. Vague evidence offers a "softer" constraint on the state of the object than does imprecise evidence, in the sense that the constraint takes the form

$$\text{"object} \in \Theta\text{"}$$

where Θ is a randomly varying subset of the state space U. Stated more precisely, suppose that the distribution of Θ is

$$\begin{array}{lll} p(RED\_BALL|\text{red ball}) & = & .3 \\ p(RED|\text{red ball}) & = & .6 \\ p(UNCERTAIN|\text{red ball}) & = & .1 \end{array}$$

Then it follows that

$$p(\text{red ball} \ \in \ \Theta | \text{red ball}) = .3 + .6 + .1 = 1$$
$$p(\text{red cube} \ \in \ \Theta | \text{red ball}) = .6 + .1 = .7$$
$$p(u \ \in \ \Theta | \text{red ball}) = .1 \qquad (\text{if } u \neq \text{ red ball, red cube})$$

and so on. In fact it is easy to show that

$$\Theta = \Sigma_x(f) = \{u \in U | \ x \leq f(u)\}$$

where the constraint is imposed by the *fuzzy subset* of U defined by the fuzzy membership function

$$f(\text{red ball}) \quad = \quad 1$$
$$f(\text{red cube}) \quad = \quad .7$$
$$f(u) \quad = \quad .1 \qquad (\text{all } u \neq \text{ red ball, red cube})$$

Ambiguous Observations in General

A "soft" constraint on states will consist of a range of hypotheses of the form

$$\mathbf{x} \in T_1 , ..., \mathbf{x} \in T_e$$

where T_i are subsets of state space and where the hypothesis "$\mathbf{x} \in T_i$" is believed to be true with degree of belief m_i. In this case ambiguous evidence can be represented as a random subset Θ of state space such that $p(\Theta = T_i) = m_i$ for all $i = 1, ..., e$.

Evidence as a Constraint on Data

The random subset Θ just described is a random subset of the *state space*, which means that ambiguous observations (imprecise, vague, etc.) constrain the state of the target directly. In general, however, many state variables may be partially hidden from the direct view of the observer (as, for example, with the singular observation model $\mathbf{z} = C\mathbf{x} + \mathbf{v}$. Accordingly, more general situations will be covered if the random subset Θ is assumed to be a random subset of *measurement space*. In this case it is not the *state* of the object that is directly constrained by the evidence, but rather the *data* generated by the object. The object state is still constrained by evidence but now it is constrained *indirectly* via the evidential constraint on data.

For example, evidence could take the form of a statement "$\mathbf{z} \in R$" where R is a subset of observation space. The state $\mathbf{x}$ is still constrained by the evidence but only indirectly in the sense that $C\mathbf{x} + \mathbf{v} \in R$. If C were nonsingular then this constraint would take the form $\mathbf{x} \in \Theta$ where

$$\Theta(\omega) \triangleq \{C^{-1}(\mathbf{r} - \mathbf{v}(\omega)) | \ \mathbf{r} \in R\}$$

for all $\omega \in \Omega$ defines a random subset of state space.

Finally, let us assume that we are dealing with more than one object. Then collected data will consist of finite subsets Z of observation space. In this case evidential constraints on data by a random subset Θ (of observation space) would take the form

$$\Theta \supseteq Z$$

That is, all observations generated by the sensor are constrained by the data.

This discussion leads to the following definition:

Definition 1 *Precise observations, also called data, are elements of the observation space* $\mathcal{R}_\odot = \mathbb{R}^n \times U \times U_\odot$. *Ambiguous observations, also called evidence, are random subsets of the observation space* $\mathcal{R}_\odot$.

7.1.2   Measurement Models for Data and Evidence

In the finite-universe case Bayesian measurement models have the form

$$
\begin{aligned}
p_{z_1,\ldots,z_m|x}(a_1,\ldots,a_m|b) &\triangleq p(z_1 = a_1, \ldots, z_m = a_m | x = b) \\
&\triangleq \frac{p(z_1 = a_1, \ldots, z_m = a_m, \, x = b)}{p(x = b)}
\end{aligned}
$$

where $z_1, \ldots, z_m$ are random variables on the measurement space such that the marginal distributions

$$p_{z_i|x}(a|b) \triangleq p(z_i = a | x = b)$$

are identical, where x is a random variable on the state space, where $a, a_1, \ldots, a_m$ are elements of measurement space, and where b is an element of state space. The distribution $p_{z_1,\ldots,z_m|x}(a_1,\ldots,a_m|b)$ expresses the likelihood of observing the sequence of observations $a_1,\ldots,a_m$ given that a target with state b is present. When multiple data is collected, such measurement models are often simplified by assuming conditional independence of the data, e.g.:

$$p_{z_1,\ldots,z_m|x}(a_1,\ldots,a_m|b) = p_{z|x}(a_1|b) \cdots p_{z|x}(a_m|b)$$

Using a Bayesian analysis, R.W. Johnson has concluded that situations in which this independence assumption "is justified (at least as an approximation) are quite common..." [2, p. 200].

When we pass to the case of observations which are imprecise, vague, etc., however, the concept of a measurement model becomes considerably more complex. In fact there are *many plausible such models*, each reflecting different assumptions about the degree to which evidence can be understood to constrain data. In what follows we will describe the following measurement models: (1) strong consistency model; (2) data-dependent model; and (3) weak consistency models.

In all cases, the following will be assumed:

(1) $Z_1, ..., Z_m$ will be subsets of a (finite) observation space which represent data;

(2) $\Theta_1, ..., \Theta_{m'}$ will be random subsets of observation space which represent ambiguous evidence;

(3) $\Sigma_1, ..., \Sigma_{m+m'}$ will be i.i.d. random subsets of observation space whose common distribution is

$$m(Z|X) = p(\Sigma_i = Z|\Gamma = X), \qquad m(X) = p(\Gamma = X)$$

If $Z_1, ..., Z_m, T_1, ..., T_m \subseteq U$ we will abbreviate

$$m(Z_1, ..., Z_e; T_1, ..., T_{m'}|X)$$
$$\triangleq \ p(\Sigma_1 = Z_1, ..., \Sigma_e = Z_e, \Theta_1 = T_1, ..., \Theta_{m'} = T_{m'}|\Gamma = X)$$

so that, in particular, $m(Z; T|X) = p(\Sigma = Z, \Theta = T|\Gamma = X)$. We will also adopt the following additional abbreviations: $Z^{(m)}$ for the sequence $Z_1, ..., Z_m$ and $\Theta^{(m')}$ for the sequence $\Theta_1, ..., \Theta_{m'}$.

7.1.3   The Strong-Consistency Measurement Model

The strong consistency model is characterized by the following definition, which requires that data must be completely consistent with evidence, but only in an *overall, probabilistic* sense:

Definition 2 *The strong-consistency measurement model for data $Z_1, ..., Z_m$ and evidence $\Theta_1, ..., \Theta_{m'}$ is*

$$m(Z_1, ..., Z_m, \Theta_1, ..., \Theta_{m'}|X)$$
$$= \ p(\Sigma_1 = Z_1, ..., \Sigma_m = Z_m, \Theta_1 \supseteq \Sigma_{m+1}, ..., \Theta_{m'} \supseteq \Sigma_{m+m'}|\Gamma = X)$$

Data and evidence are conditionally independent for this model if

$$m(Z_1, ..., Z_m, Y_1, ..., Y_{m'}; T_1, ..., T_{m'}|X)$$
$$= \ m(Z_1|X) \cdots m(Z_m|X) \, m(Y_1, T_1|X) \cdots m(Y_{m'}, T_{m'}|X)$$

for all subsets $Z_1, ..., Z_m, Y_1, ..., Y_{m'}, T_1, .., T_{m'}$ of measurement space and all subsets X of state space.

Given this definition of conditional independence it immediately follows that

$$m(Z_1, ..., Z_m, \Theta_1, ..., \Theta_{m'}|X) = m(Z_1|X) \cdots m(Z_m|X) \, \beta(\Theta_1|X) \cdots \beta(\Theta_{m'}|X)$$

for all subsets $Z_1, ..., Z_m$ of measurement space and all subsets X of state space, and where

$$\beta(\Theta|X) \triangleq p(\Sigma \subseteq \Theta|\Gamma = X)$$

is a *generalized belief.* For,

$$m(Z_1, ..., Z_m, \Theta_1, ..., \Theta_{m'}|X)$$
$$= \sum_{T_1 \supseteq Y_1, ..., T_{m'} \supseteq Y_{m'}} m(Z_1|X) \cdots m(Z_m|X) m(Y_1, T_1|X) \cdots m(Y_{m'}, T_{m'}|X)$$
$$= m(Z_1|X) \cdots m(Z_m|X) \, \beta(\Theta_1|X) \cdots \beta(\Theta_{m'}|X)$$

as claimed. Thus if we are given one data-set Z and one piece of evidence Θ, the strong-consistency model has the form

$$m(Z, \Theta|X) = p(\Sigma_1 = Z, \ \Theta \supseteq \Sigma_2 | \Gamma = X)$$

and the conditional independence assumption has the form

$$m(Z, \Theta|X) = m(Z|X) \, \beta(\Theta|X)$$

Posterior Distributions for the Strong Consistency Model

One can obviously form *posterior distributions* conditioned on both data and evidence as defined by the following proportionality:

$$m(X|Z_1, ..., Z_m, \Theta_1, ..., \Theta_{m'}) \propto m(Z_1, ..., Z_m, \Theta_1, ..., \Theta_{m'}|X) \, m(X)$$

or, using abbreviated notation,

$$m(X|Z^{(m)}, \Theta^{(m')}) \triangleq \frac{m(Z^{(m)}, \Theta^{(m')}|X) \, m(X)}{m(Z^{(m)}, \Theta^{(m')})}$$

where

$$m(Z^{(m)}, \Theta^{(m')}) = \sum_X m(Z^{(m)}, \Theta^{(m')}|X) \, m(X)$$

and where $m(X) \triangleq p(\Gamma = X)$ is a prior distribution.

Recursive Update Formulas for the Strong consistency Model

One can ask, How is the posterior distribution

$$m(X|Z^{(m)}, \Theta^{(m')})$$

related to either of the posterior distributions $m(X|Z^{(m-1)}, \Theta^{(m')})$ or $m(X|Z^{(m)}, \Theta^{(m'-1)})$ if we make suitable independence assumptions? Or, stated in other words, *how do we update the posterior* $m(X|Z^{(m)}, \Theta^{(m')})$ when are given either a new global observation Z_m or a new piece of ambiguous evidence $\Theta_{m'}$? Some simple algebra shows that

$$m(X|Z^{(m)}, \Theta^{(m')}) \ \propto \ m(Z_m|X) \, m(X|Z^{(m-1)}, \Theta^{(m')})$$
$$m(X|Z^{(m)}, \Theta^{(m')}) \ \propto \ \beta(\Theta_{m'}|X) \, m(X|Z^{(m)}, \Theta^{(m'-1)})$$

The first proportionality is a conventional recursive update formula for unambiguous observations. The second proportionality is the corresponding recursive update formula for ambiguous observations.

7.1.4   The Data-Dependent Measurement Model

The data-dependent model is more complex and, in many ways, more interesting than either of the other two measurement models that we shall consider. It is defined as follows, and specifies that *each* data set must be directly constrained by *all* evidence:

Definition 3 *The data-dependent measurement model for evidence* $\Theta_1, ..., \Theta_{m'}$ *and data* $Z_1, ..., Z_m$ *is a mass assignment function of the form*

$$m(Z_1, ..., Z_m, \Theta_1, ..., \Theta_{m'}|X)$$
$$\triangleq p(\Sigma_1 = Z_1, ..., \Sigma_m = Z_m, \Theta_1 \supseteq Z_1, ..., \Theta_i \supseteq Z_j, ..., \Theta_{m'} \supseteq Z_m|\Gamma = X)$$

Data and evidence are conditionally independent for this model if

$$m(Z_1, ..., Z_m; T_1, ..., T_{m'}|X)$$
$$\triangleq p(\Sigma_1 = Z_1, ...\Sigma_m = Z_m, \Theta_1 = T_1, ..., \Theta_{m'} = T_{m'}|\Gamma = X)$$
$$= m(Z_1|X) \cdots m(Z|X) \cdots m_{\Theta_1}(T_1|X) \cdots m_{\Theta_{m'}}(T_{m'}|X)$$

where $m_\Theta(T|X) \triangleq p(\Theta = T|\Gamma = X)$.

In particular, note that

$$m(Z_1, ..., Z_m, \Theta_1, ..., \Theta_{m'}|X) = m(Z_1, ..., Z_m, \Theta_1 \cap ... \cap \Theta_{m'}|X)$$

Suppose that data and evidence are conditionally independent. Then it can be shown that

$$m(Z_1, ..., Z_m, \Theta_1, ..., \Theta_{m'}|X) = m(Z_1|X) \cdots m(Z_m|X)\, \delta_{\Theta_1 \cap ... \cap \Theta_{m'}}(Z_1 \cup ... \cup Z_m|X)$$

where

$$\delta_\Theta(Z|X) \triangleq p(\Theta \supseteq Z|\Gamma = X)$$

is a *commonality measure*. To see this, note that since

$$m(Z_1, ..., Z_m, \Theta_1, ..., \Theta_{m'}|X) = m(Z_1, ..., Z_m, \Theta_1 \cap ... \cap \Theta_{m'}|X)$$

then we may assume $m' = 1$ without loss of generality. In this case

$$m(Z_1, ..., Z_m, \Theta|X) \quad = \sum_{T \supseteq Z_1 \cup ... \cup Z_m} m(Z_1, ..., Z_m; T|X)$$
$$= m(Z_1|X) \cdots m(Z_m|X)\delta_\Theta(Z_1 \cup ... \cup Z_m)$$

as claimed.

In particular, if we are given one data-set Z and one piece of evidence Θ the data-dependent model has the form:

$$m(Z, \Theta|X) = p(\Sigma = Z, \Theta \supseteq Z|\Gamma = X)$$

Likewise, if Θ_1, Θ_2 are random subsets representing two pieces of ambiguous evidence then

$$m(Z, \Theta_1, \Theta_2 | X) = m(Z, \Theta_1 \cap \Theta_2 | X)$$

If data and evidence are conditionally independent then $m(Z, \Theta | X) = m(Z|X)$ $\delta_\Theta(Z|X)$ and in this case Bayes' rule give us:

$$m(Z|X, \Theta) = \frac{m(Z, \Theta | X)}{\beta(\Theta | X)} = \frac{m(Z|X) \, \delta_\Theta(Z|X)}{\beta(\Theta | X)}$$

where $\beta(\Theta | X) \triangleq p(\Sigma \subseteq \Theta | \Gamma = X)$ and $m(Z|X, \Theta) \triangleq p(\Sigma = Z | \Gamma = X, \Theta \supseteq \Sigma)$, and where the second equation is true if data and evidence are conditionally independent. (If one ignores Γ in the preceding then in [4] this last quantity $m_\Sigma(Z|\Theta)$ was called the "concordance," or *generalized pignistic probability assignment*, of the random conditional event $(\Theta | \Sigma)$.)

Posterior Distributions for the Data-Dependent Model

We can form posterior distributions conditioned on both data and evidence as follows:

$$m(X | Z_1, ..., Z_m, \Theta_1, ..., \Theta_{m'}) \triangleq \frac{m(Z_1, ..., Z_m, \Theta_1, ..., \Theta_{m'} | X) \, m(X)}{m(Z_1, ..., Z_m, \Theta_1, ..., \Theta_{m'})}$$

where as usual $m(X) \triangleq p(\Gamma = X)$ and where

$$m(Z_1, ..., Z_m, \Theta_1, ..., \Theta_{m'}) \triangleq \sum_{Y \subseteq U} m(Z_1, ..., Z_m, \Theta_1, ..., \Theta_{m'} | Y) \, m(Y)$$

In the event that data and evidence are conditionally independent this becomes

$$m(X | Z^{(m)}, \Theta^{(m')}) \propto m(X | Z^{(m)}) \, \delta_{\Theta_1 \cap ... \cap \Theta_{m'}}(Z_1 \cup ... \cup Z_m | X)$$

Suppose in particular that we are given a data-set Z and a piece of evidence Θ. Then these formulas have the form

$$m(X | Z, \Theta) \propto m(Z, \Theta | X) \, m(X)$$

and, in the event of conditional independence of data and evidence,

$$m(X | Z, \Theta) \propto m(X | Z) \, \delta_\Theta(Z | X)$$

where, note, $m(X|Z)$ is the conventional posterior distribution conditioned on the data-set Z.

Recursive Update Formulas for the Data-Dependent Model

Once again we ask, How is the posterior $m(X | Z^{(m)}, \Theta^{(m')})$ related to the posteriors $m(X | Z^{(m-1)}, \Theta^{(m')})$ and , $m(X | Z^{(m)}, \Theta^{(m'-1)})$, provided that data and evidence are independent? The answer is:

$$m(X | Z^{(m)}, \Theta^{(m')}) \quad \propto \quad m_\Sigma(Z_m | X) \, m(X | Z^{(m-1)}, \Theta^{(m')})$$

$$m(X | Z^{(m)}, \Theta^{(m')}) \quad \propto \quad \delta_{\Theta_{m'}}(Z_1 \cup ... \cup Z_m | X) \, m(X | Z^{(m)}, \Theta^{(m'-1)})$$

To prove these proportionalities we begin with the fact that

$$m(X|Z^{(m)}, \Theta^{(m')}) \propto m(X|Z^{(m)}) \, \delta_{\Theta_1 \cap ... \cap \Theta_{m'}}(Z_1 \cup ... \cup Z_m|X)$$

when data and evidence are conditionally independent.. However,

$$m(X|Z^{(m)}) \propto m(Z_m|X) \, m(X|Z^{(m-1)})$$

and so

$$m(X|Z^{(m)}, \Theta^{(m')}) \propto m_\Sigma(Z_m|X) \, m(X|Z^{(m-1)}, \Theta^{(m')})$$

On the other hand, note that because of conditional independence we can write

$$
\begin{aligned}
\delta_{\Theta_1 \cap ... \cap \Theta_{m'}}(Z|X) &= p(\Theta_1 \cap ... \cap \Theta_{m'} \supseteq Z|X) = p(\Theta_1 \supseteq Z, ..., \Theta_{m'} \supseteq Z|X) \\
&= \sum_{Y_1 \supseteq Z} \cdots \sum_{Y_{m'} \supseteq Z} p(\Theta_1 = Y_1, ..., \Theta_{m'+1} = Y_{m'}|X) \\
&= \delta_{\Theta_1}(Z|X) \cdots \delta_{\Theta_{m'}}(Z|X) \\
&= \delta_{\Theta_1 \cap ... \cap \Theta_{m'-1}}(Z|X) \, \delta_{\Theta_{m'}}(Z|X)
\end{aligned}
$$

Our claim then follows.

7.1.5   Weak-Consistency Measurement Models

The weakest possible way that evidence $\Theta_1, ..., \Theta_{m'}$ can constrain data $Z_1, ..., Z_m$ is stipulated by the measurement model

$$m(Z_1, ..., Z_m, \Theta_1, ..., \Theta_{m'}|X)$$
$$\triangleq p(\Sigma_1 = Z_1, ..., \Sigma_m = Z_m, \Theta_1 \cap \Sigma_{m+1} \neq \emptyset, ..., \Theta_{m'} \cap \Sigma_{m+m'} \neq \emptyset|\Gamma = X)$$

However, the constraint on data offered by this model is *so* weak that in its place we offer the following definition, which requires that data must merely *not contradict* evidence in an overall, probabilistic sense:

Definition 4 *Let $\Sigma_1, ..., \Sigma_m$ be identically distributed random subsets of observation space and let $\Sigma_1^{(N)}, ..., \Sigma_{m'}^{(N)}$ be identically distributed random subsets of the N'th Cartesian power of observation space, where N is the number of elements in state space. Then a general weak consistency measurement model for data and evidence is*

$$m(Z_1, ..., Z_m, \Theta_1, ..., \Theta_{m'}|X)$$
$$\triangleq p(\Sigma_1 = Z_1, ..., \Sigma_m = Z_m, \Theta_1^N \cap \Sigma_{m+1}^{(N)} \neq \emptyset, ..., \Theta_{m'}^N \cap \Sigma_{m+m'}^{(N)} \neq \emptyset|\Gamma = X)$$

where $\Theta^N \triangleq \Theta \times ... \times \Theta$ is the N'th Cartesian power of Θ. Data and evidence are conditionally independent for this model if

$$m(Z_1, ..., Z_m, \Theta_1, ..., \Theta_{m'}|X) = m(Z_1|X) \cdots m(Z_m|X) \, \rho(\Theta_1|X) \cdots \rho(\Theta_{m'}|X)$$

for all subsets $Z_1, ..., Z_m$ of measurement space and all subsets X of state space, where

$$\rho(\Theta|X) \triangleq p(\Sigma^{(N)} \cap \Theta^N \neq \emptyset|\Gamma = X)$$

is a "generalized plausibility."

Posterior Distributions for Weak Measurement Models

One can form posterior distributions conditioned on both data and evidence as follows:

$$m(X|Z_1, ..., Z_m, \Theta_1, ..., \Theta_{m'}) \triangleq \frac{m(Z_1, ..., Z_m, \Theta_1, ..., \Theta_{m'}|X) \, m(X)}{m(Z_1, ..., Z_m, \Theta_1, ..., \Theta_{m'})}$$

where

$$m(Z_1, ..., Z_m, \Theta_1, ..., \Theta_{m'}) \triangleq \sum_{Y \subseteq U} m(Z_1, ..., Z_m, \Theta_1, ..., \Theta_{m'}|Y) \, m(Y)$$

Recursive Update Formulas for Weak Measurement Models

The recursive update formulas for weak measurement models are:

$$m(X|Z_1, ..., Z_m, \Theta_1, ..., \Theta_{m'}) \; \propto \; m(Z_m|X) \, m(Y|Z_1, ..., Z_{m-1}, \Theta_1, ..., \Theta_{m'})$$
$$m(X|Z_1, ..., Z_m, \Theta_1, ..., \Theta_{m'}) \; \propto \; \rho(\Theta_{m'}|X) \, m(Y|Z_1, ..., Z_m, \Theta_1, ..., \Theta_{m'-1})$$

assuming, of course, that data and evidence are conditionally independent.

7.1.6 Signature-Based Measurement Models for Evidence

Depending on the measurement model and the corresponding notion of conditional independence, the process of updating a posterior distribution using ambiguous evidence represented by a random subset Θ requires the computation of one of the quantities

$$\beta_{\Sigma|\Gamma}(\Theta|X) \; \triangleq \; p(\Sigma \subseteq \Theta|\Gamma = X)$$
$$\rho(\Theta|X) \; \triangleq \; p(\Sigma^{(N)} \cap \Theta^N \neq \emptyset|\Gamma = X)$$
$$\delta_\Theta(Z|X) \; \triangleq \; p(\Theta \supseteq Z|\Gamma = X)$$

The purpose of this section is to discuss, for the finite-universe case, construction of plausible models for ambiguous evidence. Our approach is based on the idea of *matching observed and possibly ambiguous target signatures to stored model signatures, which are also possibly ambiguous.*

The basic idea is as follows. Let V be our state space and assume that for each $v \in V$ there is an associated *nonempty* random subset Σ_v of measurement space. This random subset is to be interpreted as a *model signature* for the target type v—that is, as the typical *ambiguous observation* that one would observe if a target of type v were present in the scenario. Next, suppose that the random subset Θ of measurement space models an ambiguous observation. Then one means of determining which targets are consistent with the measurement would be to *match* the observation Θ to the model signatures Σ_v for all $v \in V$, using some comparison process

$$\Theta \Longleftrightarrow \Sigma_v$$

and then determining which targets match the observation. The two most obvious—and extreme—ways of determining degree of match are also polar opposites of each other:

$$\text{lack of contradiction} \quad : \quad \Theta \cap \Sigma_v \neq \emptyset$$

$$\text{complete consistency} \quad : \quad \Theta \supseteq \Sigma_v$$

In the first case a target type v is consistent with the evidence O only if its corresponding model signature is not in flat contradiction of Θ. In the second case v is consistent with Θ only if its corresponding model signature is in flat contradiction of the *negation* of Θ (i.e., $\Theta^c \cap \Sigma_v = \emptyset$)—that is, if it is entirely consistent with Θ.

Given this, define

$$\beta(\Theta|\{v_1, ..., v_a\}) \triangleq p(\Theta \supseteq \Sigma_{v_1} \cup ... \cup \Sigma_{v_a})$$

$$\rho(\Theta|\{v_1, ..., v_a\}) \triangleq p(\Theta^N \cap (\Sigma_{v_1} \times ... \times \Sigma_{v_a} \times U^{N-a}) \neq \emptyset)$$

$$= p(\Theta \cap \Sigma_{v_1} \neq \emptyset, ..., \Theta \cap \Sigma_{v_a} \neq \emptyset)$$

$$\delta_\Theta(Z|\{v_1, ..., v_a\}) \triangleq p(\Theta \supseteq Z | \Theta \supseteq \Sigma_{v_1} \cup ... \cup \Sigma_{v_a})$$

for all distinct $v_1, ..., v_a$ in state space, and each a less than or equal to the number of elements in state space, and where U denotes observation space. The first (strong-consistency) model stipulates that the targets which are consistent with evidence Θ are those which are strongly consistent with it. The second (weak-consistency) model insists only that targets which are consistent with evidence are those which do not flatly contradict that evidence.

The following are specific examples of measurement models that result from a signature-matching scheme such as the one proposed.

Example: Matching Vague Observations With Vague Models

Let V be our state space and assume that to each $v \in V$ we have associated a "fuzzy" model signature defined by a fuzzy membership function $f_v : U \to [0, 1]$ where U denotes observation space. Let A be a uniformly distributed random number on the unit interval $[0, 1]$ and define

$$\Sigma_v \triangleq \Sigma_A(f_v) = \{u \in U | A \leq f_v(u)\}$$

Also, assume that we have collected a piece of fuzzy evidence in the form of a fuzzy membership function $g : U \to [0, 1]$ on observation space. Define

$$\Theta \triangleq \Sigma_A(g)$$

and fuzzy membership functions $g^o, g^\bullet : V \to [0, 1]$ on V by

$$g^o(v) \triangleq \max_{u \in U} (g \wedge f_v)(u)$$

$$g^\bullet(v) \triangleq \min_{u \in U} ((1 - f_v) \vee g)(u)$$

where as usual '$\wedge$' and '$\vee$' denote the fuzzy "min/max" *AND* and *OR* operators, respectively. We will show that

$$\rho(\Theta|X) = \min_{v \in X} g^o(v) = p(\Theta^o \supseteq X)$$

$$\beta(\Theta|X) = \min_{v \in X} g^\bullet(v) = p(\Theta^\bullet \supseteq X)$$

$$\delta_\Theta(Z|X) = \min\left\{1, \frac{\min_{u \in Z} g(u)}{\min_{v \in X} g^\bullet(v)}\right\}$$

where

$$\Theta^o \triangleq \Sigma_A(g^o), \qquad \Theta^\bullet \triangleq \Sigma_A(g^\bullet)$$

are random subsets of the state space. In other words, evidential constraint of data using either the weak consistency model ρ or the strong consistency model β is the same thing as *strongly constraining* the state directly using a random subset of state space.

We begin by recalling that, given the assumptions of our signature model,

$$\beta(\Theta|X) \triangleq p(\Theta \supseteq \Sigma_v, \text{ all } v \in X)$$

$$\rho(\Theta|X) \triangleq p(\Theta \cap \Sigma_v \neq \emptyset, \text{ all } v \in X)$$

$$\delta_\Theta(Z|X) \triangleq \frac{p(\Theta \supseteq Z, \Theta \supseteq \Sigma_v, \text{ all } v \in X)}{p(\Theta \supseteq \Sigma_v, \text{ all } v \in X)}$$

First note that

$$\Theta(\omega) \cap \Sigma_v(\omega) = \Sigma_{A(\omega)}(g \wedge f_v) = \{u \in U| A(\omega) \leq (g \wedge f_v)(u)\}$$

and thus that $\Theta(\omega) \cap \Sigma_v(\omega) \neq \emptyset$ if and only if $A(\omega) \leq g^o(v)$. Accordingly,

$$\rho(\Theta|X) = p(\Theta \cap \Sigma_v \neq \emptyset, \text{ all } v \in X) = p(A \leq g^o(v), \text{ all } v \in X)$$
$$= p(A \leq \min_{v \in X} g^o(v)) = \min_{v \in X} g^o(v)$$

Also note that

$$p(\Sigma_A(g^o) \supseteq X) = p(A \leq g^o(v), \text{ all } v \in X) = \min_{v \in X} g^o(v)$$

and so

$$\rho(\Theta|X) = \min_{v \in X} g^o(v)$$

as claimed.

Next note that

$$\{u \in U| A(\omega) \leq g(u)\} = \Sigma_{A(\omega)}(g) \supseteq \Sigma_{A(\omega)}(f_v) = \{u \in U| A(\omega) \leq f_v(u)\}$$

is the same thing as saying that $A(\omega) \leq g(u)$ whenever $A(\omega) \leq f_v(u)$; or, in other words, that $A(\omega) \leq g(u)$ or $A(\omega) \leq 1 - f_v(u)$ for all $u \in U$; or that

$A(\omega) \leq \min_{u \in U}((1 - f_v) \vee g)(u) = g^{\bullet}(v)$. Thus

$$
\begin{aligned}
\beta(\Theta | X) &= p(\Theta \supseteq \Sigma_v, \text{ all } v \in X) \\
&= p(A \leq g^{\bullet}(v), \text{ all } v \in X) = p(A \leq \min_{v \in X} g^{\bullet}(v)) \\
&= \min_{v \in X} g^{\bullet}(v)
\end{aligned}
$$

The formula for $\delta_{\Theta}(Z | X)$ follows easily from the results just ostablished. By definition,

$$
\begin{aligned}
\delta_{\Theta}(Z | X) &\triangleq \frac{p(\Theta \supseteq Z, \Theta \supseteq \Sigma_v, \text{ all } v \in X)}{p(\Theta \supseteq \Sigma_v, \text{ all } v \in X)} \\
&= \frac{p(A \leq g(u), \text{ all } u \in Z, A \leq g^{\bullet}(v), \text{ all } v \in X)}{\min_{v \in X} g^{\bullet}(v)} \\
&= \frac{p(A \leq \min\{\min_{u \in Z} g(u), \min_{v \in X} g^{\bullet}(v)\})}{\min_{v \in X} g^{\bullet}(v)} \\
&= \min\left\{1, \frac{\min_{u \in Z} g(u)}{\min_{v \in X} g^{\bullet}(v)}\right\}
\end{aligned}
$$

as claimed.

Example: Matching Bayesian Observations With Bayesian Models

In this case we let $\Theta = \{a\}$ where a is a fixed (deterministic) element of observation space U. Also, we let z be a random variable on U and x a random variable on state space V. Define the random subset Σ_v of U by

$$
\Sigma_v = \{z|_{x=v}\}
$$

where $z|_{x=v}$ is a conditional random variable in measurement space such that

$$
p(z|_{x=v} = b) = p(z = b | x = v) \triangleq p_{z|x}(b|v)
$$

(see the concluding remark in this section). Note that $\Theta(\omega) \cap \Sigma_v \neq \emptyset$ if and only if $(z|_{x=v})(\omega) = a$. Therefore,

$$
\rho(\Theta | \{v\}) = p(z|_{x=v} = a) = p_{z|x}(a|v)
$$

Thus in this case the measurement model for ambiguous evidence reduces to a conventional Bayesian measurement model. The same result would be true if we used $\beta(\Theta | \{v\})$ instead.

Remark 1 *Let $(\Omega, \sigma(\Omega), p)$ be our probability space and let $x : \Omega \to U, k : \Omega \to V$ be random variables on the finite universes U, V. Recall from Section 2.3.3 of Chapter 2 that a conditional event $(S | X)$ for $S, X \subseteq \Omega$ is a subset of the product probability space $(\hat{\Omega}, \sigma(\hat{\Omega}), \hat{p})$ with $\hat{\Omega} = \Omega \times \ldots \times \Omega \times \ldots$ defined by*

$$
\begin{aligned}
(S | X) &\triangleq ((S \cap X) \times X^c \times \Omega \times \ldots) \vee (X^c \times (S \cap X) \times \Omega \times \ldots) \vee \ldots \\
&\vee (X^c \times \ldots \times X^c \times (S \cap X) \times \Omega \ldots) \vee \ldots
\end{aligned}
$$

and where $\sigma(\hat{\Omega})$ is generated by the cylinders of $\sigma(\Omega)$. In particular, given $u \in U, a \in V$ define the sets $S_u \in \sigma(\hat{\Omega})$ to be the conditional events

$$S_u \triangleq (x^{-1}(u)|k^{-1}(a))$$

Then the S_u are mutually disjoint for all $u \in U$ (since the $x^{-1}(u)$ are mutually disjoint) and $\cup_{u \in U} S_u = (\Omega|k^{-1}(a)) = (k^{-1}(a)|k^{-1}(a)) \triangleq \hat{\Omega}|_a$. Since $\hat{p}((S|S)) = p(S|S) = 1$ for all $S \subseteq \Omega$ it follows that the restriction $(\hat{\Omega}|_a, \sigma(\hat{\Omega}|_a), \hat{p})$ is also a probability space. Define $x|_{k=a} : \hat{\Omega}|_a \rightarrow U$ by

$$x|_{k=a}(\hat{\omega}) = u \qquad if \qquad \hat{\omega} \in S_u$$

Since $(x|_{k=a})^{-1}(u) = S_u \in \sigma(\hat{\Omega}|_a)$ for all $u \in U$ and since U is finite, it follows that $x|_{k=a}$ is a random variable on U. Then, notice that

$$
\begin{aligned}
\hat{p}(x|_{k=a} &= u) = \hat{p}(S_u) = \hat{p}((x^{-1}(u)|k^{-1}(a))) = p(x^{-1}(u)|k^{-1}(a)) \\
&= p(x = u|k = a) = p_{x|k}(u|a)
\end{aligned}
$$

as claimed.

7.2 Unified Data Fusion

Having illustrated the basic concept of a *measurement model* for ambiguous observations, we now turn to the problem of generalizing the results of the previous section to *hybrid continuous-discrete spaces* . This, in turn, will allow us to incorporate ambiguous evidence into the multisource, multitarget estimation approach constructed in earlier chapters.

7.2.1 Modeling Ambiguous Evidence Using DRACS

In order to extend the results of Section 7.1 to hybrid (i.e., mixed continuous-discrete) spaces we will be forced to make certain special assumptions about ambiguous evidence:

(1) Ambiguous evidence is representable by *discrete random closed subsets*; and

(2) Evidence and data are *statistically independent*.

We begin with the following generalized version of Definition 1:

Definition 5 *(a) A discrete random closed subset (DRACS) of the hybrid observation space* $\mathcal{R}_{\odot} = \mathbb{R}^n \times U \times U_{\odot}$ *is a random closed subset of* $\mathcal{R}_{\odot}$ *which takes only finitely many closed subsets* $C_1, ..., C_r$ *of* $\mathcal{R}_{\odot}$ *as values. (b) Precise observations (data) are elements of the hybrid observation space* $\mathcal{R}_{\odot}$. *Ambiguous observations (evidence) are DRACS of* $\mathcal{R}_{\odot}$.

In other words, a random closed subset Θ of $\mathcal{R}_{\odot}$ is a DRACS if there are closed subsets $C_1, ..., C_r$ of $\mathcal{R}_{\odot}$ and nonnegative numbers $q_1, ..., q_r$ such that $\sum_{i=1}^r q_i = 1$ and $p(\Theta = C_i) = q_i$ for all $i = 1, ..., r$. We will abbreviate $m_\Theta(T_i) \triangleq p(\Theta = T_i)$ for all $i = 1, ..., r$ and $m_\Theta(T) \triangleq 0$ if $T \neq T_i$ for any $i = 1, ..., r$.

7.2.2   Conditioning on Ambiguous Evidence: Single Sensor, Single Target Case

We begin by considering the situation when a single sensor interrogates a single target and, in addition, there are no missed detections or false alarms. In this case the typical observation collected by the sensor is a random singleton set of $\mathcal{R} = \mathbb{R}^n \times U$ of the form $\Sigma = \{\psi\}$ where $\psi = (\mathbf{Z}, z)$, $\mathbf{Z}$ is a state-dependent random vector in $\mathbb{R}^n$ and z is a random variable in U. Let $p_\psi(S|\zeta) = p(\psi \in S)$ be the parametrized probability measure of ψ with parameter $\zeta = (\mathbf{x}, v)$ and assume that it has a density f_ψ. Also, $\psi_1, ..., \psi_{m+m'}$ will denote i.i.d. random variables whose common density function is f_ψ.

First suppose that we are given a fixed constraint of the form $T = \mathbb{R}^n \times R$ for some $R \subseteq U$. Then T places no constraint at all on the continuous part $\mathbf{z}$ of an observation $\xi = (\mathbf{z}, u)$ but does force a constraint of the form $u \in R$ on the discrete part of the observation. Thus if R corresponds to a evidential proposition such as "$BLUE$" (which is to say: Imprecise evidence has become available which indicates that the unknown target is $BLUE$), then the imprecise evidence T states that any datum ξ which indicates that the target is *not BLUE* must surely be incorrect. Somewhat more generally, suppose that $T = B \times R$ where B is some closed subset of $\mathbb{R}^n$. This T now places a geokinematic constraint on $\mathbf{z}$ as well as a target I.D. constraint on u. That is, the subset $T \subseteq \mathcal{R}$ is equivalent to an evidential hypothesis of the form: '*The target is observable only in region B and must be $BLUE$*'. The evidence is unequivocal and thus is, in effect, assumed to originate with an infallible observer—or, at least, a supremely confident or dogmatic one. If the geokinematic measurement model has the form $\mathbf{z} = \mathbf{x} + \mathbf{w}$ where $\mathbf{w}$ is noise then this is essentially the same as saying that, according to the unequivocal evidence T, the target must be in region B and must be blue.

In general, evidence will be *equivocal* in the sense that it will consist of a range of constraining hypotheses $T_0, T_1, ..., T_a$ where, associated with each T_i, is some belief q_i that the evidential constraint T_i is valid. In fact, one of the hypotheses may be $T_0 = U$—which expresses the fact that there is a certain possibility that our evidential observation offers no constraint whatsoever and thus may be entirely in error. Let us express this equivocal evidence as a discrete closed random subset—a DRACS—Θ such that $p(\Theta = T_i) = q_i$ for all $i = 1, ..., a$.

We now define single-sensor, single-target continuous-discrete analogs of the various measurement models introduced in section 7.1.

Strong-Consistency Measurement Model

Definition 6 *The strong-consistency measurement model for data and evidence* $\Theta_1, ..., \Theta_{m'}$ *is the probability measure*

$$p(S_1, ..., S_m, \Theta_1, ..., \Theta_{m'}|\zeta)$$
$$= p(\psi_1 \in S_1, ..., \psi_m \in Z_m, \Theta_1 \ni \psi_{m+1}, ..., \Theta_{m'} \ni \psi_{m+m'})$$

Data and evidence are conditionally independent for this model if

$$p(S_1, ..., S_m, R_1, ..., R_{m'}; T_1, ..., T_{m'} | \zeta)$$
$$= \ p_\psi(S_1 | \zeta) \cdots p_\psi(S_m | \zeta) \ p(R_1; T_1 | \zeta) \cdots p(R_{m'}; T_{m'} | \zeta)$$

for all measurable subsets $S_1, ..., S_m, R_1, ..., R_{m'}, T_1, .., T_{m'}$ of measurement space and all parameters ζ in state space, where

$$p(R; T | \zeta) \triangleq p(R \ni \psi, \Theta = T)$$

It immediately follows that

$$p(S_1, ..., S_m, \Theta_1, ..., \Theta_{m'} | \zeta) = p_\psi(S_1 | \zeta) \cdots p_\psi(S_m | \zeta) \ \beta(\Theta_1 | \zeta) \cdots \beta(\Theta_{m'} | \zeta)$$

for all measurable $S_1, ..., S_m$ in observation space and where

$$\beta(\Theta | \zeta) \triangleq p(\psi \in \Theta) = \sum_T p(\psi \in T, \ \Theta = T) = \sum_T m_\Theta(T | \zeta) \ p_\psi(T | \Theta = T, \zeta)$$

is a generalized belief. If ψ, Θ are independent then, of course, we can write $\beta(\Theta | \zeta) = \sum_T m_\Theta(T | \zeta) \ p_\psi(T | \zeta)$.

Now, let $f(\xi_1, ..., \xi_m, \Theta_1, ..., \Theta_{m'} | \zeta)$ be the density corresponding to the measure $p(S_1, ..., S_m, \Theta_1, ..., \Theta_{m'} | \zeta)$. Given conditional independence, it immediately follows that

$$f(\xi_1, ..., \xi_m, \Theta_1, ..., \Theta_{m'} | \zeta) = f_\psi(\xi_1 | \zeta) \cdots f_\psi(\xi_m | \zeta) \ \beta(\Theta_1 | \zeta) \cdots \beta(\Theta_{m'} | \zeta)$$

for all elements $\xi_1, ..., \xi_m$ of measurement space and all parameters ζ of state space. Thus the strong-consistency measurement model reduces to the conventional sensor measurement model $f_\psi(\xi | \zeta)$ for unambiguous observations, and a measurement model $\beta(\Theta | \zeta)$ for ambiguous observations.

Weak-Consistency Measurement Models

Definition 7 *Let $\Sigma_1^{(\nu)}, ..., \Sigma_{m'}^{(\nu)}$ be identically distributed random subsets of the N'th Cartesian power of observation space, where ν is some integer. Then a general weak consistency measurement model for data and evidence is*

$$p(S_1, ..., S_m, \Theta_1, ..., \Theta_{m'} | \zeta)$$
$$\triangleq \ p(\psi_1 \in S_1, ..., \psi_m \in S_m, \Theta_1^\nu \cap \Sigma_{m+1}^{(\nu)} \neq \emptyset, ..., \Theta_{m'}^\nu \cap \Sigma_{m+m'}^{(\nu)} \neq \emptyset)$$

where $\Theta^\nu \triangleq \Theta \times ... \times \Theta$ is the ν'th Cartesian power of Θ. Data and evidence are conditionally independent for this model if

$$p(S_1, ..., S_m, \Theta_1, ..., \Theta_{m'} | \zeta) = p_\psi(S_1 | \zeta) \cdots p_\psi(S_m | \zeta) \ \rho(\Theta_1 | \zeta) \cdots \rho(\Theta_{m'} | \zeta)$$

for all measurable subsets $S_1, ..., S_m$ of measurement space and all parameters ζ of state space, where

$$\rho(\Theta | \zeta) \ \triangleq \ p(\Sigma^{(\nu)} \cap \Theta^\nu \neq \emptyset) = \sum_T p(\Sigma^{(\nu)} \cap T^\nu \neq \emptyset, \Theta = T)$$
$$= \ \sum_T m_\Theta(T | \zeta) \ \rho(T | \Theta = T, \zeta)$$

is a "generalized plausibility."

Let $f(\xi_1, ..., \xi_m, \Theta_1, ..., \Theta_{m'}|\zeta)$ be the density of the measure

$$p(S_1, ..., S_m, \Theta_1, ..., \Theta_{m'}|\zeta)$$

Then it follows that, given conditional independence,

$$f(\xi_1, ..., \xi_m, \Theta_1, ..., \Theta_{m'}|\zeta) = f_\psi(\xi_1|\zeta) \cdots f_\psi(\xi_m|\zeta)\, \rho(\Theta_1|\zeta) \cdots \rho(\Theta_{m'}|\zeta)$$

Once again, this measurement model decomposes into a conventional sensor measurement model $f_\psi(\xi|\zeta)$ for unambiguous observations and a measurement model $\rho(\Theta|\zeta)$ for ambiguous observations.

Data-Dependent Measurement Model

Definition 8 *The data-dependent measurement model for evidence $\Theta_1, ..., \Theta_{m'}$ and data is a probability measure of the form*

$$p(S_1, ..., S_m, \Theta_1, ..., \Theta_{m'}|\zeta)$$
$$\triangleq p(\psi_1 \in S_1, ..., \psi_m \in S_m, \Theta_1 \ni \psi_1, ..., \Theta_i \ni \psi_j, ..., \Theta_{m'} \ni \psi_m)$$

Data and evidence are conditionally independent for this model if

$$p(S_1, ..., S_m; T_1, ..., T_{m'}|\zeta)$$
$$\triangleq p(\psi_1 \in S_1, ...\psi_m \in S_m, \Theta_1 = T_1, ..., \Theta_{m'} = T_{m'})$$
$$= p_\psi(S_1|\zeta) \cdots p_\psi(S_m|\zeta)\, m_{\Theta_1}(T_1|\zeta) \cdots m_{\Theta_{m'}}(T_{m'}|\zeta)$$

where $m_\Theta(T|\zeta) \triangleq p(\Theta = T)$.

Hereafter, we will assume conditional independence. In particular, note that

$$p(S_1, ..., S_m, \Theta_1, ..., \Theta_{m'}|\zeta)$$
$$= p(\psi_1 \in S_1, ..., \psi_m \in S_m, \Theta_1 \cap ... \cap \Theta_{m'} \supseteq \{\psi_1, ..., \psi_m\})$$
$$= p(S_1, ..., S_m, \Theta_1 \cap ... \cap \Theta_{m'}|\zeta)$$
$$= p(S_1 \cap \Theta_1 \cap ... \cap \Theta_{m'}, ..., S_m \cap \Theta_1 \cap ... \cap \Theta_{m'}|\zeta)$$

Suppose that data and evidence are conditionally independent and define the *commonality measure*

$$\delta_\Theta(S|\zeta) \triangleq p(\Theta \supseteq S)$$

associated with Θ. Since

$$p(S_1, ..., S_m, \Theta_1, ..., \Theta_{m'}|\zeta) = p(S_1, ..., S_m, \Theta_1 \cap ... \cap \Theta_{m'}|\zeta)$$

it is clear that we can assume that $m' = 1$ without loss of generality. Since conditional independence is assumed,

$$p(S_1, ..., S_m, \Theta|\zeta) = p(S_1 \cap \Theta, ..., S_m \cap \Theta|\zeta)$$
$$= \sum_T p(\psi_1 \in S_1 \cap T, ..., \psi_m \in S_m \cap T, \Theta = T)$$
$$= \sum_T p_\psi(S_1 \cap T) \cdots p_\psi(S_m \cap T)\, m_\Theta(T)$$

The density function corresponding to $p(S_1, ..., S_m, \Theta|\zeta)$ is therefore

$$
\begin{aligned}
f(\xi_1, ..., \xi_m, \Theta|\zeta) &= \sum_T f_\psi(\xi_1|\zeta) 1_T(\xi_1) \cdots f_\psi(\xi_m|\zeta) 1_T(\xi_m) \, m_\Theta(T|\zeta) \\
&= f_\psi(\xi_1|\zeta) \cdots f_\psi(\xi_m|\zeta) \sum_T 1_T(\xi_1) \cdots 1_T(\xi_m) \, m_\Theta(T|\zeta) \\
&= f_\psi(\xi_1|\zeta) \cdots f_\psi(\xi_m|\zeta) \, \delta_\Theta(\{\xi_1, ..., \xi_m\}|\zeta)
\end{aligned}
$$

Thus, in general, we will get the following relationship:

$$
f(\xi_1, ..., \xi_m, \Theta_1, ..., \Theta_{m'}|\zeta) = f_\psi(\xi_1|\zeta) \cdots f_\psi(\xi_m|\zeta) \, \delta_{\Theta_1 \cap ... \cap \Theta_{m'}}(\{\xi_1, ..., \xi_m\}|\zeta)
$$

Non-Unity Probability of Detection

Let us now generalize slightly and suppose that probability of detection is $p_D(\zeta)$ and not necessarily equal to one. The belief measure of the typical observation is just

$$
\beta_\Sigma(S|\zeta) = 1 - p_D(\zeta) + p_D(\zeta) \, q_\Sigma(S|\zeta)
$$

where $q_\Sigma(S|\zeta) = p(\Sigma \subseteq S | \Sigma \neq \emptyset)$ and where we assume that $q_\Sigma(S|\zeta)$ has a density $f_\Sigma(\xi|\zeta)$. The global density is, as usual, given by the equations

$$
f_\Sigma(\emptyset) = 1 - p_D(\zeta), \qquad f_\Sigma(\{\xi\}) = p_D(\zeta) \, f(\xi|\zeta)
$$

The preceding analysis can be extended as follows. Let $\Sigma_1, ..., \Sigma_{m+m'}$ be i.i.d. random subsets whose common belief measure is β_Σ. Then:

The *strong-consistency measurement model* for data and evidence $\Theta_1, ..., \Theta_{m'}$ is the belief measure

$$
\begin{aligned}
&\beta(S_1, ..., S_m, \Theta_1, ..., \Theta_{m'}|\zeta) \\
\triangleq\; &p(\Sigma_1 \subseteq S_1, ..., \Sigma_m \subseteq Z_m, \Theta_1 \supseteq \Sigma_{m+1}, ..., \Theta_{m'} \supseteq \Sigma_{m+m'})
\end{aligned}
$$

and conditional independence for the model is defined as

$$
\begin{aligned}
&\beta(S_1, ..., S_m, R_1, ..., R_{m'}; T_1, ..., T_{m'}|\zeta) \\
=\; &\beta_\Sigma(S_1|\zeta) \cdots \beta_\Sigma(S_m|\zeta) \, \beta(R_1; T_1|\zeta) \cdots \beta(R_{m'}; T_{m'}|\zeta)
\end{aligned}
$$

where $\beta(R; T|\zeta) = \beta(R \supseteq \Sigma, \Theta = T)$. It immediately follows from independence that

$$
\beta(S_1, ..., S_m, \Theta_1, ..., \Theta_{m'}|\zeta) = \beta_\Sigma(S_1|\zeta) \cdots \beta_\Sigma(S_m|\zeta) \, \beta(\Theta_1|\zeta) \cdots \beta(\Theta_{m'}|\zeta)
$$

for all measurable $S_1, ..., S_m$ in observation space and where

$$
\beta(\Theta|\zeta) \triangleq p(\Sigma \subseteq \Theta) = \sum_T p(\Sigma \subseteq T, \, \Theta = T) = \sum_T m_\Theta(T|\zeta) \, \beta_\Sigma(T|\Theta = T, \zeta)
$$

and, if Σ, Θ are independent, $\beta(\Theta|\zeta) = \sum_T m_\Theta(T|\zeta) \, \beta_\Sigma(T|\zeta)$.

Let $f(Z_1, ..., Z_m, \Theta_1, ..., \Theta_{m'}|\zeta)$ be the global density corresponding to the belief measure $\beta(S_1, ..., S_m, \Theta_1, ..., \Theta_{m'}|\zeta)$. Then just as before,

$$f(Z_1, ..., Z_m, \Theta_1, ..., \Theta_{m'}|\zeta) = f_\Sigma(Z_1|\zeta) \cdots f_\Sigma(Z_m|\zeta) \, \beta(\Theta_1|\zeta) \cdots \beta(\Theta_{m'}|\zeta)$$

for all subsets $Z_1, ..., Z_m$ of measurement space with $|Z_i| \leq 1$ and all parameters ζ of state space.

The case for *weak-consistency models* is much the same. In this case we end up with

$$f(Z_1, ..., Z_m, \Theta_1, ..., \Theta_{m'}|\zeta) = f_\Sigma(Z_1|\zeta) \cdots f_\Sigma(Z_m|\zeta) \, p(\Theta_1|\zeta) \cdots p(\Theta_{m'}|\zeta)$$

for all $Z_1, ..., Z_m$ with $|Z_i| \leq 1$ and where as before

$$p(\Theta|\zeta) \triangleq p(\Sigma^{(\nu)} \cap \Theta^\nu \neq \emptyset) = \sum_T p(\Sigma^{(\nu)} \cap T^\nu \neq \emptyset, \Theta = T)$$

Finally, the *data-dependent model* we get

$$f(Z_1, ..., Z_m, \Theta_1, ..., \Theta_{m'}|\zeta) = f_\Sigma(Z_1|\zeta) \cdots f_\Sigma(Z_m|\zeta) \, \delta_{\Theta_1 \cap ... \cap \Theta_{m'}}(Z_1 \cup ... \cup Z_m|\zeta)$$

where $\delta_\Theta(Z|\zeta) \triangleq p(\Theta \supseteq Z) = \sum_{T \supseteq Z} m_\Theta(T|\zeta)$.

7.2.3　Conditioning on Ambiguous Evidence: Single Sensor, Multitarget Case

Let us now assume that our single sensor interrogates an unknown number of targets, that probability of detection is not necessarily unity and the false alarm rate is zero.

When multiple targets are present one must be careful about how one expresses constraints. For example, quite different evidential constraints are imposed by the two pieces of unequivocal evidence

$$T_1 = (B \times u) \cup (\mathbb{R}^n \times \{u\}^c), \qquad T_2 = B \times u$$

for some $u \in U$. The constraint $\Sigma \subseteq T_1$ expresses the belief that *any* target of type u must be in region B but that there is *no constraint of any kind* on targets of any other type. The constraint $\Sigma \subseteq T_2$, on the other hand, expresses the belief not only that any target of type u must be in region B but that, additionally, there are *no targets of any other type*.

In either case, if it is subsequently suspected that there is an additional target of type u in region B' with $B \cap B' = \emptyset$ then either $T_1' = (B' \times u) \cup (U \times \{u\}^c)$ or $T_2' = B' \times u$ will contradict both T_1 and T_2 even though two distinct targets of the same type are suspected and so there should be no contradiction. This problem is only apparent since it results from the fact that ambiguous observations, like precise observations, should be functions of the measurement model parameters $\zeta_1 = (\mathbf{x}_1, v_1), \zeta_2 = (\mathbf{x}_2, v_2)$ of the two-target system.

Suppose, then, that there are t targets with global parameter $X = \{\zeta_1, ..., \zeta_t\}$ and that $\Sigma = \Sigma_1' \cup ... \cup \Sigma_t'$ where Σ_i is the (empty or singleton) observation set corresponding to the i'th target. Let $f_\Sigma(Z|X)$ be the global density corresponding to the finite random subset Σ. Assume that $\Sigma_1', ..., \Sigma_t'$ are independent and let $\Sigma_1, ..., \Sigma_{m+m'}$ be i.i.d. finite random subsets whose common global density is $f_\Sigma(Z|X)$. We define the measurement models for this situation each in turn. The various measurement models have the same general form as specified by the formulas for the single-sensor, single-target but non-unity p_D case. That is:

Multitarget Strong Consistency Measurement Model

The measurement model is

$$\begin{aligned} & \beta(S_1, ..., S_m, \Theta_1, ..., \Theta_{m'} | X) \\ \triangleq\ & p(\Sigma_1 \subseteq S_1, ..., \Sigma_m \subseteq Z_m, \Theta_1 \supseteq \Sigma_{m+1}, ..., \Theta_{m'} \supseteq \Sigma_{m+m'}) \end{aligned}$$

and conditional independence is defined as

$$\begin{aligned} & \beta(S_1, ..., S_m, R_1, ..., R_{m'}; T_1, ..., T_{m'} | X) \\ =\ & \beta_\Sigma(S_1|X) \cdots \beta_\Sigma(S_m|X)\, \beta(R_1; T_1|X) \cdots \beta(R_{m'}; T_{m'}|X) \end{aligned}$$

where

$$\beta(R; T|X) \triangleq \beta(R \supseteq \Sigma, \Theta = T)$$

It immediately follows from independence that

$$\beta(S_1, ..., S_m, \Theta_1, ..., \Theta_{m'} | X) = \beta_\Sigma(S_1|X) \cdots \beta_\Sigma(S_m|X)\, \beta(\Theta_1|X) \cdots \beta(\Theta_{m'}|X)$$

for all measurable $S_1, ..., S_m$ in observation space and where

$$\beta(\Theta|X) \triangleq p(\Sigma \subseteq \Theta) = \sum_T p(\Sigma \subseteq T,\ \Theta = T) = \sum_T m_\Theta(T|X)\, \beta_\Sigma(T|\Theta = T, X)$$

and, if Σ, Θ are independent,

$$\beta(\Theta|X) = \sum_T m_\Theta(T|X)\, \beta_\Sigma(T|X)$$

If $\Sigma_1', ..., \Sigma_m'$ are independent and $X = \{\zeta_1, ..., \zeta_t\}$ with $|X| = t$ then, as usual,

$$\beta_\Sigma(S|X) = \beta_{\Sigma'}(S|\zeta_1) \cdots \beta_{\Sigma'}(S|\zeta_t)$$

and if $\Sigma_1', ..., \Sigma_m', \Theta$ are independent then

$$\beta(\Theta|X) = \sum_T m_\Theta(T|X)\, \beta_\Sigma(T|X) = \sum_T m_\Theta(T|X)\, \beta_{\Sigma'}(T|\zeta_1) \cdots \beta_{\Sigma'}(S|\zeta_t)$$

The corresponding global density is, therefore,

$$f(Z_1, ..., Z_m, \Theta_1, ..., \Theta_{m'} | X) = f_\Sigma(Z_1|X) \cdots f_\Sigma(Z_m|X)\, \beta(\Theta_1|X) \cdots \beta(\Theta_{m'}|X)$$

for all finite subsets $Z_1, ..., Z_m$ of measurement space and all parameter sets X of state space.

As in the finite-universe case discussion in section 7.1, we can define posterior global densities conditioned on both data and evidence. Abbreviate $Z_1, ..., Z_m$ as $Z^{(m)}$ and abbreviate $\Theta_1, ..., \Theta_{m'}$ as $\Theta^{\{m'\}}$. Then the global posterior density is:

$$f(X|Z^{(m)}, \Theta^{(m')}) \triangleq \frac{f(Z^{(m)}, \Theta^{(m')}|X)\, f(X)}{f(Z^{(m)}, \Theta^{(m')})}$$

where

$$f(Z^{(m)}, \Theta^{(m')}) \triangleq \int f(Z^{(m)}, \Theta^{(m')}|X)\, f(X) \delta X$$

and $f(X)$ is a prior distribution on parameter-sets X. Likewise, we can derive recursive update formulas:

$$f(X|Z^{(m)}, \Theta^{(m')}) \quad \propto \quad f(Z_m|X)\, f(X|Z^{(m-1)}, \Theta^{(m')})$$
$$f(X|Z^{(m)}, \Theta^{(m')}) \quad \propto \quad \beta(\Theta_{m'}|X)\, f(X|Z^{(m)}, \Theta^{(m'-1)})$$

Example 1 *For the sake of illustration we will assume that vector observations* $\mathbf{z}$ *and vector parameters* $\mathbf{x}$ *lie in the same Euclidean space* $\mathbb{R}^n$. *Suppose that evidence consists of (i) a closed region* B *of* $\mathbb{R}^n$, *expressing the fact that a target is located somewhere in* B; *and (ii) a fuzzy subset* h *of* U. *Suppose further that* h *is to be matched to a set* $f_v : v \in V$ *of fuzzy signatures (see section 7.1.6). Let* A *be a uniformly distributed random variable on* $[0, 1]$. *Then the random closed subset* $\Theta \triangleq B \times \Sigma_A(h)$ *of* $\mathcal{R}$ *models the information contained in the evidence. Also, the random subsets* $\Sigma_v \triangleq \{\mathbf{x}\} \times \Sigma_A(f_v)$ *model the concept of a target of type* v *located at* $\mathbf{x}$. *In this case we know from section 7.1.6 that*

$$\beta(\Theta|\{(\mathbf{x}_1, v_1), ..., (\mathbf{x}_t, v_t)\}) = p(\Theta \supseteq (\mathbf{x}_1 \times \Sigma_{v_1}), ..., \Theta \supseteq (\mathbf{x}_t \times \Sigma_{v_t}))$$
$$= 1_B(\mathbf{x}_1) \cdots 1_B(\mathbf{x}_t)\, \min\{h^\bullet(v_1), ..., h^\bullet(v_t)\}$$

Thus the only allowable target configurations are those in which all suspected targets are contained in the region B. *If in fact only one target is in region* B *then the only nonvanishing belief must have the form* $\beta(\Theta|\{(\mathbf{x}_1, v_1)\})$.

Multitarget Weak Consistency Models

Definition 9 *As usual, let* $\Sigma_1, ..., \Sigma_m$ *be i.i.d. random subsets of observation space and* $\Sigma_1^{(\nu)}, ..., \Sigma_{m'}^{(\nu)}$ *i.i.d. random subsets of the N'th Cartesian power of observation space, where* ν *is some integer. Then a general weak consistency measurement model for data and evidence is a joint belief measure of the form*

$$\beta(S_1, ..., S_m, \Theta_1, ..., \Theta_{m'}|X)$$
$$\triangleq p(\Sigma_1 \subseteq S_1, ..., \Sigma_m \subseteq S_m, \Theta_1^\nu \cap \Sigma_{m+1}^{(\nu)} \neq \emptyset, ..., \Theta_{m'}^\nu \cap \Sigma_{m+m'}^{(\nu)} \neq \emptyset)$$

where $\Theta^{\nu} \triangleq \Theta \times ... \times \Theta$ *is the ν'th Cartesian power of Θ. Data and evidence are conditionally independent for this model if*

$$\beta(S_1, ..., S_m, \Theta_1, ..., \Theta_{m'}|\zeta) = \beta_{\Sigma}(S_1|X) \cdots \beta_{\Sigma}(S_m|X) \, \rho(\Theta_1|X) \cdots \rho(\Theta_{m'}|X)$$

for all measurable subsets $S_1, ..., S_m$ of measurement space and all parameters ζ of state space, where

$$\rho(\Theta|X) \quad \triangleq \quad p(\Sigma^{(\nu)} \cap \Theta^{\nu} \neq \emptyset) = \sum_{T} p(\Sigma^{(\nu)} \cap T^{\nu} \neq \emptyset, \Theta = T)$$

$$= \quad \sum_{T} m_{\Theta}(T|X) \, \rho(T|\Theta = T, X)$$

Let $f(\xi_1, ..., \xi_m, \Theta_1, ..., \Theta_{m'}|X)$ be the density of the measurement-model belief measure $\beta(S_1, ..., S_m, \Theta_1, ..., \Theta_{m'}|X)$. Then given conditional independence,

$$f(Z_1, ..., Z_m, \Theta_1, ..., \Theta_{m'}|X) = f_{\Sigma}(Z_1|X) \cdots f_{\Sigma}(Z_m|X) \, \rho(\Theta_1|X) \cdots \rho(\Theta_{m'}|X)$$

for all finite subsets $Z_1, ..., Z_m$ of measurement space and all set parameters X. We can define posterior global densities in the obvious way and, from them, derive the following update formulas:

$$f(X|Z^{(m)}, \Theta^{(m')}) \quad \propto \quad f(Z_m|X) \, f(X|Z^{(m-1)}, \Theta^{(m')})$$
$$f(X|Z^{(m)}, \Theta^{(m')}) \quad \propto \quad \rho(\Theta_{m'}|X) \, f(X|Z^{(m)}, \Theta^{(m'-1)})$$

Example 2 *Make the same general assumptions as in Example 1. From section 7.1.6 we know that*

$$\rho(\Theta|\{(\mathbf{x}_1, v_1), ..., (\mathbf{x}_t, v_t)\}) \quad = \quad p(\Theta \cap (\mathbf{x}_1 \times \Sigma_{v_1}) \neq \emptyset, ..., \Theta \cap (\mathbf{x}_t \times \Sigma_{v_t}) \neq \emptyset)$$
$$= \quad \mathbf{1}_B(\mathbf{x}_1) \cdots \mathbf{1}_B(\mathbf{x}_t) \, \min\{h^{\circ}(v_1), ..., h^{\circ}(v_t)\}$$

Example 3 *Under the same general assumptions as in Example 2, suppose in addition that the constraint on the location of the target is "soft" in the sense that it is defined by a finite-level fuzzy subset g of $\mathbb{R}^n$ rather than by a hard constraint B. (The statement that g is "finite-level" means that $g(u)$ always takes its values in the ordered list $0 = l_0 < l_1 < ... < l_s < l_{s+1} \triangleq 1$ for some $l_0, ..., l_{s+1}$.) Then the random subset which models the information contained in the evidence is $\Theta \triangleq \Sigma_{A'}(g) \times \Sigma_A(h)$ where A', A are uniformly distributed random numbers on $[0,1]$. Note that Θ is a DRACS (see Definition 5) since g is finite-level. In this case,*

$$\rho(\Theta|\{(\mathbf{x}_1, v_1), ..., (\mathbf{x}_t, v_t)\})$$
$$= \quad p(\Theta \cap (\mathbf{x}_1 \times \Sigma_{v_1}) \neq \emptyset, ..., \Theta \cap (\mathbf{x}_t \times \Sigma_{v_t}) \neq \emptyset)$$
$$= \quad p(A' \leq \min\{g(\mathbf{x}_1), ..., g(\mathbf{x}_t)\}, A \leq \min\{h^{\circ}(v_1), ..., h^{\circ}(v_t)\})$$
$$= \quad p_{A', A}(\min\{g(\mathbf{x}_1), ..., g(\mathbf{x}_t)\}, \min\{h^{\circ}(v_1), ..., h^{\circ}(v_t)\})$$

where $p_{A',A}(a, b) \triangleq p(A' \leq a, A \leq b)$. *If* A', A *are independent then we will get*

$$\rho(\Theta|\{(\mathbf{x}_1, v_1), ..., (\mathbf{x}_t, v_t)\}) = \min\{g(\mathbf{x}_1), ..., g(\mathbf{x}_t)\} \ \min\{h^\circ(v_1), ..., h^\circ(v_t)\}$$

If A', A *are completely correlated (i.e.,* $A' = A$) *then*

$$\rho(\Theta|\{(\mathbf{x}_1, v_1), ..., (\mathbf{x}_t, v_t)\}) = \min\{g(\mathbf{x}_1), ..., g(\mathbf{x}_t), \ h^\circ(v_1), ..., h^\circ(v_t)\}$$

and so on.

Multitarget Data-Dependent Model

Let $\Sigma_1, ..., \Sigma_m$ be i.i.d. finite random subsets of measurement space and $\Theta_1, ...,$ $\Theta_{m'}$ discrete random closed subsets of measurement space. Then the *data-dependent measurement model* for evidence $\Theta_1, ..., \Theta_{m'}$ and data is a joint belief measure of the form

$$\beta(S_1, ..., S_m, \Theta_1, ..., \Theta_{m'}|X)$$
$$\triangleq \ p(\Sigma_1 \subseteq S_1, ..., \Sigma_m \subseteq S_m, \Theta_1 \supseteq \Sigma_1, ..., \Theta_i \supseteq \Sigma_j, ..., \Theta_{m'} \supseteq \Sigma_m)$$

Data and evidence are *conditionally independent* for this model if

$$\beta(S_1, ..., S_m; T_1, ..., T_{m'}|X)$$
$$\triangleq \ p(\Sigma_1 \subseteq S_1, ..., \Sigma_m \subseteq S_m, \Theta_1 = T_1, ..., \Theta_{m'} = T_{m'})$$
$$= \ \beta_\Sigma(S_1|X) \cdots \beta_\Sigma(S_m|X) \ m_{\Theta_1}(T_1|X) \cdots m_{\Theta_{m'}}(T_{m'}|X)$$

where $m_\Theta(T|X) \triangleq p(\Theta = T)$.

Given conditional independence we get

$$f(Z_1, ..., Z_m, \Theta_1, ..., \Theta_{m'}|X)$$
$$= f_\Sigma(Z_1|X) \cdots f_\Sigma(Z_m|X) \ \delta_{\Theta_1 \cap ... \cap \Theta_{m'}}(Z_1 \cup ... \cup Z_m|X)$$

where $\delta_\Theta(Z|X) \triangleq p(\Theta \supseteq Z) = \sum_{T \supseteq Z} m_\Theta(T|X)$.

Once again we can define posterior global densities and derive recursive update formulas for them:

$$f(X|Z^{(m)}, \Theta^{(m')}) \ \propto \ f(Z_m|X) \ f(X|Z^{(m-1)}, \Theta^{(m')})$$
$$f(X|Z^{(m)}, \Theta^{(m')}) \ \propto \ \delta_{\Theta_{m'}}(Z_1 \cup ... \cup Z_m|X) \ f(X|Z^{(m)}, \Theta^{(m'-1)})$$

Example 4 *Make the same general assumptions as in Example 3. Let*

$$Z = \{(\mathbf{z}_1, u_1), ..., (\mathbf{z}_k, u_k)\}, \qquad X = \{(\mathbf{x}_1, v_1), ..., (\mathbf{x}_t, v_t)\}$$

$$Z_0 = \{u_1, ..., u_k\}, Z_1 = \{\mathbf{z}_1, ..., \mathbf{z}_k\}, X_0 = \{v_1, ..., v_t\}, X_1 = \{\mathbf{x}_1, ..., \mathbf{x}_t\}$$

and assume that $\min_{\mathbf{x} \in X_1} g(\mathbf{x}) \neq 0$ *and* $\min_{v \in X_0} h^\bullet(v) \neq 0$. *(Since by definition* $h^\bullet(v) \triangleq \min_{u \in U} ((1 - f_v) \vee h)(u)$ *it follows that the second requirement will be satisfied if* $h > 0$ *identically or if* $f_v > 0$ *identically for all* $v \in V$. *To assume* *that $h(u) > 0$ for all u is the same as saying that U is a level subset of h and hence that the evidence h includes some allowance for the possibility that h is in completely erroneous.) Given this, from section 7.1.6 we know that*

$$\delta_\Theta(Z|X) = \frac{p(\Theta \supseteq Z, \, \Theta \supseteq \mathbf{x}_1 \times \Sigma_{v_1}, ..., \Theta \supseteq \mathbf{x}_t \times \Sigma_{v_t})}{p(\Theta \supseteq \mathbf{x}_1 \times \Sigma_{v_1}, ..., \Theta \supseteq \mathbf{x}_t \times \Sigma_{v_t})}$$

$$= \frac{p\left(\begin{array}{l} A' \le \min_{z \in Z_1} g(z), \ A \le \min_{u \in Z_0} h(u), \\ A' \le \min_{x \in X_1} g(x), \ A \le \min_{v \in X_0} h^\bullet(v) \end{array}\right)}{p(A' \le \min_{x \in X_1} g(x), \ A \le \min_{v \in X_0} h^\bullet(v))}$$

Assume that A', A are independent. Then this becomes

$$\delta_\Theta(Z|X) = \left(\min\left\{1, \frac{\min_{z \in Z_1} g(z)}{\min_{x \in X_1} g(x)}\right\}\right)\left(\min\left\{1, \frac{\min_{u \in Z_0} h(u)}{\min_{v \in X_0} h^\bullet(v)}\right\}\right)$$

If on the other hand A', A are perfectly correlated ($A' = A$) then we get

$$\delta_\Theta(Z|X) = \frac{\min\{\min_{z \in Z_1} g(z), \ \min_{x \in X_1} g(x), \ \min_{u \in Z_0} h(u), \ \min_{v \in X_0} h^\bullet(v)\}}{\min\{\min_{x \in X_1} g(x), \ \min_{v \in X_0} h^\bullet(v)\}}$$

This concludes the example.

7.2.4   Multisensor, Multitarget Case

The multisensor case will not be considered further here since its development is exactly parallel to the discussion in section 7.2.3.

7.2.5   Bayesian Characterization of Rules of Evidential Combination

Let us assume that the data-dependent measurement model is being used. Then it is clear that, just as was the case in the finite-universe case,

$$f(Z_1, ..., Z_m, \Theta_1, ..., \Theta_{m'}|X) = f(Z_1, ..., Z_m, \Theta_1 \cap ... \cap \Theta_{m'}|X)$$

The random-set intersection operator '$\cap$' may therefore be interpreted as a means of fusing multiple pieces of ambiguous evidence in such a way that posteriors conditioned on the fused evidence are identical to posteriors conditioned on the individual evidence and computed using Bayes' rule alone. Thus if we set

$$\Theta_1 = \Sigma_A(f), \qquad \Theta_2 = \Sigma_A(g)$$

where f, g are two fuzzy membership functions on $\mathcal{R}$ and A is a uniformly distributed random number on $[0, 1]$ then

$$f(X|Z_1, ... Z_m, f, g) = f(X|Z_1, ..., Z_m, f \wedge g)$$

where

$$f(X|Z_1, ..., Z_m, f, g) \triangleq f(X|Z_1, ..., Z_m, \Sigma_A(f), \Sigma_A(g))$$
$$f(X|Z_1, ..., Z_m, f \wedge g) \triangleq f(X|Z_1, ..., Z_m, \Sigma_A(f \wedge g))$$

Hence the fuzzy *AND* is a "Bayesian" rule of evidential combination in the sense that it is a means of fusing fuzzy evidence in such a way that posteriors conditioned on the fused evidence are identical to posteriors conditioned on the fuzzy evidence individually and computed using Bayes' rule alone. Similar observations apply to *any* rule of evidential combination '$*$' which bears a homomorphic relationship to the random set intersection operator. (By this we mean that there is some means of associating a random set $\Sigma(E)$ to each piece of evidence E in such a manner that $\Sigma(E * E') = \Sigma(E) \cap \Sigma(E')$.)

This leads us to the following "Bayesian characterization" of a rule of evidential combination (given that a data-dependent measurement model is assumed):

> *A rule of evidential combination is a means of fusing multiple pieces of ambiguous evidence in such a way that posteriors conditioned on the fused evidence are identical to posteriors conditioned on the evidence individually and computed using Bayes' rule alone.*

7.2.6   Nonlinear Filtering With Data and Evidence

In Section 6.4.2 of Chapter 6 we showed how *precise* data collected by multiple sensors from multiple moving targets can be fused using the *Bayesian nonlinear filtering equations*. These equations were, recall,

$$f_{\alpha+1|\alpha+1}(X_{\alpha+1}|Z^{(\alpha+1)}) = \frac{f(Z_{\alpha+1}|X_{\alpha+1}) \, f_{\alpha+1|\alpha}(X_{\alpha+1}|Z^{(\alpha)})}{\int f(Z_{\alpha+1}|Y_{\alpha+1}) \, f_{\alpha+1|\alpha}(Y_{\alpha+1}|Z^{(\alpha)}) \delta Y_{\alpha+1}}$$

$$f_{\alpha+1|\alpha}(X_{\alpha+1}|Z^{(\alpha)}) = \int f_{\alpha+1|\alpha}(X_{\alpha+1}|X_\alpha) f_{\alpha|\alpha}(X_\alpha|Z^{(\alpha)}) \delta X_\alpha$$

where as usual $Z^{(\alpha)}$ is an abbreviation for the list $Z_1, ..., Z_\alpha$ of global observations. The second of these equations is the time-update equation and the first equation is the information-update equation.

It is clear from the discussion of the preceding sections of this chapter that, given any of the measurement models that we have considered, and given suitable independence assumptions, *ambiguous evidence* can be incorporated into the nonlinear filtering process. Thus suppose, in addition, that we have a list $\Theta_1, ..., \Theta_\alpha$ of evidence, which we abbreviate as $\Theta^{(\alpha)}$. Then, regardless of the measurement model, the time-update equation will always have the form

$$f_{\alpha+1|\alpha}(X_{\alpha+1}|Z^{(\alpha)}, \Theta^{(\alpha)}) = \int f_{\alpha+1|\alpha}(X_{\alpha+1}|X_\alpha) f_{\alpha|\alpha}(X_\alpha|Z^{(\alpha)}, \Theta^{(\alpha)}) \delta X_\alpha$$

Also, if an additional global observation $Z_{\alpha+1}$ is collected then the information-update equation will always have the form

$$f_{\alpha+1|\alpha+1}(X_{\alpha+1}|Z^{(\alpha+1)}, \Theta^{(\alpha)}) = \frac{f(Z_{\alpha+1}|X_{\alpha+1}) \, f_{\alpha+1|\alpha}(X_{\alpha+1}|Z^{(\alpha)}, \Theta^{(\alpha)})}{\int f(Z_{\alpha+1}|Y_{\alpha+1}) \, f_{\alpha+1|\alpha}(Y_{\alpha+1}|Z^{(\alpha)}, \Theta^{(\alpha)}) \delta Y_{\alpha+1}}$$

regardless of which measurement model is employed. If an additional piece of evidence $\Theta_{\alpha+1}$ is collected, however, then the update equations will have the following forms:

Information-Update for Strong-Consistency Model

$$f_{\alpha+1|\alpha+1}(X_{\alpha+1}|Z^{(\alpha)}, \Theta^{(\alpha+1)}) = \frac{\beta(\Theta_{\alpha+1}|X_{\alpha+1})\, f_{\alpha+1|\alpha}(X_{\alpha+1}|Z^{(\alpha)}, \Theta^{(\alpha)})}{\int \beta(\Theta_{\alpha+1}|Y_{\alpha+1})\, f_{\alpha+1|\alpha}(Y_{\alpha+1}|Z^{(\alpha)}, \Theta^{(\alpha)}) \delta Y_{\alpha+1}}$$

Information-Update for Weak-Consistency Models

$$f_{\alpha+1|\alpha+1}(X_{\alpha+1}|Z^{(\alpha)}, \Theta^{(\alpha+1)}) = \frac{\rho(\Theta_{\alpha+1}|X_{\alpha+1})\, f_{\alpha+1|\alpha}(X_{\alpha+1}|Z^{(\alpha)}, \Theta^{(\alpha)})}{\int \rho(\Theta_{\alpha+1}|Y_{\alpha+1})\, f_{\alpha+1|\alpha}(Y_{\alpha+1}|Z^{(\alpha)}, \Theta^{(\alpha)}) \delta Y_{\alpha+1}}$$

Information-Update for Data-Dependent Model

$$f_{\alpha+1|\alpha+1}(X_{\alpha+1}|Z^{(\alpha)}, \Theta^{(\alpha+1)})$$
$$= \frac{\delta_{\Theta_{\alpha+1}}(Z_1 \cup ... \cup Z_\alpha|X_{\alpha+1})\, f_{\alpha+1|\alpha}(X_{\alpha+1}|Z^{(\alpha)}, \Theta^{(\alpha)})}{\int \delta_{\Theta_{\alpha+1}}(Z_1 \cup ... \cup Z_\alpha|Y_{\alpha+1})\, f_{\alpha+1|\alpha}(Y_{\alpha+1}|Z^{(\alpha)}, \Theta^{(\alpha)}) \delta Y_{\alpha+1}}$$

We will demonstrate how these equations are derived for only the case of the strong-consistency model. The other equations follow similarly. From the total probability theorem,

$$f_{\alpha+1|\alpha}(X_{\alpha+1}|Z^{(\alpha)}, \Theta^{(\alpha)})$$
$$= \int f_{\alpha+1|\alpha}(X_{\alpha+1}|X_\alpha, Z^{(\alpha)}, \Theta^{(\alpha)})\, f_{\alpha|\alpha}(X_\alpha|Z^{(\alpha)}, \Theta^{(\alpha)})\, \delta X_\alpha$$

and so the prediction integral follows by assuming that state transitions are independent of previous accumulated observations:

$$f_{\alpha+1|\alpha}(X_{\alpha+1}|X_\alpha, Z^{(\alpha)}, \Theta^{(\alpha)}) = f_{\alpha+1|\alpha}(X_{\alpha+1}|X_\alpha)$$

Using the same reasoning employed in Section 7.2.3 we know that, with suitable conditional independence assumptions, that

$$f_{\alpha+1|\alpha+1}(X_{\alpha+1}|Z^{(\alpha)}, \Theta^{(\alpha+1)}) \propto \beta(\Theta_{\alpha+1}|X)\, f_{\alpha+1|\alpha}(X_{\alpha+1}|Z^{(\alpha)}, \Theta^{(\alpha)})$$

from which the assertion follows. That is, from Bayes' rule and the reasoning used in that section we get

$$f_{\alpha+1|\alpha+1}(X_{\alpha+1}|Z^{(\alpha)}, \Theta^{(\alpha+1)})$$
$$\propto \beta(\Theta_{\alpha+1}|X_{\alpha+1}, Z^{(\alpha)}, \Theta^{(\alpha)})\, f_{\alpha+1|\alpha}(X_{\alpha+1}|Z^{(\alpha)}, \Theta^{(\alpha)})$$

Then, assuming that generation of current evidence does not depend on previous data or previous evidence, we get

$$\beta(\Theta_{\alpha+1}|X_{\alpha+1}, Z^{(\alpha)}, \Theta^{(\alpha)}) = \beta(\Theta_{\alpha+1}|X_{\alpha+1})$$

which yields the result claimed.

Bibliography

[1] Dubois, D. and Prade, H. (1990) Consonant approximations of belief functions, *International Journal of Approximate Reasoning*, **4**, 419-449.

[2] Johnson, R. W. (1986) Independence and Bayesian updating methods, in L.N. Kanal and J.F. Lemmer (eds.), *Uncertainty in Artificial Intelligence*, Elsevier, pp. 197-201.

[3] Li, Y. (1994) Probabilistic Interpretations of Fuzzy Sets and Systems, Ph.D. Dissertation, M.I.T. Laboratory for Information and Decision Systems, Cambridge MA, July 1994, 141 pages.

[4] Mahler, R. (1996) Combining ambiguous evidence with respect to ambiguous *a priori* knowledge, I: Boolean Logic, *IEEE Transactions on Systems, Man, and Cybernetics—Part A: Systems and Humans*, **26** (1), 27-41.

Chapter 8

Output Measurement

The term *performance evaluation* in data fusion refers to problems of the following kind:

(1) Comparing the performance of two data fusion algorithms against "ground truth", with respect to some parameter of interest

(2) Determining the effectiveness of some internal function of a data fusion algorithm

(3) Determining the performance of a data fusion algorithm relative to the current quality of data

Performance evaluation in data fusion is not as straightforward as one might assume: There is no simple "meter" that one can attach to the output of a data fusion algorithm and determine how good it is. The most common practice in data fusion is to use what we will call "local" metrics—that is, metrics which measure some particular and narrow aspect of algorithm performance: e.g., radial miss distance; probability of correct classification; probability of dropped track; correct correlation ratios; and so on. However, data fusion algorithms are typically very complex nonlinear processes whose subsystems interact with each other in often unpredictable ways. Consequently, it is not infrequently the case that optimization of a data fusion algorithm with respect to one local metric will result in performance degradation with respect to another local metric. Worse, it could be the case that the algorithm has been degraded with respect to some aspect of performance that is not even being measured.

If many performance parameters are of interest, it can be very difficult to arrive at an *over-all* interpretation of algorithm performance. For example, how does one rate one algorithm which has 80% target-localization effectiveness but 60% target-identification effectiveness, with one which is 60% effective at target localization and 80% effective at target I.D. This problem is sometimes addressed by producing *ad hoc* composite metrics—e.g. weighted sums of various local metrics. These metrics are examples of "global" metrics—i.e., metrics which attempt to measure the performance of a data fusion algorithm *as an entirety.* Such composite metrics are, however, usually even more difficult to interpret than the individual metrics from which they were constructed [9, p.

500]. More obviously, they are not invariant with respect to changes in units of measurement, so that it is difficult to determine what percentage of algorithm performance is actually due to competence with respect to one local metric, and what percentage is due to competence with respect to another local metric.

In practical performance evaluation one also frequently finds that *a priori* knowledge of ground truth is imperfect or even nonexistent. Under such circumstances, how does one determine how well a data fusion algorithm is doing? Even if one has perfect knowledge of ground truth, the quality of data provided by sensors is frequently very poor. It is often difficult, under such circumstances, to determine whether an algorithm is performing poorly against this data because the algorithm is deficient or because even a theoretically ideal algorithm would do poorly against such data.

Still another difficulty results from the fact that different end-users can have very different, and perhaps even subjective, interpretations of what constitutes good performance in an algorithm. For example, one user may be primarily interested in the effectiveness of an algorithm in target identification, rather than its effectiveness in locating targets. Similarly, some end-users will not be impressed by an algorithm which is 80% effective at identifying targets if it is ineffective at identifying the specific target types in which they happen to be interested. The underlying problem, of course, is the fact that one end-users "information" is another's "noise."

The perspective taken in this chapter is that it is better to define a single, "top-down" standard of effectiveness that applies to data fusion algorithms *as organic wholes* using so-called *global information metrics* and then "cut down" from this common measure to determine specific aspects of performance using "subglobal" metrics. The advantage of such an approach is that performance comparisons rest on a single "level playing field" that is applicable across a wide range of algorithms, and across a wide range of performance parameters.

The "level playing field" advocated in this chapter is *information*. The purpose of data fusion systems is to increase information (or, equivalently, reduce confusion) about one or more aspects of "ground truth". Suppose that we were able to measure the *total amount of information* being produced by a data fusion algorithm at any moment, and to determine how much of this total was associated with specific parameters of interest—e.g., the amount of information due to competence in target identification or in target localization. Then it would be possible to compare different algorithms, and different subsystems of the same algorithm, on an "apples with apples" basis.

Information theory is a well-understood and mathematically rigorous approach to determining the amount of information being produced by probabilistic systems. However, in order to apply it one needs the following things:

(1) The probability distributions/densities which characterize the system

(2) A concept of integration

(3) Information metrics defined in terms of integrals

Thus, for example, suppose that we wish to measure the performance of an algorithm on the basis of the output data that it produces. In a multisource, multitarget environment this output data will typically be a randomly varying finite subset Γ of "tracks," each track containing information about what is currently known about individual targets. If we know the global density function f_Γ of Γ then we can compute various information measures. For example, we can compare f_Γ to some reference density g using the "global" Kullback-Leibler discrimination measure introduced in an earlier chapter:

$$I(f_\Gamma; g) = \int f_\Gamma(X) \ln\left(\frac{f_\Gamma(X)}{g(X)}\right) \delta X$$

To compute such a measure we need to have an estimate of f_Γ. If we cannot determine f_Γ analytically then we must be able to estimate it empirically—that is, deduce it directly from some set $\Gamma = X_1, ..., \Gamma = X_m$ of sample data $X_1, ..., X_m$.

The concepts of local, global, and subglobal information metrics, and their application to practical performance estimation, were partly inspired by work of D. Fixsen and T. Wolff and apparently were first proposed in [21]. The work described in that paper was partially *ad hoc* and addressed performance estimation of multihypothesis-type data fusion engines only. The purpose of this chapter is to show how an information-theory based approach to performance estimation in data fusion can be placed on a scientifically sound basis.

In this chapter we will first show (Section 8.1) how finite-set statistics can be used to develop a "global" approach to performance evaluation of data fusion algorithms. We will also show that, for a broad class of data fusion algorithms—those which are of "multihypothesis type"—information can be computed purely analytically. In general, however, the global densities which characterize the output of a data fusion algorithm cannot be determined analytically. In this case empirical approaches must be used to determine the form of f_Γ. Thus in Section 8.2.2 we will show how a certain approach to nonparametric density estimation—the so-called "projection kernel estimator" approach—can be directly generalized to multisource, multitarget situations.

Remark 1 *It should be pointed out here that if $n = 0$, so that $\mathcal{R} = U$ is a finite universe, then a large number of information or entropy measures have been devised which are applicable to measurement of random sets. See [16], [7], [14], [10], [11], [12], [18].*

8.1 Performance Evaluation

8.1.1 Information Measured With Respect to Ground Truth

The most straightforward sort of performance estimation occurs when we have essentially perfect knowledge of "ground truth"—that is, we know the number, identities, positions, velocities, threat states, etc. of the actual targets in a given scenario. Let us suppose that the states of these known ground truth targets are $\gamma_1, ..., \gamma_t$ where $\gamma_i = (g_i, g_i) \in \mathcal{R} \triangleq \mathbb{R}^n \times V$ for all $i = 1, ..., t$. We then ask: Given the output of a data fusion algorithm at any time-instant α, *how much information is the algorithm providing about the actual state of ground truth?*

The Single-Sensor, Single-Target Case

For purposes of illustration let us begin by examining a simple case: Determining the performance of a single-sensor tracker-classifier algorithm against a single target whose actual state is $\gamma_0 = (g_0, g_0)$. To keep our illustration concrete, let us assume that the algorithm supplies two pieces of information: an estimate $\hat{x}$ of geokinematics together with an estimated Gaussian error-covariance matrix $\hat{C}$; and a probability distribution $\hat{q}$ on V. The estimate is the same as specifying a density $\hat{f}(x, v) = N_{\hat{C}}(x - \hat{x})\, \hat{q}(v)$ for all $(x, v) \in \mathcal{R}$. The information contained in $\hat{f}$ is what we must compare against ground truth.

Any actual ground truth target has a physical "footprint." That is, it has a definite physical size (i.e., volume) and it is unnecessary to try to determine the target's location to any greater degree of precision than dictated by that extent. Likewise, it is unnecessary to determine the target's speed to any greater degree of precision than some practical minimum. Similar remarks apply to other continuous physical state variables.

So, we may assume that there is some closed ball $B_r(g_0) \subseteq \mathbb{R}^n$ of radius r centered at g which reflects the minimal degree of resolution, say $L \triangleq \lambda(B_r(g_0))$, to which we wish to resolve the target's continuous state variables. Let $f_{Grnd}(x, v) \triangleq l_{Grnd}(x, v | g_0, g_0)$ where

$$
l_{Grnd}(x, v | g_0, g_0) \triangleq
\begin{cases}
L^{-1} & \text{if } v = g_0 \text{ and } x \in B_r(g_0) \\
0 & \text{if } v = g_0 \text{ and } x \notin B_r(g_0) \\
0 & \text{if } v \neq g_0
\end{cases}
$$

be a uniform density defined on state space. Then f_{Grnd} models our knowledge of actual ground truth. Also notice that f_{Grnd} is absolutely continuous with respect to $\hat{f}$ and so we can compute the Kullback-Leibler discrimination

$$
\begin{aligned}
I(f_{Grnd}; \hat{f}) &= \int_{B_0 \times g} f_{Grnd}(\zeta) \ln\left(\frac{f_{Grnd}(\zeta)}{\hat{f}(\zeta)}\right) d\lambda(\zeta) \\
&= \int_{B_r(g_0)} L^{-1} \ln\left(\frac{1}{L\,\hat{f}(x, g_0)}\right) d\lambda(x) \\
&\cong -\ln\left(L\,\hat{f}(g_0, g_0)\right) = -\ln\left(L\, N_{\hat{C}}(g_0 - \hat{x})\right) - \ln \hat{q}(g_0)
\end{aligned}
$$

provided that L is small compared to the extent of $\hat{f}$. Since L is constant we could (following the practice used in conventional information theory with respect to the differential entropy) simply dispense with it altogether and use the quantity $\tilde{I}(f_{Grnd}; \hat{f}) \triangleq -\ln \hat{f}(g_0, g_0)$ instead. The more the estimate $\hat{x}$ deviates from g_0 the greater the value of $I(f_{Grnd}; \hat{f})$; and similarly the more the maximal value of $\hat{q}$ deviates from g_0. The quantity $I(f_{Grnd}; \hat{f})$ is the *absolute entropy* of the estimate $\hat{x}, \hat{C}, \hat{q}$, i.e. the degree to which it deviates from ground truth.

Generally speaking, there will be a maximum possible value for the entropy in the problem. That is, there will be a maximal region of interest R; a maximal range of velocities that are of interest; and so on. Thus there will be a closed subset $B_\infty \subseteq \mathbb{R}^n$ consisting of all possible feasible states of interest. If

$$f_\infty(\mathbf{x}, v) \triangleq \begin{cases} \lambda(B_\infty)^{-1} \cdot |V|^{-1} & \text{if } \mathbf{x} \in B_\infty \\ 0 & \text{if otherwise} \end{cases}$$

is the corresponding uniform distribution then the incremental entropy

$$I_\infty(\hat{f}) = I(f_{Grnd}; f_\infty) - I(f_{Grnd}; \hat{f})$$

is the *absolute information* supplied by the tracker-classifier algorithm, relative to ground truth. If we have a time-sequence of estimates $\hat{f}_1, ..., \hat{f}_\alpha, ...$ then the incremental information

$$I_\alpha = I_\infty(\hat{f}_\alpha) - I_\infty(\hat{f}_{\alpha-1}) = I(f_{Grnd}; \hat{f}_{\alpha-1}) - I(f_{Grnd}; \hat{f}_\alpha)$$

measures the *increase (or decrease) in information* supplied by the tracker-classifier from time-instant $\alpha - 1$ to time-instant α.

Finally, what if we do not have essentially perfect knowledge of ground truth, as reflected in the global density f_{Grnd}? Suppose that there is some uncertainty in our knowledge of the locations and identities of targets, and perhaps uncertainty even in regard to their number? In this case it would be reasonable to proceed as before but make the uniform densities less precise. More generally, one could model one's knowledge of ground truth as a density of the form $f_{Grnd}(\mathbf{x}, v) = N_C(\mathbf{x} - \mathbf{x}_0)P(v)$.

The Multisource, Multitarget Case

From the foregoing discussion it is clear how to extend the concepts of entropy and information to the multisource, multitarget case. Suppose as before that the ground truth targets have actual states $\gamma_1, ..., \gamma_t$ where $\gamma_i = (\mathbf{g}_i, g_i)$ for all $i = 1, ..., l$. As before, for each ground truth target there will be a corresponding closed ball $B_r(\mathbf{g}_1), ..., B_r(\mathbf{g}_t)$ of $\mathbb{R}^n$ dictated by the physical extent of the targets in state space. We may assume that these balls are mutually disjoint. We also note that $\lambda(B_r(\mathbf{g}_1)) = ... = \lambda(B_r(\mathbf{g}_t)) = L$.

What is the global density which describes ground truth? The performance evaluator who has special access to ground truth can be modeled as a special, extremely high-resolution "ground truth sensor." The density which describes the (single target) observation noise for such an all-knowing evaluator/sensor is the uniform density $l_{Grnd}(\zeta|\gamma)$ defined earlier. The all-knowing evaluator is also immune to false alarms and missed detections. Consequently the multitarget global *prior* density $f_{Grnd}(X)$ for the ground truth sensor is just

$$f_{Grnd}(\{\zeta_1, ..., \zeta_t\}) = \sum_\pi l_{Grnd}(\zeta_{\pi 1}|\gamma_1) \cdots l_{Grnd}(\zeta_{\pi t}|\gamma_t)$$

for all distinct $\zeta_1, ..., \zeta_t$, where the summation is taken over all permutations π on the numbers $1, ..., t$. Let $\zeta_i = (\mathbf{x}_i, v_i)$ for all $i = 1, ..., t$. Notice that

$f_{Grnd}(\{\zeta_1, ..., \zeta_t\}) = 0$ unless there exists a permutation π such that $v_{\pi 1} = g_1, ..., v_{\pi t} = g_t$ and $\mathbf{x}_{\pi 1} \in B_r(\mathbf{g}_1), ..., \mathbf{x}_{\pi t} \in B_r(\mathbf{g}_t)$; and in this case $f_{Grnd}(\{\zeta_1, ..., \zeta_t\}) = L^{-t}$.

Suppose, then, that a data fusion algorithm supplies us with estimates about ground truth $\gamma_1, ..., \gamma_t$. Suppose that we are able, using analytic or empirical nonparametric methods, to convert this information into a global density $\hat{f}$ on state space. Generally speaking the ground truth distribution f_{Grnd} will be absolutely continuous with respect to $\hat{f}$ and so we can compute the global Kullback-Leibler discrimination

$$
\begin{aligned}
I(f_{Grnd}; \hat{f}) &= \int f_{Grnd}(X) \ln \left(\frac{f_{Grnd}(X)}{\hat{f}(X)} \right) \delta X \\
&= \frac{1}{t!} \int f_{Grnd}(\{\zeta_1, ..., \zeta_t\}) \ln \left(\frac{f_{Grnd}(\{\zeta_1, ..., \zeta_t\})}{\hat{f}(\{\zeta_1, ..., \zeta_t\})} \right) d\lambda(\zeta_1) \cdots d\lambda(\zeta_t) \\
&= \int_{B_r(\mathbf{g}_1) \times ... \times B_r(\mathbf{g}_t)} L^{-t} \ln \left(\frac{1}{L^t \hat{f}(\{(\mathbf{x}_1, g_1), ..., (\mathbf{x}_t, g_t)\})} \right) \\
&\quad \cdot d\lambda(\mathbf{x}_1) \cdots d\lambda(\mathbf{x}_t) \\
&\cong -\ln \left(L^t \hat{f}(\{(\mathbf{g}_1, g_1), ..., (\mathbf{g}_t, g_t)\}) \right)
\end{aligned}
$$

The same reasoning applied previously then extends to the multisensor, multitarget case: It is possible to define absolute information, the incremental moment-to-moment information, and so on. We are thus justified in making the following definition:

Definition 1 *Let ground truth be given by* $X_{Grnd} = \{\gamma_1, ..., \gamma_t\}$ *where* $\gamma_i = (\mathbf{g}_i, g_i) \in \mathcal{R} \triangleq \mathbb{R}^n \times U$ *for all* $i = 1, ..., t$. *Let* $\hat{f}$ *be the global density on state space supplied by a data fusion algorithm. Then the absolute global entropy of the estimate* $\hat{f}$ *with respect to ground truth is* $-\ln L^t \hat{f}(X_{Grnd})$.

Global Entropy of a Multihypothesis-Type Algorithm

It is possible to derive explicit formulas for the global entropy of a *multihypothesis-type algorithm* (see Section 6.5 of Chapter 6). Recall that the output of such an algorithm consists of a list $L = (H_1, p_{hyp}(H_1)), ..., (H_c, p_{hyp}(H_c))$ of hypotheses, where each hypothesis consists of a list of tracks, and where each track τ is drawn from a track table $\{\tau_1, ..., \tau_d\}$ and consists of a geokinematic estimate $\hat{\mathbf{x}}_\tau$, an error-covariance estimate $\hat{C}_\tau$, and an estimated distribution $\hat{q}_\tau$ on V. Let

$$
f_\tau(\mathbf{x}, v) \triangleq N_{C_\tau}(\mathbf{x} - \hat{\mathbf{x}}_\tau) \, \hat{q}_\tau(v), \qquad p_i \triangleq \sum_{H:H \ni \tau_i} p_{hyp}(H)
$$

and $f_i \triangleq f_{\tau_i}$ for all $i = 1, ..., d$. Then the absolute global entropy of the estimate $\hat{f}$ is $-\ln L^t \hat{f}(X_{Grnd})$ where:

$$
\hat{f}(X_{Grnd}) = \sum_{1 \leq i_1 \neq ... \neq i_t \leq d} \frac{(1 - p_1) \cdots (1 - p_d)}{(1 - p_{i_1}) \cdots (1 - p_{i_t})} \, p_{i_1} \cdots p_{i_t} \, f_{i_1}(\gamma_1) \cdots f_{i_t}(\gamma_t)
$$

is the global density corresponding to the algorithm output. Here, the summation is taken over all distinct $i_1, ..., i_t$ such that $1 \le i_1, ..., i_t \le d$ and where $X_{Grnd} = \{\gamma_1, ..., \gamma_t\}$.

It is possible to write a more useful formula for the above global density. Assume that no track in the track table occurs in every hypothesis of L, so that $q_i \neq 1$ for all $i = 1, ..., d$. Define the *probabilistic confusion matrix* of the algorithm output L to be the $d \times t$ matrix $\kappa(L) = (\kappa(L)_{i,j})_{i,j}$ whose components are

$$\kappa(L)_{i,j} \triangleq \frac{p_i}{1 - p_i} f_i(\gamma_j)$$

for all $i = 1, ..., d$ and $j = 1, ..., t$. (Note that the more hypotheses a track τ_i is in, the more the entry $\kappa(L)_{i,j}$ is weighted by the factor $(1 - p_i)^{-1}$.)

Then we can write

$$
\begin{aligned}
\hat{f}(X_{Grnd}) &= (1 - p_1) \cdots (1 - p_d) \sum_{1 \le i_1 \neq ... \neq i_t \le d} \kappa_{i_1.1}(L) \cdots \kappa_{i_t,t}(L) \\
&= \left(\prod_{i=1}^{d} (1 - p_i) \right) \left(\sum_{Q \subseteq \kappa(L)} perm(Q) \right)
\end{aligned}
$$

where the summation is taken over all $t \times t$ submatrices Q of $\kappa(L)$. For any square matrix $Q = (Q_{i,k})_{i,k}$ the *permanent* of Q is defined [2, Chapter 7] as:

$$perm(Q) \triangleq \sum_{\pi} Q_{1,\pi 1} \cdots Q_{t,\pi t}$$

where the summation is taken over all permutations π on the numbers $1, ..., t$.

Generally speaking, the exact computation of permanents is an NP-complete problem. However, the permanent shares many of the same properties as the determinant. For example, if Q can be written in block-diagonal form where $Q_1, ..., Q_r$ are the block square submatrices, then $perm(Q) = perm(Q_1) \cdots perm(Q_r)$. Permanents can be expanded by rows or by columns, $perm(Q) = \sum_{i=1}^{t} Q_{i,k} perm(\check{Q}_{i,k})$ for any $k = 1, ..., t$, where $\check{Q}_{i,k}$ denotes the cofactor matrix of the entry $Q_{i,k}$. Also, $perm(Q)$ is invariant with respect to interchanges of rows or interchanges of columns. All of these operations can be used to simply the exact computation of a permanent. The computation of $perm(Q)$ is less complex if Q is a sparse matrix; and Q can be always be approximated by a "sparsified" matrix $\tilde{Q}$ in which small entries of Q can be replaced by zero. See also [17] for discussion of fast algorithms for computing permanents exactly.

Example 1 *Suppose that at some instant a multihypothesis-type multitarget tracker provides us with the following information about locations of targets on a flat earth:*

$H_1 : p_{hyp}(H_1) = .7$ *One track located at* $\mathbf{x}_1$ *with covariance matrix* C_1
$H_2 : p_{hyp}(H_2) = .3$ *Two tracks, consisting of* $\mathbf{x}_1, C_1$ *and* $\mathbf{x}_2, C_2$

Let the associated Gaussian density functions be

$$f_i(\mathbf{x}) = \frac{1}{2\pi\sqrt{\det(C_i)}} \, e^{-\frac{1}{2}(\mathbf{x}-\mathbf{x}_i)^T C_i^{-1}(\mathbf{x}-\mathbf{x}_i)}$$

for $i = 1, 2$. The two tracks have track probabilities $p_1 = .7 + .3 = 1$ and $p_2 = 0 + .3 = .3$, respectively. The global density function corresponding to the tracker's output is therefore

$$f(\emptyset) = (1 - p_1)(1 - p_2) = (0)(.7) = 0$$
$$f(\{\mathbf{x}\}) = p_1(1 - p_2)f_1(\mathbf{x}) + (1 - p_1)p_2 f_2(\mathbf{x}) = .7 f_1(\mathbf{x})$$
$$f(\{\mathbf{x}, \mathbf{y}\}) = p_1 p_2 f_1(\mathbf{x})f_2(\mathbf{y}) + p_1 p_2 f_1(\mathbf{y})f_2(\mathbf{x}) = .3 f_1(\mathbf{x})f_2(\mathbf{y}) + .3 f_1(\mathbf{y})f_2(\mathbf{x})$$

Suppose for the sake of illustration that the tracker is capable of resolving the targets to the minimal resolution L, so that $L^2 = \det(C_1) = \det(C_2)$. Consider two possible situations: (1) the evaluator knows that ground truth consists of one target located at $\mathbf{g}_1$; or (2) the evaluator knows that ground truth consists of two targets located at $\mathbf{g}_1$ and $\mathbf{g}_2$. Then we can show that the respective entropies ε_1 and ε_2 for these two situations are $\varepsilon_1 = -\ln(L f(\mathbf{g}_1)) = -\ln\left(\frac{.7}{2\pi}\right) = 2.20$ and $\varepsilon_2 = -\ln(L^2 f(\{\mathbf{g}_1, \mathbf{g}_2\})) = -\ln\left(\frac{.6}{4\pi^2}\right) = 4.20$.

Information Measures With Imperfect Ground Truth

It will not infrequently be the case that ground truth will not be known to the degree of precision assumed in the preceding paragraphs. There, we assumed that our knowledge of ground truth states $\gamma_1, ..., \gamma_t$ could be described by uniform densities $l_{Grnd}(\zeta|\gamma_1), ..., l_{Grnd}(\zeta|\gamma_t)$. If this is not the case we could replace the previous uniform densities by *Gaussian* densities, each density describing our uncertainty in our knowledge of what each individual track looks like.

More seriously, we may not even be sure of the *number* of ground truth tracks. Rather than having precise knowledge of ground truth, we may only have one or more *hypotheses* about what ground truth looks like, each hypothesis consisting of varying number of hypothesized tracks. Nevertheless, the previous reasoning can still be applied—though the entropy formulas will not be nearly as simple in form. In this case the *ground truth* global density f_{Grnd} will have the same form as the density of Example 1. We will still be able to define the absolute entropy as $I(f_{Grnd}; \hat{f})$ where $\hat{f}$ is the estimate supplied by the data fusion algorithm being evaluated.

8.1.2   Relative Information

In many cases we will not have any knowledge of ground truth at all but, nevertheless, may be asked to evaluate the performance of a data fusion algorithm. This requires a concept of *relative information* as opposed to *absolute information*. In the single-sensor, single-target case a well-known approach to determining relative information exists. Suppose that $\hat{f}$ is the density estimate supplied by a tracker-classifier algorithm. Suppose further that there is a closed compact region D such that $\int_D \hat{f}(\zeta)d\lambda(\zeta) \cong 1$, so that there is very little possibility that the ground truth target could be outside of D. Let f_D be the uniform density on D. Then as a general proposition, any estimate $\hat{f}$ will typically satisfy the relationship $f_{Grnd} \ll \hat{f} \ll f_D$ where '$\ll$' denotes absolute continuity and where $I(f_{Grnd}; f_D)$ is a measure of the total possible entropy of the system. The relationship $f_{Grnd} \ll \hat{f}$ expresses a relationship in which the estimate is compared to a state of complete certainty in a scenario; whereas the relationship $\hat{f} \ll f_D$ expresses a relationship in which the estimate is compared to a state of complete *uncertainty* concerning ground truth.

The Kullback-Leibler discrimination of $\hat{f}$ with respect to f_D is

$$I(\hat{f}; f_D) = \int \hat{f}(\zeta) \ln \left(\frac{\hat{f}(\zeta)}{f_D(\zeta)} \right) d\lambda(\zeta)$$

$$\cong \ln \lambda(D) + \int \hat{f}(\zeta) \ln \hat{f}(\zeta) \, d\lambda(\zeta) = \ln \lambda(D) - \varepsilon(\hat{f})$$

The quantity $\varepsilon(\hat{f})$ is just the Shannon, or differential, entropy of $\hat{f}$. The differential entropy is a measure of the non-uniformity of $\hat{f}$. If $\hat{f}_1, ..., \hat{f}_\alpha, ...$ is a time-sequence of estimates then the incremental differential entropy

$$\Delta I_\alpha \triangleq I(\hat{f}_\alpha; f_G) - I(\hat{f}_{\alpha-1}; f_G) = \varepsilon(\hat{f}_{\alpha-1}) - \varepsilon(\hat{f}_\alpha)$$

is a measure of the decrease of differential entropy or, equivalently, the increase in information.

Now, recall from Definition 12 of Chapter 4 that the *global uniform density* $u_{D,M}$ is defined as

$$u_{D,M}(Z) = \frac{|Z|!}{M \, N^{|Z|} \lambda(D)^{|Z|}}$$

whenever $Z = \{(z_1, w_1), ..., (z_k, w_k)\}$ is such that $z_1, ..., z_k \in D$, $k < M$; and $u_{D,M}(Z) = 0$ otherwise, and where D is a closed bounded domain of $\mathbb{R}^n$ and $M > 0$ is an integer. Assume that $\hat{f}$ is absolutely continuous with respect to $u_{D,M}$ and let

$$I(\hat{f}; u_{D,M}) = \int \hat{f}(X) \ln \left(\frac{\hat{f}(X)}{u_{D,M}(X)} \right) \delta X$$

Then $u_{D,M}$ is a measure of the degree to which $\hat{f}$ deviates from uniformity. However, it is easy to show that

$$I(\hat{f}; u_{D,M}) \cong \int \hat{f}(X) \ln \hat{f}(X) \, \delta X + \ln M$$

$$+ \left(\sum_{k=0}^{M} k\hat{f}(k) \right) \ln (N\lambda(D)) - \sum_{k=0}^{M} \hat{f}(k) \ln k!$$

where $\hat{f}(k)$ is the cardinality distribution of $\hat{f}$.

Thus unlike the conventional case, *differential entropy does not arise in a rigorous manner from the Kullback-Leibler metric*

8.1.3   Components of Information

In many cases "global" information measures may be *too* global: One may be less interested in the overall performance of an algorithm than in some specific aspect of its performance—e.g., how well it is classifying, how well it is localizing targets, how well it is detecting targets, and so on. We thus require not only global performance evaluation metrics, but also metrics which are "subglobal" in that they measure a *specific component of information*—e.g., the component of information due to classification, the component due to localization, the component due to detection, and so on. We may also want to be able to answer questions such as, How much of the information supplied by the algorithm is due to classification-type data, how much is due to geokinematic-type data, and so on. If some particular component of information is supplying very little of the total information under certain circumstances, it is obviously not very useful to continue trying to collect that kind of information under those particular circumstances.

Single-Sensor, Single-Target Case

To see how to proceed, let us once again take the single-sensor, single-target case as an illustration. Once again the ground truth state is $\gamma_0 = (g_0, g_0)$ and the algorithm output density is $\hat{f}(\mathbf{x}, v) = N_{\hat{C}}(\mathbf{x} - \hat{\mathbf{x}})\, \hat{q}(v)$ for all $(\mathbf{x}, v) \in \mathcal{R}$. Suppose that we wanted to know what part of the information supplied by the sensor is due to the target-localization capabilities of the algorithm. We can filter out the target-I.D. information by "disabling" the target-I.D. part of the algorithm output as follows. The marginal distribution $\hat{f}_{loc}(\mathbf{x}) \triangleq \sum_{v \in V} \hat{f}(\mathbf{x}, v) = N_{\hat{C}}(\mathbf{x} - \hat{\mathbf{x}})$ describes the localization-only performance of the algorithm. The density $\hat{f}_{loc}(\mathbf{x}, v) \triangleq \frac{1}{N} \hat{f}_{loc}(\mathbf{x}) = \frac{1}{N} N_{\hat{C}}(\mathbf{x} - \hat{\mathbf{x}})$ for all $\mathbf{x}, v$ models an algorithm which has the same localization performance as the original algorithm, but which provides absolutely no target-I.D. information. The entropy

$$I(f_{Grnd}; \hat{f}_{loc}) \cong -\ln\left(\frac{L}{N}\, \hat{f}_{loc}(g_0)\right)$$

describes the performance of the disabled algorithm with respect to ground truth. Thus the incremental entropy

$$\Delta I_{cls} = I(f_{Grnd}; \hat{f}_{loc}) - I(f_{Grnd}; \hat{f}) \cong \ln\left(N\, \hat{q}(g_0)\right)$$

which results from disabling the localization part of the information, describes the part of $I(f_{Grnd}; \hat{f})$ which is attributable to the influence of classification data. (This influence is reflected regardless of whether it is due to synergistic interaction with the localization part of the algorithm or not. Since localization and classification data are assumed independent under our current assumptions, however, no such synergy exists under these assumptions.) It gives the same result as comparing the classifier part of the algorithm to a classifier algorithm which provides no information at all about target I.D.

Similarly, from the marginal distribution $\hat{f}_{cls}(v) \triangleq \int \hat{f}(\mathbf{x}, v) d\lambda(\mathbf{x}) = \hat{q}(v)$ we can construct an output in which localization information has been disabled:

$$\hat{f}_{cls}(\mathbf{x}, v) \triangleq \begin{cases} \lambda(D)^{-1} \hat{f}_{cls}(v) & \text{if } \mathbf{x} \in D \\ 0 & \text{if } \mathbf{x} \notin D \end{cases}$$

where D is the geokinematic region of interest for the problem. The differential entropy

$$\Delta I_{loc} = I(f_{Grnd}; \hat{f}_{cls}) - I(f_{Grnd}; \hat{f}) \cong \ln\left(\lambda(D) N_{\hat{C}}(\mathbf{g}_0 - \hat{\mathbf{x}})\right)$$

then represents the increase in information attributable to the localization performance of the algorithm.

The same ideas apply to *relative* global information. In this case we look at the relative discriminations

$$\begin{aligned} \Delta I_{cls} &= I(\hat{f}; f_R) - I(\hat{f}_{loc}; f_R) = \varepsilon(\hat{f}_{loc}) - \varepsilon(\hat{f}) = \ln N - \varepsilon(\hat{q}) \\ \Delta I_{loc} &= I(\hat{f}; f_R) - I(\hat{f}_{cls}; f_R) = \varepsilon(\hat{f}_{cls}) - \varepsilon(\hat{f}) = \lambda(D) - \varepsilon(N_{\hat{C}}) \end{aligned}$$

Remark 2 *Note that because localization and classification information are independent by assumption,*

$$\Delta I_{cls} + \Delta I_{loc} = \ln\left(M\lambda(D)\right) - \varepsilon(\hat{q}) - \varepsilon(N_{\hat{C}}) = I(\hat{f}; f_R)$$

In general, however, ΔI_{cls} and ΔI_{loc} will contain overlapping information since some classification performance will result from positive interactions between classifier and localizer; and likewise for localizer performance. In this case $\Delta I_{cls} + \Delta I_{loc} > I(\hat{f}; f_R)$ and the difference $\Delta I_{cor} = \Delta I_{cls} + \Delta I_{loc} - I(\hat{f}; f_R)$ is the information which can be attributed to interactions between the classification and localization aspects of the algorithm. Likewise, $\Delta I_{cls} - \Delta I_{cor}$ is the part of information attributable strictly to classification, and $\Delta I_{loc} - \Delta I_{cor}$ is the part attributable strictly to localization.

These ideas can be generalized to the multisensor, multitarget case as follows.

The Detection Component of Information

The *detection component* of information is that part of global information which can be attributed to the process of correctly determining the number of targets in a scenario. Let $card(\hat{f})$ denote the cardinality of $\hat{f}$, i.e. the largest k such that $\hat{f}(X) \neq 0$ for some X and $|X| = k$. Define

$$\hat{f}_{loc-cls}(X) \triangleq \frac{1}{(1 + card(\hat{f}))} \frac{\hat{f}(X)}{\hat{f}(|X|)}$$

for all X. Then $\hat{f}_{loc-cls}$ is a global density and its cardinality distribution is uniform: $\hat{f}_{loc-cls}(k) = 1 + card(\hat{f})^{-1}$ for all $0 \le k \le card(\hat{f})$ and $\hat{f}_{loc-cls}(k) = 0$

for $k > card(\hat{f})$. That is, $\hat{f}_{loc-cls}(X)$ models a data fusion algorithm which has the same localization and classification capabilities as the original algorithm, but whose ability to determine the number of targets has been completely disabled.

We can then proceed in the same manner as in the single-sensor, single-target illustration. Let f_{Grnd} denote as usual the global density which models ground truth. Then the incremental entropy

$$\Delta I_{num} = I(f_{Grnd}, \hat{f}_{loc-cls}) - I(f_{Grnd}; \hat{f})$$

describes the information, as measured against ground truth, collected by the algorithm which is attributable to the detection function of the algorithm alone. (Note that ΔI_{num} includes information that is attributable just to detection acting alone, but also detection acting in concert with both the classification and localization functions of the algorithm.) It is easy to see that

$$\Delta I_{num} \cong \ln \hat{f}(|X_{Grnd}|) + \ln(1 + card(\hat{f}))$$

Similarly, let $u_{D,M}$ denote the global uniform distribution. Then

$$\Delta I_{num} = I(\hat{f}; u_{D,M}) - I(\hat{f}_{loc-cls}; u_{D,M})$$

describes the information, as measured against a state of complete uncertainty in ground truth, attributable to the detection function alone.

The Localization Component of Information

Given a global density $\hat{f}$ produced as the output of a data fusion algorithm, define the new global density $\hat{f}_{cls-num}$ by

$$\hat{f}_{cls-num}(\{(\mathbf{x}_1, v_1), ..., (\mathbf{x}_k, v_k)\})$$

$$\triangleq \frac{1}{\lambda(D)^k} \int_{D^k} \hat{f}(\{(\mathbf{y}_1, v_1), ..., (\mathbf{y}_k, v_k)\}) \, d\lambda(\mathbf{y}_1) \cdots d\lambda(\mathbf{y}_k)$$

for all $(\mathbf{x}_1, v_1), ..., (\mathbf{x}_k, v_k) \in \mathcal{R}$ if $\mathbf{x}_1, ..., \mathbf{x}_k \in D$ are distinct and

$$\hat{f}_{cls-num}(\{(\mathbf{x}_1, v_1), ..., (\mathbf{x}_k, v_k)\}) = 0$$

otherwise. The reader may verify that $\hat{f}_{cls-num}$ is indeed a global density: $\int \hat{f}_{cls-num}(X)\delta X = 1$. In other words, $\hat{f}_{cls-num}$ models the output of a data fusion algorithm which has the same detection and classification performance as the original algorithm, but in which the localization part of the output has been completely disabled.

We can now proceed in the usual manner. The incremental entropy

$$\Delta I_{loc} = I(f_{Grnd}; \hat{f}_{cls-num}) - I(f_{Grnd}; \hat{f})$$

describes that part of the information collected by the algorithm which is attributable to the localization function of the algorithm alone. Likewise,

$$\Delta I_{loc} = I(\hat{f}; u_{D,M}) - I(\hat{f}_{cls-num}; u_{D,M})$$

describes the information, as compared to a state of complete uncertainty concerning ground truth, attributable to the localization function alone.

The Classification Component of Information

Given a global density $\hat{f}$ produced as the output of a data fusion algorithm, define the new global density $\hat{f}_{loc-num}$ by

$$\hat{f}_{loc-num}(\{(\mathbf{x}_1, v_1), ..., (\mathbf{x}_k, v_k)\}) \triangleq \frac{1}{|V|^k} \sum_{w_1,...,w_k \in U} \hat{f}(\{(\mathbf{x}_1, w_1), ..., (\mathbf{x}_k, w_k)\})$$

for all $(\mathbf{x}_1, v_1), ..., (\mathbf{x}_k, v_k) \in \mathcal{R}$ if $\mathbf{x}_1, ..., \mathbf{x}_k$ are distinct, and

$$\hat{f}_{loc-num}(\{(\mathbf{x}_1, v_1), ..., (\mathbf{x}_k, v_k)\}) = 0$$

otherwise. Then $\hat{f}_{loc-num}$ is a global density and models the output of a data fusion algorithm which has the same detection and localization performance as the original algorithm, but in which the classification part of the output has been disabled.

The incremental entropy $\Delta I_{cls} = I(f_{Grnd}; \hat{f}_{loc-num}) - I(f_{Grnd}; \hat{f})$ describes the information, as compared to ground truth, collected by the algorithm which is attributable to the classification function of the algorithm alone. Likewise, $\Delta I_{cls} = I(\hat{f}; u_{D,M}) - I(\hat{f}_{loc-num}; u_{D,M})$ describes the information, as compared to a state of complete uncertainty concerning ground truth, attributable to the classification function alone.

8.1.4   Information With Constraints

In the preceding sections we assumed that performance evaluation consists of determining how well a data fusion algorithm performs against *all targets in ground truth*. It may be the case, however, that an algorithm evaluator is interested in how well the algorithm performs against targets which are *constrained* by the existence of additional considerations. For example, the evaluator may have no interest whatsoever in targets that lie outside of some specified geographical region, or which move slower than some specified speed. Likewise, the evaluator may be interested only in how well the algorithm is performing against certain target types: all military targets; or all fighter aircraft; or all ballistic missile submarines; and so on.

The fact that the algorithm is performing well against ground truth, generally speaking, is irrelevant to an evaluator who is interested only in a *restricted subset of possible ground truths*. At the same time, targets which lie outside this subset cannot thereby be ignored by the algorithm. (For example, even though an evaluator may not care about the merchant ship below which a submarine is hiding, nevertheless any successful algorithm must resolve the merchant from the submarine in order to provide the information the evaluator regards as relevant.)

The Single-Target Case

Once again it is instructive to begin with a relatively simple illustration drawn from a single-sensor, single-target scenario: Ground truth is $\gamma_0 = (\mathbf{g}_0, g_0)$ and the algorithm output density is $\hat{f}(\mathbf{x}, v) = N_{\hat{C}}(\mathbf{x} - \hat{\mathbf{x}})\, \hat{q}(v)$ for all $(\mathbf{x}, v) \in \mathcal{R}$.

Let us begin by supposing that an end-user is interested *only* in targets γ which satisfy a *hard-and-fast constraint* such as $\gamma \in T$ for some subset $T \subseteq \mathcal{R}$. That is, targets that satisfy constraint T are of interest and targets which do not are of no interest whatsoever. An algorithm which produces information about targets in T^c is, from the perspective of *this* end-user, not producing information of any usefulness. (For example, if $T = B \times V$ for some region $B \subseteq \mathbb{R}^n$ then the end-user is concerned only about, for example, targets in a certain geographical region. If $T = \mathbb{R}^n \times T_0$ for some $T_0 \subseteq V$ then only targets of a certain type—those in T_0—are of interest.)

We can address this problem by regarding user-constrained information as a *component* of total information, much as we did in regard to parsing out the classification, detection, etc. components of information. We thus proceed by "disabling" the algorithm's competence in regard to locating and identifying targets in T, and comparing the information produced by the "disabled" algorithm to the information produced by the non-disabled one. As usual, let $\mathbf{1}_T(\zeta) = 1$ if $\zeta \in T$ and $\mathbf{1}_T(\zeta) = 0$ if $\zeta \notin T$ define the indicator function for the subset T. Define

$$\hat{f}_T(\zeta) \triangleq (1 - \mathbf{1}_T(\zeta))\hat{f}(\zeta) + \mathbf{1}_T(\zeta)p_{\hat{f}}(T)\lambda(T)^{-1}$$

for all $\zeta \in \mathcal{R}$, where $p_{\hat{f}}$ is the probability measure corresponding to $\hat{f}$, where $\lambda(T) = \sum_{v \in U} \lambda(T(v))$ is the hybrid measure of T, and where as usual $T(v) \triangleq \{\mathbf{x} \in \mathbb{R}^n |\, (\mathbf{x}, v) \in T\}$. Then $\hat{f}_T$ is a density function such that

$$\hat{f}_T(\zeta) = \begin{cases} p_{\hat{f}}(T)\,\lambda(T)^{-1} & \text{if } \zeta \in T \\ \hat{f}(\zeta) & \text{if } \zeta \notin T \end{cases}$$

In other words, $\hat{f}_T$ is identical to the original density $\hat{f}$ outside of T and is uniform on T. Thus the output of the algorithm, insofar as its ability to produce information about targets in T is concerned, has been disabled. The entropy (confusion) produced by the output-disabled algorithm, as compared to ground truth, is

$$I(f_{Grnd}; \hat{f}_T) \cong -\ln(L\, \hat{f}_T(\gamma_0)) \cong \begin{cases} -\ln(Lp_{\hat{f}}(T)\,\lambda(T)^{-1}) & \text{if } \gamma_0 \in T \\ I(f_{Grnd}; \hat{f}) & \text{if } \gamma_0 \notin T \end{cases}$$

Accordingly, the information which is attributable to algorithm competence with respect to targets in T is

$$\Delta I_T \triangleq I(f_{Grnd}; \hat{f}_T) - I(f_{Grnd}; \hat{f}) \cong \begin{cases} \ln\left(\frac{\hat{f}(\gamma_0)\lambda(T)}{p_{\hat{f}}(T)}\right) & \text{if } \gamma_0 \in T \\ 0 & \text{if } \gamma_0 \notin T \end{cases}$$

Thus the user-relevant information which is being produced is zero if the actual target does not conform to the user's definition of what is and is not informative. On the other hand, suppose that ground truth is not known and let f_D

be the uniform density describing the region of interest. Then the information produced by the output-disabled algorithm, relative to a state of complete uncertainty, is $I(\hat{f}_T; f_D) = \lambda(D) - \varepsilon(\hat{f}_T)$ where

$$\varepsilon(\hat{f}_T) = -\int_{T^c} \hat{f}(\zeta) \ln \hat{f}(\zeta) \, d\lambda(\zeta) - p_{\hat{f}}(T) \ln p_{\hat{f}}(T) + p_{\hat{f}}(T) \ln \lambda(T)$$

and so the user-relevant information which is being produced, relative to a state of total ignorance, is

$$\begin{aligned} \Delta I_T &\triangleq I(\hat{f}; f_D) - I(\hat{f}_T; f_D) \\ &= \int_T \hat{f}(\zeta) \ln \hat{f}(\zeta) \, d\lambda(\zeta) - p_{\hat{f}}(T) \ln p_{\hat{f}}(T) + p_{\hat{f}}(T) \ln \lambda(T) \end{aligned}$$

Now let us turn to the problem of dealing with the needs of an end-user whose constraint is *not* hard-and-fast but rather is *soft*. That is, either:

(1) there are two or more hard constraints, some of which are ranked as more important than others; or

(2) the constraint itself is *subjective* and thus must be regarded as a ranked list of plausible *interpretations* of what the constraint actually is.

Case I: Ranked, Multiple Hard Constraints. In the first case, we may assume that the end-user has a list $T_1, ..., T_m$ of hard constraints, along with corresponding ranks $1 = r_1 > ... > r_m > 0$ which express the relative importance of these constraints with respect to each other. (Thus if the constraint T_1 is much more important than the constraint T_2, then r_2 should be considerably smaller than 1, e.g. $r_2 = .5$.) This list can be re-expressed as an *nested* list $T_1' \subseteq ... \subseteq T_m'$ of constraints with $T_i' \triangleq T_1 \cup ... \cup T_i$ for $i = 1, ..., m$. That is, $T_1' = T_1$ is the constraint of first priority, $T_2' = T_1 \cup T_2$ reflects the fact that constraint T_1 *taken jointly with* constraint T_2 has second priority, and so on. Given this, the function h defined by $h(\zeta) = r_i$ (if $\zeta \in T_i'$) defines a fuzzy membership function on $\mathcal{R}$ whose associated canonical random set $\Sigma_h = \Sigma_A(h)$ has the mass distribution

$$\begin{aligned} p(\Sigma_h &= T_i') = \Delta r_i \triangleq r_i - r_{i+1} \qquad (i = 1, ..., m-1) \\ p(\Sigma_h &= T_m') = \Delta r_m \triangleq r_m \end{aligned}$$

(Note that $\Delta r_1 + ... + \Delta r_m = 1$ since $r_1 = 1$).

Case II: Subjective Constraints. In the second case we can *commence* by modeling the user's subjective constraint as a finite-level fuzzy membership function on $\mathcal{R}$, i.e. as $h(\zeta) \in \{r_1, ..., r_m\}$ where $1 = r_1 > ... > r_m \geq 0$. In this case the level subsets $T_i = \{\zeta \in \mathcal{R} | \ h(\zeta) = r_i\}$ form a nested sequence $T_1 \subseteq ... \subseteq T_m$ of hard constraints, each T_i being a plausible interpretation of what the user's subjective constraint is. As before, this soft constraint can be modeled as a canonical random subset $\Sigma_h = \Sigma_A(h)$.

However, the two kinds of constraint must be approached differently.

Computing Information Using Ranked Multiple Hard Constraints. In the case of a ranked list of hard constraints we can proceed as follows. Define the density function $\hat{f}_i$ as

$$\hat{f}_i(\zeta) \triangleq (1 - \mathbf{1}_{T_i'}(\zeta))\hat{f}(\zeta) + \mathbf{1}_{T_i'}(\zeta)p_{\hat{f}}(T_i')\lambda(T_i')^{-1}$$

for $i = 1, ..., m$. That is, $\hat{f}_i$ represents the output of the algorithm when this output has been disabled in regard to producing information about targets satisfying the hard constraint T_i'. Since $T_i' \subset T_{i+1}'$ it follows that $\hat{f}_{i+1}$ is "more disabled" (produces more confusion) than $\hat{f}_i$ since the former is "more uniform" than the latter.

Assume now that ground truth γ_0 is known. Then the incremental entropy $\Delta I_i \triangleq I(f_{Grnd}; \hat{f}_i) - I(f_{Grnd}; \hat{f})$ for $i = 1, ..., m$ is a measure of the amount of user-relevant information being produced by the tracker-classifier *in regard to the hard constraint* T_i alone, regardless of how important the user thinks that information actually is. We define the *user-relevant information* to be

$$\Delta I \triangleq \Delta r_1 \Delta I_1 + ... + \Delta r_m \Delta I_m$$

This quantity can be viewed from another perspective. Note that

$$\Delta I_1 = I(f_{Grnd}; \hat{f}_1) - I(f_{Grnd}; \hat{f})$$
$$\Delta I_i = [I(f_{Grnd}; \hat{f}_1) - I(f_{Grnd}; \hat{f})]$$
$$+[I(f_{Grnd}; \hat{f}_2) - I(f_{Grnd}; \hat{f}_1)] + ... + [I(f_{Grnd}; \hat{f}_i) - I(f_{Grnd}; \hat{f}_{i-1})]$$

for all $i = 2, ..., m$. Consequently,

$$\Delta I = [I(f_{Grnd}; \hat{f}_1) - I(f_{Grnd}; \hat{f})]$$
$$+r_2[I(f_{Grnd}; \hat{f}_2) - I(f_{Grnd}; \hat{f}_1)] + ... + r_m[I(f_{Grnd}; \hat{f}_i) - I(f_{Grnd}; \hat{f}_{i-1})]$$

This is in accordance with our intuitions. The rank of importance associated with information $I(f_{Grnd}; \hat{f}_1) - I(f_{Grnd}; \hat{f})$ regarding the targets in $T_1' = T_1$ is $r_1 = 1$. The rank of importance associated with information $I(f_{Grnd}; \hat{f}_2) - I(f_{Grnd}; \hat{f}_1)$ concerning competence in regard to the targets in $T_2 = T_2' - T_1'$ is r_2, and so on. Thus the information regarding targets in $T_1, .., T_m$ is ranked according to user preferences.

Computing Information Using a Subjective Constraint. On the other hand, suppose that user preferences are "soft" in the sense that the constraint is modeled as a finite-level fuzzy membership function h with level subsets $T_1 \subseteq ... \subseteq T_m$. Then a different approach must be used. Define $\hat{f}_i$ as

$$\hat{f}_i(\zeta) \triangleq (1 - 1_{T_i}(\zeta))\hat{f}(\zeta) + 1_{T_i}(\zeta)p_j(T_i)\lambda(T_i)^{-1}$$

for all $i = 1, ..., m$. Once again, $\hat{f}_i$ represents the output of the algorithm when this output has been disabled in regard to producing information about targets satisfying the crisp interpretation T of the soft constraint. Define the *average density function* by

$$\hat{f}_*(\zeta) \triangleq \Delta r_1 f_1(\zeta) + ... + \Delta r_m f_m(\zeta)$$

for all $\zeta \in \mathcal{R}$. Then $\hat{f}_*(\zeta) = \hat{f}(\zeta)$ for all $\zeta \notin T_m$ and $\hat{f}_*(\zeta)$ is uniform for $\zeta \in T_1$ with

$$\hat{f}_*(\zeta) = \frac{p_j(T_1)}{\lambda(T_1)} + ... + \frac{p_j(T_m)}{\lambda(T_m)}$$

in this case. For $\zeta \in T_m - T_1$ the density $\hat{f}_*$ has varying values which are, however, constant within each difference set $T_i - T_{i-1}$. This density models the tracker-classifier algorithm when its output has been disabled in regard to the soft constraint. It resembles the density of $\hat{f}_{T_i}$ to the same degree of plausibility Δr_i that T_i is an interpretation of the soft constraint modeled by the random subset Σ_h. Suppose, then, that we know ground truth γ_0. Then the information which is attributable to algorithm competence with respect to the soft constraint is

$$\Delta I = I(f_{Grnd}; \hat{f}_*) - I(f_{Grnd}; \hat{f}) \cong \ln\left(\frac{\hat{f}(\gamma_0)}{\hat{f}_*(\gamma_0)}\right)$$

This vanishes for $\gamma_0 \notin T_m$ and has varying values otherwise.

Similar considerations apply when we do not know ground truth. In this case we look at $\Delta I = I(\hat{f}; f_D) - I(\hat{f}_*; f_D)$.

The Multitarget Case

The approach just described can be generalized as follows. Once again, let us begin with a fixed constraint $T \subseteq \mathcal{R}$. We proceed by treating information regarding the targets satisfying constraint T as a *component* of total information. If $\hat{f}$ is the global density which describes the current output of a data fusion algorithm, we "disable" $\hat{f}$ in regard to T by computing a "marginal" density relative to T^c. Let $\zeta_1, ..., \zeta_k \in T^c$ and $\eta_1, ..., \eta_j \in T$ be distinct elements of T^c and T, respectively. Then $\hat{f}$ would be "disabled" with regard to elements in T if it were true that, for any fixed choice of $\zeta_1, ..., \zeta_k \in T^c$, the quantity $\hat{f}(\{\zeta_1, ..., \zeta_k, \eta_1, ..., \eta_j\})$ were constant regardless of the particular values of j and of $\eta_1, ..., \eta_j$. Of course, this cannot be strictly the case since the units of dimensionality of $\hat{f}(\{\zeta_1, ..., \zeta_k, \eta_1, ..., \eta_j\})$ must vary with $k + j$. Thus we can insist that this quantity be dependent upon $\eta_1, ..., \eta_j$ only in regard to the value of j. Thus we "integrate out" the influence of the targets in T by computing the marginal distribution $\int_T \hat{f}(Z \cup Y)\delta Y$ for all $Z \subseteq T^c$. Specifically, let M be the smallest positive integer such that $\hat{f}(X) = 0$ identically for all X with $|X| \geq M$. Then define the density $\hat{f}_T$ as follows: $\hat{f}_T(\emptyset) \triangleq \hat{f}(\emptyset)$ and

$$\hat{f}_T(X) \triangleq \frac{|X|!}{\lambda(T)^{|X \cap T|} M} \int_T \frac{|Y|!}{|(X \cap T^c) \cup Y|!} \hat{f}((X \cap T^c) \cup Y)\delta Y$$

for all X with $0 < |X| < M$ and $\hat{f}_T(X) = 0$ otherwise.

The set function $\hat{f}_T$ actually is a global density. For, let $X \cap T^c = \{\zeta_1, ..., \zeta_k\}$ with $|X \cap T^c| = k$ and $X \cap T = \{\eta_1, ..., \eta_j\}$ with $|X \cap T| = j$. Then the reader may verify that $\int \hat{f}_T(X)\delta X = 1$.

In particular, if $X \subseteq T$ with $|X| < M$ then

$$\hat{f}_T(X) = \frac{|X|!}{\lambda(T)^{|X|} M} \int_T \hat{f}(Y)\delta Y = \frac{|X|!}{\lambda(T)^{|X|} M} \hat{\beta}(T)$$

whereas if $X \subseteq T^c$ with $|X| < M$ then

$$\hat{f}_T(X) = \frac{|X|!}{M} \int_T \frac{|Y|!}{|X \cup Y|!} \, \hat{f}(X \cup Y) \delta Y$$

Note that $\int_T \hat{f}_T(X) \delta X = \hat{\beta}(T)$. Also notice that the definition of the global density $\hat{f}_T(X)$ is consistent with the definition of the conventional single-target density $\hat{f}_T(\zeta)$ previously described. For, suppose that $\hat{f}(X) = 0$ unless $|X| = 1$ and that $\hat{f}(\{\zeta\}) = f(\zeta)$ for some density function f on $\mathcal{R}$ with corresponding probability measure p_f. Notice that $\hat{f}((X \cap T^c) \cup Y) \neq 0$ for $Y \subseteq T$ if and only if $|Y \cap T| = 1$ or $|X \cap T^c| = 1$ or, in other words, if and only if $|X| = 1$. If $X = \{\zeta\} \subseteq T^c$ then $\hat{f}_T(\{\zeta\}) = \hat{f}(\{\zeta\} \cup \emptyset) = f(\zeta)$ and if $X = \{\eta\} \subseteq T$ then

$$\hat{f}_T(\{\eta\}) = \frac{1}{\lambda(T)} \int_T f(\eta) d\lambda(\eta) = \frac{p_f(T)}{\lambda(T)}$$

which is the same formula derived previously for the single-target case.

The global density $\hat{f}_T$ models a data fusion algorithm whose output has been "disabled" so as to render it incompetent with respect to all targets which are consistent with the hard constraint T. From this point on, therefore, we may proceed as we did in the single-target case. If ground truth $X_{Grnd} = \{\gamma_1, ..., \gamma_t\}$ is known, for example, then the incremental discriminations

$$\Delta I = I(f_{Grnd}; \hat{f}_T) - I(f_{Grnd}; \hat{f}) \cong -\ln(L^t \, \hat{f}_T(X_{Grnd})) + \ln(L^t \, \hat{f}(X_{Grnd}))$$
$$\Delta I = I(\hat{f}; u_{D,M}) - I(\hat{f}_T; u_{D,M})$$

measure the amount of information which is attributable to algorithm competence in regard to targets which satisfy the hard constraint T, depending on whether ground truth is known or not.

It should also be clear that if an end-user's constraint is "soft" in either of the senses previously described in the single-target case, then we may proceed exactly as in the single-target case—except for the fact that we will be using global densities $\hat{f}_{T_i}(X)$ rather than conventional densities $\hat{f}_{T_i}(\zeta)$.

8.2   Nonparametric Estimation

The output of a typical multisensor, multitarget data fusion algorithm will be some finite random subset $\Lambda = J(\Sigma_1, ..., \Sigma_m)$ of some parameter space, where $\Sigma_1, ..., \Sigma_m$ are i.i.d. random subsets of measurement space. We have seen (Section 6.5 of Chapter 6) that it is possible to determine the corresponding global density $f_\Lambda(X)$ analytically for data fusion algorithms which are of multihypothesis type. More generally, however, real-world data fusion algorithms are complex, highly nonlinear, and at least partially *ad hoc* constructions. Only rarely will we be able to determine the f_Λ of a specific algorithm from purely analytical considerations. For this reason we must be able to *estimate* f_Λ *empirically* from *simulated or actual data*. For example, suppose that we have a testbed facility which has the capability of generating statistically diverse simulated data (i.e., Monte Carlo simulation). If we stimulate the algorithm with a sequence $Z_1(l), ..., Z_m(l)$ of global measurement sets, for $l = 1, ..., M$, the result will be an output sequence $X_l = J(Z_1(l), ..., Z_m(l))$ of parameter sets for $l = 1, ..., M$. We must somehow construct an estimate of the global density f_Λ from the output sequence $X_1, ..., X_M$.

In conventional (i.e., single-sensor, single-target) statistics this is the purpose of *nonparametric density estimation*. Our goal in this section is to rigorously extend standard nonparametric density estimation techniques—so-called *projection-kernel estimation* in particular—to the problem of empirically estimating the form of global densities.

8.2.1 Review of Nonparametric Estimation

The purpose of this section is to summarize the basic concepts of nonparametric density estimation and, in particular, to review some specific such estimators: the histogram, moving-window, and kernel estimators. Our account follows that of Devroye [5]. The intuitive basis for empirical density estimators is the constructive definition of the Radon-Nikodým derivative. Let μ_f be a countably additive measure which is absolutely continuous with respect to Lebesgue measure λ. Then recall from Section 4.2.3 of Chapter 4 that the probability density function f corresponding to μ_f can be recovered from μ by means of the formula

$$f = \frac{d\mu_f}{d\lambda}(z) = \lim_{i \to \infty} \frac{\mu_f(E_{z;i})}{\lambda(E_{z;i})}$$

where $E_{z;i}$ is a sequence of measurable subsets (e.g. in a net or Vitali sequence) converging to the point set $\{z\}$. It follows that the density f can be approximated as a quotient of the form

$$f(z) = \frac{d\mu_f}{d\lambda}(z) \cong \frac{\mu_f(E_z)}{\lambda(E_z)}$$

where E_z is some suitably small closed ball centered at the point z.

Common Nonparametric Density Estimators

Let $z \in \mathbb{R}^n$. Then the *Dirac measure* concentrated at the vector z is defined to be: $\delta_z(S) \triangleq 1$ if $z \in S$ and $\delta_z(S) \triangleq 0$ if $z \notin S$. Let $z_1, ..., z_m$ be a set of samples of the unknown distribution. Then the *histogram measure* associated with the sample set is defined as:

$$p_{Hist}(S) \triangleq \frac{1}{m} \sum_{i=1}^{m} \delta_{z_i}(S)$$

Next, let $B_\varepsilon(z) \triangleq \{x \in \mathbb{R}^n | \; \|x - z\| \le \varepsilon\}$ denote the closed ball of radius ε centered at z. Let $z_1, ..., z_m$ be a set of samples of the unknown distribution.

Then the *moving-window estimator* is defined as:

$$f_{Win}(\mathbf{x}) \triangleq \frac{p_{Hist}(B_\varepsilon(\mathbf{x}))}{\lambda(B_\varepsilon(\mathbf{x}))} = \frac{1}{m} \sum_{i=1}^{m} \frac{\delta_{\mathbf{z}_i}(B_\varepsilon(\mathbf{x}))}{\lambda(B_\varepsilon(\mathbf{x}))}$$

where $\lambda(B_\varepsilon(\mathbf{x}))$ denotes the Lebesgue measure (i.e., hypervolume) of the ball $B_\varepsilon(\mathbf{x})$.

If m is very large and ε is very small the ratio

$$f_{Win}(\mathbf{x}) \triangleq \frac{p_{Hist}(B_\varepsilon(\mathbf{x}))}{\lambda(B_\varepsilon(\mathbf{x}))} \cong \frac{\mu_f(B_\varepsilon(\mathbf{x}))}{\lambda(B_\varepsilon(\mathbf{x}))} \cong f(\mathbf{x})$$

is a good approximation of f if the quantities ε and m are chosen correctly.

Estimators such as the histogram and moving-window estimators produce estimates which are discontinuous functions. "Kernel estimators" [5], [6] are generalizations of the moving-window estimator in which the discontinuous function

$$\frac{\delta_{\mathbf{z}_i}(B_\varepsilon(\mathbf{x}))}{\lambda(B_\varepsilon(\mathbf{x}))}$$

is "smoothed out" by replacing it with a continuous function. The kernel estimator has the form:

$$f_{Ker}(\mathbf{x}) \triangleq \frac{1}{m\varepsilon^n} \sum_{i=1}^{m} K\left(\frac{\mathbf{x} - \mathbf{z}_i}{\varepsilon}\right)$$

where n is the dimension of the space and $K(\mathbf{y})$, the kernel, is a unitless continuous function of the unitless vector variable $\mathbf{y}$ which integrates to unity: $\int K(\mathbf{y})dq(\mathbf{y}) = 1$ where $q \triangleq \lambda/\varepsilon^n$. If K is nonnegative, the kernel estimate of an unknown density is itself a density:

$$\int f_{Ker}(\mathbf{x})d\lambda(\mathbf{x}) = \frac{1}{m\varepsilon^n} \sum_{i=1}^{m} \int K\left(\frac{\mathbf{x} - \mathbf{z}_i}{\varepsilon}\right) d\lambda(\mathbf{x}) = \frac{1}{m\varepsilon^n} \sum_{i=1}^{m} \varepsilon^n = 1$$

The definition of a kernel estimator just given assumed that we are working in a vector space whose components all have the same units. In data fusion, however, both state and observation vectors usually have components with differing units. To take account of this possibility, we define

$$f_{Ker}(\mathbf{x}) \triangleq \frac{1}{m \det(E)} \sum_{i=1}^{m} K\left(E^{-1}(\mathbf{x} - \mathbf{z}_i)\right)$$

where $E = diag(\varepsilon_1, ..., \varepsilon_n)$ is a diagonal matrix. The concept of a kernel estimate can likewise be extended to hybrid continuous-discrete spaces as follows. Let $\mathcal{R} = \mathbb{R}^n \times U$ and let $\zeta_1, ..., \zeta_m \in \mathcal{R}$ be a sequence of data with $\zeta_i = (\mathbf{z}, u_i)$ for all $i = 1, ..., m$. For convenience assume that U has an even number of elements, and to each $u \in U$ assign an integer $d_u \in \{-|U|, ..., 0, ..., +|U|\}$

and a permutation π_u on the numbers $1, ..., |U|$ such that $\pi_u(d_u) = 0$. Define $k(u|v) = k(\pi_v(d_u))$ where $k(-)$ is another kernel function. Then

$$f_{Ker}(\mathbf{x}, v) = \frac{1}{m \det(E)} \sum_{i=1}^{m} K\left(E^{-1}(\mathbf{x} - \mathbf{z}_i)\right) k(v|u_i)$$

defines a kernel estimator for measurements drawn from hybrid space.

If we are given m then we must choose $\varepsilon = \varepsilon(m)$ correctly as a function of m so that a good estimate results. One also must choose the kernel function correctly (see [5, p. 126]). For more details see [5], [2], [8], [13].

Projection Kernel Estimators

How does one construct kernel estimators? One way of doing so arises from taking a functional analysis perspective based on so-called "reproducing kernel spaces" [1], [20]. The major property which allows the histogram measure p_{Hist} to approximate an unknown probability measure is its *counting property*–that is, the fact that it counts the number of samples which occur in any given subset S (i.e., it estimates the "density" of samples in the set S). This counting property derives from the *reproducing kernel behavior* of the Dirac density. That is, the Dirac density $\delta_{\mathbf{z}}(\mathbf{y}) = \delta(\mathbf{z} - \mathbf{y})$ has the well-known property that $\int \delta_{\mathbf{z}}(\mathbf{x}) f(\mathbf{x}) d\lambda(\mathbf{x}) = f(\mathbf{z})$ for any $\mathbf{z}$ and any f. In particular let $\mathbf{1}_S(\mathbf{x})$ be the indicator function of the Lebesgue-measurable subset S of $\mathbb{R}^n$: that is, $\mathbf{1}_S(\mathbf{x}) = 1$ if $\mathbf{x} \in S$ and $\mathbf{1}_S(\mathbf{x}) = 0$ otherwise. Then we can write the Dirac measure in the form:

$$\delta_{\mathbf{z}}(S) = \int_S \delta_{\mathbf{z}}(\mathbf{x}) \, d\lambda(\mathbf{x}) = \int \mathbf{1}_S(\mathbf{x}) \, \delta_{\mathbf{z}}(\mathbf{x}) \, d\lambda(\mathbf{x}) = \mathbf{1}_S(\mathbf{z})$$

That is, the reproducing kernel property of the Dirac density accounts for the fact that the Dirac measure $\delta_{\mathbf{z}}(S)$ is concentrated at the point $\mathbf{z}$. Unfortunately, the Dirac distribution is neither continuous nor differentiable in the conventional sense, so the histogram measure does not produce continuous-function estimates of the unknown probability measure.

Therefore, let us generalize the reproducing kernel property. Suppose that we have any Hilbert subspace H of the space $L^2(\mu_0)$ of functions which are square-integrable with respect to the measure μ_0. Assume further that there is a function $K : \mathbb{R}^n \times \mathbb{R}^n \to \mathbb{R}$ which has the following properties:

(1) $K(\mathbf{x}, \mathbf{y}) = K(\mathbf{y}, \mathbf{x})$ for all $\mathbf{x}, \mathbf{y} \in \mathbb{R}^n$

(2) If $K_{\mathbf{y}}(\mathbf{x}) \triangleq K(\mathbf{y}, \mathbf{x})$ for all $\mathbf{x} \in \mathbb{R}^n$ then $K_{\mathbf{y}} \in H$ for all $\mathbf{y} \in \mathbb{R}^n$

(3) The following holds: $\int K_{\mathbf{y}}(\mathbf{x}) \, f(\mathbf{x}) \, d\mu_0(\mathbf{x}) = f(\mathbf{y})$ for all $\mathbf{z} \in \mathbb{R}^n$. Then $K(-, -)$ is called a *reproducing kernel* for the Hilbert space H and in this case H is called a *reproducing kernel subspace* of $L^2(\mu_0)$ (see [1], [20]). If H has a reproducing kernel then this kernel is unique. For, suppose that $J(\mathbf{x}, \mathbf{y})$ is a reproducing kernel for H and let $J_{\mathbf{y}}(\mathbf{x}) \triangleq J(\mathbf{x}, \mathbf{y})$ for all $\mathbf{x}, \mathbf{y}$. Then

$$J(\mathbf{y}, \mathbf{z}) = J_{\mathbf{z}}(\mathbf{y}) = \int K_{\mathbf{y}}(\mathbf{x}) \, J_{\mathbf{z}}(\mathbf{x}) \, d\mu_0(\mathbf{x}) = K_{\mathbf{y}}(\mathbf{z}) = K(\mathbf{y}, \mathbf{z})$$

The following lemma suggests that there is a connection between reproducing kernels and estimation kernels:

Lemma 1 *Suppose that μ_0 has a density f_0 with respect to Lebesgue measure λ. (a) Suppose that H contains the constant functions. Then K_z integrates to unity with respect to μ_0 and $K_z f_0$ integrates to unity with respect to λ, for every $z \in \mathbb{R}^n$. (b) Suppose that all translates of the reproducing kernel K defined by $x \mapsto K(x - z, 0) = K_0(x - z)$ are contained in H. Then*

$$K(z - y, 0)\, f_0(z - y) = K(z, y)\, f_0(y)$$

for all $y, z \in \mathbb{R}^n$. (c) Suppose that the conditions of b) hold and that, in addition, H contains all contractions and dilations $\varepsilon^{-n} K_z(\varepsilon^{-1} x)$ of the reproducing kernel, for ε any unitless positive real number. Then

$$\frac{1}{\varepsilon^n} K\left(\frac{z - y}{\varepsilon}, 0\right) f_0\left(\frac{z - y}{\varepsilon}\right) = K(z, y)\, f_0(y)$$

for all $y, z \in \mathbb{R}^n$.

Proof. Easy and left to the reader.   □

Reproducing kernel spaces which satisfy all three conditions of Lemma 1 exist. An example is the space of polynomial functions in n real variables of degree less than or equal to M, for any $M \geq 1$. These facts will, in a moment, allow us to treat the function $K(x) \triangleq K(x, 0)\, f_0(x)$ as an estimating kernel. To see this let $\kappa_z(S) \triangleq \int_S K_z(x)\, d\lambda(x)$ be the additive measure corresponding to K_z. If $1_S \in H$ then we will have

$$\kappa_z(S) = \int_S K_z(x)\, d\lambda(x) = \int 1_S(x)\, K(z, x)\, d\lambda(x) = 1_S(z)$$

More generally let $f \in H$ be any square-integrable function which is concentrated at the point z—that is, which takes large values in a small neighborhood of z and small values outside of that neighborhood. Then we will likewise get $\int f(x)\, K_y(x)\, d\lambda(x) = f(y)$. This number will be large for y near z and small otherwise. Thus the measure $\kappa_z(S)$ tends, like the Dirac measure $\delta_z(S)$, to be concentrated at the point z. Accordingly, like the Dirac measure it can be used to construct a histogram-like measure which "counts" the number of occurrences in a set S of a given list of sample vectors. The approximate Radon-Nikodým derivative of this pseudo-histogram measure, in turn, can be used as an approximation to the unknown density. Unlike the histogram measure, however, this process will result in a continuous (or even differentiable, depending on how K is chosen) approximation to the unknown density.

In more detail, let f_0 be an $L^2(\lambda)$-integrable density function (called the "initial kernel"). That is, $\|f_0\|^2 = \int f_0(z)^2 d\lambda(z) < \infty$. Assume that the unknown density f is in $L^2(\mu_0)$:

$$\|f\|_0^2 = \int f(z)^2 d\mu_0(z) = \int f(z)^2\, f_0(z)\, d\lambda(z) < \infty$$

Let $\hat{\varphi}_1, ..., \hat{\varphi}_j, ...$ be an orthonormal basis for the Hilbert space $L^2(\mu_0)$. Let $\mathbf{z}_1, ..., \mathbf{z}_m$ be a set of samples drawn from the random vector which is governed by the unknown density and let $N(m)$ be some suitably chosen positive integer which functionally depends on the number of samples m. Let $H_{N(m)}$ be the subspace of $L^2(\mu_0)$ which is spanned by the functions $\hat{\varphi}_1, ..., \hat{\varphi}_{N(m)}$. Then

$$\langle f, g \rangle_0 \triangleq \int f(\mathbf{x}) g(\mathbf{x}) d\mu_0(\mathbf{x}) = \int f(\mathbf{x}) g(\mathbf{x}) f_0(\mathbf{x}) d\lambda(\mathbf{x})$$

is the inner product for the Hilbert space $H_{N(m)}$ and $\langle \hat{\varphi}_i, \hat{\varphi}_j \rangle_0 = \delta_{i,j}$ for all $i, j = 1, ..., N(m)$, where $\delta_{i,j}$ denotes the Kronecker delta.

Define the function $K_{N(m)}(\mathbf{x}, \mathbf{y})$ by

$$K_{N(m)}(\mathbf{x}, \mathbf{y}) = \sum_{j=1}^{N(m)} \hat{\varphi}_j(\mathbf{x}) \hat{\varphi}_j(\mathbf{y})$$

and define the function $K_{\mathbf{y}}(\mathbf{x}) \triangleq K_{N(m)}(\mathbf{x}, \mathbf{y})$ for all $\mathbf{x} \in \mathbb{R}^n$. Then $K_{\mathbf{y}} \in H_{N(m)}$ for all $\mathbf{y} \in \mathbb{R}^n$ and $\langle f, K_{\mathbf{y}} \rangle_0 = f(\mathbf{y})$ for all $\mathbf{y} \in \mathbb{R}^n$. For,

$$\langle f, K_{\mathbf{y}} \rangle_0 = \sum_{j=1}^{N(m)} \hat{\varphi}_j(\mathbf{y}) \langle f, \hat{\varphi}_j \rangle_0 = \left(\sum_{j=1}^{N(m)} \langle f, \hat{\varphi}_j \rangle \hat{\varphi}_j \right)(\mathbf{y}) = f(\mathbf{y})$$

In other words, the function $K_{N(m)}(\mathbf{x}, \mathbf{y})$ is the reproducing kernel function for the Hilbert space $H_{N(m)}$ and $H_{N(m)}$ is a reproducing kernel subspace of $L^2(\mu_0)$.

Given these preliminaries it can be shown [4, pp. 253-258] that the function

$$f_{Ker}(\mathbf{x}) \triangleq f_{Ker}(\mathbf{x}; \mathbf{z}_1, ..., \mathbf{z}_m) \triangleq \frac{1}{m} f_0(\mathbf{x}) \sum_{j=1}^{m} K_{\mathbf{z}_j}(\mathbf{x})$$

is an asymptotically unbiased pointwise estimator of the unknown density. It is called a *projection-kernel estimator* [4], [3].

Remark 3 *Note that f_{Ker} is rarely a nonnegative function. Also, f_{Ker} exhibits erratic behavior if the unknown density f is not in $L^2(\mu_0)$ (see [3, p. 5]).*

If we assume the validity of the conditions in the three parts of Lemma 1 above then we can write

$$f_{Ker}(\mathbf{x}) = \frac{1}{m\varepsilon^n} \sum_{j=1}^{m} K\left(\frac{\mathbf{x} - \mathbf{z}_j}{\varepsilon}\right)$$

where $K(\mathbf{x}) \triangleq K(\mathbf{x}, 0) f_0(\mathbf{x})$.

The projection-kernel estimator has many useful properties (see [4]). We will generalize three of these properties to nonparametric estimation of global densities in Propositions 7, 10, and 11 below.

8.2.2   "Global" Nonparametric Estimation

The concepts of nonparametric density estimation, and in particular the concepts of kernel estimation, can be directly extended to multisource, multitarget problems. Recall that the major intuitive idea behind conventional nonparametric density estimation was the fact that an unknown f whose probability measure is μ_f can be written as an approximate Radon-Nikodým derivative. Similar reasoning can be applied by using the fact (see Proposition 9 of Chapter 4) that global density functions can be written as *approximate set derivatives*. Let β_Σ be the belief measure of an absolutely continuous finite random subset Σ and let $\xi_1, ..., \xi_k$ be distinct elements of $\mathcal{R}$ with $\xi_j = (\mathbf{z}_j, u_j)$ for $j = 1, ...k$. Define the set function $\beta_{\Sigma,i}$ by

$$\beta_{\Sigma,i}(\{\xi_1, ..., \xi_k\}) \triangleq \beta_\Sigma((E_{\mathbf{z}_1;i} \times u_1) \cup ... \cup (E_{\mathbf{z}_k;i} \times u_k))$$

where $E_{\mathbf{z}_1;i}$ is a closed ball of radius $1/i$ centered at $\mathbf{z} \in \mathbb{R}^n$. Then recall that if $Z = \{\xi_1, ..., \xi_k\}$ with $|Z| = k$ then:

$$f_\Sigma(Z) = \lim_{i \to \infty} \frac{\sum_{Y \subseteq Z} (-1)^{|Z-Y|} \beta_{\Sigma,i}(Y)}{\lambda(E_{\mathbf{z}_1;i}) \cdots \lambda(E_{\mathbf{z}_k;i})}$$

and so we can write

$$f_\Sigma(Z) \cong \frac{\sum_{Y \subseteq Z} (-1)^{|Z-Y|} \beta_{\Sigma,i}(Y)}{\lambda(E_{\mathbf{z}_1;i}) \cdots \lambda(E_{\mathbf{z}_k;i})}$$

if i is sufficiently large.

The concept of a histogram can be extended to global densities as follows. Let $Z = \{\mathbf{z}_1, ..., \mathbf{z}_k\}$ be a finite subset of $\mathbb{R}^n$ and let S be an arbitrary subset of $\mathbb{R}^n$. Then the *Dirac belief measure*, concentrated at the finite subset Z, is the set function $\delta_Z(S)$ defined by $\delta_Z(S) = 1$ if $Z \subseteq S$ and $\delta_Z(S) = 0$ otherwise. Note that δ_Z is just the product of the usual Dirac probability measures, concentrated at the elements of Z: $\delta_Z(S) = \delta_{\mathbf{z}_1}(S) \cdots \delta_{\mathbf{z}_k}(S)$. Now, let $Z_1, ..., Z_m$ be a sequence of global samples drawn from an unknown finite random subset whose global density is to be estimated. Then the *histogram belief measure* of the sample is:

$$\beta_{Hist}(S) = \frac{1}{m} \sum_{j=1}^{m} \delta_{Z_j}(S)$$

If we treat β_{Hist} as a good approximation of the belief measure of the unknown global density, then we are led to the following definition of a *global moving-window estimator*:

$$f_{Win}(\{\mathbf{x}_1, ..., \mathbf{x}_n\}) \triangleq \frac{\sum_{Y \subseteq Z} (-1)^{|Z-Y|} \beta_{Hist,\varepsilon}(Y)}{\lambda(B_\varepsilon(\mathbf{x}_1)) \cdots \lambda(B_\varepsilon(\mathbf{x}_n))}$$

where

$$\beta_{Hist,\varepsilon}(\{\mathbf{x}_1, ..., \mathbf{x}_j\}) = \beta_{Hist}(B_\varepsilon(\mathbf{x}_1) \cup ... \cup B_\varepsilon(\mathbf{x}_j))$$

for suitably small ε.

The functions $\delta_Z, \beta_{Hist}, \beta_{Hist,\varepsilon}$ and f_{Win} can be extended from $\mathbb{R}^n$ to $\mathcal{R} = \mathbb{R}^n \times U$ in the obvious way:

$$f_{Win}(\{(\mathbf{x}_1, v_1), ..., (\mathbf{x}_n, v_n)\}) \triangleq \frac{\sum_{Y \subseteq Z} (-1)^{|Z-Y|} \beta_{Hist,\varepsilon}(Y)}{\lambda(B_\varepsilon(\mathbf{x}_1)) \cdots \lambda(B_\varepsilon(\mathbf{x}_n))}$$

$$\beta_{Hist,\varepsilon}(\{(\mathbf{x}_1, v_1), ..., (\mathbf{x}_j, v_j)\}) = \beta_{Hist}((B_\varepsilon(\mathbf{x}_1) \times v_1) \cup ... \cup (B_\varepsilon(\mathbf{x}_j) \times v_j))$$

8.2.3   Global Reproducing-Kernel Estimation

It is therefore possible to define a "global" analog of the moving-window estimator. Our intention in this chapter, however, is to adopt a less heuristic point of view and show that one can rigorously generalize the concept of a *projection-kernel estimator* to the problem of estimating global densities. We will also establish three convergence results for this estimator: Propositions 7, 10, and 11. To do this, however, we must first show that certain basic elements of functional analysis—the L^2 inner product and norm, orthonormal bases, reproducing-kernel subspaces, etc.—can be suitably generalized.

Set Integrals With Respect to a Belief Measure

We begin by slightly extending the concept of a set integral. This is necessary because, if $f(X)$ is a global density function, set integrals of the form $\int f(X)^2 \delta X$ are not well-defined (since the units of dimensionality of the global density $f(X)$ are not independent of the cardinality of X). We therefore define the concept of a set integral $\int_S \Phi(X) \delta\beta_0(X)$ with respect to a belief measure β_0. Let $f_0(X)$ be a global density function and let $\beta_0(S) = \int_S f_0(X) \delta X$ be its corresponding belief measure. Let $\Phi(X)$ be a *unitless* real-valued function of the finite-set variable X such that $\Phi(X) = 0$ identically for all $|X|$ sufficiently large. Then define the set integral of Φ with respect to the belief measure β_0 to be

$$\int_S \Phi(X)\, \delta\beta_0(X) \triangleq \int \Phi(X)\, f_0(X)\, \delta X$$

for all closed subsets S, assuming that the integral exists.

One special case is of particular interest. Let μ_0 be an ordinary probability measure with associated density function $f_0(\zeta)$ (with respect to hybrid measure) and let $f_0(k)$ be a probability distribution on the nonnegative integers (i.e., $f_0(k) = 0$ if $k < 0$ and $\sum_{k=0}^\infty f_0(k) = 1$. Then extend the conventional density function $f_0(\zeta)$ to a global density function by defining

$$f_0(X) \triangleq k!\, f_0(k)\, f_0(\zeta_1) \cdots f_0(\zeta_k)$$

for all $X = \{\zeta_1, ..., \zeta_k\}$ with $|X| = k$. Note that we have slightly extended the concept of a global density here as well, since $f_0(X)$ does not necessarily vanish identically for all X of sufficiently large cardinality. Note that

$$\beta_0(S) = \int_S f_0(X)\delta X = \sum_{k=0}^\infty f_0(k)\, \mu_0(S)^k$$

for all closed S (if, of course, the infinite sum converges). Also note that for β_0 thus defined,

$$\int_S \Phi(X)\, \delta\beta_0(X) = \sum_{k=0}^{\infty} f_0(k) \int_{S^k} \Phi(\{\zeta_1, ..., \zeta_k) d\mu_0(\zeta_1) \cdots d\mu_0(\zeta_k)$$

whore wo have defined the hybrid integral

$$\int_{S^k} h(\zeta_1, ..., \zeta_k) d\mu_0(\zeta_1) \cdots d\mu_0(\zeta_k)$$

$$\triangleq \sum_{v_1, ..., v_k \in U} \int_{S(v_1, ..., v_k)} h(\mathbf{x}_1, v_1, ..., \mathbf{x}_k, v_k) d\mu_0(\mathbf{x}_1) \cdots d\mu_0(\mathbf{x}_k)$$

$$= \sum_{v_1, ..., v_k \in U} \int_{S(v_1, ..., v_k)} h(\mathbf{x}_1, v_1, ..., \mathbf{x}_k, v_k)$$
$$\cdot f_0(\mathbf{x}_1) \cdots f_0(\mathbf{x}_k) d\lambda(\mathbf{x}_1) \cdots d\lambda(\mathbf{x}_k)$$

for any unitless μ_0-integrable function h, and where as usual

$$S(v_1, ..., v_k) \triangleq \{(\mathbf{x}_1, ..., \mathbf{x}_k) \in \mathbb{R}^{nk} |\ (\mathbf{x}_1, v_1, ..., \mathbf{x}_k, v_k) \in S\}$$

Since Φ is assumed to be unitless, set integrals of the form

$$\int \Phi(X)^2 \delta\beta_0(X) = \int \Phi(X)^2 f_0(X) \delta X$$

are well-defined. Thus we can speak of the space $L^2(\beta_0)$ of functions of a finite-set variable which are square-integrable with respect to the belief measure β_0 in the sense that $\int \Phi(X)^2 \delta\beta_0(X) < \infty$. If as usual we define the functions Φ_k by

$$\Phi_k(\zeta_1, ..., \zeta_k) \triangleq \begin{cases} \Phi(\{\zeta_1, ..., \zeta_k\}) & \text{if}\quad \zeta_1, ..., \zeta_k \text{ are distinct} \\ 0 & \text{if}\quad \text{otherwise} \end{cases}$$

then saying that Φ is square-integrable is the same as saying that the Φ_k are square-integrable for all $k \geq 1$ with respect to the product measure $\mu_0 \times ... \times \mu_0$ (product taken k times).

The set $L^2(\beta_0)$ is a linear space, as clearly follows from the inequality $(a + b)^2 \leq 2a^2 + 2b^2$ for any $a, b \in \mathbb{R}$ and the fact that the set integral is linear and preserves inequalities. Also, given two unitless square-integrable functions Φ, Ψ of a finite-set variable, the *inner product* with respect to the belief measure β_0 exists (because of the inequality $2|ab| \leq a^2 + b^2$) and is defined as:

$$\langle \Phi, \Psi \rangle_0 \triangleq \int \Phi(X)\, \Psi(X)\, \delta\beta_0(X) = \int \Phi(X)\, \Psi(X)\, f_0(X)\, \delta X$$

This inner product induces a norm $\|\Phi\|_0^2 \triangleq \langle \Phi, \Phi \rangle_0$ and it is easy to see that the Cauchy-Schwartz inequality is valid for this inner product: $\langle \Phi, \Psi \rangle_0^2 \leq \|\Phi\|_0^2 \|\Psi\|_0^2$.

Global Orthonormal Bases

For the remainder of this chapter we will assume that the global density $f_0(X)$ always has the form

$$f_0(X) = k! f_0(k) \prod_{\zeta \in X} f_0(\zeta)$$

where $f_0(k)$ is a probability distribution on the nonnegative integers, where $f_0(\zeta) = f_0(\mathbf{x}, v) = f_0(\mathbf{x}) f_0(v)$, and where $f_0(\mathbf{x})$ is a conventional density function on $\mathbb{R}^m$ with probability measure μ_0 and $f_0(v)$ is a probability distribution on U with $f_0(v) \neq 0$ for all $v \in U$. We will also use the notation μ_0 to denote the probability measure on the hybrid space $\mathcal{R}$ corresponding to $f_0(\zeta)$. (The five different usages of the notation 'f_0' and the two different usages of 'μ_0' will always be clear from context.) We will also assume that any function $\Phi(X)$ of a finite-set variable X vanishes identically for all X of sufficiently high cardinality. Accordingly, note that

$$
\begin{aligned}
\langle \Phi, \Psi \rangle_0 &= \int \Phi(X)\, \Psi(X)\, f_0(X)\, \delta X \\
&= \sum_{k=0}^{\infty} f_0(k) \int \Phi(\{\zeta_1, ..., \zeta_k\}) \Psi(\{\zeta_1, ..., \zeta_k\}) d\mu_0(\zeta_1) \cdots d\mu_0(\zeta_k)
\end{aligned}
$$

We now show how to extend any orthonormal basis of the conventional L^2-space $L^2(\mu_0)$ to an orthonormal basis of square-integrable functions of a finite-set variable. Specifically, let $\hat{\varphi}_1, ..., \hat{\varphi}_j, ...$ be an orthonormal basis for the Hilbert space $L^2(\mu_0)$ of functions on $\mathbb{R}^m$ which are square-integrable with respect to μ_0. By definition,

$$\langle \hat{\varphi}_i, \hat{\varphi}_j \rangle_0 \triangleq \int \hat{\varphi}_i(\mathbf{x})\, \hat{\varphi}_j(\mathbf{x})\, d\mu_0(\mathbf{x}) = \delta_{i,j}$$

for all $i, j \geq 0$. Extend the functions $\hat{\varphi}_j$ on $\mathbb{R}^m$ to functions $\hat{\varphi}_{j,w}$ on $\mathcal{R} = \mathbb{R}^m \times V$ by defining for all $\zeta = (\mathbf{x}, v)$:

$$\hat{\varphi}_{j,w}(\zeta) = \hat{\varphi}_{j,w}(\mathbf{x}, v) \triangleq \hat{\varphi}_j(\mathbf{x})\, \delta_{w,v} \sqrt{f_0(w)^{-1}}$$

where $\delta_{w,v} = 1$ if $w = v$ and $\delta_{w,v} = 0$ otherwise. Then it is easily verified that

$$\int \hat{\varphi}_{i,w}(\zeta)\, \hat{\varphi}_{j,u}(\zeta)\, d\mu_0(\zeta) = \delta_{w,u} \delta_{i,j}$$

for all $i, j \geq 0$ and $w, u \in U$. Thus the $\hat{\varphi}_{i,w}$ form an orthonormal basis with respect to the hybrid measure μ_0.

If a function $\psi(\zeta)$ is square-integrable with respect to the hybrid integral then

$$\int \psi(\zeta)^2 d\mu_0(\zeta) = \sum_{v \in V} \int \psi(\mathbf{x}, v)^2 f_0(\mathbf{x}) f_0(v) d\lambda(\mathbf{x}) < \infty$$

and so $\int \psi_v(\mathbf{x})^2 f_0(\mathbf{x}) d\lambda(\mathbf{x}) < \infty$ for all $v \in V$ where $\psi_v(\mathbf{z}) \triangleq \psi(\mathbf{z}, v)$. Let

$$\psi_v(\mathbf{x}) = \sum_{i=1}^{\infty} a_{i,v}\, \hat{\varphi}_i(\mathbf{x})$$

be the orthonormal expansion of ψ_v. Then

$$\psi(\mathbf{z}, v) = \sum_{i=1}^{\infty} \sum_{u \in V} a_{i,u} \sqrt{f_0(u)}\, \hat{\varphi}_{i,u}(\mathbf{x}, v)$$

is the orthonormal expansion for ψ. Accordingly, hereafter we will simplify notation by forcing an arbitrary linear ordering $u_1 \prec ... \prec u_N$ on the elements of $V = \{u_1, ..., u_N\}$ and adopting the lexicographic ordering

$$\hat{\varphi}_1 = \hat{\varphi}_{1,u_1}, ..., \hat{\varphi}_N = \hat{\varphi}_{1,u_N}, ..., \hat{\varphi}_{N(i-1)+1} = \hat{\varphi}_{i,u_1}, ..., \hat{\varphi}_{Ni} = \hat{\varphi}_{i,u_N}, ...$$

of the orthonormal basis $\{\hat{\varphi}_{i,u}\}_{i,u}$.

Next, for $k \geq 0$ and $1 \leq i_1, ..., i_k \leq k$ define

$$C_{k;i_1,...,i_k} \triangleq N_1(i_1, ..., i_k)! \; \cdots \; N_k(i_1, ..., i_k)!$$

where $N_d(i_1, ..., i_k) \triangleq \delta_{d,i_1} + ... + \delta_{d,i_k}$ denotes the number of elements of the list $i_1, ..., i_k$ which are equal to d, where $1 \leq d \leq k$.

We now turn to the problem of defining and constructing orthonormal bases for spaces of scalar-valued functions of a finite-set variable. The following definition and proposition tells us how to construct a "canonical global orthonormal basis" for spaces of finite-set functions from a conventional orthonormal basis on a space of ordinary functions.

Definition 2 *Let $X = \{\zeta_1, ..., \zeta_k\}$ be a finite subset of $\mathcal{R} = \mathbb{R}^m \times V$ with $|X| = k$ and let $\hat{\varphi}_1, ..., \hat{\varphi}_j, ...$ be an orthonormal basis for $L^2(\mu_0)$. Then the functions $\hat{\Phi}_{k;i_1,...,i_k}$ for $1 \leq i_1 \leq ... \leq i_k$ are defined as*

$$\hat{\Phi}_0(X) \triangleq \delta_{0,|X|} \frac{1}{\sqrt{f_0(0)}}$$

$$\hat{\Phi}_{k;i_1,...,i_k}(X) \triangleq \delta_{k,|X|}\, A_{k;i_1,...,i_k}\, perm(\hat{\varphi}_{k;i_1,...,i_k}(X))$$

where the constants $A_{k;i_1,...,i_k}$ and the $k \times k$ matrix $\hat{\varphi}_{k;i_1,...,i_k}(X)$ are defined by

$$A_{k;i_1,...,i_k} \triangleq \frac{1}{\sqrt{k!\, f_0(k)\, C_{k;i_1,...,i_k}}}$$

$$\hat{\varphi}_{k;i_1,...,i_k}(X)_{j,i} \triangleq \hat{\varphi}_{i_j}(\zeta_i) \qquad (i, j = 1, ..., k)$$

and where, as usual, for any $k \times k$ square matrix Q the permanentis defined by $perm(Q) \triangleq \sum_{\pi} Q_{1,\pi 1} \cdots Q_{k,\pi k}$ where the summation is taken over all permutations π on the numbers $1, ..., k$.

Note that $perm(\hat{\varphi}_{k;i_1,...,i_k}(X)) = perm(\hat{\varphi}_{k;i_{\pi 1},...,i_{\pi k}}(X))$ for all permutations π on the numbers $1, ..., k$, for each $k \geq 1$ and each list $1 \leq i_1, ..., i_k$.

The following shows that the functions $\hat{\Phi}_{k;i_1,...,i_k}$ form an orthonormal basis with respect to the inner product $\langle -, - \rangle_0$.

Proposition 2 *Let $\hat{\Phi}_{k;i_1,\ldots,i_k}$ be as just defined. Then*

$$\langle \hat{\Phi}_{k;i_1,\ldots,i_k} , \hat{\Phi}_{l;j_1,\ldots,j_l} \rangle_0 = \delta_{k,l}\, \delta_{i_1,j_1} \cdots \delta_{i_k,j_k}$$

for all $k, l \geq 0$ and all $1 \leq i_1 \leq \ldots \leq i_k$ and $1 \leq j_1 \leq \ldots \leq j_l$.

Proof. First note that $\langle \hat{\Phi}_0, \hat{\Phi}_0 \rangle_0 = f_0(0)\hat{\Phi}_0(\emptyset)\hat{\Phi}_0(\emptyset) = 1$ and that $\langle \hat{\Phi}_0, \hat{\Phi}_{k;i_1,\ldots,i_k} \rangle_0 = 0$ for $k > 0$ and $1 \leq i_1 \leq \ldots \leq i_k$. Assume now that $k, l > 0$. By definition, $\hat{\Phi}_{k;i_1,\ldots,i_k}(X) = 0$ unless $|X| = k$. Thus $\langle \hat{\Phi}_{k;i_1,\ldots,i_k} , \hat{\Phi}_{l;j_1,\ldots,j_l} \rangle = 0$ unless $k = l$. If $k = l$ then the reader may show that

$$\langle A^{-1}_{k;i_1,\ldots,i_k} \hat{\Phi}_{k;i_1,\ldots,i_k} , A^{-1}_{k;j_1,\ldots,j_k} \hat{\Phi}_{k;j_1,\ldots,j_k} \rangle_0$$

$$= f_0(k) \int \left(\sum_\pi \hat{\varphi}_{i_{\pi 1}}(\zeta_1) \cdots \hat{\varphi}_{i_{\pi k}}(\zeta_k) \right) \left(\sum_\sigma \hat{\varphi}_{j_{\sigma 1}}(\zeta_1) \cdots \hat{\varphi}_{j_{\sigma k}}(\zeta_k) \right)$$
$$\cdot d\mu_0(\zeta_1) \cdots d\mu_0(\zeta_k)$$

$$= f_0(k) \sum_\pi \sum_\sigma \langle \hat{\varphi}_{i_{\pi 1}}, \hat{\varphi}_{j_{\sigma 1}} \rangle_0 \cdots \langle \hat{\varphi}_{i_{\pi k}}, \hat{\varphi}_{j_{\sigma k}} \rangle_0$$

$$= k! f_0(k) \sum_\pi \sum_\sigma \delta_{i_{\pi 1}, j_{\sigma 1}} \cdots \delta_{i_{\pi k}, j_{\sigma k}} = k! f_0(k) \sum_\sigma \delta_{i_1, j_{\sigma 1}} \cdots \delta_{i_k, j_{\sigma k}}$$

The only nonvanishing terms in the rightmost sum will be those for which $i_1 = j_{\sigma 1}, \ldots, i_k = j_{\sigma k}$ for some permutation σ—which is to say, if the lists $i_1, \ldots, i_k$ and $j_1, \ldots, j_k$ differ from each other only up to permutation. Since $i_1 \leq \ldots \leq i_k$ and $j_1 \leq \ldots \leq j_k$, however, this means that $i_1 = j_1$ and $i_k = j_k$. Thus $i_2, \ldots, i_{k-1}$ and $j_2, \ldots, j_{k-1}$ must be identical ordered lists, which forces $i_2 = j_2$ and $i_{k-1} = j_{k-1}$; and so on. Thus

$$\langle A^{-1}_{k;i_1,\ldots,i_k} \hat{\Phi}_{k;i_1,\ldots,i_k} , A^{-1}_{k;j_1,\ldots,j_k} \hat{\Phi}_{k;j_1,\ldots,j_k} \rangle_0 = 0$$

unless $i_1 = j_1, \ldots, i_k = j_k$, in which case

$$\langle A^{-1}_{k;i_1,\ldots,i_k} \hat{\Phi}_{k;i_1,\ldots,i_k} , A^{-1}_{k;i_1,\ldots,i_k} \hat{\Phi}_{k;i_1,\ldots,i_k} \rangle_0 = k! f_0(k) \sum_\sigma \delta_{i_1, i_{\sigma 1}} \cdots \delta_{i_k, i_{\sigma k}}$$

The rightmost quantity equals the number of permutations which will transform the ordered list $i_1 \leq \ldots \leq i_k$ into itself. For each $1 \leq d \leq k$, by definition there are $0 \leq N_d(i_1, \ldots, i_k) \leq k$ numbers in the list $i_1, \ldots, i_k$ which equal d, with $\sum_{d=1}^k N_d(i_1, \ldots, i_k) = k$. In other words, the list $i_1, \ldots, i_k$ can be written as the disjoint union of sublists $T_d(i_1, \ldots, i_k)$ such that all the numbers in $T_d(i_1, \ldots, i_k)$ are equal to d. So, there are $N_1(i_1, \ldots, i_k)! \cdots N_k(i_1, \ldots, i_k)! = C_{k;i_1,\ldots,i_k}$ permutations σ which transform $i_1 \leq \ldots \leq i_k$ into itself. So

$$\langle A^{-1}_{k;i_1,\ldots,i_k} \hat{\Phi}_{k;i_1,\ldots,i_k} , A^{-1}_{k;j_1,\ldots,j_k} \hat{\Phi}_{k;j_1,\ldots,j_k} \rangle_0 = A^{-2}_{k;i_1,\ldots,i_k}$$

and we are done.   □

Remark 4 *In what follows we will simply notation by using multi-indices rather than indices. That is, we will abbreviate $\hat{\Phi}_{k;i_1,...,i_k}$ as $\hat{\Phi}_\tau$ where τ is any $(k+1)$-tuple of the form $(k, i_1, ..., i_k)$ with $k \geq 0$ and $1 \leq i_1 \leq ... \leq i_k$. We will also let $T(M, m)$ denote the set of all multi-indices $(k, i_1, ..., i_m)$ such that $1 \leq k \leq M$ and $1 \leq i_1 \leq ... \leq i_k \leq N(m)$, and will denote its cardinality by*

$$\tilde{N}(M, m) \triangleq |T(M, m)|$$

Having shown that the functions $\hat{\Phi}_{k;i_1,...,i_k}$ form an orthonormal basis, we show that it is an orthonormal basis for the space of square-integrable functions:

Proposition 3 *Let $\Psi \in L^2(\beta_0)$, i.e. $\Phi(X)$ vanishes identically for all X of sufficiently large cardinality and $\int \Phi(X)^2 \delta\beta_0(X) < \infty$. Let M be the smallest integer such that $\Psi(X) = 0$ for all X with $|X| > M$. Assume that $f_0(k) \neq 0$ for all $0 \leq k \leq M$. Then*

$$\Psi(X) = \sum_\tau \langle \Psi, \hat{\Phi}_\tau \rangle_0 \hat{\Phi}_\tau(X)$$

$$= \sum_{k=0}^\infty \sum_{1 \leq i_1 \leq ... \leq i_k < \infty} \langle \Psi, \hat{\Phi}_{k;i_1,...,i_k} \rangle_0 \, \hat{\Phi}_{k;i_1,...,i_k}(X)$$

for almost all finite subsets X.

Proof. Let $X = \{\zeta_1, ..., \zeta_e\}$ with $|X| = e$ and define the functions $\Phi_{k;i_1,...,i_k}$ by $\Phi_0(X) \triangleq \delta_{0,|X|}$ and by

$$\Phi_{k;i_1,...,i_k}(X) \triangleq \delta_{k,|X|} \sum_\pi \hat{\varphi}_{i_{\pi 1}}(\zeta_1) \cdots \hat{\varphi}_{i_{\pi k}}(\zeta_k)$$

for all finite subsets X, all $k \geq 1$ and all $i_1, ..., i_k \geq 1$. Then there will always be a permutation σ on the numbers $1, ..., e$ and a constant $C \neq 0$ such that $\Phi_{k;i_{\sigma 1},...,i_{\sigma k}} = C\hat{\Phi}_{k;i_1,...,i_k}$ and such that $i_1 \leq ... \leq i_k$. That is, the functions $\Phi_{k;i_1,...,i_k}$ generate the same Hilbert subspace of $L^2(\beta_0)$ as the orthonormal basis $\hat{\Phi}_{k;i_{\sigma 1},...,i_{\sigma k}}$. It is therefore enough to show that Ψ is in the Hilbert subspace generated by the $\Phi_{k;i_1,...,i_k}$. For, if we know that $\Psi = \sum_\tau a_\tau \hat{\Phi}_\tau$ for some $a_\tau \in \mathbb{R}$ then $\langle \Psi, \hat{\Phi}_{\tau'} \rangle_0 = \sum_\tau a_\tau \langle \hat{\Phi}_\tau, \hat{\Phi}_{\tau'} \rangle_0 = a_{\tau'}$ for all multi-indices $\tau' = (k, i_1, ..., i_k)$ with $k \geq 0$ and $1 \leq i_1 \leq ... \leq i_k$.

First note that $\langle \Psi, \Phi_0 \rangle_0 = f_0(0)\Psi(\emptyset)\Phi_0(\emptyset)$ and that

$$\langle \Psi, \Phi_{k;i_1,...,i_k} \rangle_0 = f_0(k) \int \Psi(\{\zeta_1, ..., \zeta_k\}) \left(\sum_\pi \hat{\varphi}_{i_{\pi 1}}(\zeta_1) \cdots \hat{\varphi}_{i_{\pi k}}(\zeta_k) \right)$$

$$\cdot d\mu_0(\zeta_1) \cdots d\mu_0(\zeta_k)$$

$$= f_0(k) \sum_\pi \int \Psi(\{\zeta_1, ..., \zeta_k\}) \hat{\varphi}_{i_{\pi 1}}(\zeta_1) \cdots \hat{\varphi}_{i_{\pi k}}(\zeta_k)$$

$$\cdot d\mu_0(\zeta_1) \cdots d\mu_0(\zeta_k)$$

$$= f_0(k) \sum_\pi \int \Psi(\{\zeta_{\pi 1}, ..., \zeta_{\pi k}\}) \hat{\varphi}_{i_1}(\zeta_{\pi 1}) \cdots \hat{\varphi}_{i_k}(\zeta_{\pi k})$$

$$\cdot d\mu_0(\zeta_{\pi 1}) \cdots d\mu_0(\zeta_{\pi k})$$

$$= k! f_0(k) \int \Psi(\{\zeta_1, ..., \zeta_k\}) \hat{\varphi}_{i_1}(\zeta_1) \cdots \hat{\varphi}_{i_k}(\zeta_k)$$

$$\cdot d\mu_0(\zeta_1) \cdots d\mu_0(\zeta_k)$$

So if $X = \{\eta_1, ..., \eta_k\}$ with $|X| = k$ then $\langle \Psi, \Phi_0 \rangle_0 \Phi_0(\emptyset) = f_0(0) \Psi(\emptyset) \Phi_0(\emptyset)^2$ for $k = 0$ and

$$\langle \Psi, \Phi_{k; i_1, ..., i_k} \rangle_0 \Phi_{k; i_1, ..., i_k}(X)$$

$$= k! f_0(k) \sum_\sigma \int \Psi(\{\zeta_1, ..., \zeta_k\}) \hat{\varphi}_{i_1}(\zeta_1) \cdots \hat{\varphi}_{i_k}(\zeta_k) \hat{\varphi}_{i_1}(\eta_{\sigma 1}) \cdots \hat{\varphi}_{i_k}(\eta_{\sigma k})$$

$$\cdot d\mu_0(\zeta_1) \cdots d\mu_0(\zeta_k)$$

$$= k! f_0(k) \sum_\sigma \int$$

$$\Psi(\{\zeta_1, ..., \zeta_k\}) \hat{\varphi}_{i_1}(\zeta_1) \hat{\varphi}_{i_1}(\eta_{\sigma 1}) d\mu_0(\zeta_1) \cdots \hat{\varphi}_{i_k}(\zeta_k) \hat{\varphi}_{i_k}(\eta_{\sigma k}) d\mu_0(\zeta_k)$$

for $k > 0$. Now consider the functions

$$\Psi_{k,i}(\zeta_i) \triangleq \begin{cases} \Psi(\{\zeta_1, ..., \zeta_i, ..., \zeta_k\}) & \text{if} \quad \zeta_1, ..., \zeta_i, ..., \zeta_k \text{ are distinct} \\ 0 & \text{if} \quad \text{otherwise} \end{cases}$$

for all $i = 1, ..., k$. Thus $\Psi_{k,i}(\zeta_i) = \Psi(\{\zeta_1, ..., \zeta_i, ..., \zeta_k\})$ almost everywhere. Since Ψ is square-integrable $\Psi_{k,i}$ is square-integrable and so we know that $\sum_{j=1}^\infty \langle \Psi_{k,i}, \hat{\varphi}_j \rangle_0 \hat{\varphi}_j = \Psi_{k,i}$ almost everywhere for all k, i and all η. In other words, we know that

$$\sum_{j=1}^\infty \int \Psi(\{\zeta_1, ..., \zeta_i, ..., \zeta_k\}) \hat{\varphi}_j(\zeta_i) \hat{\varphi}_j(\eta) d\mu_0(\zeta_i) = \Psi(\{\zeta_1, ..., \zeta_{i-1}, \eta, \zeta_{i+1}, ..., \zeta_k\})$$

almost everywhere. Thus

$$\sum_{i_1=1}^\infty \cdots \sum_{i_k=1}^\infty \langle \Psi, \Phi_{k; i_1, ..., i_k} \rangle_0 \Phi_{k; i_1, ..., i_k}(X)$$

$$= k! f_0(k) \sum_\sigma \sum_{i_1=1}^\infty \cdots \sum_{i_k=1}^\infty \int$$

$$\Psi(\{\zeta_1, ..., \zeta_k\}) \hat{\varphi}_{i_1}(\zeta_1) \hat{\varphi}_{i_1}(\eta_{\sigma 1}) d\mu_0(\zeta_1) \cdots \hat{\varphi}_{i_k}(\zeta_k) \hat{\varphi}_{i_k}(\eta_{\sigma k}) d\mu_0(\zeta_k)$$

$$= k! f_0(k) \sum_\sigma \Psi(\{\eta_{\sigma 1}, ..., \eta_{\sigma k}\}) = (k!)^2 f_0(k) \Psi(\{\zeta_1, ..., \zeta_k\}) = (k!)^2 f_0(k) \Psi(X)$$

almost everywhere. Therefore

$$\Psi(X) = \sum_{k=0}^M \frac{1}{(k!)^2 f_0(k)} \sum_{i_1, ..., i_k=1}^\infty \langle \Psi, \Phi_{k; i_1, ..., i_k} \rangle_0 \Phi_{k; i_1, ..., i_k}(X) = \sum_\tau a_\tau \Phi_\tau(X)$$

for suitable $a_\tau \in \mathbb{R}$. Since this holds for almost any finite subset X this completes the proof.   □

Global Reproducing Kernels

Let M be a positive integer, let

$$Z_1 = \{\zeta_{1;1}, ..., \zeta_{1;n(1)}\}, ..., Z_m = \{\zeta_{m;1}, ..., \zeta_{m;n(m)}\}$$

be a set of global samples with $|Z_i| = n(i)$ for all $i = 1, ..., m$ and let $N(m)$ be a positive integer which is a function of m, the number of global samples. Let $H_{N(m)}$ be the subspace of $L^2(\mu_0)$ generated by the $\hat{\varphi}_1, ..., \hat{\varphi}_{N(m)}$ and let $D_{N(m)}$ be the subspace of $L^2(\beta_0)$ generated by the $\hat{\Phi}_{k;i_1,...,i_k}$ for all $k = 1, ..., M$ and for all $0 \leq i_1 \leq ... \leq i_k \leq N(m)$. Then:

Definition 3 *The global reproducing kernel of the subspace $D_{N(m)}$ is*

$$K_{N(m)}(X, Y) \triangleq \sum_{\tau \in T(M,m)} \hat{\Phi}_\tau(X)\, \hat{\Phi}_\tau(Y)$$

where $T(M, m)$ is the set of all multi-indices $\tau = (k, i_1, ..., i_k)$ with $0 \leq k \leq M$ and $0 \leq i_1 \leq ... \leq i_k \leq N(m)$.

We evidently have:

Proposition 4 *The function $K_{N(m)}$ just defined is a reproducing kernel for the subspace $D_{N(m)}$. That is, if $K_Y(X) \triangleq K_{N(m)}(X, Y)$ then $K_{N(m)}(X, Y) = K_{N(m)}(Y, X)$ and $\langle \Psi, K_X \rangle_0 = \Psi(X)$ for all $\Psi \in D_{N(m)}$ and all finite subsets X.*

Proof. If $\Psi \in D_{N(m)}$ then $\Psi = \sum_\tau a_\tau \hat{\Phi}_\tau$ for some $a_\tau \in \mathbb{R}$. Thus

$$\langle \Psi, K_X \rangle_0 = \sum_{\tau \in T(M,m)} \hat{\Phi}_\tau(X)\, \langle \Psi, \hat{\Phi}_\tau \rangle_0 = \sum_{\tau,\sigma \in T(M,m)} a_\sigma \hat{\Phi}_\tau \, \langle \hat{\Phi}_\sigma, \hat{\Phi}_\tau \rangle_0$$

$$= \sum_{\tau \in T(M,m)} a_\tau \hat{\Phi}_\tau(X) = \Psi(X)$$

as desired.   □

The following shows that a global reproducing kernel function can be expressed in a simpler form:

Proposition 5 *Let $K_{N(m)}$ be the global reproducing kernel for the subspace $D_{N(m)}$ and let $X = \{\zeta_1, ..., \zeta_k\}$ and $Y = \{\eta_1, ..., \eta_l\}$ with $|X| = k$ and $|Y| = l$. Then*

$$K_{N(m)}(X, Y) = \delta_{k,l}\, \frac{1}{k! f_0(k)} \sum_\pi K_{N(m)}(\zeta_{\pi 1}, \eta_1) \cdots K_{N(m)}(\zeta_{\pi k}, \eta_k)$$

where $K_{N(m)}(\zeta, \eta) = \sum_{i=1}^{N(m)} \hat{\varphi}_i(\zeta) \hat{\varphi}_i(\eta)$ *is the reproducing kernel of the subspace* $H_{N(m)}$. *(As usual, the summation* $\sum_{\pi}$ *is taken over all permutations on the numbers* $1, ..., k$.*)*

Proof. We will use a somewhat roundabout approach requiring three lemmas. Let $E_{N(m)}$ be the subspace of all real-valued functions Ψ of a finite-set variable which satisfy the property that the functions $\Psi_{k,i}$ defined by

$$\Psi_{k,i}(\zeta_i) = \begin{cases} \Psi(\{\zeta_1, ..., \zeta_i, ..., \zeta_k\}) & \text{if } \zeta_1, ..., \zeta_i, ..., \zeta_k \text{ are distinct} \\ 0 & \text{if otherwise} \end{cases}$$

are in $H_{N(m)}$ for any $k \geq 0$ and all fixed and distinct $\zeta_1, ..., \zeta_{j-1}, \zeta_{j+1}, ..., \zeta_k$. Define the function $J(X, Y)$ by

$$J(X, Y) \triangleq \delta_{k,l} \frac{1}{k! f_0(k)} \sum_{\pi} K_{N(m)}(\zeta_{\pi 1}, \eta_1) \cdots K_{N(m)}(\zeta_{\pi k}, \eta_k)$$

Given this, we will prove the following: In Lemma A we will show that J is a reproducing kernel for the space $E_{N(m)}$. In Lemma B we will prove that the space $D_{N(m)}$ is a subspace of the space $E_{N(m)}$. In Lemma C we will show that the functions $J_Y(X) \triangleq J(X, Y)$ are contained in the space $D_{N(m)}$ for every Y. Then, finally, $J(X, Y) = K_{N(m)}(X, Y)$ will follow from the uniqueness of reproducing kernels. We begin with:

Lemma A to Proposition 5: Define $J_Y(X) \triangleq J(X, Y)$ for all X. Then: (a) $J_Y \in E_{N(m)}$ for all Y; (b) if $\Psi \in E_{N(m)}$ then $\langle \Psi, J_Y \rangle = \Psi(Y)$ for all Y. In other words, J is a reproducing kernel for the subspace $E_{N(m)}$.

Proof. (a) is obvious from the definition of J. (b) By definition,

$$\begin{aligned} \langle \Psi, J_Y \rangle_0 &= \int \Psi(X) J(X, Y) f_0(X) \delta X \\ &= \sum_{k=0}^{\infty} f_0(k) \int \Psi(\{\zeta_1, ..., \zeta_k\}) \, J(\{\zeta_1, ..., \zeta_k\}, Y) \, d\mu_0(\zeta_1) \cdots d\mu_0(\zeta_k) \\ &= \sum_{k=0}^{\infty} f_0(k) \int \Psi(\{\zeta_1, ..., \zeta_k\}) \\ &\quad \cdot \left(\delta_{k,l} \frac{1}{k! f_0(k)} \sum_{\pi} K_{N(m)}(\zeta_{\pi 1}, \eta_1) \cdots K_{N(m)}(\zeta_{\pi k}, \eta_k) \right) \\ &\quad \cdot d\mu_0(\zeta_1) \cdots d\mu_0(\zeta_k) \\ &= \frac{1}{l!} \sum_{\pi} \int \Psi(\{\zeta_1, ..., \zeta_l\}) \, K_{N(m)}(\zeta_1, \eta_{\pi 1}) \cdots K_{N(m)}(\zeta_l, \eta_{\pi l}) \\ &\quad \cdot d\mu_0(\zeta_1) \cdots d\mu_0(\zeta_l) \end{aligned}$$

By assumption the function $\Psi_{l,1}$ is an element of $H_{N(m)}$. Thus the reproducing kernel property yields

$$\langle \Psi, J_Y \rangle_0 = \frac{1}{l!} \sum_{\pi} \int \Psi(\{\eta_{\pi 1}, \zeta_2, ..., \zeta_l\}) \, K_{N(m)}(\zeta_2, \eta_{\pi 2}) \cdots K_{N(m)}(\zeta_l, \eta_{\pi l})$$

$$\cdot d\mu_0(\zeta_2) \cdots d\mu_0(\zeta_l)$$

and thus, in general,

$$\langle \Psi, J_Y \rangle = \frac{1}{l!} \sum_\pi \Psi(\{\eta_{\pi 1}, ..., \eta_{\pi l}\}) = \Psi(\{\eta_1, ..., \eta_l\}) = \Psi(Y)$$

which completes the proof of Lemma A.   □

Lemma B to Proposition 5: $D_{N(m)} \subseteq E_{N(m)}$
Proof. By definition $\Psi \in D_{N(m)}$ if and only if

$$\Psi(X) = \sum_{k=0}^{M} \sum_{1 \le i_1 \le ... \le i_k \le N(m)} a_{k;i_1,...,i_k} \hat{\Phi}_{k;i_1,...,i_k}(X)$$

for all X, for suitable $a_{k;i_1,...,i_k} \in \mathbb{R}$. Consequently, $\Psi \in E_{N(m)}$ if $\hat{\Phi}_{k;i_1,...,i_k} \in E_{N(m)}$ for all $k, i_1, ..., i_k$. By definition,

$$\hat{\Phi}_{k;i_1,...,i_k}(\{\zeta_1, ..., \zeta_e\}) = \delta_{k,e} A_{k;i_1,...,i_k} \sum_\pi \hat{\varphi}_{i_{\pi 1}}(\zeta_1) \cdots \hat{\varphi}_{i_{\pi k}}(\zeta_k)$$

The real-valued function defined by

$$\zeta_j \mapsto \delta_{k,e} A_{k;i_1,...,i_k} \sum_\pi \hat{\varphi}_{i_{\pi 1}}(\zeta_1) \cdots \varphi_{i_{\pi j}}(\zeta_j) \cdots \hat{\varphi}_{i_{\pi k}}(\zeta_k)$$

is obviously an element of $H_{N(m)}$, concluding the proof of Lemma B.   □

Lemma C to Proposition 5: $J_Y \in D_{N(m)}$ for all finite subsets Y.
Proof. Define $C_{l;i_{\pi 1},...,i_{\pi l}} \triangleq C_{l;i_1,...,i_l}$ for all permutations π on the numbers $1,..,l$. Then we will prove that

$$\sum_{i_1,...,i_l=1}^{N(m)} C_{l;i_1,...,i_l} \hat{\Phi}_{l;i_1,...,i_l}(X) \hat{\Phi}_{l;i_1,...,i_l}(Y) = |X|! \, J(X,Y)$$

from which will immediately follow that $J(X,Y)$ is a linear combination of the $\hat{\Phi}$'s for any fixed Y. Both sides of this equation vanish identically if $|X| \ne |Y|$ and so we may assume that $|X| = |Y| = e$. We then get:

$$\sum_{i_1,...,i_l=1}^{N(m)} C_{l;i_1,...,i_l} \hat{\Phi}_{l;i_1,...,i_l}(X) \hat{\Phi}_{l;i_1,...,i_l}(Y)$$

$$= \frac{1}{e! f_0(e)} \sum_{i_1,...,i_e=1}^{N(m)} \left(\sum_\pi \hat{\varphi}_{i_1}(\zeta_{\pi 1}) \cdots \hat{\varphi}_{i_e}(\zeta_{\pi e}) \right) \left(\sum_\sigma \hat{\varphi}_{i_1}(\eta_{\sigma 1}) \cdots \hat{\varphi}_{i_e}(\eta_{\sigma e}) \right)$$

$$= \frac{1}{e! f_0(e)} \sum_{\pi,\sigma} K_{N(m)}(\zeta_{\pi 1}, \eta_{\sigma 1}) \cdots K_{N(m)}(\zeta_{\pi e}, \eta_{\sigma e})$$

$$= \frac{1}{f_0(e)} \sum_\pi K_{N(m)}(\zeta_{\pi 1}, \eta_1) \cdots K_{N(m)}(\zeta_{\pi e}, \eta_e)$$

However, by definition of J we know that

$$J(X, Y) = \frac{1}{e! f_0(e)} \sum_{\pi} K_{N(m)}(\zeta_{\pi 1}, \eta_1) \cdots K_{N(m)}(\zeta_{\pi e}, \eta_e)$$

so that

$$\sum_{i_1, \ldots, i_l = 1}^{N(m)} C_{l; i_1, \ldots, i_l} \; \hat{\Phi}_{l; i_1, \ldots, i_l}(X) \; \hat{\Phi}_{l; i_1, \ldots, i_l}(Y) = e! \, J(X, Y)$$

as claimed.   □

We can now complete the proof of Proposition 5. Because J is a reproducing kernel for the space $E_{N(m)}$ it is also a reproducing kernel for the subspace $D_{N(m)} \subseteq E_{N(m)}$ provided that $J_Y \in D_{N(m)}$ for all Y (which is what we demonstrated in Lemma C). We then get

$$K_{N(m)}(X, Y) = K_Y(X) = \langle K_Y, J_X \rangle_0 = J_X(Y) = J(X, Y)$$

for all X, Y and we are done.   □

Global Projection-Kernel Estimators

In this section we will generalize the concept of a projection-kernel estimator to the case of global densities. We will establish a relatively simple formula for the estimator (Proposition 6) and, by directly generalizing results of [4, pp. 253-258], prove three convergence results regarding it. In Proposition 7 we show that the global projection-kernel estimator is an asymptotically unbiased pointwise estimator of the unknown density. In Proposition 10 we show that the (normalized) estimator converges to the unknown (normalized) density with respect to the L^2 norm, and in Proposition 11 that it converges to the unknown (normalized) density absolutely in probability.

In regard to the latter two results, the qualifier "normalized" refers to the fact that one begins with a fixed "initial global density" f_0 and then proves the results not for the unknown density f itself, but rather for the normalized density f/f_0. The reason that this is necessary is that the "global" generalizations of two common norms used in functional analysis, the L^2 and uniform-convergence norms, namely $\|g\|_0^2 \triangleq \int g(X)^2 \delta X$ and $|g|_0 \triangleq \sup_X |g(X)|$, respectively, are not well-defined if g is a global density function. The reason is, of course, that the unit of dimensionality of $g(X)$ is a function of $|X|$ and thus these two formulas do not make any sense because $g(X)$ and $g(Y)$ are incomparable if $|X| \neq |Y|$. As a result one must use *unitless* functions g of a finite-set variable X. (This is why we were forced to slightly generalize the concept of a set integral.)

Let $Z_1, \ldots, Z_m$ be a list of global samples drawn from an absolutely continuous finite random subset which has unknown global density function f. We assume that besides being a global density, f satisfies the property $f/f_0 \in L^2(\beta_0)$ or, equivalently, that $f(X) = f_0(X)\check{f}(X)$ for all X, where $\check{f}$ is a unitless real-valued function of a finite-set variable such that $\check{f} \in L^2(\beta_0)$. As in the previous sections, the initial density is assumed to have the form $f_0(X) = k! f_0(k) \prod_{\zeta \in X} f_0(\zeta)$ where $f_0(k)$ is a probability distribution on the nonnegative integers, where $f_0(\zeta) = f_0(\mathbf{x}, v) = f_0(\mathbf{x}) f_0(v)$, and where $f_0(\mathbf{x})$ is a conventional density function on $\mathbb{R}^n$ with probability measure μ_0 and $f_0(v)$ is a probability distribution on U with $f_0(v) \neq 0$ for all $v \in U$.

Let $D_{N(m)}$ and $K_{N(m)}$ be defined as in the previous section. In direct parallel to the discussion of Section 8.2.1 we define

$$f_{Ker;m}(X) \triangleq f_{Ker}(X; Z_1, ..., Z_m) \triangleq \frac{1}{m} f_0(X) \sum_{i=1}^{m} K_{N(m)}(Z_i, X)$$

It is easy to show that:

Proposition 6 *Suppose that*

$$Z_1 = \{\xi_{1;1}, ..., \xi_{1;n(1)}\}, ..., Z_m = \{\xi_{m;1}, ..., \xi_{m;n(m)}\}$$

are global observations such that $|Z_i| = n(i)$ *for all* $i = 1, ..., m$ *and such that* $X = \{\zeta_1, ..., \zeta_e\}$. *Then the global projection-kernel estimator can be written in the form*

$$f_{Ker}(X; Z_1, ..., Z_m) = \frac{1}{me! f_0(e)} f_0(X) \sum_{i=1}^{m} \delta_{e,n(i)} \, perm(K(i))$$

where $K(i)$ *is the* $e \times e$ *confusion matrix defined by* $K(i)_{j,k} \triangleq K_{N(m)}(\zeta_j, \xi_{i;k})$ *for all* $j, k = 1, ..., e$.

Proof. By Proposition 5,

$$K_{N(m)}(Z_i, X) = \delta_{n(i),e} \frac{1}{e! f_0(e)} \sum_{\pi} K_{N(m)}(\zeta_{\pi 1}, \eta_1) \cdots K_{N(m)}(\zeta_{\pi e}, \eta_e)$$

and thus

$$
\begin{aligned}
f_{Ker}(X; Z_1, ..., Z_m) &= \frac{1}{m} f_0(X) \sum_{i=1}^{m} K_{N(m)}(Z_i, X) \\
&= \frac{1}{me! f_0(e)} f_0(X) \sum_{i=1}^{m} \delta_{n(i),e} \, perm(K(i))
\end{aligned}
$$

as claimed.   □

We now prove some basic properties of the global projection-kernel estimator. The discussion is a direct parallel of [4, pp. 253-258].

Proposition 7 *(a) Let* $f = f_0 \check{f}$ *where* $\check{f} \in L^2(\beta_0)$. *Define the projection of* f *into the subspace* $D_{N(m)}$ *by*

$$f_m^{\downarrow}(X) \triangleq f_0(X) \sum_{\tau \in T(M,m)} \langle \check{f}, \, \hat{\Phi}_\tau \rangle_0 \hat{\Phi}_\tau(X)$$

for all finite subsets X, and where the summation is taken over all multi-indices $\tau = (k, i_1, ..., i_k)$ with $0 \le k \le M$ and $1 \le i_1 \le ... \le i_k \le N(m)$. Then $f_{Ker}(X; \Sigma_1, ..., \Sigma_m)$ is a pointwise unbiased estimator of $f_m^\downarrow$, i.e.

$$E[f_{Ker}(X; \Sigma_1, ..., \Sigma_m)] = f_m^\downarrow(X)$$

for almost all finite subsets X. (b) The global projection-kernel estimator is an asymptotically unbiased pointwise estimator of the unknown density f. That is, let $\Sigma_1, ..., \Sigma_m, ...$ be an i.i.d. sequence of absolutely continuous finite random subsets which have f has their common global density function. Then

$$\lim_{m \to \infty} E[f_{Ker}(X; \Sigma_1, ..., \Sigma_m)] = f(X)$$

for almost all X.

Proof. Note that $\int \check{f}(X) \delta \beta_0(X) = \int f(X) \delta X = 1$ and that

$$
\begin{aligned}
\lim_{m \to \infty} f_m^\downarrow(X) &= f_0(X) \lim_{m \to \infty} \sum_{\tau \in T(M,m)} \langle \check{f}, \hat{\Phi}_\tau \rangle_0 \hat{\Phi}_\tau(X) \\
&= f_0(X) \sum_{k=0}^{M} \sum_{1 \le i_1 \le ... \le i_k < \infty} \langle \check{f}, \hat{\Phi}_{k;i_1,...,i_k} \rangle_0 \hat{\Phi}_{k;i_1,...,i_k}(X) \\
&= f_0(X) \check{f}(X) = f(X)
\end{aligned}
$$

for almost all X, by Proposition 3. Next, by definition

$$
\begin{aligned}
E[f_{Ker}(X; \Sigma_1, ..., \Sigma_m)] &= \int f_{Ker}(X; Y_1, ..., Y_m) f(Y_1) \cdots f(Y_m) \delta Y_1 \cdots \delta Y_m \\
&= \frac{1}{m} f_0(X) \sum_{i=1}^{m} \int \\
&\quad K_{N(m)}(Y_i, X) f(Y_1) \cdots f(Y_m) \delta Y_1 \cdots \delta Y_m \\
&= \frac{1}{m} f_0(X) \sum_{i=1}^{m} \int \\
&\quad K_{N(m)}(Y_i, X) \check{f}(Y_1) \cdots \check{f}(Y_m) \delta \beta_0(Y_1) \cdots \delta \beta_0(Y_m) \\
&= \frac{1}{m} f_0(X) \sum_{i=1}^{m} \left(\sum_{\tau \in T(M,m)} \hat{\Phi}_\tau(Y_i) \hat{\Phi}_\tau(X) \right) \\
&\quad \cdot \check{f}(Y_1) \cdots \check{f}(Y_m) \delta \beta_0(Y_1) \cdots \delta \beta_0(Y_m) \\
&= \frac{1}{m} \sum_{i=1}^{m} \left(f_0(X) \sum_{\tau \in T(M,m)} \langle \hat{\Phi}_\tau, \check{f} \rangle_0 \hat{\Phi}_\tau(X) \right) \\
&= f_m^\downarrow(X)
\end{aligned}
$$

The assertion then immediately follows.   □

Proposition 7 establishes a pointwise convergence property for global projection-kernel estimators. In the remainder of this section we will establish a few additional convergence properties. We begin with the following lemma:

Lemma 8 *Abbreviate*

$$\check{f}_m(X) \triangleq \frac{f_{Ker}(X; Z_1, ..., Z_m)}{f_0(X)} - \frac{1}{m} \sum_{i=1}^{m} \sum_{\tau \in T(M,m)} \hat{\Phi}_\tau(Z_i)\hat{\Phi}_\tau(X)$$

$$\check{f}_m^\downarrow(X) \triangleq \frac{f_m^\downarrow(X)}{f_0(X)} = \sum_{\tau \in T(M,m)} \langle \check{f}, \hat{\Phi}_\tau \rangle_0 \hat{\Phi}_\tau(X)$$

Then:

$$\lim_{m \to \infty} E[\|\check{f}_m - \check{f}_m^\downarrow\|_0^2] = \lim_{m \to \infty} \frac{1}{m} \int K_{N(m)}(Y,Y) \, \check{f}(Y) \, \delta\beta_0(Y)$$

Proof. On the one hand,

$$\|\check{f}_m - \check{f}_m^\downarrow\|_0^2 = \|\check{f}_m\|_0^2 - 2\langle \check{f}_m, \check{f}_m^\downarrow \rangle_0 + \|\check{f}_m^\downarrow\|_0^2$$

Taking expectations of both sides and using the fact that $\check{f}_m$ is an unbiased pointwise estimator of $\check{f}_m^\downarrow$ by Proposition 7, we get

$$E[\|\check{f}_m - \check{f}_m^\downarrow\|_0^2] = E[\|\check{f}_m\|_0^2] - \|\check{f}_m^\downarrow\|_0^2$$

By definition

$$\|\check{f}_m\|_0^2 = \frac{1}{m^2} \left\| \sum_{\tau \in T(M,m)} \left(\sum_{i=1}^{m} \hat{\Phi}_\tau(Y_i) \right) \hat{\Phi}_\tau(X) \right\|_0^2 = \frac{1}{m^2} \sum_{\tau \in T(M,m)} \left(\sum_{i=1}^{m} \hat{\Phi}_\tau(Y_i) \right)^2$$

$$= \frac{1}{m^2} \sum_{\tau \in T(M,m)} \sum_{i \neq k} \hat{\Phi}_\tau(Y_i)\hat{\Phi}_\tau(Y_k) + \frac{1}{m^2} \sum_{\tau \in T(M,m)} \sum_{i=1}^{m} \hat{\Phi}_\tau(Y_i)^2$$

Taking expectations of both sides yields

$$E[\|\check{f}_m\|_0^2] = \frac{1}{m^2} \sum_{\tau \in T(M,m)} \sum_{i \neq k} \int \hat{\Phi}_\tau(Y_i)\hat{\Phi}_\tau(Y_k)\check{f}(Y_1) \cdots \check{f}(Y_m) \, \delta\beta_0(Y_1) \cdots \delta\beta_0(Y_m)$$

$$+ \frac{1}{m^2} \sum_{\tau \in T(M,m)} \sum_{i=1}^{m} \int \hat{\Phi}_\tau(Y_i)^2 \check{f}(Y_1) \cdots \check{f}(Y_m) \, \delta\beta_0(Y_1) \cdots \delta\beta_0(Y_m)$$

$$= \frac{1}{m^2} \sum_{\tau \in T(M,m)} \sum_{i \neq k} \left(\int \hat{\Phi}_\tau(Y)\check{f}(Y)\delta\beta_0(Y) \right)^2$$

$$+ \frac{1}{m^2} \sum_{\tau \in T(M,m)} \sum_{i=1}^{m} \int \hat{\Phi}_\tau(Y)^2 \check{f}(Y)\delta\beta_0(Y)$$

$$= \frac{m^2 - m}{m^2}\|\check{f}_m^\downarrow\|_0^2 + \frac{1}{m} \int K_{N(m)}(Y,Y)\check{f}(Y)\delta\beta_0(Y)$$

Thus

$$E[\|\check{f}_m - \check{f}_m^\perp\|_0^2] = E[\|\check{f}_m\|_0^2] - \|\check{f}_m^\perp\|_0^2 = -\frac{1}{m}\|\check{f}_m^\perp\|_0^2 + \frac{1}{m}\int K_{N(m)}(Y,Y)\check{f}(Y)\delta\beta_0(Y)$$

The lemma immediately follows from the fact that $\|\check{f}_m^\perp\|_0 \leq \|\check{f}\|_0 < \infty$. $\qquad \square$

Lemma 9 *Suppose that* $\lim_{m\to\infty} \|\check{f}_m^\perp - \check{f}\|_0 = 0$. *Then* $\lim_{m\to\infty} \|\check{f}_m - \check{f}\|_0 = 0$ *if and only if*

$$\lim_{m\to\infty}\frac{1}{m}\int K_{N(m)}(Y,Y)\,\check{f}(Y)\,\delta\beta_0(Y) = 0$$

Proof. By definition $\check{f}_m - \check{f}_m^\perp$ is in the subspace $D_{N(m)}$ (generated by the $\hat{\Phi}_\tau$ for $\tau \in T(M,m)$) and $\check{f}_m^\perp - \check{f}$ is orthogonal to this subspace. Therefore

$$\|\check{f}_m - \check{f}\|_0^2 = \|\check{f}_m - \check{f}_m^\perp\|_0^2 + \|\check{f}_m^\perp - \check{f}\|_0^2$$

Taking expectations of both sides of this equation yields

$$E[\|\check{f}_m - \check{f}\|_0^2] = E[\|\check{f}_m - \check{f}_m^\perp\|_0^2] + \|\check{f}_m^\perp - \check{f}\|_0^2$$

The left-hand side of the equation will vanish as $m \to \infty$ if and only if both terms in the right hand side vanish as $m \to \infty$. Since the second of these terms vanishes by assumption, the proposition follows from Lemma 8 $\qquad \square$

From Lemmas 9 and 8 we get the following L^2 convergence result:

Proposition 10 *Assume that* $\|\check{f}_m^\perp - \check{f}\|_0 \to 0$ *as* $m \to \infty$. *Let*

$$\tilde{N}(M,m) \triangleq |T(M,m)|$$

where as usual $T(M,m)$ *denotes the set of multi-indices* $(k,i_1,...,i_k)$ *such that* $1 \leq k \leq M$ *and* $1 \leq i_1 \leq ... \leq i_k \leq N(m)$ *and where, also as usual,* M *denotes the smallest integer such that* $\check{f}(X) = 0$ *for all* X *with* $|X| > M$. *(a) If one of the following conditions holds*

$$sup_\tau \sup_X \left|\hat{\Phi}_\tau(X)\right| < \infty \qquad or \qquad sup_X \left|\check{f}(X)\right| < \infty$$

then

$$\lim_{m\to\infty}\frac{\tilde{N}(M,m)}{m} = 0 \qquad implies \qquad \lim_{m\to\infty} E[\|\check{f}_m - \check{f}\|_0^2] = 0$$

(b) Conversely, suppose that $\inf_X \check{f}(X) > 0$. *Then*

$$\lim_{m\to\infty} E[\|\check{f}_m - \check{f}\|_0^2] = 0 \qquad implies \qquad \lim_{m\to\infty}\frac{\tilde{N}(M,m)}{m} = 0$$

Proof. (a) By definition

$$\int K_{N(m)}(Y,Y)\check{f}(Y)\delta\beta_0(Y) = \sum_{\tau \in T(M,m)} \int \hat{\Phi}_\tau(Y)^2 \check{f}(Y)\delta\beta_0(Y)$$

Given either assumption (1) or assumption (2) it follows that the integrals on the right-hand side of this equation are bounded by some constant C and therefore that

$$\frac{1}{m}\int K_{N(m)}(Y,Y)\check{f}(Y)\delta\beta_0(Y) \le \frac{1}{m}\sum_{\tau \in T(M,m)} C = C\frac{\tilde{N}(M,m)}{m}$$

The first part of the proposition then follows immediately from Lemma 9. (b) In this case we get

$$\frac{1}{m}\int K_{N(m)}(Y,Y)\check{f}(Y)\delta\beta_0(Y) = \frac{1}{m}\sum_{\tau \in T(M,m)} \int \hat{\Phi}_\tau(Y)^2\check{f}(Y)\delta\beta_0(Y)$$

$$\ge \frac{\inf_X \check{f}(X)}{m}\sum_{\tau \in T(M,m)} \int \|\hat{\Phi}_\tau\|^2$$

$$= \frac{\tilde{N}(M,m)}{m}\inf_X \check{f}(X)$$

where $\inf_X \check{f}(X) > 0$ by assumption. Part (b) then immediately follows. □

Definition 4 *Let $g(X)$ be a unitless, real-valued function of a finite-set variable X. In what follows we will write $|g|_0 \triangleq \sup_X |g(X)|$ for the uniform-convergence norm, where the supremum is taken over all finite subsets X of $\mathcal{R}$.*

We conclude this section with the following convergence result using this norm:

Proposition 11 *Suppose that $\lim_{m\to\infty} |\check{f}_m^\downarrow - \check{f}|_0 = 0$ and that $C_0 \triangleq \sup_\tau |\hat{\Phi}_\tau|_0 < \infty$. Suppose further that*

$$\sum_{m=1}^\infty \tilde{N}(M,m)\, e^{-\gamma m/\tilde{N}(M,m)^2} < \infty$$

for all positive real numbers γ, where $\tilde{N}(M,m)$ is as defined in Proposition 10. Then $\check{f}_m$ converges absolutely in probability to $\check{f}$, i.e., $\sum_{m=1}^\infty p(|\check{f}_m - \check{f}|_0 > \varepsilon) < \infty$ for all positive real numbers ε.

Proof. From the triangle inequality we know that

$$|\check{f}_m - \check{f}|_0 \le |\check{f}_m - \check{f}_m^\downarrow|_0 + |\check{f}_m^\downarrow - \check{f}|_0$$

Define $a_m \triangleq \left| \check{f}_m^{\downarrow} - \check{f} \right|_0$. Then

$$p(\left| \check{f}_m - \check{f} \right|_0 > \varepsilon) \leq p(\left| \check{f}_m - \check{f}_m^{\downarrow} \right|_0 + \left| \check{f}_m^{\downarrow} - \check{f} \right|_0 > \varepsilon) = p(\left| \check{f}_m - \check{f}_m^{\downarrow} \right|_0 > \varepsilon - a_m)$$

Since $a_m \to 0$ as $m \to \infty$ by assumption, it follows that for each integer $C > 0$ there is a number d_C such that $a_m \leq d_C$ and therefore $\varepsilon - a_m \geq \varepsilon - d_C$ for all $m \geq C$. Accordingly

$$p(\left| \check{f}_m - \check{f} \right|_0 > \varepsilon) \leq p(\left| \check{f}_m - \check{f}_m^{\downarrow} \right|_0 > \varepsilon - d_C)$$

for all $m \geq C$. If we prove that the proposition holds for the sequence $m \mapsto \left| \check{f}_m - \check{f}_m^{\downarrow} \right|_0$ we will be done. For, we will then know that $\sum_{m=1}^{\infty} p(\left| \check{f}_m - \check{f}_m^{\downarrow} \right|_0 > \delta) < \infty$ for all $\delta > 0$, therefore that

$$\sum_{m \geq C} p(\left| \check{f}_m - \check{f}_m^{\downarrow} \right|_0 > \varepsilon - d_C) < \infty$$

for all $\varepsilon > 0$, and therefore that

$$\sum_{m=1}^{C} p(\left| \check{f}_m - \check{f} \right|_0 > \varepsilon) + \sum_{m \geq C} p(\left| \check{f}_m - \check{f} \right|_0 > \varepsilon) < \infty$$

Consequently, let us prove that $\sum_{m=1}^{\infty} p(\left| \check{f}_m - \check{f}_m^{\downarrow} \right|_0 > \varepsilon) < \infty$ for all $\varepsilon > 0$. The proof will be based on the following inequality:

Bernstein-Fréchet Inequality [4, p. 41]: *Suppose that we are given real random variables $w_1, ..., w_m$ such that $a_i \leq w_i \leq b_i$ for all $i = 1, ..., m$ and some constant real numbers $a_1, ..., a_m, b_1, ..., b_m$. Then the following is true:*

$$p\left(\left| \sum_{i=1}^{m} (w_i - \mathrm{E}[w_i]) \right| \geq t \right) \leq 2 \exp \left(\frac{-2t^2}{\sum_{i=1}^{m} (b_i - a_i)^2} \right)$$

for all $t > 0$.

In preparation, note that for any $g = \sum_{\tau \in T(M,m)} a_\tau \hat{\Phi}_\tau$ and $h = \sum_{\tau \in T(M,m)} b_\tau \hat{\Phi}_\tau$ that

$$|g - h|_0 = \sup_X \left| \sum_{\tau \in T(M,m)} (a_\tau - b_\tau) \hat{\Phi}_\tau(X) \right| \leq C_0 \sum_{\tau \in T(M,m)} |a_\tau - b_\tau|$$

Therefore

$$\left| \check{f}_m - \check{f}_m^{\downarrow} \right|_0 \leq C_0 \sum_{\tau \in T(M,m)} \left| \frac{1}{m} \sum_{i=1}^{m} \hat{\Phi}_\tau(Y_i) - \int \hat{\Phi}_\tau(Y) \check{f}(Y) \delta \beta_0(Y) \right|$$

Now fix the multi-index τ and define the independent random real numbers $w_{i;m}$ for $1 \leq i \leq m$ by $w_{i;m} \triangleq \frac{1}{m} \hat{\Phi}_\tau(\Sigma_i)$ where as usual $\Sigma_1, ..., \Sigma_m$ is an i.i.d.

sequence of absolutely continuous finite random subsets whose common global density function is $f = f_0 \check{f}$. The expected value of $w_{m;i}$ is

$$E[w_{i;m}] = \frac{1}{m} \int \hat{\Phi}_\tau(Y) \check{f}(Y) \delta \beta_0(Y)$$

Consequently,

$$\sum_{i=1}^{m} (w_{i;m} - E[w_{i;m}]) = \frac{1}{m} \sum_{i=1}^{m} \hat{\Phi}_\tau(\Sigma_i) - \int \hat{\Phi}_\tau(Y) \check{f}(Y) \delta \beta_0(Y) \triangleq x_{\tau;m}$$

where we have chosen $x_{\tau;m}$ as an abbreviation to simplify notation. Also, note that

$$-\frac{C_0}{m} \leq w_{i;m} \leq \frac{C_0}{m}$$

Set $a_i = -C_0/m$, $b_i = C_0/m$ for all $i = 1, ..., m$ and note that $\sum_{i=1}^{m}(b_i - a_i)^2 = m\left(\frac{2C_0}{m}\right)^2 = \frac{4C_0^2}{m}$. Then from the Bernstein-Fréchet inequality we get:

$$p(|x_{\tau;m}| \geq t) \leq 2 \exp\left(-\frac{mt^2}{2C_0^2}\right)$$

for all $t > 0$ and for each fixed multi-index τ. Thus

$$p\left(\sum_{\tau \in T(M,m)} |x_{\tau;m}| \geq \varepsilon\right) \leq \sum_{\tau \in T(M,m)} p\left(|x_{\tau;m}| \geq \frac{\varepsilon}{\tilde{N}(M,m)}\right)$$

$$\leq 2\tilde{N}(M,m) \exp\left(-\frac{\varepsilon^2}{2C_0^2} \frac{m}{\tilde{N}(M,m)^2}\right)$$

for all $\varepsilon > 0$ which, by assumption, yields

$$\sum_{m=1}^{\infty} p\left(\sum_{\tau \in T(M,m)} |x_{\tau;m}| \geq \varepsilon\right) \leq 2 \sum_{m=1}^{\infty} \tilde{N}(M,m) \exp\left(-\frac{\varepsilon^2}{2C_0^2} \frac{m}{\tilde{N}(M,m)^2}\right) < \infty$$

for all $\varepsilon > 0$. This completes the proof.   □

Bibliography

[1] Aronszajn, N. (1943) La théorie des noyaux reproduisants et ses applications. I, *Proceedings of the Cambridge Philosophical Society*, **39**, 133-153.

[2] Brualdi, R. A. and Ryser, H. J. (1991) *Combinatorial Matrix Theory*, Cambridge University Press.

[3] Berlinet, A. (1991) Reproducing Kernels and Finite Order Kernels, in G. Roussas (ed.), *Nonparametric Functional Estimation and Related Topics*, Kluwer Academic Publishers, Dordrecht, pp. 3-18.

[4] Bosq, D. and Lecoutre, J. P. (1987) *Théorie de l'Estimation Fonctionnelle*, Economica, Paris.

[5] Devroye, L. (1987) *A Course in Density Estimation*, Birkhäuser.

[6] Devroye, L., and Györfi, L. (1985) *Nonparametric Density Estimation: The L_1 View*, J. Wiley, New York.

[7] Dubois, D. and Prade, H. (1987) Properties of Measures of Information in Evidence and Possibility Theories, *Fuzzy Sets and Systems*, **24**, 161-182.

[8] Gasser, T., and Herrmann, E. (1991) Data-Adaptive Kernel Estimation, in G. Roussas (ed.), *Nonparametric Functional Estimation and Related Topics*, Kluwer Academic Publishers, Dordrecht, pp. 67-79.

[9] Jaffee, R. J., Thode, S., Gauss, J. A., and Perry, F. A. (1992) Evaluation of Correlation and Tracking Algorithms for Transition from Research and Development to Production, *Proceedings of the 1991 Joint Service Data Fusion Symposium*, Vol. I (Unclassified), Oct. 7-11 1991, Naval Air Development Center, Warminster PA, 483-501.

[10] Klir, G.J. (1987) Where Do We Stand on Measures of Uncertainty, Ambiguity, Fuzziness, and the Like?, *Fuzzy Sets and Systems*, **24**, 141-160.

[11] Klir, G.J. (1991) Generalized information theory, *Fuzzy Sets and Systems*, **40**, 127-142

[12] Klir, G. J. and Folger, T. (1988) *Fuzzy Sets, Uncertainty and Information*, Prentice-Hall.

[13] Jones, M. C. (1991) Prospects for Automatic Bandwidth Selection in Extensions to Basic Kernel Density Estimation, in G. Roussas (ed.), *Nonparametric Functional Estimation and Related Topics*, Kluwer Academic Publishers, Dordrecht, pp. 241-249.

[14] Maeda, Y. and Ichihashi, H. (1993) An Uncertainty Measure With Monotinicity Under the Random Set Inclusion, *International Journal of Intelligent Systems*, **21**, 379-392.

[15] Mahler, R. (1995) Unified nonparametric data fusion, *SPIE Proceedings* **2484**, 66-74.

[16] Nguyen, H. T. (1987) On Entropy of Random Sets and Possibility Distributions, in J. C. Bezdek (ed.), *Analysis of Fuzzy Information, Vol. I: Mathematics an Logic*, CRC Press, Boca Raton FL, pp. 145-156.

[17] O'Neil, S. D. and Bridgland, M. F. (1991) Fast Algorithms for Joint Probabilistic Data Association, *Proceedings of the Fourth National Symposium on Sensor Fusion*, Vol. I (Unclassified), Apr. 2-4 1991, Orlando, 1991, ERIM, Ann Arbor MI, 173-189.

[18] Stephanou, H. E., and Lu, S. Y. (1984) Measuring Consensus Effectiveness By a Generalized Entropy Criterion, *Proceedings of the First IEEE Conference on Artificial Intelligence Applications*, 510-517.

[19] Thoma, H. M. Belief Function Computations, in I.R. Goodman, M.M. Gupta, H.T. Nguyen and G.S. Rogers (eds.), *Conditional Logic in Expert Systems*, North-Holland, 269-308.

[20] Yosida, K. (1980) *Functional Analysis*, Springer-Verlag, New York.

[21] Wolff, T. O., Lutes, C. L., Mahler, R., and Fixsen, D. (1992) Standards, Metrics, Benchmarks, and Monte Carlo: Evaluating Multi-Sensor Fusion Systems, *Proceedings of the 1991 Joint Service Data Fusion Symposium*, Vol. I (Unclassified), Oct. 7-11 1991, Naval Air Development Center, Warminster PA, 394-446.

Part III

Use of Conditional and Relational Events in Data Fusion

Scope of Work

In Part III here, we consider a number of topics associated with conditional and relational event algebra, focusing on potential applications to data fusion issues. The work is divided into eight basic chapters as follows:

Chapter 9 presents introductory material, with Section 9.1 providing the philosophy of approach: the presentation revolves around a number of motivating examples and conditional and relational event algebra are shown to address these issues in a more meaningful manner than previously developed approaches. Section 9.2 presents an overview of the problems involved and how an algebraic approach – in conjunction with appropriate numerical implementations – can provide more coherency and a stronger logical foundation instead of considering numerical techniques alone. Section 9.3 describes further details of the algebraic approach. Section 9.3.1 considers partitioning of information. Section 9.3.2 establishes the basic analytic framework used throughout Part III – boolean and sigma algebras and probability spaces – and introduces the concepts of deduction and "enduction"(backward deduction). Section 9.3.3 introduces the idea of algebraic combining or averaging of information, prior to taking probability evaluations. Section 9.3.3 also contains an informal introduction to the concept of a "constant-probability" event, a critical part of the development of relational event algebra (which properly begins in Chapter 15), and itself a particular type of "conditional event" (the main introduction being in Chapter 10). Section 9.4 presents the basic problem of determining similarity of events and introduces two measures of similarity (or probability distance functions), the absolute value of the difference of probability of events $d_{P,1}$, which we call the "naive probability distance" function, and the probability of the boolean symmetric difference (or sum) of events $d_{P,2}$, which we call the "natural absolute probability distance" function (due to its well-known role in the standard development of metrics spaces from probability spaces). This section also briefly discusses the actual role conditional and relational event algebra play in implementing such measures.

Chapter 10 introduces in a more formal manner conditional event algebra. In Section 10.1, Example 1 is presented to illustrate how conditional event algebra can be used to address the modeling of implicational statements and rules of inference in Expert Systems. In addition, a fundamental inequality (eq.(10.7)) is presented showing the discrepancy between using the material conditional operation and conditional probability in evaluating conditional statements. Section 10.2 introduces the basic properties all conditional event algebras should satisfy. The Fréchet-Hailperin bounds on the probability of logical combinations of events – applied to conditional events when in a boolean setting – are also discussed and applied to deriving inequalities for some measures of similarity of conditional events. Section 10.3 continues the discussion on constant-probability events begun in Section 9.3.

Chapter 11 introduces briefly the three best-investigated conditional event algebras for interpreting conditional statements. Section 11.1 provides general remarks; Section 11.2 introduces the DeFinetti-Goodman-Nguyen-Walker

(DGNW) conditional event algebra (cea); Section 11.3 introduces the Adams-Calabrese (AC) cea; Section 11.4 provides some comparisons between AC and DGNW cea's; Section 11.5 presents in some detail Lewis' negative result on forcing boolean conditional events to lie in the same boolean algebra as unconditional events (Theorem 0); and Section 11.6 presents an introduction to the only well-analyzed boolean cea, Product Space (PS).

Chapter 12 details further important development of PS cea. Section 12.1 is devoted to the determination of equivalence, ordering relations and the basic calculus of logical operations (all detailed in Theorems 1-9). Section 12.2 treats additional properties of PS, including: comparison of two natural types of events that occur in PS (Section 12.2.1), showing that though they are similar, they are non-isomorphic distinct entities; Lewis' Theorem and its relation to PS (Section 12.2.2); higher order conditionals in PS (Section 12.2.3); and other properties (Section 12.2.4), including emphasis on the fact that PS obeys all laws of probability (while DGNW and AC do not) as well as natural independence properties (which again the other two cea's do not have). Section 12.3 considers the compatibility of PS conditional events with conditioning of random variables (Theorem 10). Section 12.4 illustrates the fact that boolean cea's other than PS are possible to construct – but necessarily, in light of the characterization theorem (Theorem 11) of Section 12.5, lack certain fundamental properties PS enjoys.

Chapter 13 provides details for PS as a means for both analyzing cea issues and serving as an umbrella where all three leading cea's can be related. Section 13.1 shows direct ordering relations (Theorem 12), and a key result showing all six notions of deduction and enduction, relative to the nontrivial simple conditional events of PS can be completely characterized – quite simply and entirely ! – by the imbedded DGNW and AC operations. (See eqs.(13.28)-(13.31) and Theorems 13, and 14.) Section 13.2 returns to Example 1 introduced in Section 10.1 and shows specifically how PS can address it. Section 13.3 provides a full calculus of operations for constant probability events together with an important result on combining constant-probability events with ordinary events or other conditional events (Theorem 15).

Chapter 14 considers particular details in the testing of hypotheses for the distinctness of events when they are indirectly given via numerical functions of probabilities. Section 14.1– with Example 2 – simply illustrates classical statistical testing of hypotheses for the same event relative to different probability measures, while Section 14.2 – with Example 3 – considers the main problem to be addressed here: *testing of hypotheses of possibly different events, which, as stated above, may not be directly known a priori, relative to a common probability measure.* Section 14.3 presents a basic higher order probability assumption (Q), under which all hypotheses testing here is to be implemented and an outline of the actual test procedure is provided. Essentially, Q states that all relative atomic probabilities involved in the formulation of the problem are themselves assumed to have a joint uniform distribution over their natural domain of values as all possible probabilities and the events are varied. Also, in Section 14.3 the cumulative distribution functions associated with the two probability distance functions $d_{P,1}$ and $d_{P,2}$ are derived under assumption Q for use in testing of the hypotheses. Two additional measures of association or probability distance functions are introduced in Section 14.4: $d_{P,3}$, the "relative probability distance" function and $d_{P,4}$, the "negative form relative probability distance" function . Section 14.5 presents a brief review (Theorems 16 and 17) – with some new results apropos to our interest (Theorem 18) concerning classical (numerically-valued) metrics or distance measures and numerical subadditivity for generating additional metrics. Section 14.6 introduces algebraic metrics and algebraic subadditivity for generating additional algebraic metrics (Theorems 19 and 20). Section 14.7 connects probability with algebraic metrics by showing all probability evaluations of algebraic metrics are legitimate metrics and conversely, any binary operations on a boolean algebra which when evaluated by arbitrary probabilities yields numerical metrics is itself an algebraic metric (Theorem 21). It is also shown that three of the four measures of association previously introduced, namely: $d_{P,2}$, $d_{P,3}$, $d_{P,4}$, are actual metrics arising from corresponding algebraic metrics (Theorem 22), justifying the previous use of the term "probability distance " functions. The fourth, $d_{P,1}$, while obviously a legitimate metric, does not appear to arise as the probability evaluation of an algebraic metric. However, in a related direction, dP,1 is shown to arise as the probability evaluation of an event which in general is probability-dependent itself. See Section 16.4, eq.(16.58) and especially eqs.(16.80)-(16.91), for the construction of the associated relational events $\alpha(f_{17})$ and $\alpha(f_{18})$. Also, Theorem 23 shows that the only possible unconditional and conditional boolean algebraic metrics - and associated metrics via probability evaluations – are the same three as above. Section 14.8 provides additional relations among the four probability distance functions previously introduced, derives the cumulative distribution functions (under assumption Q) associated with $d_{P,3}$ and $d_{P,4}$, and provides in Table 1 a summary of basic properties that the four candidate probability distance functions possess or lack and, as a result, shows in Table 2 a natural preference ordering among the four for implementation. The most desirable, among the four turns out to be $d_{P,3}$ – not $d_{P,2}$ – though the latter is much more well-known in the standard development of probability. In any case, the interactive metrics, i.e., the metrics requiring knowledge of the probability of logical conjunctions (or disjunctions) of the relevant events in addition to the separate probabilities of the events, $d_{P,2}$, $d_{P,3}$, $d_{P,4}$, are all seen to be superior to the non-interactive metric $d_{P,1}$. *These interesting results, together with Theorems 19-23, show nontrivial use of PS cea in deriving metrics for probability spaces.*

Chapter 15 turns the focus on natural language and testing of hypotheses for similarities of linguistic narratives or descriptions. The basic issue is illustrated by Example 4 in Section 15.1, where it is seen how fuzzy logic can be used as a natural way to model the problem initially. Then, except for Section 15.3, the remainder of the Chapter shows how probability theory can be used to re-interpret the fuzzy logic model – including the case of exponential-form modifiers – via full homomorphic-like one-point coverage representations of appropriately constructed random subsets of the domains of the fuzzy sets involved in the modeling. In turn, the theory of hypotheses testing developed previously in Chapter 14 can be applied to this situation – see Section 15.5. To carry this out, the first tool for dealing with the problem, the dual concepts of "copula" and "cocopula" are briefly treated in Section 15.2. Before proceeding further, Section 15.3 illustrates, alternatively, a few typical standard ad hoc numerical approaches to the problem, via ordinary real analysis distance functions and fuzzy logic. But, it is seen that such approaches do not go to the heart of the issue: first, determining the natural underlying events represented by the numerical functions of probabilities, and then finding their interactive probability distances – as opposed to the non-interactive numerically-based ones for hypotheses testing. This can lead to a significantly improved specification of values over a possibly wide range of values relative to the potential use of interactive distances. The latter are seen to have a sounder theoretical basis than the non-interactive ones when only the numerical approaches are used. (See again the comparisons in Tables 1 and 2 in Section 14.8.) More details on this are provided at the end of Section 15.3. Section 15.4 provides a brief review of one-point coverage representations. Section 15.6 shows how the modeling of Example 4 can be further refined by taking into account in a nontrivial way fuzzy logic modifiers in the form of exponentials, and in turn, how to apply relational event algebra in combination with probability distance functions. Section 15.7 gives additional analysis of the problem begun in Section 15.6, where a homomorphic-like representation of the full fuzzy logic model of Example 4 can be obtained in relation to the one-point coverage random set representations.

Chapter 16, is concerned how relational event algebra – other than solely conditional event algebra – can used to treat data fusion and hypotheses testing issues from an algebraic viewpoint before utilizing numerical implementations. Section 16.1 introduces and addresses, via Example 5, a problem involving the comparison of two weighted averages of probabilities of several possibly overlapping events by different experts. Section 16.2, likewise, considers Example 6, a problem of comparing or contrasting certain types of polynomial or power series (i.e., analytic functions) of a common probability value, while Section 16.3, beginning with Example 7, treats the comparison of quadratic functions in two common probability values. Section 16.4 considers first a full mathematical formulation of the general relational event algebra problem – as treated throughout all of Part III up to this point – and proposes a weakening of the form of relational (or conditional, as a particular case) events whereby the constant-probability terms acting as coefficients are now allowed to depend on the particular probability measure used to evaluate them. (This is justified theoretically from the development of constant-probability events in general – see Section 13.3.) Such an extension of the concept of relational events allows for the representation of certain functions of probabilities, which up to this point, did not seem to be representable by relational events, including the naive probability distance function. (Again, see the last part of Section 16.4.). Finally, Section 16.5 presents concluding remarks on both conditional and relational event algebras and poses a number of general open questions which, hopefully, will be addressed by future research.

Chapter 9

Introduction to the Conditional and Relational Event Algebra Aspects of Data Fusion

9.1 Philosophy of Approach

The issue considered here is concerned with the analysis and combination of certain multi-source information which has some degree of uncertainty attached to it. When the uncertainty corresponding to each source is provided via the probability of a single unconditional event, standard probability and statistics techniques are adequate for dealing with the problem. However, it often arises that uncertainty is not provided in such a format. This can happen, e.g., when information is supplied initially through natural language descriptions which are then converted to fuzzy logic. Another occurrence is when weighted combinations, or appropriate nonlinear functions, of probabilities of contributing simpler events represent the degree of uncertainty. In such cases it is not obvious initially how to represent (or even whether such a representation is possible for) the uncertainty as a probability evaluation of a legitimate event. Further details of the problem considered here are presented in the next section. Two excellent treatments on the general topic of data fusion which not only survey various statistical and probabilistic techniques, but also consider nonstandard approaches , including fuzzy logic and Dempster-Shafer theory, may be found in Hall [7] and Waltz and Llinas [14]. However, these texts, as well as the bulk of the current approaches to data fusion and combination of evidence problems, have not addressed the problems stated above. Because the tools used in dealing with these issues, conditional and relational event algebra , lie outside of the mainstream development of probability, as well as the current development of fuzzy logic and other areas, most readers of this text, even with a strong background in probability, will find the concepts introduced here, at first, quite unfamiliar. Consequently, the approach taken here is to subdivide the topic interest through the introduction of a number of motivating examples which also show how conditional and relational event algebra can be used to address them.

9.2   Overview of the Problem and the Need for an Algebraic Basis Preceding Numerical Calculations

Multiple source information forms one of the key components of data fusion. Such information may emanate from various mechanical sensor sources such as radar or doppler systems, or it may derive from human-based sources, such as via expert opinion expressed through natural language; or the information may consist of both types of sources. In general, associated with each unit of information is a degree of uncertainty/certainty, which in the case of mechanical-based sources, is traditionally determined through the use of probability. Surprisingly, probability also plays a natural role in the case of linguistic-based information. This is because of the following: 1. It is natural to measure uncertainty initially through the use fuzzy logic [9], [11] . 2. It is now well-established that a central aspect of first order fuzzy logic can be identified (in a one-to-many relationship) with corresponding probability descriptions. This is achievable through the use of one-point coverage representations of certain classes of random sets and classical logical operations on these representations corresponding to the fuzzy sets and fuzzy logic involved [2]; [4], Section 4. Often, information uncertainty is provided in standard form, such as through unconditional probability or cumulative probability distribution functions (or via parameterized families of such). In this case, at least in theory, all relevant events are given explicitly in a boolean (or sigma-) algebra, so that the full probability space representing the uncertainty is known. Thus, one can evaluate probabilistically any desired logical combination of such events for use in decision-making, and in particular in determining the degrees of similarity or consistency between these events by use of appropriately chosen metrics, which, in general, for implementation require explicit knowledge of probability evaluations of logical conjunctions or disjunctions.

On the other hand, at times, information uncertainty is provided in a way that there appears to be no single underlying boolean event whose probability evaluation matches the prescribed uncertainty. Such uncertainty is often expressed in the form of given functions of probability evaluations of contributing simpler events. For example, when these functions are simple arithmetic division with arguments being pairs of events, each numerator argument event being a subevent of the denominator argument event, the quantitative uncertainty corresponding to each unit of information then, becomes a conditional probability. But, in general, the standard development of probability theory and statistics has not produced a way to represent conditional probabilities as single event probability evaluations, so that standard statistical decision-making techniques cannot be used in a systematic sound way here. In particular, we cannot compare, for similarities or differences, using the same techniques available when ordinary events are involved, a collection of inference rules such as "if b, then a", "if d, then c",...when the uncertainty associated with these rules are the conditional probabilities $P(a|b)$, $P(c|d)$,... For example, if only ordinary events (or propositions) a, b, c, d, e, f,... were considered for comparisons, we could use e.g., the standard probability distance function (denoted here as $d_{P,2}$ and considered in more detail, beginning with Section 9.4 – or see Kappos [8]) could be used, whereby we compute the probabilities $P(a+b)$, $P(a+c)$, $P(b+c)$,..., where $+$ here is the boolean symmetric operation (again, see Section 9.4). But, how do we compute analogously $P((\text{if } b, \text{ then } a) + (\text{if } d, \text{ then } c))$, $P((\text{if } b, \text{ then } a), (\text{if } f, \text{ then } e))$,... ? Some exceptions to this include the situation where all of the conditional probabilities represent independent information, or where all of them have an identical denominator, or where the conditional expressions are in a chained form. Recently, probability theory has been expanded to address the above problem in the form *of conditional event algebra*, which we provide in some detail, beginning with Chapter 10.

Another situation where uncertainty does not seem to correspond to the probability evaluation of a single event occurs when it is provided as a weighted sum of probabilities of more elemental events which may be overlapping (so that the total probability expansion theorem which critically depends on disjointness is not applicable here). This can arise when experts are combining probabilities subjectively or, via the connection of probability with fuzzy logic mentioned above, through the use of fuzzy partitions. Similarly, uncertainty may be provided as more complicated nonlinear functions of contributing event probabilities.

The basic issue of determining whether information from the various sources is consistent, or at least compatible in some sense, is paramount before carrying out additional analysis, such as combining/fusing, estimating common associated parameters of interest, deducing, or making further decisions based on the information present. As in the conditional probability case, the standard development of probability and statistics, has bypassed the problem of being able to implement measures of comparison and other processes which require in a key way the probability evaluation of logical combinations of events underlying the given uncertainties. The primary goal of the following sections collectively is to show how a relatively new mathematical tool, *relational event algebra*, which is an extension of conditional event algebra, can be utilized in addressing the issue of comparison and consistency of information whose uncertainty appears initially to be not representable by ordinary events in an ordinary probability space setting.

9.3   Algebraic Approach to Treating Information

9.3.1   Partitioning of Information

Following up on the overview in more detail, we first test hypotheses that the multiple source information $\{I_1, I_2, \ldots, I_n\}$ is consistent or reasonably similar vs. the information is inconsistent or dissimilar. Each unit of information I_j corresponding to source j may itself be a set of information subunits. Two important cases arise for a typical I_j:

Case 1. I_j represents a collection of events, with uncertainty [or certainty – from now on, for convenience, we use the term "uncertainty" to refer to either concept, depending on the context] given through a probability measure, such as the distribution of the location of the main body of enemy troops over possible regions of interest. This, of course, can be provided in any of a number of equivalent forms, such as cumulative distribution functions, probability functions (when the corresponding domain of values is discrete), etc.

Case 2. I_j represents a single generalized event in the following extended sense: First, this includes all situations where I_j actually represents an ordinary boolean event, possibly compounded logically or using cartesian products, from simpler contributing events. In this case the degree of uncertainty will be provided as a single probability value. Second, I_j may represent a single natural language description – possibly compounded analogously from simpler contributing linguistic components. In this situation, as mentioned before, the degree of uncertainty is naturally initially modeled by use of fuzzy logic. Third, the generalized event may also be in an indirect form whereby only its uncertainty structure is provided as a function of the probability of uncertainty for contributing simpler events. Actually, all three seemingly disparate senses of a generalized event making up Case 2 later will be shown to fit essentially the same format: namely, legitimate individual events (albeit, in possible complex form) lying in ordinary probability spaces (which, again, may possibly have a relatively complicated structure).

We partition those sources as being sufficiently similar to each other into m sets

$$\{I_1, \ldots, I_n\} = I(1) \cup \cdots \cup I(m), \qquad I(i) \cap I(j) = \emptyset, \quad \text{if } i \neq j,$$

$$I(1) = \{I_{1,1}, \ldots, I_{1,n_1}\}, \ldots, I(m) = \{I_{m,1}, \ldots, I_{m,n_m}\}, \qquad n_1 + \cdots + n_m = n.$$
$$(9.1)$$

Some of these sets $I(j)$ may be singletons, i.e., $n_j = 1$, since it may be adjudged that the corresponding information is inconsistent with all other present information. Then, for each such set $I(j)$, we may fuse or estimate the common parameters of interest, resulting in combined information $I(j, 0)$. We may consider making other decisions / actions, based on $I(j)$, such as attacking the enemy, calling for more information, or we may consider associated deduction and "enduction" classes (the latter being a coined term here to indicate the reverse of deduction – see below for details), when all the elements in $I(j)$

represent boolean events relative to a probability space $(\Omega, \mathcal{B}, P)$. In addition, we consider newly arriving information $I_{n+1}, \ldots, I_r$ in conjunction with the now-processed information $I(1, 0), \ldots, I(m, 0)$. Then, repeat the process of testing hypotheses, partitioning and estimating or combining for the collection $\{I(1, 0), \ldots, I(m, 0), I_{n+1}, \ldots, I_r\}$, etc. The procedure can be further refined by taking into account those sets – and consequent combined information – for which there the hypotheses tests do not yield low enough error levels.

9.3.2 Boolean Algebra, Probability Spaces, and Deduction and Enduction

The logical and mathematical aspect of this work is chiefly concerned with (1) spaces $\mathcal{B}$ of, equivalently: propositions, sentences, events, or sets; (2) probability measures evaluated on the spaces, in brief, *probability spaces* of the form $(\Omega, \mathcal{B}, P)$, where $\mathcal{B}$ is either a boolean or, more strongly, a sigma algebra, with $\Omega \in \mathcal{B}$ being the universal event or proposition – always true with $P(\Omega) = 1$ and $\emptyset \in \mathcal{B}$ being the null event – always false with $P(\emptyset) = 0$. We usually indicate such events or propositions by $a, b, c, \ldots \in \mathcal{B}$, or at times, by capital or Greek letters, as $A, B, C, \alpha, \beta, \gamma, \ldots \in \mathcal{B}$ with the standard boolean algebra notation for negation or complement over $\mathcal{B}$ as in a', conjunction or intersection over $\mathcal{B}$ by $\&$, $\cdot$, or even a missing symbol, when no ambiguity arises, such as $a\&b$ or $a \cdot b$ or ab; and disjunction or union over $\mathcal{B}$ by use of $\vee$ as in $a \vee b$. However, when we are working with classes or collections of events or propositions, we will usually use set/class operation notation, so that if A, B are any two such collections, i.e., $A, B \subseteq \mathcal{B}$, $A \cap B$ is used to indicate the intersection of A and B. On the other hand, when we employ the notation $A\&B$ or AB, we mean the pairwise result of using $\&$ on A and B, i.e., the collection of all ab, $a \in A$, $b \in B$. We also use the following class notation, for any collection of events or class $A \subseteq \mathcal{B}$, for given boolean or sigma algebra $\mathcal{B}$ (referring to probability space $(\Omega, \mathcal{B}, P)$):

$$\&(A) \quad = \quad \&\{a : a \in A\} = \&_{a \in A}(a) \tag{9.2}$$

$$\vee(A) \quad = \quad \vee\{a : a \in A\} = \vee_{a \in A}(a) \tag{9.3}$$

$$A^{(')} \quad = \quad \{a' : a \in A\}. \tag{9.4}$$

Because we are working with boolean (or sigma) algebras, the concepts of logical implication, deduction, and being a subevent or subset of, are all equivalent and succinctly expressed by the partial order relation $\leq$ over $\mathcal{B}$, where as usual, "a deduces b", or "b enduces a" iff $a \leq b$ iff $a = ab$ iff $b = a \vee b$ iff $a' \vee b = \Omega$. For a thorough and readable treatise on boolean algebra and its relation to logic, see Mendelson [12]. For a basic work on probability spaces, emphasizing the boolean algebra aspect, see Kappos [8]. Mention should also be made of the recent text by Hailperin on sentential probability logic [6]

We are also interested in the deduction of events e from possibly more than one event or proposition, say $\{a, b, c\}$ corresponding to – using the ideas of the previous section – a batch of information that is deemed to be consistent. But, when multiple deduction – or enduction – sources are involved, there appears to be several possible natural ways to extend the simple deduction case discussed above.

Specifically, we consider the structure of the six following subclasses of B with respect to any other subclass $A \subseteq B$, keeping in mind the application to $A = I(j)$:

disjunctive deduction class of A (with respect to B):

$$
\begin{aligned}
D(A, \leq, \vee) &= \{x \in B : A \leq x\} = \{x \in B : a \leq x \text{ for all } a \in A\} \\
&= \{x \in B : \vee(A) \leq x\}, \tag{9.5}
\end{aligned}
$$

intermediate deduction class of A:

$$
\begin{aligned}
D(A, \leq, or) &= \{x \in B : or_{a \in A}(a \leq x)\} \\
&= \{x \in B : \text{there exists } a \in A \text{ such that } a \leq x\}, \tag{9.6}
\end{aligned}
$$

conjunctive deduction class of A:

$$
D(A, \leq, \&) = \{x \in B : \&(A) \leq x\}, \tag{9.7}
$$

conjunctive enduction class of A:

$$
\begin{aligned}
D(A, \geq, \&) &= \{x \in B : A \geq x\} = \{x \in B : a \geq x, \text{ for all } a \in A\} \\
&= \{x \in B : \&(A) \geq x\}, \tag{9.8}
\end{aligned}
$$

intermediate enduction class of A:

$$
\begin{aligned}
D(A, \geq, or) &= \{x \in B : or_{a \in A}(a \geq x)\} \\
&= \{x \in B : \text{there exists } a \in A \text{ such that } a \geq x\}, \tag{9.9}
\end{aligned}
$$

disjunctive enduction class of A:

$$
D(A, \geq, \vee) = \{x \in B : \vee(A) \geq x\}. \tag{9.10}
$$

Clearly, from the above definitions, it follows that

$$
\begin{aligned}
&D(A, \leq, \vee) \subseteq D(A, \leq, or) \subseteq D(A, \leq, \&), \\
&D(A, \geq, \vee) \supseteq D(A, \geq, or) \supseteq D(A, \geq, \&) \\
&(D(A, \leq, \vee))^{(')} = D(A^{(')}, \geq, \&), \\
&(D(A, \leq, or))^{(')} = D(A^{(')}, \geq, or). \tag{9.11}
\end{aligned}
$$

Furthermore, it may be of interest to restrict all of the deduction and enduction classes to a fixed subclass C of B of interest, so that e.g., we may consider

$$
D(A, C <\geq, \vee) = \{x \in C : \vee(A) \geq x\} = D(A, \geq, \vee) \cap C. \tag{9.12}
$$

Similar comments and notation hold for the remainder of the deduction and enduction classes.

9.3.3 Algebraic Combining of Information

One reason we may wish to consider algebraic averaging or combining of events, or more generally, an algebraic analogue of standard regression – as opposed to just numerical averaging or combining – is in the same spirit as mentioned before: if single events can be found underlying such operations as numerical averaging of probabilities, a more basic approach can be established for comparing such quantities via logical operations followed by probability evaluations – now in the form of single events – through the application of well-established tools from boolean algebra and probability theory, such as the probability of boolean sums of the events (see Section 9.4).

In particular, consider weighted linear averaging for the case when information packet I_j consists of ordinary boolean events, for simplicity here, say just α, β: If appropriate "weighting events" $\theta(w)$, $\theta(1-w)$ for any choice of numerical weights $w, 1-w$, with $0 \le w \le 1$, would exist, satisfying

$$P(\theta(w)) = w, \quad P(\theta(1-w)) = 1-w, \quad \text{for all probabilities } P, \qquad (9.13)$$

then one apparently natural algebraic counterpart to numerically averaging the event probabilities, $P(\alpha)$ and $P(\beta)$, is first to compute the cartesian products $\alpha \times \theta(w)$, $\beta \times \theta(1-w)$ and then form the *algebraic average*

$$\mathrm{av}(a, \beta; w) = (\alpha \times \theta(w)) \vee (\beta \times \theta(1-w)). \qquad (9.14)$$

Clearly, eq.(9.14) implies

$$\alpha\beta \le \mathrm{av}(\alpha, \beta; w) \le \alpha \vee \beta \qquad (9.15)$$

and

$$P(\alpha\beta) \le P(\mathrm{av}(\alpha, \beta; w)) \le P(\alpha \vee \beta) \qquad (9.16)$$

and as α approaches β in probability, so do all three events in eq.(9.15) approach each other.

Note that by considering the relative atoms $\alpha\beta$, $\alpha\beta'$, $\alpha'\beta$, we can rewrite eq.(9.14) as

$$\mathrm{av}\,(\alpha, \beta; w) = (\alpha \times \theta(w)) \vee (\beta \times \theta(1-w))$$
$$= \alpha\beta \times (\theta(w) \vee \theta(1-w)) \vee \alpha\beta' \times \theta(w) \vee \alpha'\beta \times \theta(1-w). \qquad (9.17)$$

Then, taking probabilities and extending P to cartesian product as a product measure in the obvious way, we obtain, slightly abusing notation, by disjointness of all of the relative atoms and eq.(9.17) and (9.22),

$$P(\mathrm{av}(\alpha, \beta; w) = P(\alpha\beta).P(\theta(w)v\theta(1-w)) + P(\alpha\beta')w + P(\alpha'\beta)(1-w). \qquad (9.18)$$

If α and β are disjoint, then $\alpha\beta = \emptyset$, $\alpha\beta' = \alpha$ and $\alpha'\beta = \alpha$, and eq.(9.18) reduces to the desired compatibility result with the corresponding numerical averaging of the probabilities $P(\alpha)$ and $P(\beta)$, namely,

$$P(\mathrm{av}(\alpha, \beta; w)) = wP(\alpha) + (1-w)P(\beta). \qquad (9.19)$$

However, when α and β are not disjoint, then, in order to obtain the compatibility result of eq.(9.19), for all w, $0 \leq w \leq 1$, we must have either

$$\theta(1 - w) = (\theta(w))', \qquad \text{all } w, \ 0 \leq w \leq 1 \qquad (9.20)$$

or we must define $av(\alpha, \beta; w)$ somewhat differently. In any case, note first that in general, eq.(9.13) cannot be satisfied. In fact, it is easy to show that the only constant-probability events $\theta(t) \in \mathcal{B}$, the original boolean algebra containing events α and β, corresponding to probability space $(\Omega, \mathcal{B}, P)$, are the trivial ones at $w = 0$ and $w = 1$, i.e.,

$$\theta(0) = \emptyset \qquad \text{and} \qquad \theta(1) = \Omega. \qquad (9.21)$$

Nevertheless, it will be shown later (Sections 10.3 and 13.3) that constant-probability events can be found, *lying outside of probability space* $(\Omega, \mathcal{B}, P)$, i.e., there is a strictly bigger boolean algebra $\mathcal{B}_1$ which extends $\mathcal{B}$ – and hence contains imbeddings of α and β – and a probability measure P_1, extending P in a product probability form, yielding a legitimate probability space $(\Omega_1, \mathcal{B}_1, P_1)$, so that in place of eq.(9.13), we now have

$$\theta(w), \ \theta(1 - w) \in \mathcal{B}_1, \quad P_0(\theta(w)) = w, \quad P_0(\theta(1 - w)) = 1 - w \qquad (9.22)$$

for all w, $0 \leq w \leq 1$. In this construction, the counterpart of eq.(9.20) does not hold – though of course we do have the corresponding probability evaluations holding trivially, i.e.,

$$P_1(\theta(1 - w)) = 1 - w = P_1((\theta(w))'), \qquad \text{for all } w, \ 0 \leq w \leq 1.$$

Also, the averaging of the events can be reformulated so that the counterparts of eqs.(9.15), (9.16) and (9.19) all hold, by now redefining

$$av(\alpha, \beta; w) = \alpha\beta \times \Omega_1 \ \vee \ \alpha\beta' \times \theta(w) \ \vee \ \alpha'\beta \times \theta(1 - w), \qquad (9.23)$$

for all wt, $0 \leq w \leq 1$. For example, the counterpart of eq.(9.19) holds, since

$$
\begin{aligned}
P_1(av(\alpha, \beta; w)) &= P(\alpha\beta) + P(\alpha\beta')w + P(\alpha'\beta)(1 - w) \\
&= P(\alpha\beta)(w + 1 - w) + P(\alpha\beta')w + P(\alpha'\beta)(1 - w) \\
&= (P(\alpha\beta) + P(\alpha\beta'))w + (P(\alpha\beta) + P(a(b)))(1 - w) \\
&= P(\alpha)w + P(\beta)(1 - w), \text{ for all } w, \ 0 \leq w \leq 1. \quad (9.24)
\end{aligned}
$$

More generally, *collections* of constant-probability events exist, whose probability evaluations by all such P_1 match any given real values in $[0,1]$. Averaging of events will also be considered in Example 5, Section 16.1, but only after we have developed the appropriate tools for carrying this out.

It should be emphasized again that for a given probability space $(\Omega, \mathcal{B}, P)$ with events, say $a, b, c, d \in \mathcal{B}$ (boolean or sigma-algebra), the corresponding conditional events and, in particular, weighting (or constant-probability) events do not lie, in general, with the $a, b, c, d \in \mathcal{B}$, but rather, there exists another probability space $(\Omega_0, \mathcal{B}_0, P_0)$ which strictly extends $(\Omega, \mathcal{B}, P)$ in the sense that there is a natural imbedding of the former into the latter, so that the imbeddings of a, b, c, d then are $\in \mathcal{B}_0$ alongside all required conditional events made up from them (i.e., having these events occur as antecedents and consequents). In fact, Lewis has shown that, in general, all such probability spaces $(\Omega_0, \mathcal{B}_0, P_0)$ can never coincide with the original space $(\Omega, \mathcal{B}, P)$ [10]. (See also Section 11.5.)

9.4   The Algebraic Decision Theory Problem, Relational Event Algebra, and Introduction to Measures of Similarity

In this part we will concentrate mainly on Case 2, as introduced in Section 9.3.1. A number of situations arise where, in effect the certainty or uncertainty of each model is given as some prescribed function of probabilities of other events. When each model explicitly represents a single (possibly compound) event, the testing of hypotheses becomes relatively straightforward. On the other hand, situations can arise where it may not be obvious that each model can actually be represented by a single (albeit complex) event in an appropriate probability space which also contains in some sense the contributing simpler events. The standard development of probability has emphasized the numerical aspect, rather than the algebraic underpinnings. Thus, hypotheses testing and estimation has revolved around numerical quantities such as probability functions (pf's), probability density functions (pdf's) and cumulative distribution functions (cdf's), which consider ensembles of singleton events, rather than directly with the individual events. To reiterate: the goal of all of Part III is to show that a program can be implemented for a wide variety of cases which provides algebraic analogues to the standard numerical probability approach and allows for testing of hypotheses and combination or estimation of individual events in a comprehensive sound manner. Let us denote the problem of meeting such a goal as the *algebraic decision theory problem*.

In more detail, the algebraic decision theory algebra problem often reduces to the case where for each source, the corresponding information represents a single underlying event that is a known, possibly complex, logical combination of contributing simpler events. This can occur, e.g., when several competing models are proposed for the logical design of a large scale system (such as a C3 system or the running of an industrial complex) and it is desired to determine how similar or consistent these models are with respect to each other. In such cases the underlying events determining the information have explicitly known logical structure and with an appropriate choice of metric to measure their relative probabilistic distances, testing of hypotheses of similarity can be readily accomplished. For example, if a and b are two such compound events, one could use the *naive probability distance*

$$d_{P,1}(\alpha, \beta) = |P(\alpha) - P(\beta)| = |P(\alpha\beta') - P(\alpha'\beta)| \tag{9.25}$$

to test hypotheses. But clearly, the use of an algebraic operation first such as
the well-known boolean sum or symmetric difference

$$\alpha + \beta = \alpha'\beta \vee \alpha\beta' = (\alpha\beta \vee \alpha'\beta')' \tag{9.26}$$

followed by probability evaluation yielding the *natural absolute probability distance*

$$
\begin{aligned}
d_{P,2}(\alpha,\beta) &= P(\alpha+\beta) = P(\alpha'\beta) + P(\alpha\beta') = 1 - P(\alpha\beta) - P(\alpha'\beta') \\
&= P(\alpha) + P(\beta) - 2P(\alpha\beta) \\
&= 2 - P(\alpha) - P(\beta) - 2P(\alpha'\beta') \tag{9.27}
\end{aligned}
$$

provides a much more satisfactory procedure than the non-interactive approach
in eq.(9.25), which only makes use of the individual events and not their logical
relations. For more details see Section 14.2 or later sections. See also Kappos
([8], pp. 16-17 et passim) for general use of this metric in constructions involving
probability spaces. In fact this distance function is used in a standard way in
Section 12.6, where it plays a crucial role in the establishment of a particular
product probability space (PS) used to model conditional expressions.

We apply the terms "distance" or "metric" to both functions $d_{P,1}$ and $d_{P,2}$
since each satisfies the fundamental requirements of a topological pseudometric
function d over a space of events $\alpha, \beta, \gamma,...$:

$$\text{(D1)} \quad d(\alpha,\beta) = d(\beta,\alpha) \quad \text{symmetry,}$$
$$\text{(D2)} \quad d(\alpha,\alpha) = 0, \quad \text{reflexivity,}$$
$$\text{(D3)} \quad d(\alpha,\gamma) \leq d(\alpha,\beta) + d(\beta,\gamma) \quad \text{triangle inequality.} \tag{9.28}$$

with the probability-caused bounding to be in [0,1]. When we identify all events
α_1, α_2 in the equivalence classes for which $d(\alpha_1,\alpha_2) = 0$, d becomes a true
distance function. (See again [8].)

Even for the case considered above of two events, the problem of actually
combining in some optimal sense those models considered consistent with respect to each other is a more difficult problem, since we are concerned here with
the fusion or averaging in some sense of events – rather than with numerical
quantities, with which traditional regression and estimation are concerned [13].

On the other hand, when we are dealing with linguistic-based information, it
is natural to inquire if competing models – such as gleaned from the descriptions
by different experts surmising the same situation – can be treated analogously to
the standard single event probability case described above. Of course, as stated
earlier, fuzzy logic seems to be a natural tool to deal with such information
[9], [11]. But, there remains much "ad hockery" in its implementation with
little, if any justification based on probability. In response to this, it should be
stated first that it has been demonstrated that a central part of fuzzy logic can
be directly related to probability via the one-point coverages of appropriately
chosen families of random sets (see again [2];[4], Section 4). This can be shown
to lead to the conclusion that a large number of situations where competing

natural language descriptions are to be tested for similarity can also be treated as appropriately modified relational event algebra problems (as is shown in Chapter 16).

As another example of the general relational event algebra problem, let us reconsider the previously briefly mentioned situation where one wishes to test for consistency or redundancy a collection of inference rules "if a, then b", "if d, then c",..., where $a, b, c, d,...$ represent possibly logical compounds of simpler contributing events and where the uncertainty of the inference rules is given through the natural corresponding conditional probability evaluations, $P(a|b)$, $P(c|d)$,..., assuming at least the underlying probability space for $a, b, c, d,..$ is known. In this case, the common function of probabilities of the contributing events is division (since $P(a|b) = P(ab)/P(b)$, etc.). As stated earlier, in the usual probability literature the issue of determining the underlying probability space of events representing simultaneously all of these conditional probabilities has been ignored – until the recent development of conditional event algebra [1],[3],[5] – and the only available procedure has been to consider the conditional probabilities externally and compute, e.g., the naive probability distances $|P(a|b) - P(c|d)|$, rather than the analogue of the above more satisfactory natural probability distance $P((a|b) + (c|d))$, slightly abusing notation in the use of P as an extension of the original (unconditional) probability measure and (conditional) events $(a|b)$, $(c|d)$ corresponding to $P(a|b)$, $P(c|d)$.

Still another example of the relational event algebra problem arises when one wishes to test for similarity several competing models each formally representing the overall probability of, say, enemy attack tomorrow as a different weighted linear combination of contributing probabilities of events which are not necessarily disjoint, such as e.g., weather being clear, sea state being calm, and enemy build-up not exceeding a certain limit in a prescribed region. Even when these contributing events are replaced by truly disjoint events and one appeals to the standard expansion of probability in terms of weighted conditional probabilities, the event "enemy attacks tomorrow (according to expert j)" is not explicitly obtainable for testing as in the first example using the symmetric boolean difference, though obviously the less satisfactory naive probability distance is available.

Bibliography

[1] Dubois, D., Goodman, I. R., and Calabrese, P.G. (Eds.) (1994) Special Issue on Conditional Event Algebra, *IEEE Transactions on Systems, Man, and Cybernetics* **24** (12).

[2] Goodman, I. R. (1994) A new characterization of fuzzy logic operators producing homomorphic-like relations with one-point coverages of random sets, in P. P. Wang (Ed.) *Advances in Fuzzy Theory and Technology, Vol. II.*, Duke University, Durham, NC, pp. 133-160.

[3] Goodman, I. R., Gupta, M. M., Nguyen, H. T., and Rogers, G. S.(Eds.) (1991) *Conditional Logic in Expert Systems*, North-Holland, Amsterdam.

[4] Goodman, I. R. and Kramer, G. F. (1997) Extension of relational event algebra to random sets with applications to data fusion,*Proceedings of Workshop on Applications and Theory of Random Sets*, Institute for Mathematics and Its Applications (IMA), University of Minnesota, Minneapolis, August 22-24, 1996, Springer-Verlag, Berlin (to appear).

[5] Goodman, I. R., Nguyen, H. T., and Walker, E. A. (1991) *Conditional Inference and Logic for Intelligent Systems*. North-Holland, Amsterdam.

[6] Hailperin, T. (1996) *Sentential Probability Logic*, Lehigh University Press, Bethlehem, PA.

[7] Hall, D. L. (1992) *Mathematical Techniques in Multisensor Data Fusion*, Artech House, Boston, MA.

[8] Kappos, D. A. (1969) *Probability Algebras and Stochastic Spaces*, Academic Press, New York.

[9] Kosko, B. (1993) *Fuzzy Thinking*, Hyperion, New York.

[10] Lewis, D. (1976) Probabilities of Conditionals and Conditional Probabilities, *Philosophical Review*, **85**, 297-315.

[11] Mamdami, E. H. and Gaines, B. R. (1981) *Fuzzy Reasoning and Its Applications*, Academic Press, New York.

[12] Mendelson, E. (1970) *Boolean Algebra and Switching Circuits*, McGraw-Hill Book Co., New York.

[13] Rao, C. R. (1975) *Linear Statistical Inference and Its Applications*, 2nd Edition, Wiley, New York.

[14] Waltz, E. and Llinas, J. (1990) *Multisensor Data Fusion*, Artech House, Boston, MA.

Chapter 10

Potential Application of Conditional Event Algebra to Combining Conditional Information

10.1 Modeling Inference Rules

Motivating Example 1.

Because conditional event algebra is actually a special case of relational event algebra and is needed for the development of the latter, we present here an illustrative example of its use.

Consider a rule-based system consisting of a set of inference rules (if e, then a), (if f, then c),..., where each event $a, c, e, f,...$ may itself be quite a logical compound of other simpler events. Such systems have proven to be extremely useful in both civilian area use such as medical diagnosis (especially the ground-breaking MYCIN system) and fault discovery [1],[10], and potential military use such as in tracking and correlation (R. Dillard [3]). In these systems data arrives, usually in the form of the occurrence of certain events. Then, those inference rules are selected whose antecedents either match these events exactly or are deducible in some sense from them (see the discussion on deduction in Section 9.3.2). In turn, these rules are said to "fire", in which case the corresponding consequents are said to occur. (If the rules are interpreted via the classical material conditional form– see discussion below – this corresponds to taking conjunctions.) In turn, the collection of occurring consequents now plays the role of incoming data and the process is repeated until a time deadline or all possible firing of rules is exhausted. The collection of all resulting consequents, together with the original incoming data, can be considered the basic conclusion set (from which, in turn, one may perform various types of deductions or enductions –

359

again, see Section 9.3.2).

Ideally, such rule-based systems should improve in their performance when the number of inference rules is augmented, provided they are consistent in some sense. On the other hand, a system with too many rules may be over-costly, slow in response, or just become too complex. This leads to the fundamental issue of determining just when a new inference rule is not too redundant or similar with respect to any of the rules already in the system. In addition, we must choose a way of measuring the strength or validity of the inference rule based on its structure (but not to be confused with the reliability of the source giving the rule).

One basic way is to assume the rule can be modeled by use of the material conditional operation– one of the oldest ways in classical logic to interpret "if then" statements [2] – and evaluate it accordingly by probability. Thus, in this approach we identify

$$(\text{if } b, \text{ then } a) = b \Rightarrow a = b' \vee a = b' \vee ab, \tag{10.1}$$

yielding probability evaluation

$$P(\text{if } b, \text{ then } a) = P(b') + P(ab) = 1 - P(b) + P(ab). \tag{10.2}$$

The main attraction in using the material conditional with the interpretation in eqs.(10.1) and (10.2) is that quite simply one can completely determine the probability of any finite logical combination of all such rules. For example,

$$\begin{aligned}
P((\text{if } b, \text{ then } a) \,\&\, (\text{if } d, \text{ then } c)) &= P((b \Rightarrow a)(d \Rightarrow c)) \\
&= P(abcd \vee abd' \vee b'cd \vee b'd') \\
&= P(abcd) + P(abd') + P(b'cd) + P(b'd').
\end{aligned} \tag{10.3}$$

The material conditional approach, however, has the obvious drawback that when antecedents b and d above, are such that $P(b)$ and $P(d)$ are small, the value of the conjunction of inference rules is always close to unity, no matter what values $P(ab)$ or $P(cd)$ take. (A similar problem holds for disjunction.) In addition, a number of classical "paradoxes" occur with the use of the material conditional (see, e.g. [2], Chapter 8.VI).

Another approach to modeling the validity of inference rules is to use certainty factors, a particular technique approximating in a certain sense, but, distinct from, conditional probability, as developed originally in MYCIN. Both approaches to assigning values suffer from a number of problems. For controversy concerning the latter approach see e.g. [7], Section 8.3 and [11]. Yet another approach to representing conditional relations is through *imaging*. (See an exposition and comparison with conditional probability by Gärdenfors [4], Sections 5.3-5.6.) On the other hand, a natural candidate for measuring the value of an inference rule is the associated conditional probability.

$$P(\text{if } e, \text{ then } a) = P_e(a) = P(a|e) = P_e(a|e) = P(ae)/P(e), \tag{10.4}$$

where it is assumed $P(e) > 0$. None of the above issues arise with the standard use of conditional probability. However, note that in the usual development of conditional probability – unlike that for the material conditional – there is no actual corresponding "conditional event" (if e, then a), or symbolized from now on for convenience as $(a|e)$, such that eq.(10.4) holds in the sense that a, e, and $(a|e)$ all lie $\in \mathcal{B}$, for probability space $(\Omega, \mathcal{B}, P)$. In any case, let us now set up the problem as

$$\alpha = (a|e), \qquad \beta = (c|f), \tag{10.5}$$

with formal evaluations

$$P(\alpha) = P_e(a), \qquad P(\beta) = P_f(c), \tag{10.6}$$

Note that if conditional probability or the material conditional probability approach is taken for the interpretation of P(if b, then a), the following inequalities always holds (see also eqs.(10.1), (10.2)) for all $a, b \in \mathcal{B}$ $P(b) > 0$,

$$
\begin{aligned}
P(b \Rightarrow a) &= 1 - P(b) + P(ab) = 1 - P(a'b) = P(a|b) + P(a'|b) - P(a'|b)P(b) \\
&= P(a|b) + [P(b')P(a'|b)] \geq P(a|b) \geq P(ab), \quad \text{for } P(\beta) > 0;
\end{aligned}
$$

$$P(b \Rightarrow a) > P(a|b) \text{ iff } P(b')P(a'|b) > 0 \text{ iff } [P(b) < 1 \text{ and } P(ab) < P(b)]. \tag{10.7}$$

The now well-noted inequalities in eq.(10.7) (see Goodman, Nguyen, and Walker [8], Chapter 1 for a brief history of them) show a basic discrepancy between the two approaches, which can be quite large, especially when $P(b)$ is small and $P(ab)$ is even smaller, in which case $P(b \Rightarrow a)$ approaches unity, while $P(a|b)$ can take high, medium, or even low values, depending on the relative sizes of $P(ab)$ and $P(b)$.

Before continuing on, let us now consider the standard view on conditional probability for positive probability antecedents (the case for zero probability antecedents is best treated by the well-known Radon-Nikodym theory, which we have seen extended in another part of this text to the realm of random sets, but which is not necessary here for the development of our concepts). Let $(\Omega, \mathcal{B}, P)$ be an initially given probability space and let $e \in \mathcal{B}$ be any event with $P(e) > 0$. Then, the corresponding probability space e produces is $(\Omega, \mathcal{B}, P_e)$, which also may be considered in reduced or trace form as $(e, \mathcal{B}e, P_e)$, where

$$\mathcal{B}e = \{ce : c \in \mathcal{B}\}. \tag{10.8}$$

As stated above, it is not part of the standard development of probability to treat conditional probability as the legitimate unconditional probability evaluation of conditional events, so that the following relations hold, when both antecedents are identical, i.e. $e = f$ here:

$$(a|e)' = (a'|e), \quad (a|e)\&(c|e) = (ac|e), \quad (a|e) \vee (c|e) = (a \vee c|e),$$

$$P((a|e)') = P((a'|e)) = P_e(a'), \quad P((a|e)\&(c|e)) = P((ac|e)) = P_e(ac),$$

$$P((a|e) \vee (c|e)) = P((a \vee c|e)) = P_e(a \vee c). \tag{10.9}$$

So far, there is no real need to carry out such cumbersome-appearing notation. But, now suppose that $e \neq f$ and we wish to evaluate in place of the conjunction and disjunction forms in eq.(29),

$$P((a|e)\&(c|f)) =?, \qquad P((a|e)v(c|f)) =? \qquad (10.10)$$

whore not only the use of P is a formal one, but now emphatically that of operators $\&$ and $\vee$. Even if in any sensé e ls a guud approximation to f – so long as they are actually not identical – standard probability still cannot treat their logical combination. Certainly, as we have already seen, if we wish to apply a rigorous testing of hypotheses for similarity of α to β – such as in the use of $d_{P,2}(\alpha, \beta)$, we need first to be able to obtain $\alpha\beta$ and, in turn, $P(\alpha\beta)$. As discussed above, Lewis has shown that in general we cannot solve the above problem as it stands. But, if we modify it in the sense described in the next section a different story holds!

10.2   Conditional Event Algebra Problem and Connections with Similarity Measures

Let $(\Omega, \mathcal{B}, P)$ be as before. We seek a space $(\Omega_o, \mathcal{B}_o, P_o)$ (preferably, but not necessarily a probability space) – where the subscript o here is not to be confused with the event probability subscript used to indicate conditioning – which extends $(\Omega, \mathcal{B}, P)$ and an operation $(.|..)$ such that:

   (CE0) $(a|b) = (ab|b)$, for all $a, b \in \mathcal{B}$.

   (CE1)   All events $a, c, e, f, ... \in \mathcal{B}$ have isomorphic counterparts $(a|\Omega), (c|\Omega)$, $(e|\Omega), (f|\Omega) \in \mathcal{B}_o$, where, for simplicity, the (possibly boolean) logical operators for $\mathcal{B}_o$ are given the same symbols as their counterparts for $\mathcal{B}$.

   (CE2)   The compatibility relations hold:

$$P_o((a|\Omega)) = P(a), \quad P_o(c|\Omega) = P(c), \quad P_o(e|\Omega) = P(e), \quad P_o(f|\Omega) = P(f), ...$$

   (CE3)   There exist for any choice of $a, c, e, f \in \mathcal{B}$, (conditional) events $(a|e), (c|f), ... \in \mathcal{B}_o$ with compatibility relations (replacing the formal form of eq.(10.2))

$$P_o((a|e)) = P_e(a), \qquad P_o((c|f)) = P_f(c), ...$$

   (CE4)   When antecedents are identical such as $e = f$, then eq.(10.9) holds, with P replaced by P_o, but P_e left alone.

   (CE5)   Any finite logical combination of conditional events $\in \mathcal{B}_o$ is again in $\mathcal{B}_o$ and the corresponding probability evaluation via P_o of such conditional events (including any imbedding of the unconditional events as special cases) is finitely obtainable. Thus, in particular, there exist numerical-valued functions $Q_\&, Q_\vee$ such that

$$P_o((a|e)\&(c|f)) = Q_\&(P(ae), P(cf), P(acef), P(e), P(f), P(ef), ...),$$
$$P_o((a|e) \vee (c|f)) = Q_\vee(P(ae), P(cf), P(acef), P(e), P(f), P(ef), ...).$$

When the above conditions are satisfied, call $(\Omega_o, \mathcal{B}_o, P_o)$ a *conditional event algebra* extension of $(\Omega, \mathcal{B}, P)$. If $(\Omega_o, \mathcal{B}_o, P_o)$ is an actual probability space, call it a *boolean* conditional event algebra extension of $(\Omega, \mathcal{B}, P)$. For detailed histories of the problems discussed in Sections 11.1 and the above formulations of a conditional event algebra, see Goodman, Nguyen and Walker [8], Chapter 1 and Goodman [5].

Even if we cannot obtain the desired conditional event algebra and conditional events directly, but at least know they exist in a *boolean algebra setting*, then the well-known Fréchet-Hailperin bounds [10] can be applied yielding the following tightest inequalities on the computations of $P_o((a|e)\&(c|f))$, and dually $P_o((a|e) \vee (c|f))$, in terms of the individual probabilities $P_e(a)$, $P_f(c)$:

$$\begin{aligned}
\max(P_e(a) \quad &+ \quad P_f(c) - 1, 0) \leq P_o((a|e)\&(c|f)) \leq \min(P_e(a), P_f(c)) \\
&\leq \quad w \cdot P_e(a) + (1-w) \cdot P_f(c) \\
&\leq \quad \max(P_e(a), P_f(c)) \leq P_o((a|e) \vee (c|f)) \\
&\leq \quad \min(P_e(a) + P_f(c), 1),
\end{aligned} \qquad (10.11)$$

where w is any real number, $0 \leq w \leq 1$ and where, in general, all of the above inequalities are strict and because all laws of probability hold, in particular,

$$\begin{aligned}
P_o((a|e)\&(c|f)) &= P_o((a|e)) + P_o((c|f)) - P_o((a|e)\&(c|f)) \\
&= P_e(a) + P_f(c) - P_o((a|e)\&(c|f)).
\end{aligned} \qquad (10.12)$$

In fact, equality holds as follows:

$$\begin{aligned}
P_o((a|e)\&(c|f)) &= \min(P_e(a), P_f(c)) \text{ iff } P_o((a|e) \vee (c|f)) \\
&= \max(P_e(a), P_f(c)) \text{ iff } P_o((a|e) \leq (c|f)) \text{ or } (c|f) \leq (a|e)) = 1, \\
\text{i.e., } \quad P_o((a|e)\&(c|f)') &= 0 \quad \text{or} \quad P_o((a|e)'\&(c|f)) = 0,
\end{aligned} \qquad (10.13)$$

$$\begin{aligned}
P_o((a|e)\&(c|f)) &= \max(P_e(a) + P_f(c) - 1, 0) \\
\text{iff} \quad P_o((a|e) &\vee (c|f)) = \min(P_e(a) + P_f(c)), 1) \\
\text{iff} \quad &\begin{cases} P_o((a|e)\&(c|f)) = 0, & \text{when } P_e(a) + P_f(c) \leq 1 \\ P_o((a|e) \vee (c|f)) = 1, & \text{when } P_e(a) + P_f(c) \geq 1 \end{cases}
\end{aligned} \qquad (10.14)$$

Another handle can be obtained on the computation of conjunctions and disjunctions of conditional events, when it is reasonable to assume that they are independent. In this case, we have the trivial computations, with no need for any explicit conditional event algebra use

$$P_o((a|e)\&(c|f)) = P_e(a) \cdot P_f(c),$$

$$P_o((a|e) \vee (c|f)) = P_e(a) + P_f(c)) - P_e(a) \cdot P_f(c). \qquad (10.15)$$

Still another special case can arise when a chaining or modus ponens relation holds, such as

$$P_o((a|\Omega)\&(c|a)) = P(ca). \qquad (10.16)$$

Rowe ([12], especially Chapter 8) without determining a specific conditional event algebra, in effect, recognized the same problem and addressed the issue similar to that in eqs.(10.11) and (10.15). (In fact, Rowe claimed that the problem was intractable!)

In a related vein, it should be remarked that the corresponding tightest bounds derived from the Fréchet-Hailperin ones on $d_{P_o,j}(\alpha,\beta)$ for $j = 1,2$ are easily seen to be:

$$|P_e(a) - P_f(c)| = d_{P_o,1}(\alpha,\beta) \le d_{P_o,2}(\alpha,\beta)$$
$$\le \min(P_e(a) + P_f(c), 2 - (P_e(a) + P_f(c))), \quad (10.17)$$

so that if the conditional event algebra computations for $P_o(\alpha\beta)$ were unavailable, the maximum amount of probability leeway for $d_{P_o,2}(\alpha,\beta)$ is

$$\min(P_e(a) + P_f(c), 2 - (P_e(a) + P_f(c))) - |P_e(a) - P_f(c)|$$
$$= \left\{ \begin{array}{ll} 2\min(P_e(a), 1 - P_f(c)), & \text{if } P_e(a) \le P_f(c) \\ 2\min(P_f(c), 1 - P_e(a)), & \text{if } P_e(a) \ge P_f(c). \end{array} \right. \quad (10.18)$$

Clearly, such a possible wide range for the possible value of $d_{P_o,2}(\alpha,\beta)$ in eq.(10.18) shows that if the existence and choice of a well-justified, specific cea can be made, it is preferable to use the corresponding exact evaluations rather than just producing the above large range of possible values. To this end, first three different candidate cea's will be presented in Chapter 11. Later, comparisons will be made for determining the most appropriate one for given uses.

10.3 Application of Conditional Event Algebra to the Determination of Constant-Probability Events

This section is a modification of earlier work found in Goodman and Kramer [6], Section 3.1. We show here how conditional event algebra can be used to construct constant-probability events. Consider fixed measurable space $(\Omega, \mathcal{B})$ and choose any event $c \in \mathcal{B}$ and any probability measure $P : \mathcal{B} \to [0, 1]$ such that the nontriviality condition holds

$$1 > P(c) > 0. \quad (10.19)$$

Then choose any positive integers m and n with $1 \le m < n$ and form

$$\tau(m, n, c) = c^{m-1} \times c' \times c^{n-m}, \quad (10.20)$$

where c^j is the j-fold cartesian product of c with itself (see eq.(1.3.76)) with the convention that

$$c^0 \times c' = c', \qquad c' \times c^0 = c'. \quad (10.21)$$

Consider the corresponding n-fold product probability space $(\Omega^n, \mathcal{B}_n, P_n)$, all of whose single marginal probability spaces are identical to $(\Omega, \mathcal{B}, P)$. Then, assuming an appropriate conditional event algebra extension $((\Omega^n)_o, (\mathcal{B}_n)_o, (P_n)_o)$ exists relative to $(\Omega_n, \mathcal{B}_b, P_n)$, for any positive integer n and nonnegative integer m, with $0 \leq m \leq n$, define the conditional event

$$\nu(m, n, c) = \left(\bigvee_{j=1}^{m} \tau(j, n, c) \,\middle|\, \bigvee_{j=1}^{n} \tau(j, n, c) \right) \in (\mathcal{B}_n)_o, \qquad (10.22)$$

with the convention and result that

$$\nu(0, n, c) = (\emptyset_n)_o, \qquad \nu(n, n, c) = (\Omega_n)_o. \qquad (10.23)$$

It follows that for any c and P satisfying eq.(10.19), eq.(10.22) yields from the basic compatibility property and the fact that each $\tau(j, n, c)$ has the same probability evaluation, but are all mutually disjoint:

$$(P_n)_o \, (\nu(m, n, c)) = (m \cdot (P(c))^{n-1} P(c'))/(n \cdot (P(c))^{n-1} P(c')) = m/n, \quad (10.24)$$

not depending on the particular c and P chosen, so long as eq.(10.19) is satisfied. In fact, eq.(10.23) shows that the choice of m and n plays no role except through the ratio m/n. Because a given problem may have many constants to be modeled by corresponding constant probability events, we formally let $m, n \to \infty$ with m/n held fixed. With this, redefine any such $\nu(m, n, c)$ as the constant-probability event $\theta(m/n)$, suppressing the P and c. By a limit process, we can appropriately define $\theta(t)$, for any $t \in [0, 1]$, with probability space $(\Omega_1, \mathcal{B}_1, P_1)$ corresponding to the limiting form for $((\Omega_n)_o, (\mathcal{B}_n)_o, (P_n)_o)$ as above. Clearly, by use of eq.(10.22) and property (CE4) of conditional event algebras,

$$\nu(m_1, n, c) \,\&\, \nu(m_2, n, c) = \nu(\min(m_1, m_2), n, c),$$
$$\nu(m_1, n, c) \vee \nu(m_2, n, c) = \nu(\max(m_1, m_2), n, c), \qquad (10.25)$$

and thus,

$$\theta(s) \,\&\, \theta(t) = \theta(\min(s, t)), \quad \theta(s) \vee \theta(t) = \theta(max(s, t)), \quad 0 \leq s, t \leq 1, \quad (10.26)$$

whence, for all P of interest,

$$P_1(\theta(s) \,\&\, \theta(t)) = \min(s, t), \qquad P_1(\theta(s) \vee \theta(t)) = \max(s, t), \qquad (10.27)$$

justifying the earlier remarks on the properties of constant-probability events, *provided an appropriate (boolean) conditional event algebra extension of* $(\Omega, \mathcal{B}, P)$ can be found. This is indeed the case. Detailed descriptions of three basic conditional event algebras are provided, followed by some specific calculations addressing the original issue posed in eqs.(10.5), (10.6). In addition, a further development of constant-probability events introduced earlier is also presented.

Bibliography

[1] Buchanan, B. G. and Shortliffe, E. H. (1984) *Rule-Based Expert Systems*, Addison-Wesley, Reading, MA.

[2] Copi, I. M. (1986) *Introduction to Logic* (7th Ed.), Macmillan Co., New York.

[3] Dillard, R. A. (1983) *Representation of Tactical Knowledge Shared by Expert Systems*, Technical Document 632, Naval Ocean Systems Center, San Diego, CA, Oct. 1, 1983.

[4] Gärdenfors, P. (1988) *Knowledge in Flux*, MIT Press, Cambridge, MA.

[5] Goodman, I. R. (1991) Evaluation of combinations of conditioned information: a history, *Information Sciences*, **57-58**, 79-110.

[6] Goodman, I. R. and Kramer, G. F. (1996) Applications of relational event algebra to the development of a decision aid in command and control, *Proceedings 1996 Command and Control Research and Technology Symposium*, Naval Postgraduate School, Monterey, CA, June 25-28,1996, pp. 415-435.

[7] Goodman, I. R. and Nguyen, H. T. (1985) *Uncertainty Models for Knowledge-Based Systems*, North-Holland, Amsterdam.

[8] Goodman, I. R., Nguyen, H. T., and Walker, E.A. (1991) *Conditional Inference and Logic for Intelligent Systems*, North-Holland, Amsterdam.

[9] Hailperin, T. (1984) Probability logic, *Notre Dame Journal of Formal Logic* **25**(3), 198-212.

[10] Hayes-Roth, F., Waterman, D. A., and Lenat, D. B. (1983) *Building Expert Systems*, Addison-Wesley, Reading, MA.

[11] Pearl, J. (1988) *Probabilistic Reasoning in Intelligent Systems: Networks of Plausible Inference*, Morgan Kaufmann, San Mateo, CA.

[12] Rowe, N. C. (1988) *Artificial Intelligence through PROLOG*, Prentice-Hall, Englewood Cliffs, NJ.

Chapter 11

Three Particular Conditional Event Algebras

11.1 General Remarks on Conditional Event Algebra

Conditional event algebra(s) can be roughly divided into two camps: boolean and non-boolean. (Again, see the above definitions.) It appears that of the many possible conditional event algebras possible – and proposed, so far – three stand out in their significance for possible applications. All three have a number of common desirable properties – and a fundamental link concerning deduction (elaborated upon in Section 13.1), but each also has certain desirable and negative properties, the others do not possess. Two of these are non-boolean, while the third is boolean.

11.2 DeFinetti-Goodman-Nguyen-Walker Conditional Event Algebra

DeFinetti in 1935 [8], and later more expanded in [9], proposed a nonboolean cea (conditional event algebra) via the logic fragment of $[1 - (), \min, \max] \leftrightarrow$ [complement, conjunction, disjunction]-logic of Lukasiewicz' and others extensions of classical two-valued logic as in Table A below.

The form of this cea was through the indicator functions $\phi(a|b)$ of formal conditional events $(a|b)$. Though the conditional events themselves were not specifically given, the indicator function was provided in the three-valued form

369

	and = min	or = max	negation
y x	0 u 1	0 u 1	x'
0	0 0 0	0 u 1	1
u	0 u u	u u 1	u
1	0 u 1	1 1 1	0

Table A. Fragment of Lukasiewicz 3-valued
logic corresponding to DGNW

$\phi(a|b) : \Omega \to \{0, u, 1\}$, where u is a third, or indeterminate, value, in addition to the familiar 0,1 of classical or boolean logic, so that for given probability space $(\Omega, \mathcal{B}, P)$ of unconditional events $a, b, c, e, f, \ldots$

$$\phi(a|b)(\omega) = \begin{cases} 1 & \text{if } \omega \in ab \\ 0 & \text{if } \omega \in a'b \\ u & \text{if } \omega \in b'. \end{cases} \tag{11.1}$$

Thus, DeFinetti's proposed cea was that, in effect, for say any two (the argument readily extendible to any number of conditional events) $(a|b)$, $(c|d)$,

$$\phi((a|b)')(\omega) = 1 - \phi(a|b)(\omega),$$
$$\phi((a|b)\&(c|d))(\omega) = \min(\phi(a|b)(\omega), \phi(c|d)(\omega)),$$
$$\phi((a|b) \vee (c|d))(\omega) = \max(\phi(a|b)(\omega), \phi(c|d)(\omega)), \forall \omega \in \Omega, \tag{11.2}$$

where, as before the same symbol $()'$, $\&$, $\vee$ indicate the appropriate extensions of the usual boolean counterparts, and where it is assumed we have the formal ordering and identification (see Table A)

$$0 < u < 1, \qquad 1 - u = u. \tag{11.3}$$

However, as far as can be seen, DeFinetti did not show how to compute explicit probability evaluations of the above logical combinations. Goodman [12] and Goodman, Nguyen, and Walker [15] obtained explicit representations for conditional events, as well as their logical combinations and probability evaluations of such combinations, compatible (in fact, isomorphic)with DeFinetti's proposals for the indicator function form in eqs.(11.2),(11.3). In this case, any conditional event $(a|b)$ is identifiable as either the ordered pair (ab, b), the interval of events

$$[ab, b \Rightarrow a] = \{c \in \mathcal{B} : ab \le c \le b \Rightarrow a\}, \tag{11.4}$$

where the usual material conditional operator $\Rightarrow$ is as in eq.(10.1); or the coset

$$\Omega b' \vee ab = \{xb' \vee ab : x \in \Omega\} \tag{11.5}$$

$\in$ quotient boolean algebra

$$\Omega/\Omega b' = \{\Omega b' \vee ab : a \in \mathcal{B} \text{ arbitrary}\}, \tag{11.6}$$

noting the actual identity

$$[ab, b \Rightarrow a] = \Omega b' \vee ab. \tag{11.7}$$

The Goodman-Nguyen-Walker proposal showed that by interpreting &, $\vee$, $()'$ component-wise relative to the conditional events in interval of event, or equivalently, in coset form (for possibly varying antecedents, and hence possibly different quotient boolean algebras), one obtained

$$\mathcal{B}_o = \{(a|b) : a, b \in \Omega\} \tag{11.8}$$

as being closed with respect to the operations $()', \&, \vee$, yielding for one and two arguments (the multiple argument cases readily obtainable also) the specific forms

$$
\begin{aligned}
(a|b)' &= (a'|b) \\
(a|b)\&(c|d) &= ((a|b)' \vee (c|d)')' \\
&= (abcd \mid a'b \vee c'd \vee bd) = (abcd \mid a'b \vee c'd \vee abcd), \\
(a|b) \vee (c|d) &= (ab \vee cd \mid ab \vee cd \vee bd) = (abcd \mid ab \vee cd \vee a'bc'd).
\end{aligned}
\tag{11.9}
$$

The corresponding probability evaluations, in light of the fundamental compatibility conditions (CE2) – which can be shown to hold consistently here ([15], Chapters 0,1), yield

$$
\begin{aligned}
P_o((a|b)') &= P(a'|b), \\
P_o((a|b)\&(c|d)) &= P(abcd \mid a'b \vee c'd \vee bd) = P(abcd \mid a'b \vee c'd \vee abcd), \\
P_o((a|b) \vee (c|d)) &= P(ab \vee cd \mid ab \vee cd \vee bd) \\
&= P(ab \vee cd \mid ab \vee cd \vee a'bc'd),
\end{aligned}
\tag{11.10}
$$

provided none of the above denominators are zero. Thus, (CE5) holds. It is easily seen that (CE2) holds as a special case of eq.(11.10), and in fact, (CE0), (CE1), (CE4) also hold, so that this system is a legitimate conditional event algebra, which we will call DGNW (DeFinetti-Goodman-Nguyen-Walker). Also, obviously from the above definitions,

$$(a|b) = (c|d) \text{ iff } ab = cd \text{ and } b = d, \quad (a|\Omega) = \{a\} \text{ (identified with } a\text{), } \tag{11.11}$$

$$(a|\emptyset) = (\emptyset|\emptyset) = 2^\Omega, \quad (a|a) = \Omega a' \vee a = \Omega \vee a \text{ (filter class on } a\text{). } \tag{11.12}$$

Also, the standard partial order $\leq$ over $\mathcal{B}$ can be extended to $\mathcal{B}_o$, where now

$$(a|b) \leq (c|d) \quad \text{iff} \quad (a|b) = (a|b)\&(c|d), \tag{11.13}$$

whence, it can be shown (see again [15] for proofs of all the following statements concerning DGNW cea)

$$(a|b) \leq (c|d) \text{ iff } (a|b) = (a|b)\&(c|d) \text{ iff } (c|d) = (a|b) \vee (c|d)$$
$$\text{iff } [ab \leq cd \text{ and } c'd \leq a'b] \text{ iff } [ab \leq cd \text{ and } b \Rightarrow a \leq d \Rightarrow c]. \tag{11.14}$$

In addition, it can be shown that the DGNW cea is not only a fully distributed DeMorgan lattice/algebra (thus, &, $\vee$ are associative, commutative, idempotent, and mutually absorbing), but is also relatively pseudocomplemented, and in fact is a Stone algebra and, analogous to Schay's characterization of his proposed cea's (see [19] and comments in Section 12.3), a Stone Representation theorem has been obtained for DGNW ([15], Chapter 4; [22]). Nevertheless it is still not a full boolean algebra, because ()' is not a true complementation. This means some of the standard laws of probability do not hold here, i.e., the function P_o extending probability measure P is itself not a probability measure. For example, the modular expansion fails, resulting in the inequality

$$P_o((a|b) \vee (c|d)) \neq P(a|b) + P(c|d) - P_o((a|b)\&(c|d)). \tag{11.15}$$

Moreover, if $bd = \emptyset$, the undesirable property that

$$P_o((a|b)\&(c|d)) = 0, \qquad P_o((a|b) \vee (c|d)) = 1 \tag{11.16}$$

holds. On the other hand, the DGNW characterization of partial ordering is a very natural one, in that for any a, b, c, d with $\emptyset < a < b$, $\emptyset < c < d$ ([15], pp. 48, 49, 66, 67)

$$P(a|b) \leq P(c|d), \text{ for all possible well-defined } P$$
$$\text{iff } (a|b) \leq (c|d) \text{ in DGNW sense}$$
$$\text{iff } ab \leq cd \text{ and } c'd \leq a'b, \tag{11.17}$$

as in eq.(11.14). In a related vein, we also have for a, b, c, d as above

$$P(a|b) = P(c|d), \text{ for all possible well-defined } P$$
$$\text{iff } (a|b) = (c|d) \text{ in DGNW sense}$$
$$\text{iff } ab = cd \text{ and } b = d. \tag{11.18}$$

A few special conjunctions and disjunctions:

$$\begin{aligned}
&(a|b)\&(c|\Omega) = (abc \mid b \vee c') \quad \text{(generalized modus ponens form)}, \\
&(b|b)\&(c|d) = (bcd \mid bd \vee c'd), \qquad (b|b) \vee (c|d) = (b \vee cd \mid b \vee cd), \\
&(\Omega|\Omega)\&(c|d) = (c|d), \qquad (\Omega|\Omega) \vee (c|d) = (\Omega|\Omega), \\
&(\emptyset|b)\&(c|d) = (\emptyset|b \vee c'd), \qquad (\emptyset|b) \vee (c|d) = (cd|cd \vee bd), \\
&(\emptyset|\Omega)\&(c|d) = (\emptyset|\Omega), \qquad (\emptyset|\Omega) \vee (c|d) = (c|d), \\
&(\emptyset|\emptyset)\&(c|d) = (\emptyset|c'd), \qquad (\emptyset|\emptyset) \vee (c|d) = (cd|cd). \tag{11.19}
\end{aligned}$$

11.3   Adams-Calabrese (AC) Conditional Event Algebra

The other important non-boolean cea is the AC (Adams-Calabrese) cea, introduced originally by Adams [1],[2], and later independently rediscovered by Calabrese [5], [6]. Schay [19] also introduced two different non-DeMorgan conditional event algebras, where if the conjunction operation from one and the disjunction operation from the other were put together, the result would be the AC cea. Schay also proposed a complement operation which is the same as for all other proposed conditional event algebras. However, though Schay presented an in-depth analysis of his proposed candidates, as they stand, they appear to have certain properties that preclude them as reasonable candidates for use. (See [15], Chapter 1.)

In fact, Calabrese's work inspired a modern resurgence of interest in cea, beginning with [12], as well as [14], [15], and culminating in a special issue of the IEEE Transactions on Systems, Man, and Cybernetics devoted entirely to the subject of cea [10]. Although Adams did not specify what specific forms could be used to represent conditional events, he did propose a calculus of "quasi-" logical operations. Calabrese, in effect, proposed an interpretation of conditional events as ordered pairs and produced the same calculus as Adams'. Neither originally tied in their operations with three-valued logic, but it was pointed out in [15] that not only does DGNW correspond isomorphically to a particular (1-( ), min, max) three-valued logic, but also AC corresponds isomorphically to a fragment of three-valued logic extending ordinary two-valued logic (Sobocinski's indifference logic – see Table B below or the discussion in [18]). In fact, it is shown in [15], Chapter 3.4 that *all possible truth functional three-valued logics are isomorphic to all appropriately corresponding cea's such that their logical operations can be considered to result in corresponding boolean functions for consequents and antecedents*, i.e., for the binary case, e.g., where g is either a conjunction or disjunction type operation

$$g((a|b),(c|d)) = (h_g(ab,b,cd,d) \mid k_g(ab,b,cd,d)), \quad \text{all } a,b,c,d \in \mathcal{B}, \quad (11.20)$$

where h_g and k_g are ordinary (unconditional) boolean operators.

	and	or	negation
y x	0 u 1	0 u 1	x'
0	0 0 0	0 0 1	1
u	0 u 1	0 u 1	u
1	0 1 1	1 1 1	0

Table B. Fragment of Sobocinski's 3-valued
logic corresponding to DGNW

In brief, the basic logical operations for AC at the unary and binary level are:

$$(a|b)' \;=\; (a'|b),$$
$$(a|b)\&(c|d) \;=\; ((a|b)' \vee (c|d)')' = ((b \Rightarrow a)(d \Rightarrow c) \mid b \vee d)$$
$$\;=\; (abcd \vee abd' \vee b'cd \mid b \vee d),$$
$$(a|b) \vee (c|d) \;=\; (ab \vee cd \mid b \vee d), \tag{11.21}$$

where eqs.(11.4)-(11.7), (11.11), (11.12) all remain valid here. One can also show that if conditional events are considered in the form of cosets, or equivalently, intervals of events, disjunction defined in eq.(11.21) is the same as the class intersection $(a|b) \cap (c|d)$, provided they have a non-empty intersection ([15], p. 55). The corresponding probability evaluations to eq.(68) are (for all P well-defined)

$$P_o((a|b)') \;=\; P(a'|b),$$
$$P_o((a|b)\&(c|d)) \;=\; P((b \Rightarrow a)(d \Rightarrow c)|b \vee d)$$
$$\;=\; P(abcd \vee abd' \vee b'cd|b \vee d),$$
$$P_o((a|b) \vee (c|d)) \;=\; P(ab \vee cd|b \vee d). \tag{11.22}$$

AC can also be shown to be a legitimate cea. It is also not distributive and though it is a DeMorgan semilattice with respect to & and $\vee$ separately (and hence &, $\vee$ are associative, commutative and idempotent), it is not a full lattice (absorption identities – and hence full compatibility between $\leq$ and &, $\vee$ operations fail), and is thus not boolean. Hence again, as in the case of DGNW, certain laws of probability fail here, such as the modular expansion (thus inequality (11.15) holds). Thus, as in the case of DGNW, the function P_o here extending probability measure P is not a probability measure. Also, the usual partial order $\leq$ over $\mathcal{B}$ can be extended to $\mathcal{B}_o$ for AC, where one defines for AC

$$(a|b) \leq_\& (c|d) \quad \text{iff} \quad (a|b) = (a|b)\&(c|d),$$
$$(a|b) \leq_\vee (c|d) \quad \text{iff} \quad (c|d) = (a|b) \vee (c|d), \tag{11.23}$$

noting that, unlike DGNW, the partial orders $\leq_\&$ and $\leq_\vee$ do not coincide in general. The two partial orders in eq.(11.23) are readily characterized separately as

$$(a|b) \leq_\& (c|d) \quad \text{iff} \quad [b \Rightarrow a \leq d \Rightarrow c \text{ and } d \leq b], \tag{11.24}$$

$$(a|b) \leq_\vee (c|d) \quad \text{iff} \quad [ab \leq cd \text{ and } b \leq d]. \tag{11.25}$$

On the other hand, AC cea has many desirable properties, such as v actually being a type of weighted average, since we can rewrite the bottom of eq.(11.22) in terms of AC operations as

$$(a|b) \vee (c|d) = (a|b)\&(b \mid b \vee d) \;\vee\; (c|d)\&(d \mid b \vee d), \tag{11.26}$$

with corresponding probability evaluation for $bd = \emptyset$, being

$$P_o((a|b) \vee (c|d)) = P(a|b)P(b \mid b \vee d) + P(c|d)P(d \mid b \vee d), \qquad (11.27)$$

which is useful in the analysis of the well-known Simpson's "paradox" that weighted averages of portions of one collection of data can be dominated by weighted averages of corresponding parts of another collection of data, yet the overall weighted average of the first collection may at times exceed the overall weighted average of the second [7]. On the other hand, AC cca has certain drawbacks. For example if we consider $P_o((a|b) \vee (c|\Omega))$, where $0 < P(a) < P(b)$, with $P(b)$ small, but $P(a|b)$ close to unity, and with $P(c)$ also small, we obtain from eq.(11.21)

$$P_o((a|b) \vee (c|\Omega)) = P(ab \vee c) \le P(ab) + P(c), \qquad (11.28)$$

a small value, yet intuitively, since one part of the disjunction, $P(a|b)$ is large, it would seem natural that disjunction should be large. In fact, eq.(11.28) shows the failure of monotonicity of $\vee$ for AC. Furthermore, neither of the two AC semilattice orders in eq.(11.23) are compatible with the natural probability evaluation order in eq.(11.17), as is DGNW.

A few special conjunctions and disjunctions for AC:

$$(a|b)\&(c|\Omega) = (abc \vee b'c|\Omega) \quad \text{(generalized modus ponens)}$$
$$(b|b)\&(c|d) = (bcd \vee bd \vee cdb' \mid b \vee d), \qquad (b|b) \vee (c|d) = (b \vee cd \mid b \vee d),$$
$$(\Omega|\Omega)\&(c|d) = (d \Rightarrow c \mid \Omega), \qquad (\Omega|\Omega) \vee (c|d) = (\Omega|\Omega),$$
$$(\emptyset|b)\&(c|d) = (b'cd \mid b \vee d), \qquad (\emptyset|b) \vee (c|d) = (cd \mid b \vee d),$$
$$(\emptyset|\Omega)\&(c|d) = (\emptyset|\Omega), \qquad (\emptyset|\Omega) \vee (c|d) = (cd|\Omega),$$
$$(\emptyset|\emptyset)\&(c|d) = (c|cd), \qquad (\emptyset|\emptyset) \vee (c|d) = (c|d). \qquad (11.29)$$

Again, inspection of eq.(11.29) shows the failure of AC $\&$ and $\vee$ to be monotonic, even with respect to the unity and zero elements

$$(\Omega|\Omega)\&(c|d) = (d \Rightarrow c|\Omega) > (c|d) > (cd|\Omega) = (\emptyset|\Omega) \vee (c|d), \quad \text{in general.} \qquad (11.30)$$

On the positive side of things, unlike the triviality-producing property of DGNW operations for disjoint antecedents as in eq.(11.16), when $bd = \emptyset$, we obtain

$$(a|b)\&(c|d) = (a|b) \vee (c|d) = (ab \vee cd \mid b \vee d), \qquad (11.31)$$

with corresponding probability evaluation (see also related eq.(11.27))

$$P_o((a|b)\&(c|d)) = P_o((a|b) \vee (c|d)) = P(ab \vee cd|b \vee d).$$

The above corresponds to an important class of conditional statements in natural language where disjoint alternate choices (the number need not be restricted to two) constitute the antecedents, and (necessarily) disjoint statements constitute the consequents.

11.4   Some Comparisons of DGNW and AC Conditional Event Algebras

Both AC and DGNW also possess the *chaining* and *modus ponens* properties in the form:

$$(a|bc)\&(b|c) = (ab|c), \quad \text{all } a, b, c \in \mathcal{B} \quad \text{(chaining)},$$

which reduces to

$$(a|b)\&(b|\Omega) = (ab|\Omega) \quad \text{(modus ponens)}, \tag{11.32}$$

with the usual compatible probability counterpart involved

$$P_o((a|bc)\&(b|c)) = P(a|bc)P(b|c) = P_o((a|bc)) = P(a|bc). \tag{11.33}$$

It can be shown that no possible cea based on truth-functional three-valued logic or the equivalent operator forms in eq.(11.20) can possibly be boolean and that the only cea which is a Stone algebra is DGNW. (See, e.g., [21].) In addition to not being boolean – and consequently certain standard laws of probability fail – as seen in inequality (11.15) – both DGNW and AC possess other shared drawbacks, such as the possible existence (which the reader can easily verify) of events a, b, c, d for which ab, b is P-independent of cd, d, i.e., the four-way independence relations hold

$$P(abcd) = P(ab)P(cd), \qquad P(abd) = P(ab)P(d),$$
$$P(bcd) = P(b)P(cd), \qquad P(bd) = P(b)P(d), \tag{11.34}$$

yet $(a|b)$ and $(c|d)$ are not P_o-independent, i.e., in general,

$$P_o((a|b)\&(c|d)) \neq P(a|b)P(c|d), \tag{11.35}$$

for both DGNW and AC cea's.

Also, both DGNW and AC cea's have associated higher order conditionals, i.e., objects of the form $((a|b)|(c|d))$. In both cea's, the higher order conditional events are forced to be in the same space $\mathcal{B}_o$ as the conditionals $(a|b)$, $(c|d)$.

$$DGNW: \quad ((a|b)|(c|d)) = (a \mid b\&(a'd' \vee cd)),$$
$$AC: \quad ((a|b)|(c|d)) = (a \mid b\&(d \Rightarrow c)), \tag{11.36}$$

with both definitions coinciding when either $d = b$ or $d = \Omega$, yielding the common form $(a|bc)$, the first called in logic literature the import-export property and the second in abstract algebra literature the second isomorphism property:

$$((a|b)|(c|\Omega)) = (a|bc) = ((a|b)|(c|b)). \tag{11.37}$$

However, note that using eqs.(11.10) and (11.22), both higher order conditional event definitions fail to satisfy the natural relations

$$P_o((a|b)|(c|d)) = P_o((a|b)\&(c|d))/P(c|d), \tag{11.38}$$

as can be verified directly.

11.5 Lewis' Negative Result Concerning Forced Boolean Conditional Event Algebras

Recall that in eq. (10.7) it was shown that a major difference exists in general between $P(a|b)$ and $P(b \Rightarrow a)$. In fact, Calabrese considered this issue in [4] and later showed in [5] that this difference can be extended to taking the probability of any of the 16 possible binary boolean operators acting on a, b and comparing this with conditional probability $P(a|b)$. However, independent of all of this, Lewis earlier showed in [16] that no binary operation (boolean or otherwise) over the basic boolean algebra of events can have a probability evaluation identical to $P(a|b)$ for all $a, b \in \mathcal{B}$, and all well-defined P. Lewis' negative result concerning conditioning was briefly mentioned earlier (Section 9.3 and the Introduction to this book). At this point, because of both the importance of the result – and its profound impact on anyone in the logical or philosophical communities seeking possibly to construct cea's (see, e.g. Eels and Skyrms [11]) – as well as the simplicity of the theorem and its proof, we present the result in a somewhat modified form here.

Theorem 0. (See Lewis [16]) *Let $(\Omega, \mathcal{B}, P)$ be any probability space such that there are events $a, b \in \mathcal{B}$ with $\emptyset < a < b < \Omega$ and $0 < P(a) < P(b) < 1$. Then, it is impossible to have conditional event $(a|b)$ also $\in \mathcal{B}$, so that for all possible choices of P $(P(b) > 0)$, the compatibility condition (CE3) holds, i.e. here*

$$P((a|b)) = P(a|b).$$

Proof: By the assumptions and the total probability expansion,

$$
\begin{aligned}
P(a)/P(b) &= P(a|b) = P((a|b)) = P_a((a|b))P(a) + P_{a'}((a|b))P(a') \\
&= P_a(a|b)P(a) + P_{a'}(a|b)P(a') \\
&= [(P(aba)/P(a))/(P(ab)/P(a))]P(a) \\
&\quad + [(P(aba')/P(a'))/(P(a'b)/P(a'))]P(a') \\
&= 1 \cdot P(a) + 0 \cdot P(a') = P(a), \quad\quad\quad (11.39)
\end{aligned}
$$

a contradiction to the assumptions. □

Note that Lewis' theorem is not applicable to AC nor DGNW, since their conditional events do not lie in the same space as the original unconditional events, noting that, e.g., though the isomorphic imbedding of $a, b \in \mathcal{B}$, are $(a|\Omega), (b|\Omega) \in \mathcal{B}_o$, however, $(a|\Omega), (b|\Omega) \notin \mathcal{B}$ (and $a \neq (a|\Omega)$, $b \neq (b|\Omega)$). In addition, both spaces of conditional events are non-boolean. On the other hand, although both DGNW and AC cea's force the higher order conditional to be back in $\mathcal{B}_o$, since both cea's are not boolean, the total probability theorem is still not applicable. We will, however, return to Lewis' theorem for the next construction.

11.6　Introduction to Product Space Conditional Event Algebra

All of the above negative results set the stage for considering possible boolean cea. Although Van Fraasen originally introduced such a cea [20] , the conditional events dubbed "Stalnaker Bernoulli conditionals", later, in effect McGee obtained the same cea via a utility / betting scheme approach [17], and independently – inspired by a remark of Bamber that all conditional probabilities are formally the same as a certain sequential coin-tossing experiment [3]- Goodman and Nguyen [13] also developed a boolean cea identical in structure to Van Fraasen (McGee obtained the same operations, but did not specify what specific forms the conditional events should take). In short, this common boolean cea is described below and is designated PS (product space) cea:

Let $(\Omega, \mathcal{B}, P)$ be a given probability space of unconditional events $a, b, c, d, \ldots$ Following standard practice, construct a product probability space $(\Omega_o, \mathcal{B}_o, P_o)$ out of $(\Omega, \mathcal{B}, P)$, where the product space has a countable infinite number of marginal-factor spaces, each identical to $(\Omega, \mathcal{B}, P)$. Thus,

$$\Omega_o = \Omega \times \Omega \times \cdots, \tag{11.40}$$

$\mathcal{B}_o$ is the boolean or σ-algebra spanned by $\mathcal{B} \times \mathcal{B} \times \cdots$

P_o is the product probability measure with identical

$$\text{marginal-factor probability } P. \tag{11.41}$$

Next, making another important use of probabilistic distance functions, identify all events in the product space as being equivalent in the usual way: any two events $\alpha, \beta \in \mathcal{B}_o$ are the same iff

$$d_{P,2}(\alpha, \beta) = P_o(\alpha + \beta) = 0. \tag{11.42}$$

If $d_{P,2}(\alpha, \beta) = 0$, for all (well-defined) P, call α and β *strongly equivalent*. Then, we define the form

$$(a|b) = (ab \times \Omega_o) \vee (b' \times ab \times \Omega_o) \vee (b' \times b' \times ab' \times \Omega_o) \vee \ldots, \tag{11.43}$$

noting that the P_o evaluation of the above disjoint infinite disjunction series is

$$\begin{aligned} P_o((a|b)) &= P(ab) + P(b')P(ab) + (P(b'))^2 P(ab) + \ldots. \\ &= P(ab) \cdot \sum_{j=0}^{\infty} (P(b'))^j = P(ab)\frac{1}{1 - P(b')} = P(a|b), \end{aligned} \tag{11.44}$$

provided $P(b) > 0$. In fact, the evaluation in eq.(11.43) is the simplest way to express division of probabilities of one subevent of another in terms of additions of multiplications. The definition in eq.(11.37) can also be considered in recursive form as

$$(a|b) = \bigvee_{i=0}^{j} \left((b')^i \times ab \times \Omega_o \right) \vee \left((b')^{j+1} \times (a|b) \right), \quad j = 0, 1, 2, \ldots \tag{11.45}$$

and, in particular, for $j = 0$,

$$(a|b) = ab \times \Omega_o \; \vee \; b' \times (a|b). \tag{11.46}$$

Similar comments hold for the probability evaluation, such as for $j = 0$, the well-known relation

$$P(a|b) = P(ab) + P(b')P(a|b). \tag{11.47}$$

Furthermore, let us define an operation which resembles conditioning, but is not true conditioning, since the latter via the above product space construction properly extends the original probability space, whereas this operation stays in $\mathcal{B}_o$: For any $b \in \mathcal{B}$ and $\alpha \in \mathcal{B}_o$,

$$
\begin{aligned}
[\alpha|b] &= [\alpha\&(b|\Omega)\,|\,b] \\
&= \alpha\&(b|\Omega) \vee b' \times \alpha\&(b|\Omega) \; \vee b' \times b' \times \alpha\&(b|\Omega) \vee ... \in \mathcal{B}_o
\end{aligned} \tag{11.48}
$$

with recursive form analogous to eqs.(11.13),(11.14)

$$[\alpha|b] = \bigvee_{i=0}^{j} (b')^i \times \alpha\&(b|\Omega) \; \vee \; (b')^{j+1} \times [\alpha|b], \quad j = 0,1,2,..., \tag{11.49}$$

$$[\alpha|b] = \alpha\&(b|\Omega) \; \vee \; b' \times [\alpha|b]. \tag{11.50}$$

The probability evaluations of $[\alpha|b]$ produces, analogous to eq.(11.43),

$$
\begin{aligned}
P_o([\alpha|b]) &= P_o(\alpha\&(b|\Omega)) \sum_{j=0}^{\infty} (P(b'))^j = P_o(\alpha\&(b|\Omega)) \frac{1}{1 - P(b')} \\
&= P_o(\alpha\&(b|\Omega))/P(b) = P_o(\alpha|(b|\Omega)) \\
&= (P_o)_o(([\alpha|(b|\Omega)]),
\end{aligned} \tag{11.51}
$$

though $[\alpha|b] \neq (\alpha|b \times \Omega_o)$, the later $\in (\mathcal{B}_o)_o$, where probability space $((\Omega_o)_o, (\mathcal{B}_o)_o, (P_o)_o)$ is constructed as a countable infinite product space formed properly out of $(\Omega_o, \mathcal{B}_o, P_o)$, as the latter is formed out of $(\Omega, \mathcal{B}, P)$. Note, however, the easily proven reductions

$$[(a|\Omega)|b] = [a \times \Omega_o|b] = (a|b), \quad [\alpha|\Omega] = \alpha, \quad \text{for all } a,b \in \mathcal{B}, a \in \mathcal{B}_o. \tag{11.52}$$

We can show directly that

$$(a|\Omega) = [(a|\Omega)|\Omega] = a \times \Omega_o, \qquad \text{for all } a \in \mathcal{B} \tag{11.53}$$

and in the above equations we could use $(ab \times \Omega_o)$ in place of $(ab|\Omega)$. These equations also show, as well, that basic cea properties (CE0)-(CE4) hold. (Property (CE5) will follow from results below.) Finally, we mention an interesting strengthening of the basic inequality $P(b \Rightarrow a) \geq P(a|b) \geq P(ab)$ in eq.(10.7) in the PS partial ordering form

$$(b \Rightarrow a \mid \Omega) \geq (a|b) \geq (ab|\Omega), \quad \text{for all } a,b \in \mathcal{B}. \tag{11.54}$$

A simple proof of this equation follows from observing the recursive form in eq.(11.45) and using standard properties of cartesian products and the definition of $b \Rightarrow a$.

Bibliography

[1] Adams, E. W. (1966) Probability and the Logic of Conditionals, in J. Hintikka and P. Suppes (Eds.), *Aspects of Inductive Logic*, North-Holland, Amsterdam, pp. 265-316.

[2] Adams, E. W. (1975) *The Logic of Conditionals* D. Reidel Co., Dordrecht, Holland.

[3] Bamber, D. E. (1992) Personal Communications, Naval Command Control Ocean Systems Center, San Diego, CA

[4] Calabrese, P. G. (1975) The Probability that p implies q, Preliminary Report, Notices American Mathematical Society **22**(3), April, 1975, pp. A430-A431.

[5] Calabrese, P. G. (1987) An algebraic synthesis of the foundations of logic and probability, *Information Sciences*, **42**, 187-237.

[6] Calabrese, P. G. (1994) A theory of conditional information with applications, *IEEE Transactions on Systems, Man, and Cybernetics*, **24**(12), 1676-1684.

[7] Calabrese, P. G. (19106) Personal Communication.

[8] DeFinetti, B. (1935) La Logique de la Probabilité, *Actes du Congres Intern. de Philosophie Scientifique.* Hermann et Cie Editeurs, Paris, pp. IV1-IV9.

[9] DeFinetti, B. (1974) *Theory of Probability*, Vols. I & II, Wiley, New York.

[10] Dubois, D., Goodman, I. R., and Calabrese, P. G. (Eds.) (1994) Special Issue on Conditional Event Algebra, *IEEE Transactions on Systems, Man, and Cybernetics*, **24**(12), December, 1994.

[11] Eells, E. and Skyrms, B. (Eds.) (1994) *Probability and Conditionals*, Cambridge University Press, Cambridge, U.K.

[12] Goodman, I. R. (1987) A measure-free approach to conditioning, *Proceedings of Third American Association for Artificial Intelligence Workshop on Uncertainty in AI*, University of Washington at Seattle, July 10-12, 1987, pp. 270-277.

[13] Goodman, I. R. and Nguyen, H. T. (1994) A theory of conditional information for probabilistic inference in intelligent systems: II, Product space approach; III, Mathematical appendix., *Information Sciences*, **76**(1,2), 13-42; **75**(3), 253-277, December, 1993.

[14] Goodman, I. R., Gupta, M. M., Nguyen, H. T., and Rogers, G.S. (Eds.) (1991) *Conditional Logic in Expert Systems*, North-Holland, Amsterdam.

[15] Goodman, I. R., Nguyen, H. T., and Walker, E. A. (1991) *Conditional Inference and Logic for Intelligent Systems*, North-Holland, Amsterdam.

[16] Lewis, D. (1976) Probabilities of conditionals and conditional probabilities, *Philosophical Review*, **85**, 2107-315.

[17] McGee, V. (1989) Conditional probabilities and compounds of conditionals, *Philosophical Review*, **4**, 485-541.

[18] Rescher, N.(1969) *Many-Valued Logic*, McGraw-Hill Book Co., New York.

[19] Schay, G. (1968) An algebra of conditional events, *Journal of Mathematical Analysis and Applications*, **24**, 334-344.

[20] Van Fraasen, B. (1976) Probabilities of conditionals, in W. L. Harper and C. A. Hooker (Eds.), *Foundations of Probability Theory, Statistical Inference, and Statistical Theories of Science*, D. Reidel, Dordrecht, Holland, pp. 261-300.

[21] Walker, E.A. (1994) Stone algebras, conditional events, and three valued logic, *IEEE Transactions on Systems, Man, and Cybernetics*, **24**(12), 1699-1707.

Chapter 12

Further Development of Product Space Conditional Event Algebra

12.1 Equivalence, Partial Ordering, and Calculus of Logical Operations

For purposes of both self-containment and in order to show a new more improved approach to this product space algebra, we present a number of key results which also may be found in Goodman and Nguyen [1], [2], but with streamlined proofs here:

Theorem 68 *Fundamental recursive criterion. Let $a, b \in \mathcal{B}$ and $\alpha, \beta \in \mathcal{B}_o$ be arbitrary. Then,*
(I) If $b > \emptyset$,
    (i) $\beta = \alpha \& (b|\Omega) \vee b' \times \beta$ *iff* $\beta = [\alpha|b]$.
    (ii) $\beta = (ab|\Omega) \vee b' \times \beta$ *iff* $\beta = (a|b)$.
(II) If $b = \emptyset$,
    (i) $\beta = \alpha \& (b|\Omega) \vee b' \times \beta$ *iff* $\beta = \Omega \times \beta$ *iff* $[\beta = \emptyset_o$ *or* $\beta = \Omega_o]$.
    (ii) $\beta = (ab|\Omega) \vee b' \times \beta$ *iff* $\beta = \Omega \times \beta$ *iff* $[\beta = \emptyset_o$ *or* $\beta = \Omega_o]$.

Proof. (Ii): If the rhs(right-hand side) of (i) holds, then recursive eq.(11.49) yields lhs(left-hand side) of (i). If the lhs(i) holds, by making repeated substitutions, we obtain

$$\beta = \gamma_j \vee (b')^{j+1} \times \beta, \quad j = 1, 2, ..., \tag{12.1}$$

where

$$\gamma_j = \bigvee_{k=0}^{j} (b')^k \times (\alpha \& (b|\Omega)), \qquad j = 1, 2, \ldots. \tag{12.2}$$

Then, disjoining both sides of eq.(12.2) with

$$\gamma_j = \bigvee_{k=j+1}^{\infty} (b')^k \times (\alpha \& (b|\Omega)) = (b')^{j+1} \times [\alpha|b] \quad j = 1, 2, \dots . \tag{12.3}$$

and noting the relation

$$[\alpha|b] = \gamma_j \vee \delta_j, \qquad j = 1, 2, \dots, \tag{12.4}$$

yields

$$\beta \vee \delta_j = [\alpha|b] \vee (b')^{j+1} \times \beta,$$

which, using absorption and the boolean symmetric operator $+$ can be written as

$$\beta + \beta' \& \delta_j = [\alpha|b] + \left([\alpha|b]' \& ((b')^{j+1} \times \beta) \right),$$

which, using basic ring properties of $+$, simplifies to,

$$\begin{aligned} \beta + [\alpha|b] &= \beta' \& \delta_j + \left([\alpha|b]' \& ((b')^{j+1} \times \beta) \right) \\ &= ((b')^{j+1} \times \Omega_o) \& \eta_j, \quad j = 1, 2, \dots, \end{aligned} \tag{12.5}$$

where

$$\eta_j = \beta' \& (\Omega^{j+1} \times [\alpha|b]) + \left([\alpha|b]' \& (\Omega^{j+1} \times \beta) \right), \quad j = 1, 2, \dots . \tag{12.6}$$

Hence, for any probability measure P with $P_o([\alpha|b])$ well-defined, i.e., $P(b) > 0$, applying P_o to eq.(12.5) shows

$$\begin{aligned} P_o(\beta + [\alpha|b]) &= P_o(((b')^{j+1} \times \Omega_o) \& \eta_j) \\ &\leq P_o((b')^{j+1} \times \Omega_o) = (P(b'))^{j+1}, \quad j = 1, 2, \dots . \end{aligned} \tag{12.7}$$

Since $P(b') < 1$, eq.(12.7) shows

$$P_o(\beta + [\alpha|b]) < \varepsilon, \text{ for any arbitrary } \varepsilon > 0, \text{ and all well-defined } P,$$

i.e.,

$$P_o(\beta + [\alpha|b]) = 0, \quad \text{for all well-defined } P. \tag{12.8}$$

Eq.(12.8) shows β and $[\alpha|b]$ are strongly identifiable and the rhs(i) holds.

(Iii): Proof follows similarly to the proof of (Ii).

(II): If either $\beta = \emptyset_o$ or $\beta = \Omega_o$, clearly $\beta = \Omega \times \beta$. Conversely, consider

$$\beta = \Omega \times \beta. \tag{12.9}$$

By making repeated substitutions in eq.(11.28), we obtain

$$\beta = \Omega^j \times \beta, \quad j = 1, 2, \dots \tag{12.10}$$

If $\beta > \emptyset$, then eq.(12.10) means that the projection of β onto every possible factor space is Ω. Hence, β must be Ω_o. $\square$

Theorem 69 *Basic characterization of trivial conditional events.*
For all $a, b, c, d \in \mathcal{B}$,
(i) $(a|b) = \emptyset_o$, *iff* $ab = \emptyset$, $[\alpha|b] = \emptyset_o$ *iff* $\alpha \& (b|\Omega) = \emptyset_o$.
Hence, $[\alpha|\emptyset] = (a|\emptyset) = \emptyset_o$, *where* $\emptyset_o$ *is the null event in* $\mathcal{B}_o$.
(ii) $(a|a) = \Omega_o$ *iff* $a > \emptyset$.
(iii) $(b|b) \& (c|d) = (\Omega|\Omega) \& (c|d) = (c|d)$, *for* $b > \emptyset$,
$\qquad (b|b) \vee (c|d) = (\Omega|\Omega) \vee (c|d) = (\Omega|\Omega) = \Omega_o$, *for* $b > \emptyset$,
$\qquad (\emptyset|b) \& (c|d) = (\emptyset|\Omega) \& (c|d) = (\emptyset|\emptyset) \& (c|d) = (\emptyset|\Omega) = \emptyset_o$,
$\qquad (\emptyset|b) \vee (c|d) = (\emptyset|\Omega) \vee (c|d) = (\emptyset|\emptyset) \vee (c|d) = (c|d)$.

Proof. (i): If $ab = \emptyset$, clearly eq.(11.42) shows $(a|b) = \emptyset_o$. If $(a|b) = \emptyset$, using eq.(11.45), we have

$$(ab|\Omega) \vee \; b' \times (a|b) = \emptyset_o,$$

which by the property of disjunctions in boolean algebra $\mathcal{B}_o$, implies

$$(ab|\Omega) = \emptyset_o,$$

i.e.,

$$ab = \emptyset.$$

Similar remarks hold for $[\alpha|b]$.

(ii): First, $(a|a) = \Omega_o$ while $a = \emptyset$ is impossible by (i). The converse proof will be shown following the presentation of Theorem 3.

(iii): Follows immediately from (i) and (ii) and fact that $\mathcal{B}_o$ is a boolean algebra with a zero and unity element. □

Theorem 70 *For all $a, b, c, d \in \mathcal{B}$ with $b, d > \emptyset$ and all $a, b \in \mathcal{B}_o$,*
(i) $[\alpha|b]' = [\alpha'|b]$, $(a|b)' = (a'|b)$,
(ii) $[\alpha|b] \leq [\beta|d]$ *iff* $[\beta|d]' \leq [\alpha|b]'$ *iff* $[\beta'|d] \leq [\alpha'|b]$,
(iii) $(a|b) \leq (c|d)$ *iff* $(c|d)' \leq (a|b)'$ *iff* $(c'|d) \leq (a'|b)$.

Proof. (i): By the recursive form in eq.(11.49),

$$
\begin{aligned}
[\alpha|b]' &= (\alpha \& (b|\Omega)) \vee b' \times [\alpha|b])' \\
&= ((a' \& (b|\Omega)) \vee (b'|\Omega)) \& (b' \times [\alpha|b]' \vee (b|\Omega) \\
&= a' \& (b|\Omega) \vee b' \times [\alpha|b]'. \quad (12.11)
\end{aligned}
$$

But, applying Theorem 1 to eq.(12.11) shows lhs(i). The proof for rhs(i) is similar.

(ii),(iii): These follow by the usual properties of ordering in a boolean algebra in conjunction with (i). □

The proof of Theorem 2(ii) is completed by noting that for $a > \emptyset$, using Theorem 3(i) and Theorem 2(i),

$$(a|a)' = (a'|a) = (a'a|a) = (\emptyset|a) = \emptyset_o. \quad (12.12)$$

Hence, eq.(12.12) shows

$$(a|a) = (a|a)'' = \emptyset'_o = \Omega_o, \qquad (12.13)$$

the desired result.   □

Theorem 71 *For any* $a, b, c, d \in \mathcal{B}$, *any* $\alpha, \beta \in \mathcal{B}_o$,
  (i) $[\alpha|b] \leq \beta \& (d|\Omega) \vee d' \times [\alpha|b]$ *implies* $[\alpha|b] \leq [\beta|d]$,
  (ii) $(a|b) \leq (cd|\Omega) \vee d' \times (a|b)$ *implies* $(a|b) \leq (c|d)$,

Proof. (i): This proof is similar to that for Theorem 1. By repetitive substitution in lhs(i),

$$[\alpha|b] \leq \gamma_j \vee (d')^{j+1} \times [\alpha|b], \qquad (12.14)$$

where

$$\gamma_j = \bigvee_{k=0}^{j} (d')^k \times (\beta \& (d|\Omega), \quad j = 1, 2, \dots \qquad (12.15)$$

Then, disjoining both sides of eq.(12.14) with

$$\delta_j = \bigvee_{k=j+1}^{\infty} ((d')^k) \times (\beta \& (d|\Omega) = (d')^{j+1} \times [\beta|d], \quad j = 1, 2, \dots \qquad (12.16)$$

and noting the relation

$$[\beta|d] = \gamma_j \vee \delta_j, \qquad j = 1, 2, \dots, \qquad (12.17)$$

yields

$$[\alpha|b] \vee \delta_j \leq [\beta|d] \vee (d')^{j+1} \times [\alpha|b],$$

which implies, by the usual properties of boolean $+$ and $\leq$,

$$[\alpha|b] + [\alpha|b]'\&\delta_j = ([\alpha|b] + ([\alpha|b]'\&\delta_j))$$
$$\& ([\beta|d] + ([\beta|d]'\&((d')^{j+1} \times [\alpha|b]))). \qquad (12.18)$$

In turn, eq.(12.18) shows

$$[\alpha|b] + [\alpha|b]\&[\beta|d] = (d')^{j+1}\&\eta_j, \qquad j = 1, 2, \dots \qquad (12.19)$$

where for all j

$$\eta_j = (\Omega^{j+1} \times ([\alpha|b]' \& [\beta|d]))\&[\alpha|b]\&[\beta|d]'$$
$$+(\Omega^{j+1} \times [\alpha|b]) \& [\alpha|b]\&[\beta|d]'. \qquad (12.20)$$

Hence, applying P_o to eq.(12.19) yields

$$P_o([\alpha|b] + ([\alpha|b]\&[\beta|d])) = P_o((d')^{j+1}\&\eta_j) \leq P_o((d')^{j+1}) = (P(d'))^{j+1},$$

$j = 1, 2, \dots$, whence, analogous to eq.(12.8),

$$P_o([\alpha|b] + [\alpha|b]\&[\beta|d]) = 0, \quad \text{for all well-defined } P, \qquad (12.21)$$

i.e., $[\alpha|b]$ and $[\alpha|b]\&[\beta|d]$ are strongly identitiable and thus rhs(i) holds.
  (ii): Proof is similar to that for (i) above.   □

Theorem 72 *For any $a, b, c, d \in \mathcal{B}$ and $\alpha, \beta \in \mathcal{B}_o$,*
(i) $[\alpha|b] \leq [\beta|d]$ iff $[\alpha|b] = [\alpha|b]\&[\beta|d]$ iff $[\beta|d] = [\alpha|b] \vee [\beta|d]$,
(ii) $(a|b) \leq (c|d)$ iff $(a|b) = (a|b)\&(c|d)$ iff $(c|d) = (a|b) \vee (c|d)$.
Also, making the nontriviality assumption $\emptyset < a < b$ and $\emptyset < c < d$, the same criterion (under the above restriction) formally as in eq.(11.14) for DGNW (see also, eq.(11.17)),

$$(a|b) \leq (c|d) \text{ iff } [ab \leq cd \text{ and } c'd \leq a'b] \text{ iff } [ab \leq cd \text{ and } b \Rightarrow a \leq d \Rightarrow c].$$

(iii)   If $\emptyset < a < b$ and $\emptyset < c < d$, then $[(a|b) = (c|d)$ iff $ab = cd$ and $b = d]$, the same criterion (under the above restriction) formally as in eq.(11.11) for DGNW or AC and compatible with the identity of conditional probabilities criterion (eq.(11.18)).
(iv) A. If $ab = cd$ and $b = d$, then $(a|b) = (c|d)$.
(iv) B. If $\alpha\&(b|\Omega) = \beta\&(d|\Omega)$ and $b = d$, then $[\alpha|b] = [\beta|d]$.

Proof. (i) and unrestricted part of (ii): Follows from standard (lattice) properties of boolean algebra $\mathcal{B}_o$.

Restricted part of (ii): First, if either equivalent bottom part of (ii) holds, using the recursive form in eq.(11.45) and absorption property of $\mathcal{B}_o$,

$$\begin{aligned}
(a|b) &= (ab|\Omega) \vee b' \times (a|b) \leq (cd|\Omega) \vee (d' \vee cd) \times (a|b) \\
&= cd \times \Omega_o \vee cd \times (a|b) \vee d' \times (a|b) = (cd|\Omega) \vee d' \times (a|b) \quad (12.22)
\end{aligned}$$

Then, applying Theorem 4(ii) to eq.(12.22), shows

$$(a|b) \leq (c|d). \tag{12.23}$$

Conversely, if eq.(12.23) holds, apply recursive eq.(11.45):

$$(ab|\Omega) \vee b' \times (a|b) \leq (cd|\Omega) \vee d' \times (c|d). \tag{12.24}$$

Then, conjoin eq.(12.24) through by $(b(cd)'|\Omega)$ to yield

$$(ab(cd)'|\Omega) \leq bd' \times (c|d). \tag{12.25}$$

But, inspection of eq.(12.24), using basic properties of cartesian products and the restricting assumptions, shows we must have $ab(cd)' = \emptyset$, i.e.,

$$ab \leq cd. \tag{12.26}$$

Next, using Theorem 3(iii), eq.(12.23) implies

$$(c'|d) \leq (a'|b). \tag{12.27}$$

Now, applying eq.(11.45) to eq.(12.27) yields

$$(c'd|\Omega) \vee d' \times (c'|d) \leq (a'b|\Omega) \vee b' \times (a'|b). \tag{12.28}$$

Then, conjoin eq.(12.28) through by $(d(a'b)'|\Omega)$ to yield

$$((c'd)(a'b)'|\Omega) \le b'd \times (a'|b). \tag{12.29}$$

But inspection of eq.(12.29) and use of the restricting assumptions shows we must have $(c'd)(a'b)' = \emptyset$, i.e.,

$$c'd \le a'b, \tag{12.30}$$

Thus, eqs.(12.26) and (12.30) show the bottom part of (ii) holds.

(iv): A, B are obvious from the very definitions of $(.|..)$ and $[.|..]$.

(iii): rhs(iii) implies lhs(iii) by (iv)A. Next, assume lhs(iii) holds. Then, substitute recursive eq.(11.45) in lhs(C):

$$a \vee b' \times (a|b) = c \vee d' \times (c|d). \tag{12.31}$$

First, conjoin eq.(12.31) through by ad. This yields immediately, by the non-triviality assumptions, that

$$a = c. \tag{12.32}$$

Substitute eq.(12.32) into eq.(12.31) and cancel the common term $a = c$ because of mutual disjointness. Thus,

$$b' \times (a|b) = d' \times (a|d). \tag{12.33}$$

Next, conjoin eq.(12.33) through by b, yielding directly $\emptyset_o = bd' \times (a|d)$, which, under the nontriviality assumptions, implies

$$b \le d. \tag{12.34}$$

Similarly, by conjoining eq.(12.33) through by d, we obtain

$$d \le b. \tag{12.35}$$

Obviously, eqs.(12.32), (12.34), and (12.35) give rhs(iii). Alternatively, (iii) can be proven using the criterion in eq.(11.18). □

However, unlike a number of the above results where the $[.|..]$ analogue of $(.|..)$ holds, it is not true in general that the $[.|..]$ analogue of Theorem 5(ii) fully holds. The best we can say at this point is provided in the next three theorems:

Theorem 73 *For any* $\alpha \in B_o$ *and* $b \in B$,

$$[\alpha|b] = \alpha \quad iff \quad (b|\Omega) \ge \alpha + (\Omega \times \alpha). \tag{12.36}$$

Proof. lhs(12.36) holds, using recursive form in eq.(11.49)

$$\begin{aligned}
\text{iff } \alpha\&(b'|\Omega) + b' \times [\alpha|b] = \alpha \quad &\text{iff} \quad \emptyset = \alpha + \alpha\&(b|\Omega) + (b' \times [\alpha|b]) \\
&\text{iff} \quad \emptyset = (b'|\Omega)\&(\alpha + (\Omega \times \alpha)) \\
&\text{iff} \quad (b|\Omega) \ge \alpha + (\Omega \times \alpha). \qquad \square
\end{aligned}$$

Theorem 74 *For any* $\alpha, \beta \in \mathcal{B}_o$, $b, d \in \mathcal{B}$,

$$[\alpha|b] = [\beta|d] \text{ iff}$$
$$[[\alpha|b]\&(d|\Omega) = \beta\&(d|\Omega) \text{ and } (d|\Omega) \geq [\alpha|b] + (\Omega \times [\alpha|b])]. \quad (12.37)$$

Proof. If lhs(12.37) holds, then using eq.(11.49) and conjoining through with $(d|\Omega)$ directly yields

$$[\alpha|b]\&(d|\Omega) = \beta\&(d|\Omega). \quad (12.38)$$

Then, substituting eq.(12.38) into lhs(12.37),

$$[\alpha|b] = [\beta|d] = [\beta\&(d|\Omega) \mid d] = [[\alpha|b]\&(d|\Omega) \mid d] = [[\alpha|b] \mid d]. \quad (12.39)$$

But, Theorem 6 is now applicable to eq.(12.39) with a replaced by $[\alpha|b]$ and b by d, to yield

$$(d|\Omega) \geq [\alpha|b] + (\Omega \times [\alpha|b]). \quad (12.40)$$

Combining eqs. (12.38) and (12.40) produces rhs(12.37). Conversely, if rhs(12.37) holds, then Theorem 6 shows this is the same as

$$[\alpha|b]\&(d|\Omega) = \beta\&(d|\Omega) \text{ and } [\alpha|b] = [[\alpha|b]|d]. \quad (12.41)$$

Hence, eq.(12.41) shows

$$[\beta|d] = [\beta\&(d|\Omega)|d] = [[\alpha|b]\&(d|\Omega)|d] = [[\alpha|b]|d] = [\alpha|b]. \quad (12.42)$$

Eq.(12.42) is the same as the lhs(12.37) holding. $\qquad \square$

Theorem 75 *Let* $\alpha, \beta \in \mathcal{B}_o$ *and* $b, d \in \mathcal{B}$ *be arbitrary such that if the* $[.|..]$ *analogue of Theorem 5(ii) bottom holds, i.e.,*

$$\alpha\&(b|\Omega) \leq \beta\&(d|\Omega) \text{ and } \beta'\&(d|\Omega) \leq \alpha'\&(b|\Omega), \quad (12.43)$$

i.e., equivalently,

$$\alpha\&(b|\Omega) \leq \beta\&(d|\Omega) \text{ and } (b|\Omega) \Rightarrow \alpha \leq (d|\Omega) \Rightarrow \beta, \quad (12.44)$$

then,

$$[\alpha|b] \leq [\beta|d]. \quad (12.45)$$

Proof. Using the recursive forms in eq.(11.49) and the hypothesis here,

$$
\begin{aligned}
[\alpha|b] &= \alpha\&(b|\Omega) \vee b' \times [\alpha|b] = \alpha\&(b|\Omega) \vee (b'|\Omega)\&(\Omega \times [\alpha|b]) \\
&\leq \beta\&(d|\Omega) \vee ((d|\Omega) \Rightarrow \beta)\&(\Omega \times [\alpha|b]) \\
&= \beta\&(d|\Omega) \vee \beta\&(d|\Omega)\&(\Omega \times [\alpha|b]) \vee d' \times [\alpha|b] \\
&= \beta\&(d|\Omega) \vee d' \times [\alpha|b]. \quad (12.46)
\end{aligned}
$$

Then, apply Theorem 4(i) to eq(12.46) and obtain directly eq.(12.45) $\qquad \square$

Theorem 76 *Calculus of binary conjunction and disjunction relations for PS cea. For any $a, b, c, d \in B$ and $\alpha, \beta \in B_o$, the following holds:*

(i) $(a|b)\&(c|d) = [(abcd \mid \Omega) \vee abd' \times (c|d) \vee b'cd \times (a|b) \mid b \vee d]$,

(ii) $(a|b) \vee (c|d) = [(ab \vee cd \mid \Omega) \vee a'bd' \times (c|d) \vee b'c'd \times (a|b)|b \vee d]$,

(iii) $[\alpha|b]\&[\beta|d] = [\alpha\beta\&(bd|\Omega) \vee \alpha\&(bd' \times [\beta|d]) \vee \beta\&(b'd \times [\alpha|b])|b \vee d]$,

(iv) $[\alpha|b] \vee [\beta|d]$
$= [\alpha\&(b|\Omega) \vee \beta\&(d|\Omega) \vee \alpha'\&(bd' \times [\beta|d]) \vee \beta'\&(b'd \times [\alpha|b])|b \vee d]$.

Assuming that all appropriate denominators are positive, for any choice of probability measure P over B,

(v) $P_o((a|b)\&(c|d)) = (P(abcd) + P(abd')P(c|d) + P(b'cd)P(a|b))/P(b \vee d)$,

(vi) $P_o((a|b)\vee(c|d)) = (P(ab \vee cd) + P(a'bd')P(c|d) + P(b'c'd)P(a|b))/P(b \vee d)$
$= P(a|b) + P(c|d) - P_o((a|b)\&(c|d))$,

(vii) $P_o([\alpha|b]\&[\beta|d])$
$= (P_o(\alpha\beta\&(bd|\Omega)) + P_o(\alpha\&(bd' \times [\beta|d])) + P_o(\beta\&(b'd \times [\alpha|b])))/P(b \vee d)$,

(viii) $P_o([\alpha|b]v[\beta|d])$ $= (P_o(\alpha\&(b|\Omega) \vee \beta\&(d|\Omega)) + P_o(\alpha'\&(bd' \times [\beta|d]))$
$+ P_o(\beta'\&(b'd \times [\alpha|b])))/P(b \vee d)$,
$= P_o([\alpha|b]) + P_o([\beta|d] - P_o([\alpha|b]\&[\beta|d])$.

(ix) $(a|b)\&(c|\Omega) = (abc|\Omega) \vee b'c \times (a|b)$ *(generalized modus ponens)*,

$$P_o((a|b)\&(c|\Omega)) = P(abc) + P(b(c)P(a|b)),$$

Proof. (i): Using the recursive forms in eq.(11.45), expanding out all terms

$$(a|b)\&(c|d) = ((ab|\Omega) \vee b' \times (a|b))\&((cd|\Omega) \vee d' \times (c|d))$$
$$= \gamma \vee b'd' \times ((a|b)\&(c|d)), \tag{12.47}$$

where

$$\gamma = (abcd|\Omega) \vee abd' \times (c|d) \vee b'cd \times (a|b). \tag{12.48}$$

Since obviously

$$\gamma \le (b \vee d|\Omega), \tag{12.49}$$

we can rewrite eq.(12.47) as

$$(a|b)\&(c|d) = \gamma\&(b \vee d|\Omega) \vee (b \vee d)' \times ((a|b)\&(c|d)). \tag{12.50}$$

Applying Theorem 1(I) to eq.(12.50), with $\beta = (a|b)\&(c|d)$, $\alpha = \gamma$, and b replaced by $b \vee d$, directly yields the desired result in (i).

(ii) Using the recursive forms in eq.(11.45) and expanding out all terms,

$$(a|b) \vee (c|d) = ((ab|\Omega) \vee b' \times (a|b)) \vee ((cd|\Omega) \vee d' \times (c|d))$$
$$= (ab \vee cd|\Omega) \vee d' \times (c|d) \vee b' \times (a|b)$$
$$= (ab \vee cd|\Omega) \vee bd' \times (c|d) \vee b'd \times (a|b) \vee b'd' \times ((a|b) \vee (c|d))$$
$$= (ab \vee cd|\Omega) \vee ((ab \vee cd)'|\Omega)\&(bd' \times (c|d)$$
$$\vee b'd \times (a|b) \vee b'd' \times ((a|b) \vee (c|d))$$
$$= \kappa\&(b \vee d) \vee (b \vee d)' \times ((a|b) \vee (c|d))), \tag{12.51}$$

where

$$\kappa = (ab \vee cd|\Omega) \vee a'bd' \times (c|d) \vee b'c'd \times (a|b) \le (b \vee d|\Omega). \qquad (12.52)$$

Then, applying Theorem 1(I) to eq.(12.52), where $\beta = (a|b) \vee (c|d)$, $\alpha = \kappa$, and b is replaced by $b \vee d$, yields (ii).

(iii),(iv): Proofs follow analogously to (i),(ii), respectively, beginning with recursive forms in eq.(11.49) and ending with the use of Theorem 1(I). Note, once again, that because all standard laws of probability are applicable to $(\Omega_o, \mathcal{B}_o, P_o)$, the alternate forms via modular expansions in eqs.(vi) and (viii) also hold.

(v)-(viii): Follows directly from the basic properties of P_o and the fact that all terms in the consequent parts of the large-form conditional events in (i)-(iv) are mutually disjoint.

(ix): Obvious by specializing (i) and (v).   □

Remark. All of the results in Theorem 9 can be extended via a recursive technique to any number of conditional event or [.|..] operator arguments. For example, for conjunction with respect to (.|..), using the recursive forms in eq.(11.45), we obtain for any positive integer n and $a_j, b_j \in \mathcal{B}$, with at least one $b_j > \emptyset$, $j = 1, \ldots, n$, letting $I = \{1, 2, \ldots, n\}$,

$$\&_{j \in I}(a_j|b_j) = \&_{j \in I}\left((a_j b_j|\Omega) \vee b_j' \times (a_j|b_j)\right)$$

$$= \bigvee_{K \subseteq I}\left((\&_{j \in I-K}(a_j b_j)) \&_{k \in K}(b_k' \times (a_k \mid b_k))\right),$$

by straightforward combinatorics

$$\&_{j \in I}(a_j|b_j) = \bigvee_{K \subseteq I}(\&_{j \in I-K}(a_j b_j|\Omega)\&(\&_{k \in K}(b_k' \times (a_k \mid b_k))))$$

$$= \bigvee_{K \subseteq I}(\alpha_K(\bar{a}, \bar{b}) \times (\&_{k \in K}(a_k|b_k))), \qquad (12.53)$$

where

$$\alpha_K(\bar{a}, \bar{b}) = \&_{j \in I-K}(a_j b_j)\& (\&_{k \in K}b_k'), \quad \text{for all } (\emptyset \subseteq)K \subseteq I. \qquad (12.54)$$

Breaking the disjunction in the rhs(12.53) into two parts, one corresponding to all $K \subset I$ and the other to $K = I$, yields

$$\&_{j \in I}(a_j|b_j) = \beta(\bar{a}, \bar{b}) \vee \vee (\&_{k \in I}b_k') \times \&_{j \in I}(a_j|b_j), (\&_{k \in I}b_k') \times \&_{j \in I}(a_j|b_j), \qquad (12.55)$$

where

$$\beta(\bar{a}, \bar{b}) = \bigvee_{K \subset I}(\alpha_K(\bar{a}, \bar{b}) \times (\&_{k \in K}(a_k|b_k))). \qquad (12.56)$$

Since

$$\beta(\bar{a}, \bar{b}) \le (\alpha_K(\bar{a}, \bar{b})|\Omega) \le (\vee_{k \in I}b_j|\Omega), \quad \text{for any each } K \subset I, \qquad (12.57)$$

and by hypothesis, $\vee_{k\in I} b_j > \emptyset$, we can apply Theorem 1(I) to eq.(12.55) to yield, finally, the recursive form

$$\&_{j\in I}(a_j|b_j) = \left[\beta(\bar{a},\bar{b})|\vee_{k\in I} b_j\right],\qquad (12.58)$$

noting that $\beta(\bar{a},\bar{b})$ is a function of all of the K-fold conjunctions of the $(a_j|b_j)$ for $K \subset I$. In turn, the probability evaluation of I-fold conjunction in eq.(12.58), using the product form of P_o and observing all distinct $\alpha_K(\bar{a},\bar{b})$ are mutually disjoint, is

$$P_o\left(\&_{j\in I}(a_j|b_j)\right) = P_o(\beta(\bar{a},\bar{b}))/P_o\left(\vee_{k\in I}b_j\right),\qquad (12.59)$$

where

$$P_o(\beta(\bar{a},\bar{b})) = \sum_{K\subset I} P_o(\alpha_K(\bar{a},\bar{b})P_o\left(\&_{k\in K}(a_k|b_k)\right).\qquad (12.60)$$

Thus in obtaining the full I-fold conjunction here for either eq.(12.58) or (12.59), we first obtain all 2-fold conjunctions (same formula, except for appropriate substitutions), then all 3-fold (as a function of all 2-fold and trivial 1-fold), etc.,.., until the I-fold level is reached. (See [1], Theorems 9, 10 for related results.) For simplicity we present the case for the probability evaluation of the conjunction of three conditional events:

$$\begin{aligned}
P_o((a|b)\&(c|d)\&(e|f)) &= [P(abcdef) + P(abcdf')P(e|f)\\
&+P(abd'ef)P(c|d) + P(b'cdef)P(a|b)\\
&+P(abd'f')P_o((c|d)\&(e|f)) + P(b'cdf')P_o((a|b)\&(e|f))\\
&+P(b'd'ef)P_o((a|b)\&(c|d))]/P(b\vee d\vee f),\qquad (12.61)
\end{aligned}$$

where the terms $P_o((c|d)\&(e|f))$, $P_o((a|b)\&(e|f))$, $P_o((a|b)\&(c|d))$ are all obtainable from Theorem 9(v).   □

The above also shows that property (CE5) is indeed satisfied for PS. Hence, all of the required cea properties (CE0)-(CE5) are satisfied and PS is a bona fide cea. Furthermore, it is readily verified from Theorem 9(i) that PS possesses the chaining and modus ponens properties. In addition, the generalized modus ponens form for PS (corresponding to that for DGNW in top(11.19) and AC in top(11.29) is, with evaluation,

$$\begin{aligned}
(a|b)\&(c|\Omega) &= (abc|\Omega) \vee b'c \times (a|b),\\
P_o((a|b)\&(c|\Omega)) &= P(abc) + P(b'c)P(a|b).\qquad (12.62)
\end{aligned}$$

Note also, that not only are the forms of conditional events for the PS (product space) cea different from the coset or interval of event forms for the DGNW and AC cea's, but also the corresponding set indicator function $\phi(a|b) : \Omega_o \to 0,1$ is no longer three-valued, but two valued where for any

$$\omega = (\omega_1, \omega_2, \omega_3, \ldots) \in \Omega_o,$$

$$\phi(a|b)(\omega) = \begin{cases} 1, & \text{if } \omega_1 \in ab \\ 0, & \text{if } \omega_1 \in a'b \\ 1, & \text{if } \omega_1 \in b', \omega_2 \in ab \\ 0, & \text{if } \omega_1 \in b', \omega_2 \in a'b \\ 1, & \text{if } \omega_1 \in b', \omega_2 \in b', \omega_3 \in ab \\ 0, & \text{if } \omega_1 \in b', \omega_2 \in b', \omega_3 \in a'b \\ \cdots\cdots\cdots\cdots \end{cases} \quad (12.63)$$

Finally, note that while all three cea's DGNW, AC and PS have obviously distinct logical operations, in general, they all share the same criterion for equality of nontrivial events (see Theorem 5(iv) for PS), which is the same as the characterization for the identity of conditional probabilities for all evaluations (see eq.(11.18), namely:

$$(a|b) = (c|d) \text{ iff } [a = c \text{ and } b = d], \text{ provided } \emptyset < a < b, \emptyset < c < d. \quad (12.64)$$

Also, DGNW and PS also share for nontrivial events the common partial ordering, equivalent to the corresponding natural inequality characterization for all conditional probabilities (see eq.(11.17)), namely:

$$(a|b) \le (c|d) \text{ iff } [a \le c \text{ and } c'd \le a'b], \text{ provided } \emptyset < a < b, \emptyset < c < d. \quad (12.65)$$

However, by inspection of eqs.(11.24) and (11.25), neither of the two different AC semilattice orders in general are equivalent to the natural inequality characterization for all conditional probabilities.

12.2 Additional Important Properties of PS

12.2.1 Comparison of [.|..] and (.|..) Type of Events

It should be noted that, in general, though the reduction in eq.(11.51) holds for special higher order conditionals to simple conditionals , for events of the type $[(a|b)|c]$, – though their probability values are the same (eq.(11.50)), in general, not only do we have

$$[(a|b)|c] \ne ((a|b)|(c|\Omega)) \quad (12.66)$$

(they lie in different level spaces, the lhs in $\mathcal{B}_o$, while the rhs in $\mathcal{B}_{oo}$) – though they are bijective – but we also do *not* have a probability-preserving isomorphism with respect to &, or ∨ as can be verified as follows: Using Theorem 9(iii) and simplifying terms,

$$[(a|b)|c]\&[(a|b)|d] = [\lambda|c \vee d], \quad (12.67)$$

where

$$\lambda = (a|b)\&(cd|\Omega) \vee (a|b)\&(cd' \times [(a|b)|d]) \vee (a|b)\&(c'd \times [(a|b)|c]). \quad (12.68)$$

Then, using the recursive forms in eqs.(11.45), (11.49),

$$(a|b)\&(cd' \times [(a|b)|d]) = abcd' \times [(a|b)|d] \vee b'cd' \times \rho, \tag{12.69}$$

$$\rho = (a|b)\&[(a|b)|d] = (abd|\Omega) \vee (abd' \vee b'd \vee b'd') \times (a|b) \tag{12.70}$$

and analogously,

$$(a|b)\&(c'd \times \lceil(a|b)|c]) = abc'd \times [(a|b)|c] \vee b'c'd \times \tau, \tag{12.71}$$

$$\tau = (a|b)\&[(a|b)|c] = (abc|\Omega) \vee (abc' \vee b'c \vee b'c') \times (a|b). \tag{12.72}$$

Thus, eqs.(12.68)-(12.72) show, by appropriate rearrangement,

$$\lambda = \mu \vee b'cd' \times \rho \vee b'c'd \times \tau, \tag{12.73}$$

where

$$\mu = (a|b)\&(cd|\Omega) \vee abcd' \times [(a|b)|d] \vee abc'd \times [(a|b)|c]. \tag{12.74}$$

On the other hand, Theorem 9(i), applied to $(\Omega_{oo}, \mathcal{B}_{oo}, P_{oo})$, etc., shows

$$((a|b)|(c|\Omega))\&((a|b)|(d|\Omega)) = (\gamma|(c \vee d|\Omega)), \tag{12.75}$$

where

$$
\begin{aligned}
\gamma &= ((a|b)(cd|\Omega)|\Omega_o) \vee ((a|b)\&(cd'|\Omega)) \times ((a|b)|(d|\Omega)) \\
&\quad \vee ((a|b)\&(c'd|\Omega)) \times ((a|b)|(c|\Omega)) \\
&= \nu \vee b'cd' \times \psi \vee b'c'd \times \kappa, \tag{12.76}
\end{aligned}
$$

$$
\begin{aligned}
\nu = ((a|b)\&(cd|\Omega)|\Omega_o) \quad &\vee \quad (abcd'|\Omega) \times ((a|b)|(d|\Omega)) \\
&\vee \quad (abc'd|\Omega) \times ((a|b)|(c|\Omega)), \tag{12.77}
\end{aligned}
$$

$$\psi = (a|b) \times ((a|b)|(d|\Omega)), \tag{12.78}$$

$$\kappa = (a|b) \times ((a|b)|(c|\Omega)). \tag{12.79}$$

Then, if we consider the obvious bijective correspondence

$$[(a|b)|c] \longleftrightarrow ((a|b)|(c|\Omega)), \text{ for all } a, b, c \in \mathcal{B}, \tag{12.80}$$

in order for there to be an isomorphism between the two types of operations as in eq.(12.70), with respect to the standard $\&$ and $\vee$ operations in $\mathcal{B}_o$ and $\mathcal{B}_{oo}$, we must have at least the correspondence of the special case

$$[(a|b)|c]\&[(a|b)|d] \longleftrightarrow ((a|b)|(c|\Omega))\&((a|b)|(d|\Omega)), \text{ for all } a, b, c, d \in \mathcal{B}. \tag{12.81}$$

But, first comparing eqs.(12.67), (12.70), (12.72)- (12.74), relative to the left-hand side of eq.(12.81), with the corresponding eqs.(12.75), (12.78), (12.79), (12.76), (12.77), relative to the right-hand side of eq.(12.81), shows that we have similar natural correspondences between: the antecedents $c \vee d$ in eq.(12.67) and $(c \vee d \mid \Omega)$ in eq.(12.75); μ in eq.(12.73) (as given in eq.(12.74) and ν in eq.(12.76) (as given in eq. (12.77). However, the factor terms of the common expressions $b'cd'$ and $b'c'd$ appearing in eqs. (12.73) and (12.76), respectively, i.e., (ρ, τ) and (ψ, κ), do not correspond in the same way, as is easily seen by inspection of eqs.(12.70),(12.72) and (12.78),(12.79). Thus, we do not have an isomorphic relation for even this restricted case of conjunction for (12.81).

12.2.2   Lewis' Theorem and PS

Returning to Lewis' theorem, it is clear that again the conditional event $(a|b)$, and the imbeddings of the unconditional events, as conditional events themselves, $(a|\Omega)$, $(b|\Omega)$ all lie $\in \mathcal{B}_o$, but a, b remain $\in \mathcal{B}$. Thus, the key eq.(11.39) in the proof of Lewis' Theorem must be modified. Using eq.(12.3) and the fact that $\emptyset < a < b$:

$$
\begin{aligned}
P(a|b) &= P_o((a|b)) = (P_o)_{(a|\Omega)}((a|b))(P_o)((a|\Omega)) \\
&\quad + (P_o)_{(a'|\Omega)}((a|b))(P_o)((a'|\Omega)) \\
&= \frac{P_o((a|b)\&(a|\Omega))}{P_o((a|\Omega))} P_o((a|b))) + \frac{P_o((a|b)\&(a'|\Omega))}{P_o((a'|\Omega))} P_o((a'|\Omega)) \\
&= P_o((a|b)\&(a|\Omega)) + P_o((a|b)\&(a'|\Omega)) \\
&= P_o\left([aba\Omega \vee a\Omega b' \times (a|b) \vee ab\Omega' \times (a|\Omega)|b \vee \Omega]\right) \\
&\quad + P_o\left([aba'\Omega \vee a'\Omega b' \times (a|b) \vee ab\Omega' \times (a'|\Omega)|b \vee \Omega]\right) \\
&= P(a) + P(b')P(a|b) = P(a|b), \quad\quad\quad\quad\quad (12.82)
\end{aligned}
$$

a harmless identity. Furthermore, note the differences in general

$$
(P_a)_o \neq (P_o)_{(a|\Omega)}, \qquad (P_{a'})_o \neq (P_o)_{(a'|\Omega)}. \quad\quad\quad (12.83)
$$

For example,

$$
(P_{a'})_o((a|b)) = P_{a'}(a|b) = (P(a'ab)/P(a'))/(P(a'b)/P(a')) = 0, \quad (12.84)
$$

while

$$
\begin{aligned}
(P_o)_{(a'|\Omega)}((a|b)) &= P_o((a'|\Omega)\&(a|b))/P_o((a'|\Omega)) \\
&= P(b')P(a|b)/P(a'). \quad\quad\quad\quad\quad (12.85)
\end{aligned}
$$

Similar remarks hold for the operation $[.|..]$, even though it produces conditional-like expressions back into $\mathcal{B}_o$ (not $\mathcal{B}_{oo}$). In summary, no contradiction along Lewis' lines can hold since the basic probability space $(\Omega_o, \mathcal{B}_o, P_o)$ properly extending $(\Omega, \mathcal{B}, P)$ is a well-defined construction with well-defined conditional events in $\mathcal{B}_o$, not in $\mathcal{B}$.

12.2.3   Higher Order Conditionals for PS

Unlike AC and DGNW, PS cea leads to a properly extended higher order cea where each higher order conditional event, say $((a|b)|(c|d)) \in (\mathcal{B}_o)_o$, is *not* in space ($\mathcal{B}_o$ for PS) of first level conditionals $(a|b)$, $(c|d)$,... for $a, b, c, d \in \mathcal{B}$. Also, unlike that for both CA and DGNW – the natural relation holds:

$$
(P_o)_o \left(((a|b)|(c|d))\right) = P_o((a|b)|(c|d)) = P_o((a|b)\&(c|d))/P(c|d), \quad (12.86)
$$

provided, of course, that $P(c|d) > 0$. Higher order conditionals for PS also have the interesting property that setting $d = \Omega$, the following weighted import-export property holds (but not import-export as in the cases of AC and DGNW):

$$
\begin{aligned}
(P_o)_o((a|b)|(c|\Omega)) &= P_o((a|b)|(c|\Omega)) = P_o(abc \vee cb' \times (a|b))/P(c) \\
&= (P(abc) + P(cb')P(a|b))/P(c) \\
&- P(a|bo)P(b|c) + P(a|b)P(b'|c). \quad (12.87)
\end{aligned}
$$

Also, setting $d = b$.

$$
P_o((a|b)|(c|b)) = P_o((a|b)\&(c|b))/P(c|b) = P(ac|b)/P(c|b) = P(a|bc),
$$

so that the second isomorphism property shared by DGNW and AC is also possessed by PS. Finally, note that, setting $b = \Omega$,

$$
\begin{aligned}
P_o((a|\Omega)|(c|d)) &= P_o((a|\Omega)\&(c|d))/P(c|d) \\
&= [P(acd) + P(ad')P(c|d)]/P(c|d) \\
&= P(a|cd)P(d) + P(ad'). \quad (12.88)
\end{aligned}
$$

(See [1], pp. 281-283, 301, 302 for additional properties of higher order conditioning for PS.)

12.2.4   Other Properties of PS

As is obvious, since PS conditional event algebra is a boolean one, all standard laws of probability are applicable (such as seen with the modular expansion in Theorem 9 and thus one can now apply the Frćhet-Hailperin tightest bounds as in eq.(10.11) validly to PS (but not to AC nor DGNW, because of their non-boolean structure). It is a simple exercise to verify indeed that all of the bounds are compatible with the exact computations for the logical combinations given in Theorem 9. Thus, all of the tightest bounds derived for the $d_{P,j}(a,b)$ also are valid for PS, such as in eqs.(10.17),(10.18).

Another important property that PS possesses distinct from DGNW and AC is the product form it takes when antecedents are disjoint: Theorem 9 immediately shows

$$
\text{If } bd = \emptyset, \quad \text{then} \quad P_o((a|b)\&(c|d)) = P(a|b)P(c|d), \quad (12.89)
$$

as opposed to DGNW yielding a trivial result and AC yielding a weighted average (see above). In a similar vein, unlike the failure of both AC and DGNW to preserve independence in the sense that if ab, b is pairwise P- independent of cd, d (see eq.(11.34)), then correspondingly $(a|b)$ and $(c|d)$ are P_o-independent (as opposed to eq.(11.35)). Still another special independence property that PS possesses – this time shared by AC and DGNW (but provided that the interpretation of independence is via conditioning, not factoring, since for these cea the two types of independence differ) is the following: Consider any positive integer n, any events $a_j, b_j \in \mathcal{B}$, $j = 1, ..., n$. Then

$$
(a_1|b_1)\& \cdots \&(a_n|b_n) \quad \text{and} \quad (b_1 \vee \cdots \vee b_n|\Omega) \text{ are } P_o\text{-independent.} \quad (12.90)
$$

(Since PS obeys all laws of probability, factored and conditional forms of independence are equivalent.)

12.3 Conditional Events and Their Relation to Conditioning of Random Variables for PS

This section, in part is based on results in [2]. PS can be used also – again apparently because of its compatibility with standard probability laws, as well as its desirable structure – to define random variables or measurable mappings involving conditional events which relate to the standard development of conditioning of random variables: In brief, suppose the symbol

$$(\Omega_1, \mathcal{B}_1, P_1) \xrightarrow{Z} (\Omega_2, \mathcal{B}_2, P_1 \circ Z^{-1}) \tag{12.91}$$

indicates that $Z : \Omega_1 \to \Omega_2$ is a $(\mathcal{B}_1, \mathcal{B}_2)$-measurable mapping which induces probability space $(\Omega_2, \mathcal{B}_2, P_1 \circ Z^{-1})$ from probability space $(\Omega_1, \mathcal{B}_1, P_1)$. When some P_2 is used in place of $P_1 Z^{-1}$, it is understood that $P_2 = P_1 Z^{-1}$. Also, define the PS extension of Z to be the mapping $Z_o : (\Omega_1)_o \to (\Omega_2)_o$, where

$$Z_o(\omega) = (Z(\omega_1), Z(\omega_2), ...), \quad \text{for all } \bar{\omega} = (\omega_1, \omega_2, ...) \in (\Omega_1)_o. \tag{12.92}$$

Lemma (See Lemma 2.1, [2]). *If* $(\Omega_1, \mathcal{B}_1, P_1) \xrightarrow{Z} (\Omega_2, \mathcal{B}_2, P_1 \circ Z^{-1})$ *holds, then so does* $((\Omega_1)_o, (\mathcal{B}_1)_o, (P_1)_o) \xrightarrow{Z_o} ((\Omega_2)_o, (\mathcal{B}_2)_o, (P_1 \circ Z^{-1})_o)$ *hold (where we can naturally identify* $(P_1)_o Z_o^{-1}$ *with* $(P_1 Z^{-1})_o$ *) and say that* $()_o$ *lifts* Z *to* Z_o.

Next, replace Z by joint measurable mapping (X, Y) in eq.(12.91), where $(\Omega_1, \mathcal{B}_1, P_1)$ is simply $(\Omega, \mathcal{B}, P)$, Ω_2 is replaced by $\Omega_1 \times \Omega_2$, $\mathcal{B}_2$ by sigma algebra generated by $(\mathcal{B}_1 \times \mathcal{B}_2)$, and define

$$(X, Y)(\omega) = (X(\omega), Y(\omega)), \quad \text{for any } \omega \in \Omega. \tag{12.93}$$

Thus, we have the following commutative diagram:

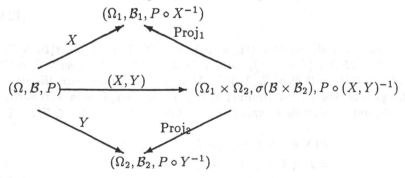

$$\tag{12.94}$$

which is part of the standard development of unconditional probability. Consider now the conditional probability

$$P(X \in a|Y \in b) = P(X^{-1}(a)|Y^{-1}(b)), \tag{12.95}$$

for any choice of $a \in \mathcal{B}_1$, $b \in \mathcal{B}_2$, where $(\Omega_j, \mathcal{B}_j)$ are any measurable spaces, $j - 1, 2$. (In particular, we could choose Ω_j as n_j-dimensional euclidean space and $\mathcal{B}_o$ as the corresponding Borel algebra, $j = 1, 2$.) We also assume here, as before that $P(Y^{-1}(b)) > 0$ and we do not treat the Radon-Nikodym case for $P(Y^{-1}(b)) = 0$. There does not exist in general any single random variable, say Z (including $Z = (X, Y)$) with Z not depending on P, with $(\Omega, \mathcal{B}, P) \xrightarrow{Z} (\Omega_1 \times \Omega_2, \sigma(\mathcal{B}_1, \mathcal{B}_2), P_o(Z^{-1}))$ holding, so that for any choice of $a \in \mathcal{B}_1$, $b \in \mathcal{B}_2$, there exists $\gamma(a, b) \in \sigma(\mathcal{B}_1 \times \mathcal{B}_2)$, also not depending on P, such that

$$P(Z \in \gamma(a, b)) = P(Z^{-1}(g(a, b))) = P(X^{-1}(a)|Y^{-1}(b)). \tag{12.96}$$

In fact, Lewis' theorem above can be used directly to show the impossibility of eq.(12.96). But, by utilizing PS cea the above can be accomplished as an imbedding:

Theorem 77 *(See [2], Theorem 2.3). The commutative diagram of arbitrary joint measurable mappings in eq.(12.94) lifts to the commutative diagram*

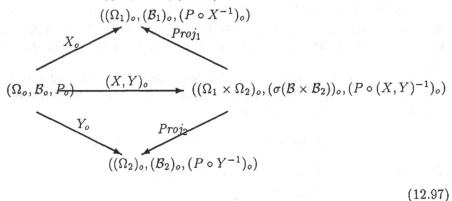

$$\tag{12.97}$$

where, we can naturally identify $(\Omega_1 \times \Omega_2)_o$ with $(\Omega_1)_o \times (\Omega_2)_o$, $(\sigma(\mathcal{B}_1 \times \mathcal{B}_2))_o$ with $\sigma((\mathcal{B}_1)_o \times (\mathcal{B}_2)_o)$, $(P \circ X^{-1})_o$ with $P_o \circ X_o^{-1}$, $(P \circ Y^{-1})_o$ with $P_o \circ Y_o^{-1}$, and $(P \circ (X, Y)^{-1})_o$ with $P_o \circ (X_o, Y_o)^{-1}$. Moreover, the basic compatibility relations always hold between conditioning of measurable mappings in unconditional events and joint measurable mappings in conditional events, all $a \in \mathcal{B}_1$, $b \in \mathcal{B}_2$:

$$P(X \in a|Y \in b) \quad (= P(X^{-1}(a)|Y^{-1}(b)))$$
$$= P_o((X, Y)_o \in (a \times b|\Omega_1 \times b))$$
$$(= P_o((X, Y)_o^{-1}((a \times b|\Omega_1 \times b)))). \tag{12.98}$$

12.4 Boolean Conditional Event Algebras Distinct from PS

Other boolean cea's differing from PS are possible to construct even for the same product probability space as $(\Omega_o, \mathcal{B}_o, P_o)$ in eq.(11.40). This can be accomplished by appropriately changing the order in which the terms appear in the definition for PS conditional events (in general choosing for each antecedent such as d a corresponding permutation of places given in eq.(11.42). For example, at its simplest, suppose we now define for a fixed $d \in \mathcal{B}$, $\emptyset < d < \Omega$ and $c \in \mathcal{B}$ arbitrary

$$(c|d)_1 = (cd|\Omega) \vee d' \times \Omega \times cd \times \Omega_o \vee d' \times cd \times d' \times \Omega_o \vee d' \times d' \times d' \times (c|d), \quad (12.99)$$

where, as usual $(c|d)$ is given in eq.(11.42). Next, we can show that

$$(c|d)_1' = (c'|d)_1, \quad (12.100)$$

where $(c'|d)$ is as in eq.(12.99), but with c replaced by c'. On the other hand, for any $a, b \in \mathcal{B}$, with $b \neq d$, let us define

$$(a|b)_1 = (a|b), \quad (12.101)$$

with $(a|b)$ as in eq.(11.42). In turn, consider the resulting collection $\{(a|b)_1 : a, b \in \mathcal{B}\}$ as defined in eqs.(12.99), (12.101). It is clear that for any choice of probability measure P over $\mathcal{B}$,

$$P_o((a|b)_1) = P_o((a|b)) = P(a|b), \qquad \text{for } b \neq d, \quad (12.102)$$

provided $P(b) > 0$. In turn, it is easily seen that eqs.(12.99)-(12.102) generate a legitimate cea which is boolean and obeys modus ponens. But, we show that it leads to a different cea than PS. In particular, for $b \neq d$, $\emptyset < a < b < \Omega$, $\emptyset < c < d$, we can show

$$|P_o((a|b)_1 \& (c|d)_1) - P_o((a|b) \& (c|d))| > 0. \quad (12.103)$$

Proof. First, note (see also eq.(14.6) for a similar form)

$$|P_o((a|b)_1 \& (c|d)_1) - P_o((a|b) \& (c|d))|$$
$$= |P_o((a|b)_1 \& (c|d)_1 \& ((a|b) \& (c|d))') - P_o(((a|b)_1 \& (c|d)_1)' \& (a|b) \& (c|d))|$$
$$= |P_o((a|b) \& (c|d)_1 \& (c'|d)) - P_o((a|b) \& (c'|d)_1 \& (c|d))|, \quad (12.104)$$

using eqs.(12.100), (12.101) and basic properties of PS. Letting

$$\tau_1(c, d) = d' \times \Omega \times cd \vee d' \times cd \times d', \qquad \tau(c, d) = d' \times cd \vee d' \times d' \times cd, \quad (12.105)$$

we can write (using eqs.(11.44) and (12.99))

$$(c|d) = cd \times \Omega_o \vee \tau(c, d) \vee (d')^3 \times (c|d).$$
$$(c|d)_1 = cd \times \Omega_o \vee \tau_1(c, d) \vee (d')^3 \times (c|d), \quad (12.106)$$

noting from eqs.(12.100), (12.101) and the properties of PS,

$$(c|d)' = (c'|d) = c'd \times \Omega_o \vee \tau(c', d) \vee (d')^3 \times (c'|d),$$
$$(c|d)'_1 = (c'|d)_1 = c'd \times \Omega_o \vee \tau_1(c', d) \vee (d')^3 \times (c'|d). \quad (12.107)$$

Then, after simplifying, using eqs.(12.106), (12.107),

$$(c|d)_1 \& (c'|d) = [d' \times c'd \times cd \mid \Omega],$$
$$(c|d) \& (c'|d)_1 = [d' \times cd \times c'd \mid \Omega]. \quad (12.108)$$

Hence, using the recursive form for $(a|b)$ in eq.(11.44),

$$(a|b) \& (c|d)_1 \& (c'|d) = (a|b) \& [d' \times c'd \times cd | \Omega]$$
$$= (ab \times \Omega_o \vee b' \times ab \times \Omega_o \vee b' \times b' \times ab \times \Omega_o \vee b' \times b' \times b' \times (a|b))$$
$$\& (d' \times c'd \times cd \times \Omega_o)$$
$$= abd' \times c'd \times cd \times \Omega_o \vee b'd' \; abc \times cd \times \Omega_o$$
$$\vee b'd' \times b'c'd \times abcd \times \Omega_o \vee b'd' \times b'c'd \times b'cd \times (a|b). \quad (12.109)$$

Similarly,

$$(a|b) \& (c|d) \& (c'|d)_1 = abd \times cd \times c'd \times \Omega_o \vee b'd' \times abcd \times c'd \times \Omega_o$$
$$\vee b'd' \times b'cd \times abc'd \times \Omega_o \vee b'd' \times b'cd \times b'c'd \times (a|b). \quad (12.110)$$

Hence, applying product measure P_o to eqs.(12.109), (12.110),

$$P_o((a|b) \& (c|d)_1 \& (c'|d))$$
$$= P(abd')P(c'd)P(cd) + P(b'd')P(abc'd)P(cd)$$
$$+ P(b'd')P(b'c'd)P(abcd) + P(b'd')P(b'c'd)P(b'cd)P(a|b), \quad (12.111)$$

and

$$P_o((a|b) \& (c|d) \& (c'|d)_1)$$
$$= P(abd')P(cd)P(c'd) + P(b'd')P(abcd)P(c'd)$$
$$+ P(b'd')P(b'cd)P(abc'd) + P(b'd')P(b'cd)P(b'c'd)P(a|b). \quad (12.112)$$

Hence, eqs.(12.111), (12.112) and (12.104) imply

$$|P_o((a|b)_1 \& (c|d)_1) - P_o((a|b) \& (c|d))|$$
$$= |P(b'd')(P(abc'd)P(cd) - P(abcd)P(c'd))$$
$$+ P(b'c'd)P(abcd)P(b'cd)P(abc'd)|$$
$$= |P(b'd')(P(abc'd)(P(cd) - P(b'cd)) + P(abcd)(P(b'c'd) - P(c'd)))|$$
$$= |P(b'd')(P(abc'd)P(bcd) - P(abcd)P(bc'd))|$$
$$= |P(b'd')(P(abc'd)(P(abcd) + P(a'bcd))$$
$$- P(abcd)(P(abc'd) + P(a'bc'd)))|$$
$$= |P(b'd')(P(abc'd)P(a'bcd) - P(abcd)P(a'bc'd))|. \quad (12.113)$$

Since the terms in eq.(12.113), $P(abc'd)$, $P(a'bcd)$, $P(abcd)$, $P(a'bc'd)$, in general are probabilities of distinct atoms, all disjoint from $b'd'$, one can choose (infinitely many) nontrivial P such that

$$P(abc'd)P(a'bcd) - P(abcd)P(a'bc'd) \neq 0 \text{ and } P(b'd') > 0, \qquad (12.114)$$

thereby yielding for many nontrivial P,

$$|P_o((a|b)_1 \& (c|d)_1) - P_o((a|b) \& (c|d))| > 0. \qquad (12.115)$$

Thus, the cea determined by eqs.(12.99)-(12.102) is distinct from PS cea, though both lie in the same space $\mathcal{B}_o$.

12.5   Fundamental Characterization of PS

In the last part we saw how to construct boolean cea's within the same product space setting as PS, but quite distinct from the latter. Here, Theorem 11 shows that if we require that candidate boolean cea's also satisfy the special independence property (see eq.(12.90)), then we cannot make such a construction.

Theorem 78 *Characterization of PS cea ([1], Theorem 16). Consider any measurable space $(\Omega, \mathcal{B})$. Suppose procedure Ψ is such that for any choice of P and hence corresponding probability space $(\Omega, \mathcal{B}, P)$, Ψ produces a boolean conditional event algebra extension, say $(\Omega_1, \mathcal{B}_1, P_1)$ of $(\Omega, \mathcal{B}, P)$ which satisfies modus ponens and the special independence property. Then, up to all probability evaluations of all finite well-defined logical combinations of corresponding conditional events, Ψ must coincide with PS cea.*

Proof. The complete proof is given in [1]. A sketch for the case of two arguments follows by first noting we can write for any $a, b, c, d \in \mathcal{B}$ with $(\Omega, \mathcal{B}, P)$ arbitrarily given and assuming the hypothesis, any candidate conditional events $(a|b)_1$, $(c|d)_1 \in \mathcal{B}_1$, using basic properties of all cea's, the unity identity and distributivity properties of a boolean algebra and the modus ponens assumption,

$$(a|b)_1 = (a|b)_1 \& ((b|\Omega)_1 \vee (b'|\Omega)_1) = (ab|\Omega)_1 \vee (a|b)_1 \& (b(|\Omega)_1,$$
$$(c|d)_1 = (c|d)_1 \& ((d|\Omega)_1 \vee (d'|\Omega)_1) = (cd|\Omega)_1 \vee (c|d)_1 \& (d(|\Omega)_1. \quad (12.116)$$

Then conjoining eq.(12.116) and using the basic properties of all cea's we have the result,

$$(a|b)_1 \& (c|d)_1 = \alpha \vee (b'd'|\Omega)_1 \& (a|b)_1 \& (c|d)_1, \qquad (12.117)$$

where

$$\alpha = (abcd|\Omega)_1 \vee (abd'|\Omega)_1 \& (c|d)_1 \vee (a|b)_1 \& (b'cd|\Omega)_1. \qquad (12.118)$$

Next, by the special independence assumption and the fact we are working in a cea,

$$P_1((a|b)_1 \& (c|d)_1 \& (b \vee d|\Omega)_1) = P_1((a|b)_1 \& (c|d)_1) P(b \vee d). \qquad (12.119)$$

Since eq.(12.119) must hold for all choices of a, b, c, d, first replace a by abd' and b by Ω, yielding

$$P_1\left((abd'|\Omega)_1 \& (c|d)_1\right) = P(abd')\,P(c|d). \tag{12.120}$$

Similarly, by replacing c by $b'cd$ and d by Ω in eq.(12.119), we obtain

$$P_1\left((a|b)_1 \& (b'cd|\Omega)_1\right) = P(a|b)P(b'cd). \tag{12.121}$$

Also, eq.(12.119) is obviously equivalent to the form

$$P_1\left((b'd'|\Omega)_1 \& (a|b)_1 \& (c|d)_1\right) = P(b'd')\,P_1((a|b)_1 \& (c|d)_1). \tag{12.122}$$

Next, noting all terms in eqs.(12.117) and (12.118) are mutually disjoint, evaluate these equations through by P_1 and use eqs.(12.119)-(12.122) to obtain

$$P_1((a|b)_1 \& (c|d)_1) = P_1(\alpha) + P(b'd)\,P_1((a|b)_1 \& (c|d)_1), \tag{12.123}$$

where

$$P_1(\alpha) = P(abcd) + P(abd')P(c|d) + P(a|b)P(b'cd). \tag{12.124}$$

Solving for $P_1((a|b)_1 \& (c|d)_1)$ in eq.(12.123) shows finally that

$$P_1((a|b)_1 \& (c|d)_1) = P_1(\alpha)/P(b \vee d), \tag{12.125}$$

with $P_1(\alpha)$ evaluated as in eq.(12.124). Comparison of eqs.(12.125), (12.124) with Theorem 9(v) shows the same evaluation. Since we are in a boolean algebra setting, all probability evaluations of disjunctions must also coincide. The coincidence of probabilities of negation is of course trivial.   □

Bibliography

[1] Goodman, I. R. and Nguyen, H. T. (1995) Mathematical foundations of conditionals and their probabilistic assignments, *International Journal of Uncertainty, Fuzziness and Knowledge-Based Systems*, **3**(3), 247-339.

[2] Goodman, I. R. and Kramer, G. F. (1997) Extension of relational event algebra to random sets with applications to data fusion, *Proceedings of Workshop on Applications and Theory of Random Sets*, Institute for Mathematics and Its Applications (IMA), University of Minnesota, Minneapolis, August 22-24, 1996, Springer-Verlag, Berlin (to appear).

Chapter 13

Product Space Conditional Event Algebra as a Tool for Further Analysis of Conditional Event Algebra Issues

13.1 Direct Connections between PS and both DGNW and AC via Partial Ordering and Deduction-Enduction Relations

So far the only connections between PS, DGNW, and AC have been through those properties that are shared (or those properties they differ on). In Section 11.6, we mentioned that McGee, independently , in effect, obtained the PS logical operations without explicitly recognizing the form of the conditional events themselves. Extending McGee's work, Goodman and Nguyen have also shown relations between DGNW and PS via "fixed point" evaluation of the third or indeterminate value for the three-valued (min,max) logic corresponding to DGNW. (See [3],Section 4.)

In the following, we proceed in another direction and show some interesting and useful direct relations involving partial ordering of conditional events. Next, (see also [3], Section 5.1), we will imbed the DGNW and AC operations in a PS setting. First, recall that conditional events for both the DGNW and AC cea's are identifiable as ordered pairs or intervals of events (or even cosets), but written for convenience using the symbol (.|..), with the fundamental equality characterization (eq.(11.11))

$$(a|b) = (c|d) \qquad \text{iff} \qquad ab = cd \text{ and } b = d. \tag{13.1}$$

On the other hand, the equality characterization between conditional events for PS differs with respect to the trivial events (Theorem 5(iii) and Theorem 2).
    If, $ab = \emptyset$ or $cd = \emptyset$, then

$$(a|b) = (c|d)\,(= \emptyset_o) \qquad \text{iff} \qquad ab = cd = \emptyset. \tag{13.2}$$

If $ab = b > \emptyset$ or $cd = d > \emptyset$, then

$$(a|b) = (c|d)\,(= \Omega_o) \quad \text{iff} \quad ab = d > \emptyset \text{ and } cd = d > \emptyset. \tag{13.3}$$

Thus, only for the *nontriviality conditions*

$$\emptyset < a < b \qquad \text{and} \qquad \emptyset < c < d, \tag{13.4}$$

does the criterion for equality coincide for DGNW and AC and PS conditional events, formally as in eq.(13.1). However, this does not preclude the possibility that resulting events produced by the imbedded nontrivial DGNW or AC operations themselves at times may well themselves be trivial – and are evaluated in PS appropriately. This situation does not give rise to any ambiguity of probability evaluation in PS and, in fact, the probability evaluations coincide always with the corresponding probability evaluations in DGNW or AC, though, of course, the actual trivial conditional event forms may differ from the PS space interpretation). *The real difficulty in attempting to extend the DGNW and AC operations to trivial events involves being well-defined at trivial conditional event arguments.* For example, suppose the AC & operation is extended to possible trivial arguments in PS and consider the formal operation (see eq.(11.30))

$$(\emptyset|b)\&_{AC}(c|d) = (b'cd|b \vee d), \tag{13.5}$$

with formal probability evaluation

$$P_o((\emptyset|b)\&_{AC}(c|d)) = P(b'cd|b \vee d). \tag{13.6}$$

But, eq.(13.4) relative to the PS setting must be the same as

$$\emptyset_o\&_{AC}(c|d) = \emptyset_o = (\emptyset|\Omega), \quad \text{for all } b \in \mathcal{B}, \tag{13.7}$$

since for PS,

$$(\emptyset|b) = \emptyset_o. \tag{13.8}$$

However, by choosing any $b, d \in \mathcal{B}$ such that $d > \emptyset$ and $b'cd > \emptyset$, not only do we have, via eqs.(13.6) and (13.7)

$$(\emptyset|b)\&_{AC}(c|d) \neq \emptyset_o\&_{AC}(c|d), \tag{13.9}$$

but the lhs(13.9) varies in general as b varies. One way of remedying this situation is to pick any event $e_o \in \mathcal{B}$, dependent, in general, upon the choice of P, such that $e_o > \emptyset$ and $P(e_o) = 0$, where preferably, $e_o \& (b \vee d) = \emptyset$. If this is not possible, choose $e_o \notin B$ and we can always imbed B in a strictly bigger boolean algebra B_1 (in the spirit of the standard compactification of sets) containing now e_o as an event, and we extend probability space (Ω, B, P) to a bigger probability (Ω_1, B_1, P_1), so that P_1 restricted to the imbedding of B is as before and $P_1(e_o) = 0$. Then, in light of eqs.(11.19) (for DGNW) and (11.29) (for AC), consider any trivial conditional events such as $(\emptyset|b)$, or $(b|b)$. Replace these with $(e_o|b \vee e_o)$ and $(b|b \vee e_o)$, respectively, assuming for simplicity $b > \emptyset$, where we slightly abuse notation in using the symbol b, $c, d,...$ now to indicate, when required, the imbedding of the corresponding original elements $b, c, d \in B$, now as elements in B_1. A similar replacement can be carried out for any DGNW or AC logical operation involving any number of trivial events. Though, of course

$$P_1(e_o|b \vee e_o) = 0, \quad P_1(b|b \vee e_o) = 1, \quad \text{provided } P(b) > 0, \qquad (13.10)$$

conditional events $(e_o|b \vee e_o)$ and $(b|b \vee e_o)$ are *not* trivial conditional events and may be considered to be $\in (B_1)_o$ relative to PS. Returning back to the original example, consider now – assuming $\emptyset < c < d$ and $b > \emptyset$ with $P(b \vee d) > 0$ – instead of eq.(13.5), we have conditional events and now (via eq.(11.21)) well-defined operation

$$(e_o|b \vee e_o), (c|d), (e_o|b \vee e_o)\&_{AC}(c|d) = (e_o bcd \vee e_o bd \vee b'e_o'cd|b \vee d \vee e_o), \quad (13.11)$$

replacing the original events and not well-defined operation, respectively,

$$(\emptyset|b), (c|d), (\emptyset|b)\&_{AC}(c|d) = (b'cd|b \vee d). \qquad (13.12)$$

In turn, we now have the matching probability evaluations – using $P(e_o) = 0$, etc.,

$$P_1(e_o|b \vee e_o) = 0 = P(\emptyset|b), \quad P_1(c|d) = P(c|d),$$
$$P_1((e_o|b \vee e_o)\&_{AC}(c|d)) = P(b'cd|b \vee d). \qquad (13.13)$$

Similarly, we can take any of the four DGNW and AC $\&$ and $\vee$ operations – or any combination of them, as defined formally to include trivial events – for any number of arguments, and by identifying all trivial conditional event arguments in either the form $(e_o|b \vee e_o)$ or $(b|b \vee e_o)$, for appropriate events $b \in B$ and choice of special event e_o, identify all four operations as essentially always acting upon only all nontrivial arguments, though, as mentioned before, any number of them may well have zero or unity values for any given probability P.

In addition, all four operations can be considered auxiliary non-boolean ones with respect to the possible PS boolean operations – though, of course, the four DGNW and AC operations have the form of conditionals of boolean operations over the original boolean algebra of unconditional events (see eq.(11.20)). Hence, for simplicity and without loss of generality, we can always restrict ourselves to AC and DGNW operations restricted to only nontrivial arguments – and hence being completely well-defined and probability-compatible with the original probability evaluations of these operations in their respective spaces. Let us denote the operations corresponding to AC and DGNW with the appropriate subscripts: $\&_{AC}$, $\vee_{AC}$, denote AC conjunction and disjunction, while $\&_{DGNW}$, $\vee_{DGNW}$ denote DGNW conjunction and disjunction, respectively. Making this convention, consider first any $a, b, c, d, e, f \in B$, where relative to PS and its partial ordering $\leq$ (or, equivalently – due to the restriction to nontrivial conditional events and results – relative to the partial ordering of DGNW),

$$(e|f) \leq (a|b), (c|d). \tag{13.14}$$

By the characterizing form in Theorem 5(iii), eq.(13.14) (under the nontriviality assumptions ($\emptyset < a < b, \emptyset < c < d, \emptyset < e < f$) is equivalent to

$$ef \leq ab \text{ and } a'b \leq e'f \text{ and } ef \leq cd \text{ and } c'd \leq e'f,$$

i.e.,

$$ef \leq abcd \quad \text{and} \quad a'b \vee c'd \leq e'f. \tag{13.15}$$

Now, it is readily shown by inspection of eq.(11.9) and use of the criterion in Theorem 5(ii), that under the nontriviality assumptions,

$$(e|f) \leq (a|b)\&_{DGNW}(c|d)$$

iff $[ef \leq abcd\&(a'b \vee c'd \vee df) \text{ and } (abcd)'\&(a'b \vee c'd \vee bd) \leq e'f]$

iff $[ef \leq abcd \quad \text{and} \quad a'b \vee c'd \leq e'f], \tag{13.16}$

noting that for (13.16) to hold here, we must have $abcd > \emptyset$.

Comparing eqs.(13.15) and (13.16), shows for the nontrivial situation

$$(e|f) \leq (a|b), (c|d) \quad \text{iff} \quad (e|f) \leq (a|b)\&_{DGNW}(c|d). \tag{13.17}$$

However, since PS is a full lattice, the left-hand side of eq.(13.17) is equivalent to the usual infimum characterization of $\leq$, so that $(a|b)\&(c|d) \leq (a|b), (c|d)$ and for any $\alpha \in B_o$, if $\alpha \leq (a|b), (c|d)$, then

$$\alpha \leq (a|b)\&(c|d). \tag{13.18}$$

Thus, eqs.(13.17) and (13.18) show, where we choose $(e|f) = (a|b)\&_{DGNW}(c|d)$ and assume $\emptyset < abcd$, noting the nontriviality condition $\emptyset < e < f$ here is the same as $\emptyset < abcd < a'b \vee c'd \vee abcd$, which holds because of the nontriviality assumption $\emptyset < a < b$ and $\emptyset < c < d$ and $\emptyset < abcd$:

$$(a|b)\&_{DGNW}(c|d) \leq (a|b)\&(c|d). \tag{13.19}$$

On the other hand for the nontriviality conditions holding, but $abcd = \emptyset$, eq.(13.19) itself holds trivially, since in this case lhs(13.19) $= \emptyset$. Thus, eq.(13.19) holds for the nontriviality conditions for a, b, c, d alone. Dually, using the De-Morgan relations for both DGNW and PS operations, eq.(13.19) shows also for the nontriviality assumption for a, b, c, d alone,

$$(a|b) \vee (c|d) \leq (a|b) \vee_{DGNW} (c|d). \tag{13.20}$$

Next, by inspection of the operations in eqs.(11.21) and Theorem 9(ii), it is clear that $\&_{AC}$ is produced formally from PS $\&$ by omitting all of the proper cartesian product factors, while $\vee_{AC}$ is produced from PS $\vee$ by formally replacing all of the proper cartesian product factors by Ω_o, noting both AC and PS cea's have corresponding operations producing the same antecedents. Thus, by a relatively simple argument, $\&_{AC}$, produces conditional events $\geq$ corresponding PS $\&$ operation and similarly for $\vee_{AC}$ being dominated by PS $\vee$. Thus, for both nontrivial and even trivial cases,

$$(a|b)\&(c|d) \leq (a|b)\&_{AC}(c|d) \text{ and } (a|b) \vee_{AC} (c|d) \leq (a|b) \vee (c|d). \tag{13.21}$$

Next, by direct inspection of the expressions for $(a|b)\&_{AC}(c|d)$ and $(a|b) \vee_{AC} (c|d)$ in eq.(11.21), clearly both have the same antecedent $b \vee d$ and the consequents satisfy

$$abcd \vee abd' \vee cdb' \leq ab \vee cd. \tag{13.22}$$

But, a basic property of PS is that for any $A, B, C \in \mathcal{B}$, with $B > \emptyset$,

$$(A|B) \leq (C|B) \qquad \text{iff } AB \leq CB. \tag{13.23}$$

Hence, applying this to eq.(13.22) shows directly that for the nontrivial case

$$(a|b)\&_{AC}(c|d) \leq (a|b) \vee_{AC} (c|d). \tag{13.24}$$

In fact, eq.(13.24) even holds for all trivial cases.

Finally, combining eqs.(13.19)-(13.21) and (13.24), using the fact that complementation is the same for all three cea's and all three cea's are DeMorgan, we obtain an improved proof of [3], Theorem 21:

Theorem 79 *Relative to PS, for the nontrivial situation* $\emptyset < a < b, \emptyset < c < d$,

$$(a|b)\&_{DGNW}(c|d) \leq (a|b)\&(c|d) \leq (a|b)\&_{AC}(c|d)$$
$$(a|b) \vee_{AC} (c|d) \leq (a|b) \vee (c|d) \leq (a|b) \vee_{DGNW} (c|d), \tag{13.25}$$

where, as usual,$\&$, $\vee$ without subscripts indicate PS operations. The corresponding probability evaluations of P_o over eq.(13.25) become

$$P_o(a|b)\&_{DGNW}(c|d)) \leq P_o((a|b)\&(c|d)) \leq P_o((a|b)\&_{AC}(c|d))$$
$$\leq P_o((a|b) \vee_{AC} (c|d)) \leq P_o((a|b) \vee (c|d))$$
$$\leq P_o((a|b) \vee_{DGNW} (c|d)), \tag{13.26}$$

where strict inequality holds in general for both eqs.(13.25) and (13.26). $\qquad \square$

Remark. Another further direct tie-in between PS and AC and DGNW is concerned with deduction and enduction, as defined in Section 9.3.2. The following development is also an improvement in both proof derivation and form over a similar result given in [3], Section 5.2 for the case of two arguments. Again, we confine ourselves to two conditional event arguments, but the results are extendible to any number of arguments. Consider $(\Omega, \mathcal{B}, P)$, its PS cea extension via $(\Omega_s, \mathcal{B}_s, P_s)$, and choose any $a, b, c, d \in \mathcal{B}$. Let $A = \{(a|b), (c|d)\}$ be a set of two nontrivial conditionals (i.e., $\emptyset < a < b$ and $\emptyset < c < d$), such as inference rules in Example 1. We then wish to conclude deductively and enductively from the information given in A. Here, we let the basic class of interest be the class of all nontrivial conditional events $C = \{(e|f) : \emptyset < e < f \in \mathcal{B}\} \subset \mathcal{B}_o$. Then, for this nontrivial situation, using the following special interval of event notation,

$$(\alpha, \beta]_C = \{(e|f) \in C : \alpha < (e|f) \leq \beta\},$$
$$[\alpha, \beta)_C = \{(e|f) \in C : \alpha \leq (e|f) < \beta\}, \quad \text{for all } \alpha, \beta \in \mathcal{B}_o. \ (13.27)$$

The conjunctive enduction class of A with respect to C here is, using (13.17),

$$
\begin{aligned}
D(A, C, \geq, \&) &= \{(e|f) \in C : (a|b)\&(c|d) \geq (e|f)\} \\
&= (\emptyset_o, (a|b)\&_{DGNW}(c|d)]_C.
\end{aligned}
\tag{13.28}
$$

The disjunctive deduction class of A with respect to C, dually, here is

$$
\begin{aligned}
D(A, C, \leq, \vee) &= \{(e|f) \in C : (a|b) \vee (c|d) \leq (e|f)\} \\
&= [(a|b) \vee_{DGNW} (c|d)], \Omega_o)_C.
\end{aligned}
\tag{13.29}
$$

The intermediate enduction class of A with respect to C is simply, by its very definition,

$$D(A, C, \geq, or) = (\emptyset, (a|b)]_C \cup (\emptyset, (c|d)]_C, \tag{13.30}$$

while, dually, the intermediate deduction class of A with respect to C is

$$D(A, C, \leq, or) = [(a|b), \Omega_o)_C \cup [(c|d), \Omega_o)_C. \tag{13.31}$$

This leaves two deduction-related classes remaining to determine: conjunctive deduction and disjunctive enduction. $\qquad\square$

We first need the following result:

Theorem 80 *Relative to PS, for $a, b, c, d, e, f \in \mathcal{B}$, with the nontrivial case holding:*
(i) Suppose that

$$not((a|b) \leq (e|f)) \quad and \quad not((c|d) \leq (e|f)). \tag{13.32}$$

Then,

$$
\begin{aligned}
(a|b)\&(c|d) \leq (e|f) \quad &iff \quad (a|b)\&_{AC}(c|d) \leq (e|f) \\
&iff (abcd \vee abd' \vee b'cd \mid b \vee d) \leq (e|f). \quad (13.33)
\end{aligned}
$$

(ii) Suppose that

$$not((a|b) \geq (e|f)) \quad and \quad not((c|d) \geq (e|f)). \tag{13.34}$$

Then,

$$(a|b) \vee (c|d) \geq (e|f) \quad iff \quad (a|b) \vee_{AC} (c|d) \geq (e|f)$$
$$iff \quad (ab \vee cd \mid b \vee d) \geq (e|f), \tag{13.35}$$

Proof. (i): First note that if rhs(12.33) holds, then by eq.(13.21), it follows that lhs(13.33) must also hold. Conversely, suppose that lhs(13.33) holds. Thus, from Theorem 9(i),

$$[\alpha|b \vee d] \leq (e|f), \tag{13.36}$$

where

$$\alpha = (abcd \mid |\Omega) \vee abd' \times (c|d) \vee b'cd \times (a|b). \tag{13.37}$$

Then, using the recursive form in eqs.(11.45) and (11.49) and noting again that

$$\alpha \leq (b \vee d|\Omega), \tag{13.38}$$

then eq.(13.36) is equivalent to

$$\alpha \vee b'd' \times [\alpha|b \vee d] \leq (ef \mid \Omega) \vee f' \times (e|f). \tag{13.39}$$

Next, conjoin eq.(13.39) through by $((b \vee d)(ef)' \mid \Omega)$, using eq.(13.38) to yield

$$\alpha \& ((ef)'|\Omega) \leq (b \vee d)f' \times (e|f). \tag{13.40}$$

But, using eq.(13.37),

$$\alpha \& ((ef)'|\Omega) = (abcd(ef)' \mid \Omega) \vee abd'(ef)' \times (c|d) \vee b'cd(ef)' \times (a|b). \tag{13.41}$$

Hence, combining eqs.(13.40) and (13.41) shows

$$(abcd(ef)'|\Omega) \leq (b \vee d)f' \times (e|f), \quad abd'(ef)' \times (c|d) \leq (b \vee d)f' \times (e|f),$$
$$b'cd(ef)' \times (a|b) \leq (b \vee d)f' \times (e|f). \tag{13.42}$$

In turn, by the basic properties of cartesian products and the hypothesis, eq.(13.32), then eq.(13.42) implies

$$\emptyset = abcd(ef)' = abd'(ef)' = b'cd(ef)'. \tag{13.43}$$

In turn, eq.(13.43) shows

$$abcd \leq ef, \quad abd' \leq ef, \quad b'cd \leq ef,$$

i.e.,

$$abcd \vee abd' \vee b'cd \leq ef. \tag{13.44}$$

Next, note that (see, e.g., Theorem 3), eq.(13.36) is equivalent to

$$(e'|f) \leq [\alpha'|b \vee d]. \tag{13.45}$$

In turn, using again recursive eqs.(11.45) and (11.49), eq.(13.45) is the same as

$$(e'f|\Omega) \vee f' \times (e'|f) < \alpha'\&(b \vee d|\Omega) \vee b'd' \times [\alpha'|b \vee d]. \tag{13.46}$$

Now, from Theorem 9 and the DeMorgan properety of $\mathcal{B}_o$,

$$[\alpha'|b \vee d] = (a'|b) \vee (c'|d) = [\alpha'|b \vee d], \tag{13.47}$$

where

$$\alpha'\&(b \vee d|\Omega) = (a'b \vee c'd|\Omega) \vee abd' \times (c'|d) \vee b'cd \times (a'|b). \tag{13.48}$$

Next, conjoin eq.(13.46) through by $(f|\Omega)\&\alpha$,

$$(e'f|\Omega)\&\alpha \leq (fb'd' \times [\alpha'|b \vee d])\&\alpha. \tag{13.49}$$

Now, by eq.(13.37)

$$(e'f|\Omega)\&\alpha = (abcde'f|\Omega) \vee abd'e'f \times (c|d) \vee b'cde'f \times (a|b), \tag{13.50}$$

and

$$(fb'd' \times ([\alpha'|b \vee d])\&\alpha = \emptyset_o \vee \emptyset_o \vee \emptyset_o = \emptyset_o. \tag{13.51}$$

Combining eqs.(13.49)-(13.51) yields

$$\emptyset_o = (abcde'f|\Omega) = abd'e'f \times (c|d) = b'cde'f \times (a|b). \tag{13.52}$$

Thus, eq.(13.52) shows, analogous to eq.(13.43), that

$$\emptyset = abcde'f = abd'e'f = b'cde'f, \tag{13.53}$$

which implies

$$e'f \leq (abcd)', \quad e'f \leq (abd')', \quad e'f \leq (b'cd)'. \tag{13.54}$$

Hence, eq.(13.54) shows, after simplifications

$$e'f \leq (abcd)'\&(abd')'\&(b'cd)' = a'b \vee c'd \vee b'd'. \tag{13.55}$$

Next, conjoin eq.(13.49) through by $(fb'd'|\Omega)$ to yield

$$(e'fb'd'|\Omega) \leq fb'd' \times [\alpha'|b \vee d]. \tag{13.56}$$

In turn, by the use of standard cartesian product properties, eq.(13.56) implies $e'fb'd' = \emptyset$, i.e.,

$$e'f \leq b \vee d. \tag{13.57}$$

Combining eqs.(13.55) and (13.57),

$$e'f \le (a'b \vee c'd \vee b'd')\&(b \vee d) = a'b \vee c'd \qquad (13.58)$$

Next, note from eq.(11.21) that

$$(a|b)\&_{AC}(c|d) = (\beta|b \vee d), \qquad (13.59)$$

where

$$\beta = abcd \vee abd' \vee b'cd \le b \vee d. \qquad (13.60)$$

Now, consider the possible validity of $(a|b)\&_{AC}(c|d) \le (e|f)$, i.e., rhs(13.33), under the nontriviality assumption for a, b, c, d, e, f, for the following possible logically exhaustive cases.

Case 1. $\beta = \emptyset$. This case is the same as $\emptyset = abcd = abd' = b'cd$, i.e.,

$$\emptyset = abcd \quad \text{and} \quad ab \le d \quad \text{and} \quad cd \le b. \qquad (13.61)$$

In this case, $(a|b)\&_{AC}(c|d) \le (e|f)$ holds trivially in the form $\emptyset < (e|f)$. Before we go to the next case, note the following identities

$$b \vee d = \beta \vee a'b \vee c'd \ge \beta \qquad (13.62)$$

and

$$\beta'(b \vee d) = a'b \vee c'd. \qquad (13.63)$$

Case 2. $\beta = b \vee d$ ($> \emptyset$, by nontriviality assumptions). Using eq.(13.62), this case is the same as $a'b \vee c'd = \emptyset$, i.e.,

$$a'b = c'd = \emptyset. \qquad (13.64)$$

But, the nontriviality conditions for a, b, c, d preclude eq.(13.64) from holding, and thus this case cannot hold, though eqs. (13.62) and (13.63) are always valid.

Case 3. $\emptyset < b < b \vee d$. This is the same as

$$\emptyset < abcd \qquad \text{or} \qquad \emptyset < abd' \qquad \text{or} \qquad \emptyset < b'cd. \qquad (13.65)$$

Then, because the nontriviality condition $\emptyset < e < f$ also holds throughout the proof, we can apply the criterion of Theorem 5(ii) here and obtain

$$(a|b)\&_{AC}(c|d) = (\beta|b \vee d) \le (e|f)$$
$$\text{iff } [\beta\&(b \vee d) \le ef \text{ and } e'f \le \beta'\&(b \vee d)]. \qquad (13.66)$$

Eqs.(13.62) and (13.63) show that eq.(13.66) is the same as

$$(a|b)\&_{AC}(c|d) \le (e|f) \text{ iff } [abcd \vee abd' \vee b'cd \le ef \text{ and } e'f \le a'b \vee c'd]. \qquad (13.67)$$

Now, let us return to what we have deduced so far from the assumption that lhs(13.33) holds: eqs.(13.44) and (13.58) hold, i.e.,

$$abcd \vee abd' \vee b'cd \le ef \quad \text{and} \quad e'f \le a'b \vee c'd. \qquad (13.68)$$

Then, apply the logical possibilities of Cases 1-3 in conjunction with eq.(13.68) and the possible validity of rhs(13.33). Thus, if Case 1 holds (which is perfectly compatible with eq.(13.68)), then rhs(13.33) holds trivially. Case 2 is logically impossible. The remaining possibility is Case 3, where eq.(13.65) holds (also, completely compatible with eq.(13.68)). Thus, the equivalence in eq.(13.67) is valid. Comparing eq.(13.67) with eq.(13.68) shows that lhs(13.67) holds. But this is the same as rhs(13.33), and the proof of (i) is completed.

(ii): This simply the DeMorgan dual of (i). □

Theorem 81 *Let $a, b, c, d, e, f \in \mathcal{B}$ with the nontrivial case holding, i.e., $\emptyset < a < b$, $\emptyset < c < d$, and $\emptyset < e < f$. Again, letting $A = \{(a|b), (c|d)\}$ and $C = \{(e|f) : \emptyset < e < f \in \mathcal{B}\} \subset \mathcal{B}_o$, as in the Remark following Theorem 12, then:*

(i) Conjunctive deduction class of A with respect to C is

$$D(A, C, \leq, \&) = \{(e|f) \in C : (a|b)\&(c|d) \leq (e|f)\}$$
$$= [(a|b), \Omega_o)_C \cup [(c|d), \Omega_o)_C \cup [(a|b)\&_{AC}(c|d), \Omega_o)_C.$$

(ii) Disjunctive enduction class of A with respect to C is

$$D(A, C, \geq, \vee) = \{(e|f) \in C : (a|b) \vee (c|d) \geq (e|f)\}$$
$$= (\emptyset_o, (a|b)]_C \cup (\emptyset_o, (c|d)]_C \cup (\emptyset_o, (a|b) \vee_{AC} (c|d)]_C.$$

Proof. (i): Since a standard property of $\&$ over boolean algebra $\mathcal{B}_o$ is obviously $(a|b)\&(c|d) \leq (a|b), (c|d)$, then by basic logic

$$(a|b)\&(c|d) \leq (e|f) \quad \text{iff} \quad [((a|b) \leq (e|f) \text{ or } (c|d) \leq (e|f)) \text{ or } R], \quad (13.69)$$

where relation R is defined as

$$R = (\text{not}((a|b) \leq (e|f)) \text{ and } \text{not}((c|d) \leq (e|f)) \text{ and } (a|b)\&(c|d) \leq (e|f)). \quad (13.70)$$

Then, applying Theorem 13(i) to R yields

$$(a|b)\&(c|d) \leq (e|f)$$

iff

$$[((a|b) \leq (e|f) \text{ or } (c|d) \leq (e|f)) \text{ or } (\text{not}((a|b) \leq (e|f))$$
$$\text{and } \text{not}((c|d) \leq (e|f)) \text{ and } (a|b)\&_{AC}(c|d) \leq (e|f))] \quad (13.71)$$

(ii): This is the DeMorgan dual of (i) and apply analogously Theorem 13(ii). □

Remark. The results of Theorem 14 extend to any number of conditional event arguments in a straightforward way, though the cases for conjunctive deduction, and, dually, disjunctive enduction, become much more complex. For example, consider conjunctive deduction for three arguments: $A = (a|b), (c|d), (e|f)$, with

C as before (with $(g|h)$'s replacing $e|f)$'s as dummy variables). Then, analogous to key equivalent equations (13.69), (13.71) in the proof of Theorem 14, we have, assuming the usual nontriviality conditions,

$$(a|b)\&(c|d)\&(e|f) \le (g|h) \quad \text{iff}$$
$$[((a|b)\&(c|d) \le (g|h) \text{ or } (a|b)\&(e|f) \le (g|h) \text{ or } (c|d)\&(e|f) \le (g|h))$$
$$\text{or } (\text{not}((a|b)\&(c|d) \le (g|h)) \text{ and } \text{not}((a|b)\&(e|f) \le (g|h)) \text{ and}$$
$$\text{not}((c|d)\&(e|f) \le (g|h)) \text{ and } (a|b)\&(c|d)\&(e|f) \le (g|h))], \qquad (13.72)$$

is equivalent to

$$(a|b)\&(c|d)\&(e|f) \le (g|h) \quad \text{iff}$$
$$[((a|b)\&(c|d) \le (g|h) \text{ or } (a|b)\&(e|f) \le (g|h) \text{ or } (c|d)\&(e|f) \le (g|h))$$
$$\text{or } (\text{not}((a|b)\&(c|d) \le (g|h)) \text{ and } \text{not}((a|b)\&(e|f) \le (g|h)) \text{ and}$$
$$\text{not}((c|d)\&(e|f) \le (g|h)) \text{ and } (a|b)\&_{AC}(c|d)\&_{AC}(e|f) \le (g|h))] .(13.73)$$

That is, we can replace $(a|b)\&(c|d)\&(e|f) \le (g|h)$ in eq.(13.72) by the tractable relation $(a|b)\&_{AC}(c|d)\&_{AC}(e|f) \le (g|h)$, where, using

$$(a|b)\&_{AC}(c|d)\&_{AC}(e|f) = ((b \Rightarrow a)\&(d \Rightarrow c)\&(f \Rightarrow e)|b \vee d \vee f), \qquad (13.74)$$

we have readily

$$(a|b)\&_{AC}(c|d)\&_{AC}(e|f) \le (g|h) \qquad \text{iff}$$
$$[(abcdef \vee abcdf' \vee abefd' \vee cdefb' \vee abd'f' \vee cdb'f' \vee efb'd' \le gh)$$
$$\text{and } (g'h \le a'b \vee c'd \vee e'f)] . \qquad (13.75)$$

In addition, we can fully realize eq.(13.73) by noting that the binary conjunction partial orderings, such as $(a|b)\&(c|d) \le (g|h)$ and its negation, are each characterized through eq.(13.71) (because of Theorem 14), where it is seen to depend only on the tractable typical binary relation $(a|b)\&_{AC}(c|d) \le (g|h)$, as given in eq.(13.67) (with $(e|f)$ replaced by $(g|h)$ here), and its negation. It also depends on the unary relations such as $(c|d) \le (e|f)$ and its negation, via the criterion as in Theorem 5(ii).

Calabrese ([1], Section IE) from a different viewpoint has also considered and characterized various types of deduction in a conditional event algebra context.

13.2   Direct Application of PS to Motivating Example 1

With much of the machinery of DGNW, AC and PS cea firmly established and streamlined, we now can not only readily address Example 1, posed at the beginning of Section 10.1, but we can analyze the situation much further, such as obtaining in closed form all relevant deduction-related classes (via Theorem 14 and the Remark preceding Theorem 13). As for the original basic issue, recall the identifications in eq.(10.5) can now be written as legitimate conditional events representing the inference rules as $\alpha = (a|b)$, $\beta = (c|d)$, which we can consider, e.g., to be with respect to the PS setting and corresponding product probability space $(\Omega_o, \mathcal{B}_o, P_o)$ with $\alpha, \beta \in \mathcal{B}_o$. We then first can obtain any of the basic probability distance functions, since we now know how to compute the conjunction probability of conditional events (Theorem 9(v)), or alternatively, we can compute the probability distances via any relative atom probability form, again using Theorem 9. In particular, we can write, as before, in its simplest trivial form,

$$d_{P_o,1}(\alpha, \beta) = |P(a|b) - P(c|d)|, \tag{13.76}$$

without reference to any particular choice of cea. One could, of course, compute the rhs(9.25), using e.g., PS,

$$d_{P_o,1}(\alpha, \beta) = |P_o(\alpha\beta') - P_o(\alpha'\beta)|. \tag{13.77}$$

using Theorems 3 and 9 to obtain:

$$P_o(\alpha\beta') = P_o((a|b)\&(c'|d))$$
$$(P(abc'd) + P(abd')P(c'|d) + P(b'c'd)P(a|b))/P(b \vee d),$$
$$P_o(\alpha'\beta) = P_o((a'|b)\&(c|d))$$
$$(P(a'bcd) + P(a'bd')P(c|d) + P(b'cd)P(a'|b))/P(b \vee d), \tag{13.78}$$

which could be substituted back into eq.(13.77). But, this is patently absurd to use, since it requires knowledge of $P(a|b)$ and $P(c|d)$, as well as having a far more complex computational form. Finally, as a check, note that by adding $P(\alpha\beta)$ to both expressions in eq.(13.78), the evaluation in eq.(13.77) remains the same with these substitutions. But,

$$\begin{aligned} P_o(\alpha\beta) &= P_o((a|b)\&(c|d)) \\ &= (P(abcd) + P(abd')P(c|d) + P(b'cd)P(a|b))/P(b \vee d) \end{aligned} \tag{13.79}$$

which added to eq.(13.78) directly yields

$$\begin{aligned} P_o(\alpha\beta') + P_o(\alpha\beta) &= (P(abc'd) + P(abd')P(c'|d) + P(b'c'd)P(a|b))/P(b \vee d) \\ &\quad + (P(abcd) + P(abd')P(c|d) + P(b'cd)P(a|b))/P(b \vee d) \\ &= (P(abd) + P(abd') + P(b'd)P(a|b))/P(b \vee d) \\ &= (P(ab) + P(b'd)P(ab)/P(b))/(P(b)P(b \vee d)) \\ &= P(ab)(P(b) + P(b(d)))/(P(b)P(b \vee d)) \\ &= P(ab)P(b \vee d)/(P(b)P(b \vee d)) = P(a|b). \end{aligned} \tag{13.80}$$

Similarly, by direct computations

$$P_o(\alpha'\beta) + P_o(\alpha\beta) = P(c|d). \tag{13.81}$$

Clearly, eqs.(13.80) and (13.81) are completely compatible with standard probability results, as well as eqs.(13.76),(13.77).

On the other hand, the choice of cea for obtaining $d_{P_o,2}(\alpha,\beta)$ is relevant. For PS this becomes in its simplest form with respect to eq.(9.27),

$$d_{P_o,2}(\alpha,\beta) = P(a|b) + P(c|d) - 2P_o(\alpha\&\beta), \tag{13.82}$$

with $P_o(\alpha\&\beta)$ given in eq.(13.79) above.

Actual testing of hypotheses using the above functions will be postponed until Example 3, Section 14.2, where two additional probability distance functions will be considered, both of which, as in the case of $d_{P_o,2}(\alpha,\beta)$, require in their simplest forms, only knowledge of $P(a|b)$, $P(c|d)$ and $P_o(\alpha\&\beta)$ (or equivalently, $P_o(\alpha \vee \beta)$, if desired).

13.3 Rigorous Formulation of Constant Probability Events and Intervals within a PS Framework

This section contains an improved formulation and proofs for corresponding work first presented in [2], Sections 6.3 and 7.3. We provide here a more general basis for the concept of a constant-probability event begun in eqs.(10.19)-(10.27). First, for any probability space $(\Omega, \mathcal{B}, P)$ and any choice of $c \in \mathcal{B}$, with $1 > P(c) > 0$, and integers q, m, n, with $0 \leq q \leq m \leq n$, define

$$\nu(q,m,n,c) = \left(\bigvee_{j=q+1}^{m} \left(\tau(j,n,c) \middle| \bigvee_{j=1}^{n} \tau(j,n,c) \right) \right) \tag{13.83}$$

where, by convention

$$\nu(0,0,n,c) = \emptyset_o. \tag{13.84}$$

Thus, analogous to eq.(10.24),

$$(P_n)_o(\nu(q,m,n,c)) = (m-q)/n, \tag{13.85}$$

not depending on c, nor particular P. Consider then relation $\Psi(q,m,n;r,s)$, for any choice of real constants, $0 \leq r, s \leq 1$,

$$\Psi(q,m,n,r,s)) : [q,m,n \to \infty, \text{ with } \tfrac{q}{n} \to r \text{ and } \tfrac{m}{n} \to s \text{ uniformly}], \tag{13.86}$$

where we indicate the corresponding limit relative to a given quantity by the symbol $\lim_{\Psi(q,m,n,r,s)}(\ )$. Thus, the limiting analogue of the constant-probability event $\theta(s)$ is

$$\theta(r,s) = \lim_{\Psi(q,m,n,r,s)} \nu(q,m,n,c), \tag{13.87}$$

with convention

$$\theta(r,s) = \emptyset_o, \quad \text{if } r \geq s. \tag{13.88}$$

The corresponding evaluation, using eq.(13.85) and the previous notation $(\Omega_1, \mathcal{B}_1, P_1)$ for the product probability space encompassing both the limiting form of conditional events and $(\Omega_o, \mathcal{B}_o, P_o)$, is

$$P_1(\theta(r, s)) = \max(s - r, 0), \qquad (13.89)$$

whence we call $\theta(r, s)$ a constant-probability interval. In particular,

$$\theta(0, s) = \theta(s), \qquad \text{all } 0 \leq s \leq 1. \qquad (13.90)$$

By a straightforward argument applied to eqs.(13.83)-(13.85), the following properties also hold for all real $0 \leq r, s, r_j, s_j \leq 1, j = 1, 2$. First,

$$\theta(s)' = \theta(s, 1), \qquad (13.91)$$

not the result originally assumed in eq.(9.20) (q.v.), with probability evaluation

$$P_1(\theta(s)') = 1 - s. \qquad (13.92)$$

Next, when logically combining expressions which contain distinct constant-probability intervals, we make the following assumption given for simplicity for the case of two arguments, but obviously extendable to any number of finite arguments:

Assumption Q: In the limit relation in eq.(13.86) forming $\theta(r_1, s_1)$ and $\theta(r_2, s_2)$, assume that integers $0 \leq q_j \leq m_j \leq n$ and real constants $0 \leq r_j, s_j \leq 1$ are all arbitrary such that the limit of $\nu(q_j, m_j, n, c)$ with respect to $\Psi(q_j, m_j, n, r_j, s_j)$ holds for the same n and c for both $j = 1$ and $j = 2$. Then, under Assumption Q,

$$\theta(r_1, s_1) \ \& \ \theta(r_2, s_2) = \theta(\max(r_1, r_2), \min(s_1, s_2)), \qquad (13.93)$$

$$\theta(r)' \ \& \ \theta(s) = \theta(r, 1)\&\theta(0, s) = \theta(r, s), \qquad (13.94)$$

$$\theta(r) \ \& \ \theta(s) = \theta(\min(r, s)), \qquad \theta(r) \vee \theta(s) = \theta(\max(r, s)), \qquad (13.95)$$

with corresponding probability evaluations

$$P_1\left(\theta(r_1, s_1) \ \& \ \theta(r_2, s_2)\right) = \max\left(\min(s_1, s_2) - \max(r_1, r_2), 0\right), \qquad (13.96)$$

$$P_1\left(\theta(r)' \ \& \ \theta(s)\right) = \max(s - r, 0), \qquad (13.97)$$

$$P_1\left(\theta(r) \ \& \ \theta(s)\right) = \min(r, s), \qquad P_1\left(\theta(r) \vee \theta(s)\right) = \max(r, s). \qquad (13.98)$$

In later examples, in developing relational event algebra, we will consider for a given probability space $(\Omega, \mathcal{B}, P)$, infinite or finite disjoint disjunctions of expressions, each of which will be in the "mixed" form of a cartesian product of ordinary events, say $a_j \in \mathcal{B}, j = 1, ..., n$ producted with a coefficient term in the form of a constant-probability event or interval, say $\theta(r, s)$, i.e., each term appears in the typical form $a_1 \times \cdots \times a_n \times \theta(r, s)$ (with n, r, s, and the a_j possibly varying from term to term). However, this expression is actually an abuse of notation, since before even taking limits to achieve $\theta(r, s)$, via eqs.(13.86), (13.87) – and Assumption Q, if appropriate – at any stage n, the approximating expression to $\theta(r,s)$, $\nu(q,m,n,c)$, as in eq.(10.22) lies $\in (\mathcal{B}_n)_o$ (see again the discussion following eq.(10.20)), as opposed to the $a_j \in \mathcal{B}$. In order to make things consistent, we should actually replace each a_j here with either $a_j \times \Omega^{n-1}$ or for the limit situation as $n \to \infty$, by $a_j \times \Omega_o$ (where as usual $\Omega_o = \Omega \times \Omega \times \cdots$). However, for convenience, we make an abuse of notation in the following, bearing in mind the above comments.

Next, we will show how to combine conjunctively different degree cartesian products of ordinary events with constant-probability or interval event coefficients.

Theorem 82 *Consider any positive fixed integers M, N with $M \leq N$ and $a_j, b_k \in \mathcal{B}$, for $j = 1, ..., M$ and $k = 1, ..., N$, and let $0 \leq r_i, s_i \leq 1$ be arbitrary real constants, $i = 1, 2$. Then, making Assumption Q,*

$$(a_1 \times \cdots \times a_M \times \theta(r_1, s_1)) \& (b_1 \times \cdots \times b_N \times \theta(r_2, s_2))$$
$$= (a_1 b_1 \times \cdots \times a_M b_M) \times b_{M+1} \times \cdots \times b_N$$
$$\times (\theta(r_1, s_1) \& \theta(r_2, s_2)), \tag{13.99}$$

where $\theta(r_1, s_1) \& \theta(r_2, s_2)$ is obtained from eq.(13.92). In turn, the obvious common probability evaluation in eq.(13.99) is

$$P_1 ((a_1 \times \cdots \times a_M \times \theta(r_1, s_1)) \& (b_1 \times \cdots \times b_N \times \theta(r_2, s_2)))$$
$$= P(a_1 b_1) \cdots P(a_M b_M) P(b_{M+1}) \cdots P(b_N)$$
$$\cdot P_1 (\theta(r_1, s_1) \& \theta(r_2, s_2)), \tag{13.100}$$

where $P_1(\theta(r_1, s_1) \& \theta(r_2, s_2))$ is given in eq.(13.94).

Proof. Obviously, if $M = N$, the result trivially follows with $b_{M+1} \times \cdots \times b_N$ omitted. Consider, thus, only $M < N$. Also for all variables involved in the above hypothesis (and assumption Q), make the following definitions:

$$\zeta = (A|B) \& (b_{M+1} \times \cdots \times b_N \times \kappa), \quad \iota = \theta(r_1, s_1) \& (b_{M+1} \times \cdots \times b_N \times \theta(r_2, s_2)), \tag{13.101}$$

$$A = \bigvee_{i=q_1+1}^{m_1} \tau(i, n, c), \quad B = \bigvee_{i=1}^{n} \tau(i, n, c)$$
$$\nu(q_j, m_j, n, c) = (A|B), \quad \kappa = \nu(q_2, m_2, n, c), \tag{13.102}$$

$$\eta = [(A|\Omega) \& (b_{M+1} \times \cdots \times b_N \times \kappa)]$$
$$\vee [(B' \& b_{M+1}) \times ((A|\Omega) \& (b_{M+2} \times \cdots \times b_N \times \kappa))]$$
$$\vee [(B' \& b_{M+1}) \times (B' \& b_{M+2}) \times ((A|\Omega) \& (b_{M+3} \times \cdots \times b_N \times \kappa))] \vee \cdots$$
$$\vee [(B' \& b_{M+1}) \times \cdots \times (B' \& b_{N-1}) \times ((A|\Omega) \& (b_N \times \kappa))]. \tag{13.103}$$

$$\chi = (B' \& b_{M+1}) \times \cdots \times (B' \& b_N) \times ((A|B) \& \kappa) \tag{13.104}$$

$$\rho = b_{M+1} \times \cdots \times b_N \times ((A|B) \& \kappa) + (\theta(r_1, s_1) \& \theta(r_2, s_2)), \tag{13.105}$$

$$\lambda = b_{M+1} \times \cdots \times b_N \times ((A|B) \& \kappa). \tag{13.106}$$

Now, note that lhs(13.99) can be written first as

$$\varphi = (a_1 \times \cdots \times a_M \times \theta(r_1, s_1)) \& (b_1 \times \cdots \times b_N \times \theta(r_2, s_2))$$
$$= (a_1 b_1 \times \cdots \times a_M b_M) \times \iota. \tag{13.107}$$

Also, define rhs(13.99) by

$$\gamma - (a_1 b_1 \times \cdots \times a_M b_M) \wedge b_{M+1} \times \cdots \times b_N \times (\theta(r_1, s_1) \,\&\, \theta(r_2, s_2)). \tag{13.108}$$

In turn, consider, the finite analogue δ of φ before taking limits with respect to assumption Q, where

$$\delta = (a_1 \times \cdots \times (A|B)) \& (b_1 \times \cdots \times b_N \times \kappa) = (a_1 b_1 \times \cdots \times a_M b_M) \times \zeta. \tag{13.109}$$

Next, apply the recursive form in eq.(11.44) to ζ, yielding

$$\zeta = [(A|\Omega o) \vee (B' \times (A|B)) \vee \cdots \vee ((B')^{N-M-1} \times (A|\Omega o))$$
$$\vee ((B')^{N-M} \times (A|B))] \& (b_{M+1} \times \cdots \times b_N \times \kappa)$$
$$= \zeta + \chi. \tag{13.110}$$

Furthermore, we also have the simple identity

$$(b_{M+1} \times \cdots \times b_N \times (\theta(r_1, s_1) \,\&\, \theta(r_2, s_2)) = \rho + \lambda. \tag{13.111}$$

Then, utilizing eqs.(13.110), (13.111), and standard properties of the boolean operation $+$ and probability,

$$P_1(\varphi + \gamma) = P_1[(a_1 b_1 \times \cdots \times a_M b_M)$$
$$\times [\iota + (b_{M+1} \times \cdots \times b_N \times (\theta(r_1, s_1) \,\&\, \theta(r_2, s_2)))]]$$
$$\leq P_1[\iota + (b_{M+1} \times \cdots \times b_N \times (\theta(r_1, s_1) \,\&\, \theta(r_2, s_2)))]$$
$$\leq P_1(\iota + \zeta) + P_1[\zeta + (b_{M+1} \times \cdots \times b_N \times (\theta(r_1, s_1) \& \theta(r_2, s_2)))]$$
$$= P_1(\iota + \zeta) + P_1(\eta + \chi + \rho + \lambda)$$
$$\leq P_1(\iota + \zeta) + P_1(\eta) + P_1(\rho) + P_1(\chi + \lambda). \tag{13.112}$$

Now, taking limits as $q_j, m_j, n \to \infty$ with respect to Assumption Q, using eqs.(13.102) and (10.20), shows

$$P_1(A) \leq P_1(B) = n(P(c))^{n-1} P(c'). \tag{13.113}$$

By standard results (since $1 > P(c) > 0$), eq.(13.113) shows

$$\lim_{\Psi(q_1, m_1, n, r_1, s_1)} P_1(A) = 0 \quad \text{and} \quad \lim_{\Psi(q_2, m_2, n, r_2, s_2)} P_1(B') = 1. \tag{13.114}$$

Hence, applying eq.(13.114) to eqs.(13.112) and (13.105),

$$\lim_{\substack{\Psi(q_1, m_1, n, r_1, s_1), \\ \Psi(q_2, m_2, n, r_2, s_2)}} P_1(\eta) = \lim_{\substack{\Psi(q_1, m_1, n, r_1, s_1), \\ \Psi(q_2, m_2, n, r_2, s_2)}} P_1(\rho) = 0. \tag{13.115}$$

Also,

$$\lim_{\substack{\Psi(q_1, m_1, n, r_1, s_1),\\ \Psi(q_2, m_2, n, r_2, s_2)}} P_1(\chi + \lambda)$$

$$= \lim_{\substack{\Psi(q_1, m_1, n, r_1, s_1),\\ \Psi(q_2, m_2, n, r_2, s_2)}} P_1\left[((B'\&b_{M+1}) \times \cdots \times (B'\&b_N))\right.$$

$$+ (b_{M+1} \times \cdots \times b_N)] \quad \lim_{\substack{\Psi(q_1, m_1, n, r_1, s_1),\\ \Psi(q_2, m_2, n, r_2, s_2)}} P_1\left((A|B)\&\kappa\right)$$

$$= \lim_{\Psi(q_2, m_2, n, r_2, s_2)} P_1\left[\left(((B')^{N-M})'\&\kappa\right)\&(b_{M+1} \times \cdots \times b_N)\right]$$

$$\cdot P_1(\theta(r_1, s_1)\&\theta(r_2, s_2))$$

$$\leq \lim_{\Psi(q_2, m_2, n, r_2, s_2)} P_1\left[((B')^{N-M})'\right] = 0 \tag{13.116}$$

By similar reasoning,

$$\lim_{\substack{\Psi(q_1, m_1, n, r_1, s_1),\\ \Psi(q_2, m_2, n, r_2, s_2)}} P_1(\iota + \zeta) = 0. \tag{13.117}$$

Finally, combining eqs.(13.115)-(13.117) with eq.(13.112) yields

$$P_1(\varphi + \gamma) = \lim_{\substack{\Psi(q_1, m_1, n, r_1, s_1),\\ \Psi(q_2, m_2, n, r_2, s_2)}} P_1(\varphi + \gamma) = 0 \tag{13.118}$$

Eq.(13.118) shows the equivalence desired between φ and γ.   □

Remark. So far, constant-probability events and intervals have been constructed which do not depend on the particular probability P chosen for evaluation nor generating event c – except of course that $1 > P(c) > 0$ – as in Section 10.3 and in this section. But, by inspection of all of these equations, especially, the limiting forms, as in eq.(13.87), there is no reason, to preclude numerical dependence on P, for each choice of P. Of course, it is clear that when it is possible, such dependence should be avoided. But, as will be seen in Section 16.4 certain functions of probabilities appear to require dependency of the constant-probability events on particular choices of P.

Bibliography

[1] Calabrese, P. G. (1994) A Theory of conditional information with applications, *IEEE Transactions on Systems, Man and Cybernetics*, **24**(12), 1676-1684.

[2] Goodman, I. R. and Kramer, G. F. (1997) Extension of relational event algebra to random sets with applications to data fusion, *Proceedings of Workshop on Applications and Theory of Random Sets*, Institute for Mathematics and Its Applications (IMA), University of Minnesota, Minneapolis, August 22-24, 1996, Springer-Verlag, Berlin (to appear).

[3] Goodman, I. R. and Nguyen, H. T. (1995) Mathematical foundations of conditionals and their probabilistic assignments, *International Journal of Uncertainty, Fuzziness and Knowledge-Based Systems*, **3**(3), 247-339.

Chapter 14

Testing of Hypotheses for Distinctness of Events and Event Similarity Issues

General Remarks

As we have stated previously, the standard development of probability theory and applications has been overwhelmingly numerically-oriented. Algebraic aspects, when involved, are still fundamentally directed toward numerical optimization criteria, such as the well-developed areas of transformation group-invariant estimation and hypotheses testing. In fact, the very foundations of modern information theory is almost entirely numerically-based as e.g. in [7]. This is not only borne out by any survey of the general literature, but also by considering any of the leading efforts devoted to combining or comparison of evidence, such as in [8] or the survey [2]. In this chapter we concentrate on determining a fundamentally more algebraically-oriented approach to the problem of testing hypotheses for the distinctness of events and related similarity issues – in the sense of "algebraic metrics" which replace, but are compatible with, all probability evaluations and numerical metrics (see Sections 14.6 and 14.7). This allows us to discover additional probability distance functions ($d_{P,3}$ and $d_{P,4}$), one of which ($d_{P,3}$) is apparently superior in a number of ways to the traditional one ($d_{P,2}$). (See Section 14.8.) PS cea is seen to play a major role in the above results. Preliminary aspects of this work can be found in [3].

14.1   Classical Testing of Statistical Hypotheses and Estimation / Regression Applied to the Comparison of Different Probability Distributions

Motivating Example 2.

This is a typical problem (actually an example of Case 1 in Section 9.3.1) which can be satisfactorily treated by the basic development of probability. Three differently located statistically independent radar systems are queried on the probability of an enemy ship being in a specified area. Assuming for simplicity that each describes the (supposed) common target(s) of interest through three-dimensional gaussian probability distributions $N(\mu_j, \Lambda_j)$, with estimated mean μ_j and covariance matrix Λ_j, corresponding to information I_j, $j = 1, 2, 3$. Using standard testing of hypotheses [9], [13], employing, e.g., weighted distances between the population means, it can be determined, up to some error level, that the sets of consistent information are, e.g., $I(1) = \{I_1, I_3\}$ and $I(2) = \{I_2\}$. Alternatively, a basic suboptimal procedure is to test pair-wise hypotheses of similarity for I_1 vs. I_2, I_1 vs. I_3, and I_2 vs. I_3, with an additional procedure for resolving quandaries, such as the case where I_1 is declared sufficiently similar to I_2, I_1 is also declared as being sufficiently similar to I_3, but I_2 is not sufficiently similar to I_3. Then, using standard regression procedures (see again, [9], e.g.), one obtains $I(1, 0) = N(\mu(1, 0), \Lambda(1, 0))$, $I(2, 0) = N(\mu_2, \Lambda_2)$, where typically

$$\mu(1, 0) = (\Lambda_1^{-1} + \Lambda_3^{-1})^{-1}(\Lambda_1^{-1}\mu_1 + \Lambda_3^{-1}\mu_3),$$
$$\Lambda(1, 0) = (\Lambda_1^{-1} + \Lambda_3^{-1})^{-1}. \tag{14.1}$$

Newly arriving information can be in the form of a fourth independent (possibly repeated from one of the previous three sensor sources) source yielding $I_4 = N(\mu_4, \Lambda_4)$, which is to be first tested against $I(1, 0)$ and $I(2, 0)$, and then combined with the appropriate distribution if the tests declares a positive association for the relevant information.

Finally, we remark that the metric $d_{P,1}$ or $d_{P,2}$ can be utilized quite easily here to determine the degree of similarity between any events α, β with respect to any given $I(j)$, such as for $j = 2$, where we denote the corresponding pdf as f_{μ_2, Λ_2}. For example, referring back to eq.(9.25),

$$d_{P,1}(\alpha, \beta) = \left| \int_{x \in \alpha\beta'} f_{\mu_2, \Lambda_2}(x)dx - \int_{x \in \alpha'\beta} f_{\mu_2, \Lambda_2}(x)dx \right|. \tag{14.2}$$

14.2   Testing Hypotheses of Different Events Relative to a Common Probability Measure

Motivating Example 3.

Two competing models of a single complex event – as compared to the situation in Example 2 – are to be compared for similarity, and if declared sufficiently compatible, are to be combined. More specifically, consider two independent models for the design of a data fusion processing system, relative to its logical structure. Suppose the following events are considered, where regions A, B, C are in general overlapping and processing of information in all the areas can be carried on simultaneously:

$$a = \text{arriving information in region } A \text{ is now being processed,}$$
$$b = \text{arriving information in region } B \text{ is now being processed,}$$
$$c = \text{arriving information in region } C \text{ is now being processed.} \quad (14.3)$$

Also suppose that all relevant probabilities of logical combinations of a,b,c occurring or not occurring are known via probability measure P. Based upon geographical, economic, and practical considerations, and using standard boolean logic notation, the two models of design for simultaneous processing are

$$\alpha = ab' \vee c, \qquad \beta = a'b \vee c \qquad (14.4)$$

with corresponding probabilities of certainty being

$$I_1 = P(\alpha) = P(ab') + P(c) - P(ab'c),$$
$$I_2 = P(\beta) = P(a'b) + P(c) - P(a'bc). \qquad (14.5)$$

One basic measure of similarity between α and β is simply the naive probability distance $d_{P,1}(\alpha, \beta)$, introduced in eq.(9.25), which here becomes

$$d_{P,1}(\alpha, \beta) = |P(a'bc') - P(ab'c')|, \qquad (14.6)$$

since

$$P(\alpha'\beta) = P(a'bc'), \qquad P(\alpha\beta') = P(ab'c'). \qquad (14.7)$$

But, it is clear that $d_{P,1}(\alpha, \beta)$ can be quite small, yet α and β may represent quite different events, in the sense that both $P(\alpha'\beta)$ and $P(\alpha\beta')$ may be quite large: By inspection of eq.(14.7), situations can arise where $P(a'bc')$ and $P(ab'c')$, are both close to a half, thus making $d_{P,1}(\alpha, \beta)$ in eq.(14.6) close to zero. A much more satisfactory measure (though not necessarily the best – see later results) is to use the natural absolute probability distance $d_{P,2}(\alpha, \beta)$, as introduced in eq.(9.27), where correspondingly

$$\begin{aligned} d_{P,2}(\alpha, \beta) &= P(\alpha + \beta) = P(\alpha\beta') + P(\alpha'\beta) \\ &= 1 - P(\alpha\beta) - P(\alpha'\beta') = P(\alpha) + P(\beta) - 2P(\alpha\beta). \quad (14.8) \end{aligned}$$

In the case here, we simply have

$$\alpha + \beta = ab'c' \vee a'bc' \qquad (14.9)$$

and thus

$$d_{P,2}(\alpha, \beta) = P(ab'c') + P(a'bc'). \qquad (14.10)$$

It is obvious that it is impossible, to have both $P(\alpha\beta')$ and $P(\alpha'\beta)$ being relatively large (such as close to a half) with $P(\alpha + \beta)$ small. Thus, $d_{P,2}(\alpha, \beta)$ will be sufficiently small iff so are $P(\alpha'\beta)$ and $P(\alpha\beta')$, in which case we can then consider the combining of α and β. (See again Section 9.3.3.) Note also that the above equations show that $d_{P,1}(\alpha, \beta)$ and $d_{P,2}(\alpha, \beta)$ differ as to adding together $P(\alpha\beta')$ and $P(\alpha'\beta)$ as compared to subtracting them in absolute value, so that in general we have

$$d_{P,1}(\alpha, \beta) \leq d_{P,2}(\alpha, \beta) \quad \text{and} \quad d_{P,2}(\alpha, \beta) - d_{P,1}(\alpha, \beta) = 2\min(P(\alpha\beta'), P(\alpha'\beta))$$
$$(14.11)$$

with strict inequality holding iff

$$P(\alpha \leq \beta) = 1 \qquad \text{or} \qquad P(\beta \leq \alpha) = 1, \tag{14.12}$$

i.e.,

$$P(\alpha\beta') = 0 \qquad \text{or} \qquad P(\alpha'\beta) = 0. \tag{14.13}$$

In the example considered here, in general, eq.(14.13) does not hold and eq.(14.11) shows the difference between use of $d_{P,2}(\alpha, \beta)$ and $d_{P,2}(\alpha, \beta)$ is $2\min(P(ab'c'), P(a'bc'))$, which can be quite large.

14.3   Testing Hypotheses Under A Higher Order Probability Assumption On The Relative Atoms

First, some comments on actually implementing hypotheses testing and measurement of similarity here: One course of action for the degree of similarity is simply to use, as is, the computed value for $d_{P,2}(\alpha, \beta)$ in eq.(14.10). (Actually, $d_{P,2}(\alpha, \beta)$ directly measures the dissimilarity.) However, we cannot use this value by itself to test hypotheses say

$$H_0 : \alpha \neq \beta \qquad \text{vs.} \qquad H_1 : \alpha = \beta, \tag{14.14}$$

since we do not know how significant the probability value $d_{P,2}(\alpha, \beta)$ is compared to the totality of situations, where possibly probability measures and events other than P acting on α and β might be employed. In order to account for this, we could attempt to assume formally a "higher order probability distribution" jointly over the possible relevant choices of probability measures P and events α, β. Of course such a "distribution" in general is intractable; but one natural approach to this problem is first to note that for the two events in question α, β, considered now generically, the possible relative atomic events are $\alpha\beta$, $\alpha\beta'$, $\alpha'\beta$, $\alpha'\beta'$, with corresponding relative atomic event probabilities being $P(\alpha\beta)$, $P(\alpha\beta')$, $P(\alpha'\beta)$, $P(\alpha'\beta')$. Noting the obvious constraints

$$0 \leq P(\alpha\beta), P(\alpha\beta'), P(\alpha'\beta), P(\alpha'\beta') \leq 1,$$
$$1 = P(\alpha\beta) + P(\alpha\beta') + P(\alpha'\beta) + P(\alpha'\beta'), \tag{14.15}$$

and assuming that "all prior things can happen equally" (a form of the so-called "Lagrange assumption"), leads us to the following:

Supposition I for H_o: At a higher order, assume P and α, β are formally joint random variables so that they induce joint random variables $X = P(\alpha\beta)$, $Y = P(\alpha\beta')$, $Z = P\alpha'\beta)$ – and hence determine functionally, the remaining possible random variable $W = P(\alpha'\beta') = 1 - X - Y - Z$ – and X, Y, Z are jointly uniformly distributed over the simplex Q_o of their possible values, where

$$Q_o = \{(x, y, z) : 0 \leq x, y, z \leq 1, x + y + z \leq 1\}. \tag{14.16}$$

Thus, integrating over Q_o yields a value of $\mathrm{vol}(Q_o) = 1/6$, whence the joint pdf f of (X, Y, Z) at any outcomes (x, y, z) is given as

$$f(x, y, z) = \begin{cases} 6 & \text{if } x, y, z \in Q_o \\ 0 & \text{if not}(x, y, z \in Q_o). \end{cases} \tag{14.17}$$

Hence, integrating out the variable Z (for $0 \leq z \leq 1-x-y$) in eq.(14.17) readily yields the joint pdf f for joint marginal random variables (X, Y) – which, by symmetry, formally the same with appropriate substitutions for joint marginal random variables (X, Z) or (Y, Z) – at any outcomes x, y, z as:

$$f(x, y) = \begin{cases} 6(1 - x - y) & \text{if } 0 \leq x, y \leq 1, x + y \leq 1 \\ 0 & \text{if } x, y \text{ is otherwise.} \end{cases} \tag{14.18}$$

In turn, integrating out the y-term above (for $0 \leq y \leq 1 - x$) yields the pdf f for the marginal random variable X (formally the same with appropriate substitutions for marginal random variables Y and Z) as:

$$f(x) = \begin{cases} 3(1 - x)^2 & \text{if } 0 \leq x \leq 1 \\ 0 & \text{if } x \text{ is otherwise.} \end{cases} \tag{14.19}$$

with obviously corresponding cdf F over [0,1] being

$$F(x) = 1 - (1 - x)^3. \tag{14.20}$$

Thus, for example, if we choose a metric $d_{P,1}$, as in eq.(9.25), because it can now be expressed as the random variable

$$D_1(Y, Z) = |Y - Z|, \tag{14.21}$$

application of the standard transformation of probability technique to Supposition I (e.g. [13], Section 2.8), where first we let $S = Y - Z$, $Z = Z$, whence $Y = S + Z$, $Z = Z$, yields the possible non-zero joint pdf g of S and Z at any outcomes s, z as

$$g(s, z) = f(s + z, z) \left| \frac{\partial(y, z)}{\partial(s, z)} \right|, \tag{14.22}$$

for $0 \leq y, z \leq 1$, $y + z \leq 1$. In terms of s and z, this is equivalent to

$$0 \leq s + z, z \leq 1, \quad s + 2z \leq 1, \quad -1 \leq s \leq 1,$$

i.e.,

$$\max(-s, -s/2) \leq z \leq \min(1, 1 - s, (1 - s)/2), \qquad -1 \leq s \leq 1,$$

whence,

$$\text{if } -1 \leq s \leq 0, \quad \text{then } |s| \leq z \leq (1 + |s|)/2,$$
$$\text{if } 0 \leq s \leq 1, \quad \text{then } 0 \leq z \leq (1 - s)/2. \tag{14.23}$$

Also,

$$\left| \frac{\partial(y, z)}{\partial(s, z)} \right| = \left| \begin{array}{cc} \partial y/\partial s & \partial y/\partial z \\ \partial z/\partial s & \partial z/\partial z \end{array} \right| = \left| \begin{array}{cc} 1 & 1 \\ 0 & 1 \end{array} \right| = 1. \tag{14.24}$$

Then, eqs.(14.18), (14.22)-(14.24) yield for possible non-zero values of g,

$$g(s, z) = 6(1 - s - 2z), \text{ for } \left\{ \begin{array}{ll} |s| \leq z \leq (1 + |s|)/2, & \text{if } -1 \leq s \leq 0, \\ 0 \leq z \leq (1 - s)/2, & \text{if } 0 \leq s \leq 1. \end{array} \right. \tag{14.25}$$

In turn, eq.(14.25) shows that the marginal pdf g of S at any nontrivial producing outcome s is

$$g(s) = \left\{ \begin{array}{ll} \int_{|s|}^{(1+|s|)/2} 6(1 + |s| - 2z)dz, & \text{if } -1 \leq s \leq 0 \\ \int_{0}^{(1-s)/2} 6(1 - s - 2z)dz, & \text{if } 0 \leq s \leq 1 \end{array} \right.$$
$$= \frac{3}{2}(1 - s)^2, \qquad \text{for all } -1 \leq s \leq 1. \tag{14.26}$$

Finally, letting $T = |S|$, we easily see that the pdf h of T, using eq.(14.26) at any nontrivial outcome t, is simply

$$h(t) = g(-t) + g(t) = 3(1 - t)^2, \qquad 0 \leq t \leq 1, \tag{14.27}$$

with corresponding cdf

$$H(t) = 1 - (1 - t)^3, \qquad 0 \leq t \leq 1. \tag{14.28}$$

Combining eqs. (14.18) and (14.28) shows that:

Any marginal random variable X, Y, or Z , as well as $D_1(Y, Z)$, under H_0, is identically (but certainly not independently) distributed with common cdf F_1, given as

$$F_1(t) = 1 - (1 - t)^3, \qquad 0 \leq t \leq 1. \tag{14.29}$$

Next, eq.(9.27) shows that the random variable corresponding to the choice of metric $d_{P,2}$ is

$$D_2(Y, Z) = Y + Z. \tag{14.30}$$

Again, application of the standard transformation of probability technique to Supposition I, where now we let $S = Y + Z$, $Z = Z$, whence

$$Y = S - Z, \qquad Z = Z,$$

yields the possible non-zero joint pdf g of S and Z at any outcomes s, z as

$$g(s, z) = f(s - z, z)\left|\frac{\partial(y, z)}{\partial(s, z)}\right|. \tag{14.31}$$

for $0 \leq y, z \leq 1$, $y + z \leq 1$. In terms of s and z, this is equivalent to

$$0 \leq s - z, z \leq 1, \qquad 0 \leq s \leq 1,$$

i.e.,

$$0 \leq z \leq s \leq 1. \tag{14.32}$$

Also, here

$$\left|\frac{\partial(y, z)}{\partial(s, z)}\right| = \left|\begin{array}{cc} \partial y/\partial s & \partial y/\partial z \\ \partial z/\partial s & \partial z/\partial z \end{array}\right| = \left|\begin{array}{cc} 1 & -1 \\ 0 & 1 \end{array}\right| = 1. \tag{14.33}$$

Combining eqs.(14.18), (14.32), (14.33) shows for the possible non-zero values of g,

$$g(s, z) = 6(1 - s), \qquad \text{for } 0 \leq z \leq s \leq 1. \tag{14.34}$$

Hence, the marginal pdf g of S at any nontrivial outcome s is

$$g(s) = \int_0^s 6(1 - s)dz = 6(1 - s)s, \qquad \text{for } 0 \leq s \leq 1.$$

Hence, after integrating out, *the corresponding cdf F_2 of $D_2(Y, Z)$ under H_0 is*

$$F_2(t) = t^2(3 - 2t), \qquad \text{for } 0 \leq t \leq 1. \tag{14.35}$$

Returning back to testing hypotheses, following standard practice (such as in [13], Chapter 13), if we choose, e.g., $d_{P,2}$ we first compute numerically $d_{P,2}(\alpha, \beta)$ as in eq.(14.5). Then, for any choice of significance level γ, $0 < \gamma < 1$, we then:

accept H_0 (and reject H_1)
  iff observed/ computed value $d_{P,2}(\alpha, \beta) > C_\gamma$,
accept H_1(and reject H_0)
  iff observed/ computed value $d_{P,2}(\alpha, \beta) \leq C_\gamma$,   (14.36)

where threshold C_γ is determined by the type one error relation, using eq.(14.35), as

$$\begin{aligned} \gamma &= P(\text{reject } H_0 \text{ using } d_{P,2}(\alpha, \beta)|H_0 \text{ holds}) \\ &= P(D_2(Y, Z) \leq C_\gamma|H_o) = F_2(C_\gamma), \end{aligned} \tag{14.37}$$

whence explicitly,

$$C_\gamma = F_2^{-1}(\gamma). \tag{14.38}$$

Substituting eq.(14.38) back into eq.(14.36) shows the equivalent form for observation/computation $d_{P,2}(\alpha, \beta)$ which avoids requiring explicit knowledge of threshold C_γ to carry out the test:

$$\text{accept } H_0 \text{ (and reject } H_1) \text{ iff } d_{P,2}(\alpha, \beta) > F_2^{-1}(g)$$
$$\text{iff } F_2(d_{P,2}(\alpha, \beta)) > \gamma,$$
$$\text{accept } H_1 (\text{and reject } H_0) \text{ iff } d_{P,2}(\alpha, \beta) \leq F_2^{-1}(y)$$
$$\text{iff } F_2(d_{P,2}(\alpha, \beta)) \leq \gamma. \tag{14.39}$$

Furthermore, instead of first picking γ and determining (at least, theoretically) threshold C_γ, alternatively, we can simply compute $F_2(d_{P,2}(\alpha, \beta))$ as the basic significance level of the test in the following sense: For $d_{P,2}(\alpha, \beta)$ now fixed as a number – not a random variable – and considering any possible new situation where a new outcome value, say $(d_{P,2}(\alpha, \beta))_o$ were observed and computed from random variable $D_2(Y, Z)$, we could carry out the test of hypotheses with (old) $d_{P,2}(\alpha, \beta)$ replaced by (new) $(d_{P,2}(\alpha, \beta))_o$, in eq.(14.36), or equivalently, eq.(14.39), but now with γ pre-chosen as

$$\gamma = F_2(d_{P,2}(\alpha, \beta)) \quad \text{(old)}, \tag{14.40}$$

whence the corresponding threshold for the test becomes

$$C_\gamma = C_{F_2(d_{P,2}(\alpha,\beta))} = d_{P,2}(\alpha, \beta) \quad \text{(old)}, \tag{14.41}$$

Because F_2 is a cubic polynomial (eq.(14.35)), eq.(14.38) shows C_γ will have a complicated form, as a function of prechosen significance level γ, in general. On the other hand, the basic significance level and corresponding threshold is readily obtained as

$$\gamma = F_2(d_{P,2}(\alpha, \beta)) = (d_{P,2}(\alpha, \beta))^2 (3 - 2d_{P,2}(\alpha, \beta)), \quad C_\gamma = d_{P,2}(\alpha, \beta).$$

If we choose $d_{P,1}$ as the metric, then all of the above results on testing of hypotheses remain valid with the appropriate substitutions, using eq.(14.29) and $d_{P,1}(\alpha, \beta)$ in place of $d_{P,2}(\alpha, \beta)$. C_γ is readily explicitly obtainable for any choice of γ as

$$C_\gamma = F_1^{-1}(\gamma) = 1 - (1 - \gamma)^{1/3}. \tag{14.42}$$

In this case, the basic significance level and corresponding threshold are

$$\gamma = F_1(d_{P,1}(\alpha, \beta)) = 1 - (1 - d_{P,1}(\alpha, \beta))^3, \quad C_\gamma = d_{P,1}(\alpha, \beta). \tag{14.43}$$

14.4   Probability Distance Functions And PS Conditional Event Algebra

In this and the subsequent sections, we will investigate in much more depth the concept of distance, or degree of association, between events.

So far, we have considered two fundamental measures of association in the form of probabilistic distance functions, the naive probability distance $d_{P,1}$, which does not depend on conjunction probabilities, and the natural absolute probability distance $d_{P,2}$, which explicitly uses conjunction probabilities. The latter is seen also (eq.(9.27)) to be the probability evaluation of a boolean event. However, note that, though $d_{P,1}$ is a legitimate distance function, it is not such an evaluation. In actuality, it will be shown later (see Theorem 22, Section 14.7) that two additional probability distance functions exist as legitimate metrics in the sense of eq.(9.28). These are the probability evaluations of conditional events: the relative probability distance function $d_{P,3}$ and the negative form relative probability distance function $d_{P,4}$, where

$$
\begin{aligned}
d_{P,3}(\alpha,\beta) &= \frac{d_{P,2}(\alpha,\beta)}{P(\alpha \vee \beta)} = P(\alpha + \beta | \alpha \vee \beta) = \frac{P(\alpha+\beta)}{P(\alpha \vee \beta)} \\
&= \frac{P(\alpha) + P(\beta) - 2P(\alpha\beta)}{P(\alpha) + P(\beta) - P(\alpha\beta)} = \frac{P(\alpha'\beta) + P(\alpha\beta')}{P(\alpha'\beta) + P(\alpha\beta') + P(\alpha\beta)}, \\
&= (1 + [P(\alpha\beta)/(P(a'b) + P(ab'))])^{-1}, \quad\quad\quad (14.44)
\end{aligned}
$$

$$
\begin{aligned}
d_{P,4}(\alpha,\beta) &= \frac{d_{P,2}(\alpha,\beta)}{P(\alpha' \vee \beta')} = P(\alpha + \beta | \alpha' \vee \beta') = \frac{P(\alpha+\beta)}{P(\alpha' \vee \beta')} \\
&= \frac{P(\alpha) + P(\beta) - 2P(\alpha\beta)}{1 - P(\alpha\beta)} = \frac{P(\alpha'\beta) + P(\alpha\beta')}{1 - P(\alpha\beta)}. \quad (14.45)
\end{aligned}
$$

Remark. Using PS cea, we can represent $d_{P,3}(\alpha,\beta)$ and $d_{P,4}(\alpha,\beta)$ in terms of the double conditional operation and related operations, where for any $\alpha, \beta \in \mathcal{B}$, for any given probability space $(\Omega, \mathcal{B}, P)$, using PS operations (see Theorem 9, etc.)

$$
\begin{aligned}
(\alpha|\beta)\&(\beta|\alpha) &= (\alpha\beta|\alpha \vee \beta), \\
(\alpha|\beta)\&(\beta|\alpha)' &= (\alpha' \vee \beta'|\alpha \vee \beta) = (\alpha + \beta|\alpha \vee \beta), \quad\quad (14.46)
\end{aligned}
$$

whence

$$
P(((\alpha|\beta)\&(\beta|\alpha))') = 1 - P((\alpha|\beta)\&(\beta|\alpha)) = d_{P,3}(\alpha,\beta). \quad\quad (14.47)
$$

Also,

$$
((\alpha'|\beta')\&(\beta'|\alpha'))' = (\alpha|\beta') \vee (\beta|\alpha') = (\alpha' + \beta'|\alpha' \vee \beta') = (\alpha + \beta|\alpha' \vee \beta'),
$$

whence

$$
P(((\alpha'|\beta')\&(\beta'|\alpha'))') = 1 - P((\alpha'|\beta')\&(\beta'|\alpha') = d_{P,4}(\alpha,\beta). \quad\quad (14.48)
$$

Note also that since $a' + b' = a + b$,

$$
d_{P,4}(\alpha,\beta) = d_{P,3}(\alpha',\beta'). \quad\quad\quad (14.49)
$$

14.5   Numerical-Valued Metrics on a Probability Space

First, let us review some basic facts concerning distances or metrics in general. (The primary definition and a brief discussion were already presented in Section 9.4.) For background, see e.g. [1], [5], [10], [11], and various articles in the three volume work edited by Shisha [12], among many other sources.

Let $D \subseteq \mathbb{R}$ (real line) and $f : D \to \mathbb{R}$ be any function. Call f subadditive over $E \subseteq D$ iff

$$f(x + y) \leq f(x) + f(y) \quad \text{for all } x, y \in E \text{ such that } x + y \in E. \qquad (14.50)$$

Recall that for any function $f : D \to \mathbb{R}$, f is *concave* iff

$$f(sx + (1 - s)y) \geq sf(x) + (1 - s)f(y), \text{ for all } x, y \in D, \text{ all } s \in [0, 1]$$

iff, assuming f second order differentiable over D

$$d^2 f(x)/d^2 x \leq 0, \text{ all } x \in D. \qquad (14.51)$$

Theorem 83 *(See [10]-[12].) Basic sufficient conditions for subadditivity. Let $r > 0$ be any fixed real number or $r = \infty$ and let $f : [0, r) \to \mathbb{R}$ with $f(0) \geq 0$. Then*

(i) If f is concave, then $f(x)/x$ is a non-increasing function of x for all $0 < x < r$.

(ii) If f is such that $f(x)/x$ is a non-increasing function of x for all $0 < x < r$, then f is subadditive over $[0, r)$.

A prime example of a concave nondecreasing function f over $[0,2)$ with $f(j) = j$, $j = 0, 1$ – and hence subadditive function – is $f(x) = x^r$, for all $x \in [0, 2]$ and any fixed $r \in (0, 1]$. Another example is $f(x) = (1/\log(2))\log(1 + x)$, for $x \in [0, 2]$. Many other examples can be constructed by graphical means. See again [10]-[12] for other sufficient conditions for subadditivity. The next two theorems show the key role subadditivity plays in generating metrics for our needs:

Theorem 84 *(See [10]-[12].) Let $\mathcal{B}$ be any boolean algebra, $d : \mathcal{B}^2 \to [0, 1]$ any (pseudo-) metric, and $f : [0, 2] \to \mathbb{R}^+$ (non-negative real line) with $f(j) = j$, $j = 0, 1$. Then, if f is subadditive and nondecreasing, then the functional composition $f \circ d : \mathcal{B} \to [0, 1]$ is also a (pseudo-)metric.*

Proof. $f \circ d$ is symmetric (property (D1)) because d is. It is reflexive (property (D2)) because $0 = f(0) = f(d(a, a))$, for any $a \in \mathcal{B}$. Finally, the triangle inequality (property (D3)) also holds for $f \circ d$ because it holds for d and f is nondecreasing. $\square$

Later, we will show rigorously that all four probability distance functions $d_{P,j}$, $j = 1, 2, 3, 4$ are legitimate (pseudo-)metrics (satisfying axioms (D1), (D2), (D3)), but for the present we need the well-known fact that specifically, $d_{P,2}$ is

(see, e.g., Kappos [6], Chapter 1). Also, from now on, we omit the distinction between pseudo-metrics and metrics, apropos to the previous comments on replac-
ing each element $a \in \mathcal{B}$, with the corresponding probability space $(\Omega, \mathcal{B}, P)$, by
the equivalence class of elements containing a in the form $\{b \in \mathcal{B} : P(b+a) = 0\}$
and considering the resulting probability space.

Theorem 85 *Partial converse of Theorem 17 (new result). Let* $f : (0, 2] \to \mathbb{R}$
be nondecreasing. Then, if for all choices of probability spaces $(\Omega, \mathcal{B}, P)$ *and for
at least the metric* $d_{P,2}$, $f \circ d_{P,2}$ *is also a metric, then necessarily* $f(0) = 0$ *and
f is subadditive over* $[0,1]$ *for all* $x, y \in [0, 1]$ *with* $x + y \in [0, 1]$.

Proof. For any $(\Omega, \mathcal{B}, P)$ and any $a, b, c \in \mathcal{B}$, first note that for any $a \in \mathcal{B}$, by
the reflexivity of $f \circ d_{P,2}$ and that of $d_{P,2}$, for any $a \in \mathcal{B}$,

$$0 = f \circ d_{P,2}(a, a) = f(d_{P,2}(a, a)) = f(0). \tag{14.52}$$

Next, since $f \circ d_{P,2}$ obeys the triangle inequality, we have for any choice of
$a, b, c \in \mathcal{B}$,

$$f(P(a + b)) + f(P(b + c)) \geq f(P(a + c)). \tag{14.53}$$

Letting

$$x = P(ac'), \qquad y = P(a'c) \tag{14.54}$$

and choosing $b = ac$ produces from eq.(14.53),

$$f(x) + f(y) \geq f(x + y), \quad \text{for all } x, y \text{ such that relation } R(x, y) \text{ holds}, \tag{14.55}$$

where

$$R(x, y) \text{ holds iff [eq.(14.54) holds, for any } a, c \in \mathcal{B}, (\Omega, \mathcal{B}, P) \text{ arbitrary]}. \tag{14.56}$$

But, a straightforward argument shows that rhs(14.56) is the same as

$$\text{all } x, y \text{ such that } 0 \leq x, y \leq 1 \text{ and } x + y \leq 1. \tag{14.57}$$

Combining eqs.(14.57) with (14.55) shows

$$f(x) + f(y) \geq f(x + y), \text{ all } x, y \text{ such that } 0 \leq x, y \leq 1 \text{ and } x + y \leq 1. \tag{14.58}$$

Eq.(14.48) finally shows the desired result.   □

Remark. One basic way to generate a metric from two other metrics is to
average them: For any choice of s, $0 \leq s \leq 1$, and any metrics d_1, d_2, it is
easily verified that $sd_1 + (1 - s)d_2$ is also a metric. Another way to generate
a new metric $d_{(d_1, 0)}$ from a given metric d_1 is to truncate the metric d_1 in the
following way: Define $d_{(d_1, 0)}$, for any $a, b \in \mathcal{B}$ as

$$d_{(d_1, 0)}(a, b) = \begin{cases} 0 & \text{if } d_1(a, b) = 0 \\ 1 & \text{if } d_1(a, b) > 0 \end{cases} \tag{14.59}$$

Clearly, $d_{(d_1,0)}(a,b)$ is both reflexive and symmetric. The only way the triangle inequality could fail here is for

$$d_{(d_1,0)}(a,b) = d_{(d_1,0)}(b,c) = 0 \quad \text{and} \quad d_{(d_1,0)}(a,c) = 1. \tag{14.60}$$

But, eq.(14.60) is the same as

$$d_1(a,b) = d_1(b,c) = 0, \qquad d_1(a,o) > 0, \tag{14.61}$$

violating the triangle inequality for $d_1(a,c)$ which must be 0 from the lhs(14.60).

Note, that any attempt here to generalize the 0-truncation level for the definition in eq. (14.59) by replacing 0 by some real t, $0 < t < 1$, as in

$$d_{(d_1,t)}(a,b) = \begin{cases} 0 & \text{if } d_1(a,b) \leq t \\ 1 & \text{if } d_1(a,b) > t, \end{cases} \tag{14.62}$$

is doomed to failure as a metric because the triangle inequality will be violated in general, since it is possible to have

$$d_1(a,b), d_1(b,c) \leq t, \qquad t < d_1(a,c) \leq d_1(a,b) + d_1(b,c) \leq 2t, \tag{14.63}$$

corresponding to

$$d_{(d_1,t)}(a,b) = d_{(d_1,t)}(b,c) = 0, \qquad d_{(d_1,t)}(a,c) = 1. \tag{14.64}$$

14.6   Algebraic Metrics on a Probability Space

On the other hand, we will see below that it is possible to obtain a related class of truncated metrics through an algebraic analogue of eq.(14.58). Let us introduce some additional concepts which involve *new algebraic counterparts to the standard numerically-based concept of a metric* so far considered.

Let $\mathcal{B}_1, \mathcal{B}_2$ be any boolean algebras and $d : \mathcal{B}_1^2 \to \mathcal{B}_2$ such that the following algebraic analogue of the properties (D1),(D2),(D3) for eq.(21) hold for all $a, b, c \in \mathcal{B}_1$,

(AD1) $d(a,b) = d(b,a)$ symmetry,

(AD2) $d(a,a) = 0$ reflexivity,

(AD3) $d(a,b) \vee d(b,c) \geq d(a,c)$ triangle inequality.   (14.65)

Then, call such a d an (unconditional) *algebraic metric or algebraic distance function*. Denote by d_+ the boolean symmetric operation $+$ (the reason for using this apparently more cumbersome notation will become later self-evident). In Theorem 22, independent of Theorems 19-21, it will be shown that d_+ is an algebraic metric. If $\mathcal{B}_1 = \mathcal{B}_2$ and d is a boolean operation, call d a (unconditional) boolean algebraic metric. Next, let $f : \mathcal{B}_1^2 \to \mathcal{B}_2$ and $g : \mathcal{B}_1^2 \to \mathcal{B}_2$ be any two functions. Consider the measurable space part $((\Omega_2)_o, (\mathcal{B})_o)$ of the PS construction and the conditional event operator $(.|..)$ with range $\in (\mathcal{B}_2)_o$. Then, define $d_{f,g} : \mathcal{B}_1^2 \to (\mathcal{B}_2)_o$, where for any $a, b \in \mathcal{B}_1$,

$$d_{f,g}(a, b) = (f(a, b)|g(a, b)). \tag{14.66}$$

Then, if $d_{f,g}$ is an algebraic metric, call it a *conditional algebraic metric*. If also, $\mathcal{B}_1 = \mathcal{B}_2$ and both f and g are (binary) boolean operations over $\mathcal{B}_1$, call $d_{f,g}$ a *conditional boolean algebraic metric*. Furthermore, analogous to the numerical definition in eq.(14.50), call $f : \mathcal{B}_1 \to \mathcal{B}_2$ *algebraically subadditive* iff

$$f(a \vee b) \leq f(a) \vee f(b) \qquad \text{for all } a, b \in \mathcal{B}_1. \tag{14.67}$$

For convenience, in the ensuing development, we slightly abuse notation by use of the same symbols for corresponding boolean operations and relations for possibly different boolean algebras. The following algebraic analogue of Theorem 17 holds:

Theorem 86 *If $f : \mathcal{B}_2 \to \mathcal{B}_3$ is any algebraically subadditive function so that $f(\emptyset) = \emptyset$, f is nondecreasing with respect to the natural partial orders $\leq$ over $\mathcal{B}_2$ and $\leq$ over $\mathcal{B}_3$, and $d : \mathcal{B}_1 \to \mathcal{B}_2$ is any algebraic metric, then the functional composition $f \circ d : \mathcal{B}_1^2 \to \mathcal{B}_3$ is also an algebraic metric.*

Proof. Clearly, symmetry and reflexivity are immediate, analogous to the numerical counterpart. The triangle inequality (AD3) follows also easily using the nondecreasing property of f and the triangle inequality property for d.   □

The following result is the converse of Theorem 19 and is the algebraic counterpart of Theorem 18:

Theorem 87 *Let $f : \mathcal{B}_1 \to \mathcal{B}$ be any nondecreasing function with respect to the partial orders $\leq$ for $\mathcal{B}_1$ and $\mathcal{B}_2$. If, for at least the algebraic metric d_+, $f \circ d_+$ is also an algebraic metric, for all choices of boolean algebras $\mathcal{B}_1, \mathcal{B}_2$, then necessarily $f(\emptyset_1) = \emptyset_2$ and f is algebraically subadditive over $\mathcal{B}_1$.*

Proof. Since, $f \circ d_+$ is reflexive and d_+ is reflexive, then for any $a \in \mathcal{B}_1$,

$$\emptyset_2 = (f \circ d_+)(a, a) = f(d_+(a, a)) = f(\emptyset_1). \tag{14.68}$$

Next, choose any $a, b, c \in \mathcal{B}_1$ and using the triangle inequality property of $f \circ d_+$, obtain

$$f(a + b) \vee f(b + c) \geq f(a + c), \tag{14.69}$$

which with $b = ac$ yields

$$f(ac') \vee f(a'c) \geq f(ac' \vee a'c) \qquad \text{for all } a, c \in \mathcal{B}_1. \tag{14.70}$$

Next, note that for any $a, c \in \mathcal{B}_1$, we can write

$$a \vee c = a \vee a'c = a(a'c)(\vee a'(a'c). \tag{14.71}$$

Now, by the arbitrariness of $a, c \in \mathcal{B}_1$ in eq.(14.70), we can let a alone and replace c by $(a'c)$ to yield:

$$f(a(a'c)') \vee f(a'(a'c)) \geq f(a(a'c) \vee a'(a'c)) \qquad \text{all } a, c \in \mathcal{B}_1. \qquad (14.72)$$

But, using eq.(14.71) in eq.(14.72) yields

$$f(a) \vee f(a'c) \geq f(a \vee c) \qquad \text{for all } a, c \in \mathcal{B}_1 \qquad (14.73)$$

Finally, using the nondecreasing property of f,

$$f(a) \vee f(c) \geq f(a) \vee f(a'c) \geq f(a \vee c) \qquad \text{for all } a, c \in \mathcal{B}_1. \qquad (14.74)$$

Hence, eq.(14.67) shows f is algebraically subadditive.   □

Remark. Some examples of algebraic metrics and algebraic subadditivity.

Theorem 22 will show that the algebraic analogues of $d_{P,2}$, $d_{P,3}$, $d_{P,4} - d_+$, $d_{+,\vee}$, $d_{+,(.)'\vee(..)'}$, respectively – are boolean algebraic metrics; indeed the last two are also conditional boolean algebraic metrics, where for all $a, b \in \mathcal{B}$,

$$d_+(a, b) = a + b, \qquad (14.75)$$
$$d_{+,\vee}(a, b) = (a + b | a \vee b), \qquad (14.76)$$
$$d_{+,(.)'\vee(..)'}(a, b) = (a + b | a' \vee b'). \qquad (14.77)$$

The (non-boolean) algebraic analogue of the fractional exponent function $f(x) = x^r$ (see the discussion following Theorem 16), is for arbitrary real constant r, $0 < r < 1$, and any $C \in \mathcal{B}$,

$$C^r = \bigvee_{j=0}^{\infty} (c')^j \times c \times \theta(t_{j,r}) \in (\mathcal{B})_1 \qquad (14.78)$$

(see eq.(15.67)) with the compatibility property (eq.(15.72)) that

$$P_1(C^r) = (P(C))^r \qquad \text{for all } P, \qquad (14.79)$$

where for the constant-probability events $\theta(t_{j,r})$, the values $t_{j,r}$ are given in eq.(15.66). Consider next any $a, b \in \mathcal{B}$. Then, eq.(14.78) shows

$$
\begin{aligned}
(a \vee b)^r &= \bigvee_{j=0}^{\infty} (a'b')^j \times a \times \theta(t_{j,r}) \vee \bigvee_{j=0}^{\infty} (a'b')^j \times b \times \theta(t_{j,r}) \\
&\leq \bigvee_{j=0}^{\infty} (a')^j \times a \times \theta(t_{j,r}) \vee \bigvee_{j=0}^{\infty} (b')^j \times b \times \theta(t_{j,r}) \\
&= a^r \vee b^r. \qquad (14.80)
\end{aligned}
$$

Thus, analogous to the numerical counterpart x^r, the algebraic function C^r as a function of C, for r arbitrary fixed, $0 < r \leq 1$ (the case $C^1 = C$, is of course trivial to show) is algebraically subadditive. Hence, by Theorem 19, if

$d : \mathcal{B}_1^2 \to \mathcal{B}$ is an algebraic metric, so is the exponentiation $d^r : \mathcal{B}_1^2 \to (\mathcal{B}_2)_1$ (a necessarily non-boolean algebraic metric, in general).

Consider, next, the function $C^2 = C \times C$. In this case – analogous to the numerical counterpart – we do not have subadditivity, since for any $a, b \in \mathcal{B}$, in general,

$$(a \vee b)^2 = a^2 \vee b^2 \vee a \times b \vee b \times a > a^2 \vee b^2. \tag{14.81}$$

However, the cartesian sum function of identical arguments, defined for any $C \in \mathcal{B}$ as

$$((C')^2)' = C' \times C \vee C \times \Omega = C' \times C \vee C \times C' \vee C \times C, \tag{14.82}$$

yields for any $a, b \in \mathcal{B}$,

$$
\begin{aligned}
(((a \vee b)')^2)' &= a'b' \times (a \vee b) \vee (a \vee b) \times \Omega \\
&\leq a' \times a \vee a \times \Omega \vee b' \times b \vee b \times \Omega \\
&= ((a')^2)' \vee ((b')^2)',
\end{aligned} \tag{14.83}
$$

thereby establishing the algebraic subadditivity of the cartesian sum function of identical arguments. Hence, by Theorem 19, if d is an algebraic metric, so is $((d')^2)'$.

Next, consider the algebraic analogue of the failed attempt at establishing a t-level truncation metric from a given metric – see eq.(14.62). In the definition, replace the fixed numerical value t, $0 \leq t < 1$, by the algebraic analogue, any fixed event $c \in \mathcal{B}$, with $c < \Omega$ (c can possibly be $\emptyset$): For any given algebraic metric $d_1 : \mathcal{B}_1^2 \to \mathcal{B}$, define $d_{(d_1,c)} : \mathcal{B}_1^2 \to \mathcal{B}$ by, for any $a, b \in \mathcal{B}_1$,

$$d_{(d_1,c)}(a,b) = \begin{cases} \emptyset_2 & \text{if } d_1(a,b) \leq c \\ \Omega_2 & \text{if not}[d_1(a,b) \leq c] \end{cases} \tag{14.84}$$

Obviously, symmetry and reflexivity hold for $d_{(d_1,c)}$. For the triangle inequality to fail, we must have

$$d_{(d_1,c)}(a,b) = d_{(d_1,c)}(b,e) = \emptyset_2 \quad \text{and} \quad d_{(d_1,c)}(a,e) = \Omega_2. \tag{14.85}$$

But, eq.(14.85) is the same as

$$d_1(a,b) \leq c, \quad d_1(b,e) \leq c, \quad \text{not}[d_1(a,e) \leq c], \tag{14.86}$$

which is impossible since d_1 satisfies the triangle inequality and

$$d_1(a,e) \leq d_1(a,b) \vee d_1(b,e) \leq c \vee c = c,$$

contradicting eq.(14.86). Thus, $d_{(d_1,c)}$. is an algebraic metric, which can be called the *c-level truncated algebraic metric of d_1*. □

14.7   Development of Probability Distance Functions using Algebraic Metrics

The next result plays a crucial role in our analysis:

Theorem 88 *Fundamental connection between algebraic and numerical metrics.*

(i) Suppose that $(\Omega_2, \mathcal{D}_2, P_2)$ is any probability space and $d : \mathcal{B}_1^2 \to \mathcal{B}$ is an algebraic metric. Define $d_{P_2,d} : \mathcal{B}_1^2 \to [0,1]$ by the probability evaluation

$$d_{P_2,d}(a, b) = P_2(d(a, b)). \tag{14.87}$$

Then, $d_{P_2,d}$ is a metric over $\mathcal{B}_1$.

(ii) Let $d : \mathcal{B}_1^2 \to \mathcal{B}$ be a given function such that for all choices of probabilities P_2 over $\mathcal{B}_2$, $d_{P_2,d} : \mathcal{B}_1^2 \to [0,1]$ defined formally the same as in eq.(14.87) is a metric over $\mathcal{B}_1$. Then, d is an algebraic metric.

Proof. (i): Again, checking that $d_{P_2,d}$ is symmetric and reflexive is simple. The triangle inequality follows because, for any $a, b, c \in \mathcal{B}_1$,

$$d_{P_2,d}(a, b) + d_{P_2,d}(b, c) \geq P_2(d(a, b) \vee d(b, c)),$$

by basic property of probabilities with respect to disjunctions,

$$= P_2(d(a, c)) = d_{P_2,d}(a, c), \tag{14.88}$$

by the monotone property of probabilities and the fact that d obeys the (algebraic)triangle inequality.

(ii): To show this, first consider the following lemma:

Lemma 1. *For any boolean algebra $\mathcal{B}$ and $a, b, c \in \mathcal{B}$, the following two statements are equivalent:*

(I) $a \vee b \geq c$.
(II) $P(a) + P(b) \geq P(c)$, for all possible choices of probabilities P over $\mathcal{B}$.

Proof. (I) implies (II): If (I) holds, by the standard properties of probabilities, for any P,

$$P(a) + P(b) \geq P(a \vee b) \geq P(c), \tag{14.89}$$

and (II) holds.

(II) implies (I): Determine the collection of all relative atomic probabilities for (II):

$$P(a) = P(abc) + P(abc') + P(ab'c) + P(ab'c'),$$
$$P(b) = P(abc) + P(abc') + P(a'bc) + P(a'bc'),$$
$$P(c) = P(abc) + P(ab'c) + P(a'bc) + P(a'b'c). \tag{14.90}$$

Substituting eq.(14.90) into (II) and cancelling out terms yields

$$P(abc) + 2P(abc') + P(ab'c') + P(a'bc') \geq P(a'b'c), \quad \text{for all } P. \tag{14.91}$$

Suppose now

$$a'b'c > \emptyset. \tag{14.92}$$

Then, at least one of the four disjoint events abc, abc', $ab'c'$, $a'bc'$ must be $> \emptyset$, otherwise, the lhs(14.91) would be identically zero for all P, so that any positive assignment of probability to $a'b'c$ would contradict (II). Pick any one of the above four disjoint events which is not $\emptyset$ and call it α. Then, choosing P such that

$$P(a'b'c) = 3/4 \quad \text{and} \quad P(\alpha) = 1/4, \tag{14.93}$$

we have, for P as in eq.(14.93),

$$\text{lhs}(14.91) \leq 2(1/4) = 1/2 < 3/4 = \text{rhs}(14.91). \tag{14.94}$$

However, eq.(14.94) contradicts eq.(14.91). Thus, the assumption in eq.(14.92) cannot hold. That is, we must have $a'b'c = \emptyset$, i.e.,

$$a \vee b \geq c, \tag{14.95}$$

and (I) must hold. $\qquad\square$

Returning back to the proof of (ii), since $d_{P_2,d}(a, b)$ is a metric for all P, it must obey the triangle inequality for all P:

$$P_2(d(a, b)) + P_2(d(b, c)) \geq P_2(d(a, c)), \text{ for all } P_2 \text{ over } \mathcal{B}_2, \text{ all } a, b, c \in \mathcal{B}_2. \tag{14.96}$$

Applying Lemma 1 to eq.(14.96) shows

$$d(a, b) \vee d(b, c) \geq d(a, c) \quad \text{for all } a, b, c \in \mathcal{B}_2. \tag{14.97}$$

But, eq.(14.97) is precisely the (algebraic) triangle inequality property for d.

The verification that d is also symmetric follows by the application of a basic cancellation property of probability (see, e.g., [4], p. 48) , i.e.,

$$P(d(a, b) = P(d(b, a)) \quad \text{for all } P \text{ implies} \quad d(a, b) = d(b, a). \tag{14.98}$$

Reflexivity of d, is as usual, an immediate consequence of the hypothesis. $\qquad\square$

Apropos to Theorem 21, call $d_{P_2,d}$ the probability metric or probability distance function generated by the (corresponding) algebraic metric d.

Theorem 89 *The functions* d_+, $d_{+,\vee}$, $d_{+,(.)'\vee(..)'}$ *are each algebraic metrics and* $d_{P,2}$, $d_{P,3}$, $d_{P,4}$ *are all probability metrics generated, respectively, by* d_+, $d_{+,\vee}$, $d_{+,(.)'\vee(..)'}$,*for any choice of* P.

Proof. Clearly, again by inspection, symmetry and reflexivity holds for d_+, $d'_{+,\vee}$, $d_{+,(.)'\vee(..)'}$. We now verify the triangle inequality for all three:

d_+: For any $a, b, c \in \mathcal{B}$,

$$(a + b) \vee (b + c) = ab' \vee a'b \vee bc' \vee b'c$$
$$= ab'c \vee ab'c' \vee a'bc \vee a'bc' \vee abc' \vee a'bc' \vee ab'c \vee a'b'c, \tag{14.99}$$

$$a + c = ac' \vee a'c = abc' \vee ab'c' \vee a'bc \vee a'b'c. \qquad (14.100)$$

Thus, comparing eqs.(14.99) and (14.100), we see that

$$(a + b) \vee (b + c) \geq a + c, \qquad (14.101)$$

and the algebraic triangle inequality is established for d_+.

$d_{+,\vee}$: First, note that for any $a, b, c \in \mathcal{B}$ and possible $\emptyset$ valued antecedents of $d_{+,v}(a, b)$, $d_{+,\vee}(b, c)$, $d_{+,\vee}(a, c)$, corresponding to non-well-defined values for $d_{P,3}(a, b)$, $d_{P,3}(b, c)$, $d_{P,3}(a, c)$, we must have either $a \vee b = \emptyset$ or $b \vee c = \emptyset$ or $a \vee c = \emptyset$. If $a \vee b = \emptyset$, then $a = b = \emptyset$ and

$$d_{+,\vee}(a, b) \vee d_{+,\vee}(b, c) = (\emptyset|\emptyset) \vee (c|c) = \emptyset \vee (c|c) = (c|c) = d_{+,\vee}(a, c). \quad (14.102)$$

Similarly, if $b \vee c = \emptyset$, $b = c = \emptyset$ and

$$d_{+,\vee}(a, b) \vee d_{+,\vee}(b, c) = (a|a) = d_{+,\vee}(a, c). \qquad (14.103)$$

Finally, if $a \vee c = \emptyset$, then $a = c = \emptyset$ and

$$d_{+,\vee}(a, b) \vee d_{+,\vee}(b, c) = (b|b) \geq \emptyset = (\emptyset|\emptyset) = d_{+,\vee}(a, c). \qquad (14.104)$$

Hence, eqs.(14.102)-(14.104) show that the triangle inequality holds even for the possible non-well-defined case for $d_{P,3}(a, b)$, $d_{P,3}(b, c)$, $d_{P,3}(a, c)$. Thus, from now on in this proof, assume

$$a \vee b > \emptyset \quad \text{and} \quad b \vee c > \emptyset \quad \text{and} \quad a \vee c > \emptyset. \qquad (14.105)$$

Next, from standard boolean algebra manipulations

$$A = (a + b)'(a \vee b)(b \vee c)' = (ab \vee a'b')(a \vee b)(b'c') = \emptyset,$$
$$B = (a \vee b)'(b + c)'(b \vee c) = a'b'(bc \vee b'c')(b \vee c) = \emptyset. \qquad (14.106)$$

But, Theorem 9(ii) and eq.(14.106) show for all $a, b, c \in \mathcal{B}$, using eq.(14.106),

$$(a + b \mid a \vee b) \vee (b + c \mid b \vee c)$$
$$= [((a + b) \vee (b + c) \mid \Omega) \vee A \times (b + c \mid b \vee c)$$
$$\vee B \times (a + b \mid a \vee b)|(a \vee b) \vee (b \vee c)]$$
$$= [(a + b) \vee (b + c)|\Omega) \mid a \vee b \vee c]$$
$$= ((a + b) \vee (b + c) \mid a \vee b \vee c). \qquad (14.107)$$

Hence, eq.(14.107) shows

$$d_{+,\vee}(a, b) \vee d_{+,\vee}(b, c) \geq d_{+,\vee}(a, c)$$
$$\text{iff } ((a + b) \vee (b + c) \mid a \vee b \vee c) \geq (a + c \mid a \vee c). \qquad (14.108)$$

Next, apropos to eq.(14.108), we take care of all possible trivial and nontrivial cases for $((a + b) \vee (b + c)|a \vee b \vee c)$ and $(a + c|a \vee c)$. Using Theorem 2, this means we consider the possible cases where one or the other quantity is $\emptyset_o$ or Ω_o, but with eq.(14.105) holding.

Case 1. $((a+b) \vee (b+c) \mid a \vee b \vee c) = \Omega_o$ or $(a+c \mid a \vee c) = \emptyset$. This case obviously always yields

$$((a+b) \vee (b+c) \mid a \vee b \vee c) \geq (a+c \mid a \vee c). \qquad (14.109)$$

Case 2. $((a+b) \vee (b+c) \mid a \vee b \vee c) = \emptyset_o$. Then, as a disjunction from eq.(14.107), $(a+b \mid a \vee b) \vee (b+c \mid b \vee c) = \emptyset_o$, i.e.,

$$\emptyset_o = (a+b \mid a \vee b) = (b+c \mid b \vee c),$$

which by eq.(14.105) and Theorem 2 is equivalent to

$$a+b = b+c = \emptyset. \qquad (14.110)$$

Thus, $\emptyset = ab' = a'b = bc' = b'c$ implying, in turn, that

$$\emptyset = ab'c = ab'c' = a'bc = a'bc' = abc' = a'bc' = ab'c = a'b'c. \qquad (14.111)$$

Hence, using eq.(14.111),

$$a+c = ac' \vee a'c = abc' \vee ab'c' \vee a'bc \vee a'b'c = \emptyset, \qquad (14.112)$$

while

$$a \vee c = (a+c) \vee ac = abc' \vee a'b'c \vee abc. \qquad (14.113)$$

In any case, eq.(14.112) shows that

$$(a+c \mid a \vee c) = (\emptyset \mid a \vee c) = \emptyset_o. \qquad (14.114)$$

Thus, we have here from the hypothesis for Case 2 here and eq.(14.114)

$$((a+b) \vee (b+c) \mid a \vee b \vee c) = (a+c \mid a \vee c) = \emptyset_o. \qquad (14.115)$$

Case 3. $(a+c \mid a \vee c) = \Omega_o$. Thus, we must have $a+c = a \vee c$, i.e.,

$$ac = \emptyset. \qquad (14.116)$$

But, eq.(14.116) implies

$$abc = ab'c = \emptyset. \qquad (14.117)$$

Now,

$$(a+b) \vee (b+c) = ab' \vee a'b \vee bc' \vee b'c$$
$$= ab'c \vee ab'c' \vee a'bc \vee a'bc' \vee abc' \vee a'bc' \vee ab'c \vee a'b'c. \qquad (14.118)$$

Applying eq.(14.117) to eq.(14.118),

$$(a+b) \vee (b+c) = ab'c' \vee a'bc \vee a'bc' \vee abc' \vee a'b'c. \qquad (14.119)$$

On the other hand,

$$a \vee b \vee c = ab'c' \vee bc' \vee c$$
$$= ab'c' \vee abc' \vee a'bc' \vee ac \vee a'bc \vee a'b'c. \quad (14.120)$$

Applying eq.(14.117) to eq.(14.120) and comparing with eq.(14.119) shows

$$a \vee b \vee c = ab'c' \vee abc' \vee a'bc' \vee a'bc \vee a'b'c = (a+b) \vee (b+c). \quad (14.121)$$

Hence, eq.(14.105), the hypothesis of Case 2 and eqs.(14.119) and (14.121) show that for this case

$$((a+b) \vee (b+c) \mid a \vee b \vee c) = (a+c \mid a \vee c) = \Omega_o. \quad (14.122)$$

With all of the trivial cases for $((a+b) \vee (b+c) \mid a \vee b \vee c)$ and $(a+c \mid a \vee c)$ taken into account, we finally have apropos to rhs(14.108):

Case 4. Nontrivial case: $\emptyset < (a+b) \vee (b+c) < (a \vee b \vee c)$ and $\emptyset < a+c < a \vee c$. We can then apply the criterion in Theorem 5(ii):

$$((a+b) \vee (b+c) \mid a \vee b \vee c) \geq (a+c \mid a \vee c) \text{ iff}$$
$$[(a+b) \vee (b+c) \geq a+c \text{ and}$$
$$(a+c)((a \vee c) \geq ((a+b) \vee (b+c)')(a \vee b \vee c)]. \quad (14.123)$$

Now,

$$(a+c)'(a \vee c) \geq ((a+b) \vee (b+c))'(a \vee b \vee c)$$
$$\text{iff} \quad (ac \vee a'c')(a \vee c) \geq (a+b)'(b+c)'(a \vee b \vee c)$$
$$\text{iff} \quad ac \geq (ab \vee a'b')(bc \vee b'c')(a \vee b \vee c) = abc, \quad (14.124)$$

which is always true. Also, eq.(14.101) shows $(a+b) \vee (b+c) \geq a+c$. Thus, rhs(14.122) is true. Hence, lhs(14.123) must also be true. Thus,

$$((a+b) \vee (b+c) \mid a \vee b \vee c) \geq (a+c \mid a \vee c). \quad (14.125)$$

Hence, in summary, eq.(14.124) not only holds for Case 4, but for Case 1, by eq.(14.109); it holds for Case 2, by eq.(14.115); and it holds for Case 3, by eq.(14.122). Hence, for all possible cases, eq.(14.125) holds. Combining this with eq.(14.107) and the non-well-defined case up to eq.(14.105) shows

$$(a+b \mid a \vee b) \vee (b+c \mid b \vee c) \geq (a+c \mid a \vee c) \quad \text{for all } a, b, c \in \mathcal{B}, \quad (14.126)$$

and, via eq.(14.108), the triangle inequality is established for $d_{+,\vee}$.

$d_{+,(.)'\vee'}$: Simply use eq.(14.49) and the validity of the triangle inequality for $d_{+,\vee}$ as above in eq.(14.126), for all a, b, c, substituting a' for a, b' for b and c' for c.

Thus, the (algebraic) triangle inequality holds for d_+, $d_{+,\vee}$, $d_{+,(.)'\vee'}$ and hence, they are all algebraic metrics. Finally, apply Theorem 21 to this fact noting the definitions in eqs.(14.75)-(14.77). This yields the fact that $d_{P,2}$, $d_{P,3}$,

$d_{P,4}$ are the probability metrics generated, respectively, by d_+, $d_{+,\vee}$, $d_{+,(.)'\vee(..)}$ for any choice of P (well-defined). □

Remark. Thus, Theorems 21 and 22 fully justify the previous use of the expression "probability distance function" applied to $d_{P,2}$, $d_{P,3}$, $d_{P,4}$. In fact, it is clear that d_+, is a boolean algebraic metric while $d_{+,\vee}$, $d_{+,(.)'\vee'}$ are conditional boolean algebraic metrics. As stated before, $d_{P,1}$ is a legitimate metric (from the basic properties of the absolute value function), but it is apparently not a true probability metric, since no corresponding algebraic metric can be found, not dependent on the choice of the relevant probability measure used to evaluate it. However, when the form of the underlying algebraic metric is allowed to depend on the probability measure in some appropriate way, $d_{P,1}$ also becomes a modified type of probability metric. (For more details, see Section 16.4.)

Next, it is fundamental to inquire if any of the types of algebraic metric classes defined earlier can be characterized. The problem of determining all non-boolean unconditional or conditional algebraic metrics is an open one, as is the problem of characterizing the class of all metrics. However, the following shows that for both unconditional and conditional boolean algebraic metrics, a relatively simple characterization can be obtained:

Theorem 90 *Characterization of boolean algebraic metrics.*
(i) The only non-constant boolean algebraic (unconditional) metric over a given boolean algebra $\mathcal{B}$ is the one corresponding to the standard probability distance $d_{P,2}$, i.e., d_+.
(ii) The only non-constant conditional boolean algebraic metrics corresponding to a given boolean algebra $\mathcal{B}$ are $d_{+,\Omega}$, $d_{+,\vee}$, $d_{+,(.)'\vee'}$, where $d_{+,\Omega}$ is the conditional event imbedded form of d_+, i.e., for all $a, b \in \mathcal{B}$,

$$d_{+,\Omega}(a,b) = (d_+(a,b) \mid \Omega) = (a + b | \Omega). \qquad (14.127)$$

Proof. (i): There are only 16 possible binary boolean operations over $\mathcal{B}$. Table C shows which operations obey any of the requisite three algebraic metric properties in eq.(14.65). By inspection of this table, only the constant function $\emptyset$ and $a + b$, as functions of $a, b \in \mathcal{B}$, satisfy all three metric properties. Thus, $a + b$ is the only non-constant one.

(ii): If we let $d_{f,g}$ be any candidate conditional boolean algebraic metric where, without loss of generality we assume $f(a,b) \le g(a,b)$ for all $a, b, \in \mathcal{B}$, it follows from the very definition in eq.(14.66) that necessarily we must have both f and g being symmetric with f having the reflexive property. From Table C this leaves as possible candidates for f, $f(a,b) = a + b$ or $f(a,b) = \emptyset$ identically. The latter only yields $d_{f,g}(a,b) = \emptyset_o$ identically, and hence we must have $f(a,b) = a + b$. Thus, $g(a,b) \ge a + b$ and is symmetric. Table C shows the only possibilities for this (without even referring to the triangle inequality) are $g(a,b) = \Omega$, or $g(a,b) = a \vee b$ or $g(a,b) = a' \vee b'$, identically in all $a, b \in \mathcal{B}$. But, clearly, via eqs.(14.75)-(14.77), these choices correspond to the three basic conditional algebraic metrics. □

Remark. A step toward characterizing all possible algebraic metrics – and by Theorem 21, all possible probability metrics – is to re-examine the algebraic triangle inequality in eq.(14.65). For the conditional form, $d_{f,g}$, this becomes, assuming without loss of generality, $f \leq g$, for any such $f, g : \mathcal{B}_1 \to \mathcal{B}_2$ ($\mathcal{B}_1$, $\mathcal{B}_2$ boolean algebras),

$$(f(a,b) \mid g(a,b)) \vee (f(b,c) \mid g(b,c)) \geq (f(a,c)|g(a,c)) \quad \text{for all } a, b, c \in \mathcal{B}_1. \tag{14.128}$$

In turn, using the disjunctive enduction characterization in Theorem 14(ii), the above equation is the same as, for those a, b, c for which the nontriviality condition

$$\emptyset < f(a,b) < g(a,b), \quad \emptyset < f(b,c) < g(b,c), \quad \emptyset < f(a,c) < g(a,c) \tag{14.129}$$

holds:

$$(f(a,c)|g(a,c)) \leq (f(a,b)|g(a,b)) \text{ or } (f(a,c)|g(a,c)) \leq (f(b,c)|g(b,c)) \text{ or}$$
$$[\text{not}((f(a,c)|g(a,c)) \leq (f(a,b)|g(a,b)))$$
$$\text{and not}((f(a,c)|g(a,c)) \leq (f(b,c)|g(b,c)))$$
$$\text{and } ((f(a,b)|g(a,b)) \vee_{AC} (f(b,c)|g(b,c)) \geq (f(a,c)|g(a,c)))] . \tag{14.130}$$

Binary boolean f at any $a, b \in \mathcal{B}$	Symmetric?	Reflexive?	Triangle Inequality?	$f(a,b) \geq a + b$?
Ω	yes	no	yes	yes
ϕ	yes	yes	yes	no
a	no	no	yes	no
b	no	no	yes	no
a'	no	no	yes	no
b'	no	no	yes	no
ab	yes	no	no	no
ab'	no	yes	yes	no
$a'b$	no	yes	yes	no
$a'b'$	yes	no	no	no
$a \vee b$	yes	no	yes	yes
$b \Rightarrow a$	no	no	yes	no
$a \Rightarrow b$	no	no	yes	no
$a \Leftrightarrow b$	yes	no	no	no
$a + b$	yes	yes	yes	yes
$a' \vee b'$	yes	no	yes	yes

Table C. Algebraic metric-related properties for the
16 binary boolean operations.

We can apply the criterion in Theorem 5(ii) to determine when the following individual inequalities or their negations hold:

$$(f(a,c)|g(a,c)) \leq (f(a,b)|g(a,b)), \quad (f(a,c)|g(a,c)) \leq (f(b,c)|g(b,c)).$$

Similarly, we can apply this criterion to the determination of the bottom inequality in eq.(14.130), by using the following lemma which provides a somewhat simplified form:

Lemma 2. *For any boolean algebra $\mathcal{B}$ and $A, B, C, D, E, F \in \mathcal{B}$ satisfying the nontriviality conditions, $\emptyset < A < B, \emptyset < C < D, \emptyset < E < F$,*

$$(A|B) \vee_{AC} (C|D) \geq (E|F) \text{ iff}$$
$$[(A \vee C = B \vee D(\text{ in which case, } (A|B) \vee_{AC} (C|D) \geq (E|F)$$
$$\text{become the trivially true relation } \Omega_o \geq (E|F)))$$
$$\text{or } (B \vee D > A \vee C \geq E \text{ and } F \geq E \vee (B \vee D)A'C')]. \qquad (14.131)$$

Proof. By a logical decomposition, using Theorems 5(ii) and 2,

$$(A \vee C \mid B \vee D) \geq (E|F) \text{ iff}$$
$$[(A \vee C = B \vee D \text{ (in which case, } (A|B) \vee_{AC} (C|D) \geq (E|F)$$
$$\text{become the trivially true relation } \Omega_o \geq (E|F)))$$
$$\text{or } (A \vee C < B \vee D \text{ and } (A \vee C|B \vee D) \geq (E|F))]. \qquad (14.132)$$

In turn, bottom (14.132), by the criterion in Theorem 5(ii) Now being applicable, is the same as

$$A \vee C < B \vee D \text{ and } A \vee C \geq E \text{ and } B'D' \vee A \vee C \geq F' \vee E. \qquad (14.133)$$

But, the two right conjunctions in eq.(14.133) can be expressed equivalently as

$$A \vee C \geq E \quad \text{and} \quad B'D' \vee E \vee (A \vee C)E' \geq F' \vee E,$$

and since E appears disjointly on both sides of the inequality, it can be cancelled to yield the equivalent relation ,

$$A \vee C \geq E \quad \text{and} \quad B'D' \vee (A \vee C)E' \geq F'. \qquad (14.134)$$

Finally, take the equivalent complement form of the rhs(14.134) to yield

$$A \vee C \geq E \text{ and } F \geq (B \vee D)\&(E \vee A'C') = E \vee (B \vee D)A'C'. \qquad (14.135)$$

Substituting the equivalent form in eq.(14.135) for that in bottom(14.132) yields the desired result.   □

Finally, note that letting in Lemma 2, $A = f(a, b)$, $B = g(a, b)$, $C = f(b, c)$, $D = g(b, c)$, $E = f(a, c)$, $F = g(a, c)$, bottom(14.131) shows that part of the criterion for $d_{f,g}$ to be an algebraic metric implies that f must also be an algebraic metric (at least for all a, b, c satisfying the non-triviality criterion).

14.8   Additional Relations among the Basic Probability Distance Functions and Open Issues

Under Supposition I, note that for random variables $X = P(\alpha\beta)$, $Y = P(\alpha\beta')$, $Z = P(\alpha'\beta)$ above, the random variables correponding to $d_{P,3}(\alpha,\beta)$ and $d_{P,4}(\alpha,\beta)$, using eqs.(14.44), (14.45) are, respectively,

$$D_3(X,Y,Z) = (Y+Z)/((Y+Z+X) = 1/(1+(X/(Y+Z))),$$
$$D_4(X,Y,Z) = (Y+Z)/(1-X). \tag{14.136}$$

Again, by use of standard transformation of probability techniques – as in the derivation of the cdf's of $D_1(Y,Z)$ and $D_2(Y,Z)$ above leading to eqs.(14.29) and (14.35) – the cdf's F_j corresponding to $D_j(X,Y,Z)$, under H_0, for $j = 3,4$ are obtained as follows:

First, let $X = X$, $S = Y + Z$, $Z = Z$, whence

$$X = X, \qquad Y = S - Z, \qquad Z = Z,$$

yielding the possible non-zero joint pdf g of X,S,Z at any outcomes x,s,z as

$$g(x,s,z) = f(x,s-z,z)\left|\frac{\partial(x,y,z)}{\partial(x,s,z)}\right| \tag{14.137}$$

for $0 \leq x,y,z \leq 1$, $x+y+z \leq 1$, i.e.,

$$0 \leq x,z,s \leq 1, \qquad x+s \leq 1. \tag{14.138}$$

Also,

$$\left|\frac{\partial(x,y,z)}{\partial(x,s,z)}\right| = \begin{vmatrix} \partial x/\partial x & \partial x/\partial s & \partial x/\partial z \\ \partial y/\partial x & \partial y/\partial s & \partial y/\partial z \\ \partial z/\partial x & \partial z/\partial s & \partial z/\partial z \end{vmatrix} = \begin{vmatrix} 1 & 0 & 0 \\ 0 & 1 & -1 \\ 0 & 0 & 1 \end{vmatrix} = 1. \tag{14.139}$$

Combining eqs.(14.137)-(14.139), the possible non-zero value form for g is given by

$$g(x,s,z) = 6 \qquad \text{for } 0 \leq x,z,s \leq 1,\ x+s \leq 1,\ z \leq s, \tag{14.140}$$

whence, integrating out z in eq.(14.140), we obtain the possible non-zero joint pdf g of X and S as

$$g(x,s) = 6s \qquad \text{for } 0 \leq x,s \leq 1,\ x+s \leq 1. \tag{14.141}$$

Next, we let $T = X/S$, $S = S$, whence

$$X = ST, \qquad S = S,$$

yielding the possible non-zero joint pdf h of T and S as

$$h(t,s) = g(st,s) = \left|\frac{\partial(x,s)}{\partial(t,s)}\right| \tag{14.142}$$

for $0 \le x, s \le 1$, $x + s \le 1$, i.e.,

$$0 \le s \le 1/(1+t), \qquad 0 \le t. \tag{14.143}$$

Also, here

$$\left| \frac{\partial(x,s)}{\partial(t,s)} \right| = \left| \begin{array}{cc} \partial x/\partial t & \partial x/\partial s \\ \partial s/\partial t & \partial s/\partial s \end{array} \right| = \left| \begin{array}{cc} s & t \\ 0 & 1 \end{array} \right| = s. \tag{14.144}$$

Hence, combining eqs.(14.141)-(14.144), we obtain the possible non-zero h as

$$h(t,s) = 6s^2, \qquad \text{for } 0 \le s \le 1/(1+t), 0 \le t \le \infty. \tag{14.145}$$

Thus, the marginal pdf h of T is obtained by integrating out s in eq.(14.145) to yield

$$h(t) = 2/(1+t)^3, \qquad \text{for } 0 \le t \le \infty. \tag{14.146}$$

Next, we consider $W = 1/(1+T)$, i.e.,

$$T = (1/W) - 1,$$

whence by again transforming pdf's, the pdf k of W at any positive outcome w is

$$k(w) = h((1/w) - 1)|dt/dw| = (2/(1/w)^3)|-1/w^2| = 2w, \quad \text{for } 0 \le w \le 1. \tag{14.147}$$

Hence, the corresponding cdf to k in eq.(14.137) yields the desired *cdf F_3 of* $D_3(X,Y,Z)$ is

$$F_3(t) = t^2, \qquad \text{for } 0 \le t \le 1. \tag{14.148}$$

Next, consider the distribution of $D_4(X,Y,Z)$. Now from above, we know the joint distribution of X and $S = Y + Z$, given in eq.(14.141) through the pdf g. Thus, consider the transforms $R = 1 - X$, $S = S$, whence

$$X = 1 - R, \qquad S = S,$$

which yields the joint pdf f of random variables R, S at any non-zero producing outcomes r, s as

$$f(r,s) = g(1 - r, s) \left| \frac{\partial(x,s)}{\partial(r,s)} \right| = 6s \left| \begin{array}{cc} -1 & 0 \\ 0 & 1 \end{array} \right| = 6s, \quad \text{for } 0 \le s \le r \le 1. \tag{14.149}$$

Next, let $U = S/R$, $R = R$, i.e.,

$$S = RU, \qquad R = R.$$

Hence, the joint pdf h of R and U at any possible non-zero producing outcomes r, u is

$$h(r,u) = f(r,ru) \left| \begin{array}{cc} 1 & 0 \\ u & r \end{array} \right| = 6ur^2, \tag{14.150}$$

for $0 \le ru \le r \le 1$, i.e.,

$$0 \le u, r \le 1. \tag{14.151}$$

Thus, the marginal pdf h of U, from eqs. (14.150) and (14.151) is.

$$h(u) = \int_0^1 6ur^2 dr = 2u \qquad \text{for } 0 \le u < 1. \tag{14.152}$$

yielding finally the *cdf F_4 of $U = D_4(X,Y,Z)$ is*

$$F_4(t) = t^2, \qquad 0 \le t \le 1, \tag{14.153}$$

the same formally as F_3 in eq.(14.148) for $D_3(X,Y,Z)$.
Note also the easily shown relations between all four $d_{P,j}(\alpha, \beta)$:

$$0 \le d_{P,1}(\alpha, \beta) \le d_{P,2}(\alpha, \beta) \le \left\{ \begin{array}{c} d_{P,3}(\alpha, \beta) \\ d_{P,4}(\alpha, \beta) \end{array} \right. \le 1. \tag{14.154}$$

In light of the previous comments concerning the apparent superiority of $d_{P,2}$ over $d_{P,1}$ (see again Example 2), it is natural to inquire which basic desirable (or undesirable) properties the d_j possess. This leads to the table of properties given in Table 1, where it should be noted that the first three properties are desirable properties for a distance function to have, while the second three are negative ones. Thus, only $d_{P,3}$ possesses all positive and no negative properties of the six offered. Also, each of the four distances can be distinguished from the others by fulfilling or not the appropriate properties. In fact, based on these positive and negative properties, we can then rank order the $d_{P,j}$ in Table 2, with $d_{P,1}(\alpha, \beta)$ least preferred, followed by $d_{P,4}(\alpha, \beta)$ more preferred, followed by $d_{P,2}(\alpha, \beta)$, still more preferred, ending with $d_{P,3}(\alpha, \beta)$ as most preferred. Of course, other criteria could possibly be added to Table 1, changing the preference order in Table 2, but the list of six there appears as a reasonable start. We can also extend the concept of probability distance functions by weakening the requirements in eq.(9.28), in particular, by omitting the triangle inequality, calling all such functions similarity functions. In the same vein, we can ask the following research questions:

1. Characterize the class of similarity functions satisfying a particular property or properties, such as 3 or not 5.

2. What properties guarantee uniqueness ?

3. What other meaningful properties can be chosen to distinguish between candidate similarity or actual metrics ?

4. What are the connections between similarity and/ metrics as applied here and pattern recognition techniques ?

5. Determine characterizations for the class of all metrics; or certain subclasses such as all algebraic metrics or all conditional algebraic metrics, etc.

Possible Property of d_P	Specific d_P			
	$d_{P,1}$	$d_{P,2}$	$d_{P,3}$	$d_{P,4}$
1 $d_P(\alpha,\beta)$ is a (pseudo) metric	Y	Y	Y	Y
2 $d_P(\alpha,\beta)$ nontrivial function of $P(\alpha), P(\beta)$ and $P(\alpha\beta)$	N	Y	Y	Y
3 $d_P(\alpha,\beta)$ monotone decreasing in $P(\alpha\beta)$ for fixed $P(\alpha), P(\beta)$	N	Y	Y	Y
4 There exists α,β such that $P(\alpha), P(\beta)$ medium, $P(\alpha\beta)=0$, but $d_P(\alpha,\beta)$ small	Y	N	N	N
5 There exists α,β such that $P(\alpha), P(\beta)$ small, $P(\alpha\beta)=0$, but $d_P(\alpha,\beta)$ small	Y	Y	N	Y
6 There exists α,β such that $P(\alpha\beta')$, $P(\alpha'\beta)$ small, $P(\alpha\beta)$ large, but also $d_P(\alpha,\beta)$ large	N	N	N	Y

Table 1. Comparison of Properties of the Four Basic
Probability Distance Functions

PROBABILITY DISTANCE FUNCTIONS			
$d_{P,1}(\alpha,\beta)$	$d_{P,4}(\alpha,\beta)$	$d_{P,2}(\alpha,\beta)$	$d_{P,3}(\alpha,\beta)$
Y to pos.	Y to pos.	Y, to pos.	Y to pos.
property 1	properties 1,2,3	properties 1,2,3	properties 1,2,3
N to pos.	N to neg	N to neg.	N to neg.
properties 2,3,	property 4,	properties 4,6	properties 4,5,6
N to neg.	Y to neg.	Y to neg.	
property 6	properties 5,6	property 5	
Y to neg. properties 4,5			
LEAST PREFER	$\longrightarrow$	MOST PREFER	

Table 2. Preference ordering of Probability
Distance Functions Utilizing Table 1.

Bibliography

[1] Blumenthal, L. (1953) *Theory and Applications of Distance Geometry*, Oxford University Press, Oxford, U.K.

[2] Genest, C. and Zidek, J. V. (1986) Combining Probability Distributions: A Critique and an Annotated Bibliography, *Statistical Science*, 1, 114-148.

[3] Goodman, I. R. and Kramer, G. F. (1996) Applications of relational event algebra to the development of a decision aid in command and control, *Proceedings 1996 Command and Control Research and Technology Symposium*, Naval Postgraduate School, Monterey, CA, June 25-28, 1996, pp. 415-435.

[4] Goodman, I. R., Nguyen, H. T., and Walker, E. A. (1991) *Conditional Inference and Logic for Intelligent Systems*, North-Holland, Amsterdam.

[5] Hardy, G. H., Littlewood, J. E., and Polya, G. (1959) *Inequalities*, Cambridge University Press, Cambridge, U.K.

[6] Kappos, D. A. (1969) *Probability Algebras and Stochastic Spaces*, Academic Press, New York.

[7] Kullback, S. (1968) *Information Theory and Statistics*, Dover Publications, New York.

[8] National Research Council (1992) *Combining Information: Statistical Issues and Opportunities for Research*, Report by National Research Council National Academy Press, Washington, D.C.

[9] Rao, C. R. (1975) *Linear Statistical Inference and Its Applications*, 2nd Edition, New York.

[10] Rathore, S. S. (1965) On subadditive and superadditive functions, *American Mathematical Monthly*, **72**(2), 653-654.

[11] Saaty, T. L. (1981) *Modern Nonlinear Equations*, Dover Publications, New York, especially pp. 151-155.

[12] Shisha, O., Ed. (1967) *Inequalities, Volumes I, II, III*, Academic Press, New York.

[13] Wilks, S. S. (1963) *Mathematical Statistics*, Wiley, New York.

Chapter 15

Testing Hypotheses And Estimation Relative To Natural Language Descriptions

15.1 Motivating Example 4

In this case, two expert observers independently (or perhaps with some coordination or influence) provide their opinions concerning the same situation of interest: namely the description of an enemy ship relative to length and visible weaponry of a certain type. Suppose that Experts 1 and 2 have stated the following information which is obviously already intermingled with uncertainty:

$I_1 : \alpha =$ "Ship A is very long, or has quite a (i.e., between moderately

and very) large number of q-type weapons on deck",

$I_2 : b =$ "Ship A is extremely long, or has a moderate number

of q-type weapons on deck".   (15.1)

As in Example 1, we seek to determine the degree of similarity and test hypotheses that the two statements in eq.(15.1) are consistent with each other – or are at least sufficiently similar to be combined.

First, note that both sentences are "or" type or disjunctions, so that we can write symbolically

$$\alpha = a \text{ or } b, \qquad \beta = c \text{ or } d, \qquad (15.2)$$

where, dropping quotes,

$a =$ ship A is very long,

$b =$ ship A has quite a large number of q-type weapons on deck,

455

c = ship A is extremely long,

d = ship A has a moderate quantity of q-type weapons on deck (15.3)

In the absence of any clear-cut boolean events actually representing a, b, c, d and any a priori probability distributions measuring their uncertainties, one natural approach to the further evaluation of this problem is via fuzzy logic, whereby a, b, c, d now represent entire fuzzy set membership functions (or equivalently, fuzzy set membership functions evaluated at generic points) as follows:

$$a \text{ corresponds to } f_a : D_a \to [0, 1],$$
$$b \text{ corresponds to } f_b : D_b \to [0, 1],$$
$$c \text{ corresponds to } f_c : D_c \to [0, 1],$$
$$d \text{ corresponds to } f_d : D_d \to [0, 1], \qquad (15.4)$$

where $[0,1]$ is the unit interval. Here, D_a is the natural domain of values for a, say lengths in feet appropriately discretized for convenience, such as

$$D_a = \{60(ft), 60.5(ft), 1.0(ft), ..., 100.0(ft), 100.5(ft), ...300(ft)\}, \qquad (15.5)$$

60 ft and 300 ft representing absolute lower and upper bounds on possible ship lengths of relevance and $t(a)$, the uncertainty (or truth or possibility) of a – or perhaps more appropriately, the degree of compatibility of length(A) with respect to the attribute (or fuzzy set) "very long" – is provided by $f_a(\text{lngth}(A))$, where variable lngth(A), the true, but unknown length of A, can vary over D_a.

It is assumed that with the above interpretations, f_a can be numerically modeled via prior empirical considerations such as through interviews with qualified individuals concerning the common meaning of "very long" in the context of ships (or more restrictively, of ships of certain classes, etc.). One can utilize parameters, if necessary – analogous to the modeling of families of probability functions. Typically, f_a may represent a certain type of unimodal function, perhaps symmetric about 160(ft) where it generally achieves the highest value, unity, and decreasing in value away from 160. Unlike probability functions, fuzzy set membership functions in general may exceed one in sum, as is obviously the case here. (The fuzzy logic literature is replete with modeling techniques for fuzzy set membership functions. For some basic techniques see, e.g., [4], pp. 253-264.) Similar comments hold for b, c, d, with obviously differently shaped fuzzy set membership functions holding for b, c, d. Thus, briefly, we have the uncertainty evaluations

$$t(a) = f_a(\text{lngth}(A)), \qquad t(b) = f_b(no.q(A)),$$
$$t(c) = f_c(\text{lngth}(A)), \qquad t(d) = f_d(no.q(A)), \qquad (15.6)$$

with the obvious measurement abbreviation $no.q(A)$ for the (true) number of q-type weapons on board A, and where, as in the case of a – and clearly c – variable $no.q(A)$ can run over $D_b = D_d$, where for example, by prior knowledge

$$D_b = \{0, 1, 2, ..., 30\}. \qquad (15.7)$$

Note also, here $D_a = D_c$. (In the general case, when a, b, c, d do not refer to attributes over common domains, D_a, D_b, D_c, D_d are all distinct.)

Finally, a truth-functional fuzzy logic operator is chosen to interpret "or", i.e., disjunction, usually in the form of a cocopula or t-conorm (in general, a different concept from the former) such as max(imum) or probsum (explained in the next section). Dually, if an "and" or linguistic conjunction operation were present, one would seek to model it through a copula or a t-norm, such as min(imum) or prod(uct). If it is reasonable to assume the fuzzy set membership functions are "independent" in some reasonable sense, then one can use prod for the copula and probsum for the cocopula. On the other hand, a traditional pair of fuzzy logic operators in use when no such assumptions are made is (min, max). Both pairs of operators above are also t-norm, t-conorm pairs. Many other fuzzy logic conjunction-disjunction pairs can be chosen. (See the next Subsection for some further explanations).

15.2   Copulas, Cocopulas, and Fuzzy Logic Operators

In this section, we present a brief review of copulas, cocopulas, t-norms, and t-conorms. For more detailed background, see [3], [4], [7], [11], [12].

First, t-norms are certain associative operations $f : [0, 1]^m \to [0, 1]$, which for one argument, by convention coincide with the identity function, for two arguments are assumed to be associative, commutative, nondecreasing, with values for any zero or unity-valued argument reducing in a similar way as min or prod does, and which are extended to any finite number of arguments by a natural recursion. On the other hand, copulas are defined as operations denoted generically by cop: $[0, 1]^m \to [0, 1]$, which are the joint cdf of any joint collection of random variables which are marginally uniformly distributed over $[0,1]$, where we use the notation cop for both its m-argument form and for all of its k-argument forms, $1 \leq k \leq m$. Also, t-conorms are operators over $[0,1]$ having similar properties to t-norms, except that their values for any zero or unity valued arguments reduce in a similar way as max or probsum does (see below for the definition of the latter). Cocopulas, denoted generically as cocop, are each defined as the DeMorgan dual of a corresponding copula, where

$$\text{cocop}(t_1, ... t_m) = 1 - \text{cop}(1 - t_1, ..., 1 - t_m), \text{ for all } 0 \leq t_j \leq 1, j = 1, .., m. \tag{15.8}$$

It can be shown that a very useful class of t-conorms arises similarly by taking the DeMorgan dual of particular t-norms of interest. Examples of fuzzy logic operator pairs representing conjunction and disjunction via DeMorgan duals which are both t-norm, t-conorm pairs as well as copula, cocopula pairs are:

$$\text{(min, max),   (prod, probsum),   (maxsum, bndsum),} \tag{15.9}$$

where, for all $s, t \in [0, 1]$,

$$\text{probsum}(s, t) = 1 - ((1 - s)(1 - t)) = s + t - st, \tag{15.10}$$

$$\text{bndsum}(s,t) = min(s+t,1), \tag{15.11}$$

$$\text{maxsum(s,t)} = \max(s+t-1,0). \tag{15.12}$$

An important class of simultaneous DeMorgan t-norm, t-conorm pairs which are also copula, cocopula pairs that are conveniently parameterized which include the three pair of operators in eq.(15.9) is the modular, associative, archimedean copula class, a part of the family of operators Frank has investigated [5], [6]. In particular, the copula-t-norm part of this family has the form for $0 \leq s \leq \infty$ and all $t_j \in [0,1]$, $j = 1, ..., m$,

$$\text{cop}_s(t_1, ..., t_m) = \log_s[1 + ((s^{t_1}-1)...(s^{t_m}-1)/(s-1)^{m-1})], \tag{15.13}$$

with the limiting cases for $s = 0$, $s = 1$, $s = \infty$ being

$$\text{cop}_0 = \min, \quad \text{cop}_1 = \text{prod}, \text{cop}_\infty = \text{maxsum}. \tag{15.14}$$

Most important to our usage here, copulas furnish a natural measure of association among one-dimensional random variables or corresponding cdf's as shown in Sklar's copula theorem which we will clarify later. (See eq.(15.30) and for background, in particular, [11] and [12].)

Returning to Example 4, whatever the choice of the fuzzy logic disjunction operator, denoted symbolically as, say, $\bar{V}$- in complete binary functional form as $\bar{V}$: $[0,1]^2 \rightarrow [0,1]$ – as opposed to the un-barred symbol V for ordinary boolean disjunction, we have for the full uncertainty (or truth or possibility, etc.) evaluation of the two sentences in eq.(15.1):

$$I_1 = t(\alpha)(x,y) = \bar{V}(t(a)(x), t(b)(y)) = t(a)(x) \ \bar{V} \ \bar{V}t(b)(y),$$
$$I_2 = t(t(b)(y),$$
$$I_2 = t(b)(x,y) = \bar{V}(t(c)(x), t(d)(y)) = t(c)(x) \ \bar{V}t(b)(y),$$
$$I_2 = t(\bar{V} \ \bar{V}t(b)(y),$$
$$I_2 = t(t(d)(y), \tag{15.15}$$

employing standard binary operator notation on the right-hand sides of eq.(15.10), where $t(a)$, $t(b)$, $t(c)$, $t(d)$ are given in eq.(15.6) and where we identify variables $x = \text{lngth}(A) \in D_a$, $y = no.q(A) \in D_b$. Hence, the truth evaluations of the two expert opinions are two functions $t(a)$, $t(b)$ – in fact, also (compound) fuzzy set membership functions – in function space $F = [0,1]^{D_a \times D_b}$.

15.3   Numerically-Based Measures of Similarity and Metrics for the Problem

At this point, a number of standard real analysis and fuzzy logic and procedures are available to test whether functions $t(\alpha)$ and $t(\beta)$ (of variables x, y) are sufficiently close to be considered consistent. From a standard function space viewpoint, one simply chooses a distance or norm d over the entire function space $F = [0,1]^{D_a \times D_b}$ such as the well-established supremum norm or L^2-norm, given respectively as

$$d_1(t(\alpha), t(\beta)) = \sup_{x \in D_a, y \in D_b} (|t(\alpha)(x,y) - t(\beta)(x,y)|),$$

$$d_2(t(\alpha), t(\beta)) = \left(\sum_{x \in D_a, y \in D_b} [t(\alpha)(x,y) - t(\beta)(x,y)]^2 \right)^{1/2}, \quad (15.16)$$

where, as before, we identify variables $x = \text{lngth}(A)$, $y = no.q(A)$. (See any standard text such as [10] for background.) On the other hand, to take into account the specific form of F as a space of fuzzy membership functions, researchers in the field have proposed a number of other measures of closeness or similarity of fuzzy set membership functions. For example, a basic similarity measure proposed by Zadeh and motivated by his extension principle (see [13] for more details) is to form the fuzzy set membership function $t(\gamma) : [0,1] \to [0,1]$, where for any $r \in [0,1]$,

$$t(\gamma)(r) = \sup[t(\alpha)((t(\beta))^{-1}(r))] = \sup_{\{x \in D_a, y \in D_b : t(\beta)(x,y) = r\}} (t(a)(x,y)). \quad (15.17)$$

Obviously, in general, when α and β are interchanged, $t(\gamma)$ changes. Measures of similarity which are invariant to interchanging α and β have also been proposed. (See, e.g., [4], especially pp. 38-40 for more details of a basic fuzzification of distance between fuzzy sets.)

In any case, all of the above measures of closeness integrate out, in a sense, the argument variables x, y. On the other hand, it is also *natural to consider measures of similarity for individual relevant argument values*. In this case, a basic measure of similarity, sensitive to any given value of x, y, is simply the absolute difference

$$t(\delta)(x,y) = |t(\alpha)(x,y) - t(\beta)(x,y)|. \quad (15.18)$$

Eq.(15.18) is of course analogous to the absolute probability distance $d_{P,1}$ in Example 3, given in eq.(9.25), which we have seen is inferior to the choice of the natural probability distance $d_{P,2}$. But, if such a metric is to be used, we must be able to model the algebraic underpinnings of Example 3. If not, recall the basic Fréchet-Hailperin inequalities considered in Section 10.2 applied there to the situation where we compared the explicit use of conditional events and their operations through some choice of a cea vs. no application of cea, only external use of the individual conditional probabilities. Thus, if we can model Example 4 through a reasonable choice of (relational or conditional) events, we can potentially improve significantly by utilizing metrics such as $d_{P,2}, d_{P,3}$, or $d_{P,4}$, which nontrivially use the distance between events in an interactive way, i.e., which require explicit knowledge of the conjunction (or disjunction) of the events in question – as opposed to the non-interactive $d_{P,1}$, or other metrics which use only the marginal, and hence, purely numerical, probability evaluation of the events separately. (In particular, for the case of the use of $d_{P,2}$, see again eq.(10.18).)

15.4   One-Point Random Set Coverage Representations of Fuzzy Sets and Fuzzy Logic

It was stated earlier that a basic relation existed between certain random sets and fuzzy sets and this extended to parts of fuzzy logic. As we have seen in other sections of this monograph, random sets provide a powerful tool for analyzing information, since they involve the concept of not just a single point varying randomly, but entire sets (although when these sets become singletons, they reduce to ordinary random variables). Though all probability concepts are completely preserved – in particular, one and only one outcome set of a random set can occur at a time – random set outcomes need only be distinct, not disjoint, and hence overlaps can now be incorporated into modeling. The connection between such possible overlaps and fuzzy set membership functions is spelled out as follows: In brief, we can show (see, e.g., [8], Section 4) that there exist probability spaces $(\Omega, \mathcal{B}, P)$ with associated marginal random sets $S(a) : \Omega \to P(D_a)$, $S(b) : \Omega \to P(D_b)$, $S(c) : \Omega \to P(D_c)$, $S(d) : \Omega \to P(D_d)$, i.e., for all outcomes

$$S(a) \subseteq D_a, \quad S(b) \subseteq D_b, \quad S(c) \subseteq D_c, \quad S(d) \subseteq D_d, \tag{15.19}$$

recalling $D_a = D_c$, $D_b = D_d$ such that the *one-point coverage relations* hold:

$$t(a)(x) = P(x \in S(a)), \quad t(b)(y) = P(y \in S(b)),$$
$$t(c)(x) = P(x \in S(c)), \quad t(d)(y) = P(y \in S(d)), \tag{15.20}$$

for all $x \in D_a$, $y \in D_b$. We must also consider the *joint behavior* of such one-point coverage type random sets, both with respect to each pair $(S(a), S(b))$, corresponding to statement α and $(S(c), S(d))$, corresponding to statement β in eq.(15.2), as well as to both pairs jointly, i.e., the overall joint behavior of $(S(a), S(b), S(c), S(d))$. At times, in order to avoid both ambiguity and identifying the behavior of one pair with the other, we will add the appropriate subscript 1 or 2 to S.

One particular choice of S is provided through the nested random set (or, alternatively, called the level random set) form. In that case, a fixed random variable U is first considered which is uniformly distributed over [0,1] and for any choice of attribute or fuzzy set q corresponding to fuzzy set membership function, say, $f_q : D_q \to [0, 1]$, one lets

$$S(q) = f_q^{-1}[U, 1] = \{z \in D_q : f_q(z) \geq U\} \tag{15.21}$$

for $q = a, b, c, d$. Or, alternatively, one could replace the common use of U for all four quantities by choosing U_1, U_2 statistically independent, identically distributed as U , with U_1 used to produce both $f_a^{-1}[U_1, 1]$ and $f_b^{-1}[U_1, 1]$, while U_2 is employed for $f_c^{-1}[U_2, 1]$ and $f_d^{-1}[U_2, 1]$. More generally, the entire solution set of possible S here is actually determined by first picking an arbitrary function in a class of copulas and, dually, cocopulas. Then, one can construct in a straightforward way random sets corresponding to joint distribution functions of

0-1 valued random variables indexed by $D_a \times D_b$ whose hitting probabilities (i.e., probabilities of taking on unit values) match the fuzzy set membership functions. It can be shown that the choice of U above corresponds to picking the copula min, while the use of independent U_1 and U_2 correspond to a different copula. In all of the above, it is clear that the one-point coverage relations are somewhat analogous to the weak specification of a random variable through a measure of its central tendency, such as its expectation. Here, the one point coverage function of a random set only determines the random set's local behavior. It is unfortunate that even today much controversy and antagonism exists between the fuzzy logic and probability communities because of the failure of both to recall the now well-documented fundamental relations described above between the core of both fields. (See, e.g., two recent rounds of polemics on this topic in well-known journals [1], [9].)

Returning to the general situation, the underlying boolean events here corresponding to fuzzy set membership functions are

$$(x \in S(a)), (y \in S(b)), (x \in S(c)), (y \in S(d)), x \in D_a, y \in D_b, \qquad (15.22)$$

noting that the x, y values vary nonrandomly – only the quantities $S()$ are random here. Alternatively, and equivalently, these events can be expressed in inverse functional notation as

$$(S(a))^{-1}(F_x), (S(b))^{-1}(F_y), (S(c))^{-1}(F_x), (S(d))^{-1}(F_y), \qquad (15.23)$$

$x \in D_a$, $y \in D_b$, respectively, all lying in boolean or sigma-algebra $\mathcal{B}$, where F_x and F_y are the filter classes on x in D_a and on y in D_b, respectively. For example,

$$F_x = \{C \subseteq D_a : x \in C\}. \qquad (15.24)$$

Next, consider any one-point coverage event such as $(x \in S(a))$ and its complement or negation

$$(x \in S(a))' = (x \notin S(a)) = (x \in (S(a))') = Fx^{-1}((S(a))'), \qquad (15.25)$$

etc. At this point, it is convenient to introduce the alternative notation

$$(x \in_1 S(a)) \text{ for } (x \in S(a)), \quad (x \in_0 S(a)) \text{ for } (x \notin S(a)), \qquad (15.26)$$

with similar notation holding relative to the coverages or non-coverages for $S(c)$ and $S(d)$. Clearly, the pair of events $((x \in S(a)), (x \in (S(a))')$ is bijective to the pair of 0-1-valued events $(\phi(S(a))(x) = 1, \phi(S(a))(x) = 0)$, where ϕ indicates the standard set membership functional. That is, the marginal random quantity one-point coverage - non-one-point coverage, $(x \in (\notin)S(a))$, bijectively corresponds to the marginal 0-1 -valued random variable $\phi(S(a))(x)$, for all x in D_a, with the compatibility conditions

$$P(x \in S(a)) = P(\phi(S(a))(x) = 1) = t(a)(x),$$
$$P(x \notin S(a)) = P(\phi(S(a))(x) = 0) = 1 - t(a)(x), \qquad (15.27)$$

for all $x \in D_a$. In the same sense as above, the remaining three one-point coverage events $(y \in S(b))$, $(x \in S(c))$, $(y \in S(d))$, and their negations, are bijective to the corresponding marginal 0-1-valued random variables $\phi(S(b))(y)$, $\phi(S(c))(x)$, $\phi(S(d))(y)$, respectively. Furthermore, when we consider the joint behavior within the pair of events $(x \in S(a))$ vs. $(x \notin S(a))$ in conjunction with events $(y \in S(c))$ vs. $(y \notin S(a))$, there is a bijection with the pair of 0-1-valued random variables $(\phi(S(a))(x), \phi(S(b))(y))$ acting jointly, so that

$$P((x \in_i S(a)) \& (y \in_j S(b))) = P((\phi(S(a))(x) = i) \& (\phi(S(a))(x) = j)), \quad (15.28)$$

for all $i, j \in \{0, 1\}$, all $x \in D_a$, $y \in D_b$.

Similar remarks hold for the joint behavior within the pair $((x \in_i S(c)), (y \in_j S(d)))$ with respect to joint 0-1-valued random variables $(\phi(S(c))(x), \phi(S(d))(y))$, for all $x \in D_c$, $y \in D_d$. Finally, the overall joint behavior of one-point coverages vs. non-one-point coverages with respect to $S(a)$, $S(b)$, $S(c)$, $S(d)$ is bijective with respect to the overall joint behavior of 0-1-valued random variables $(\phi(S(a))(x), \phi(S(b))(y), \phi(S(c))(x), \phi(S(d))(y))$, with the basic relation holding:

$$P\left[((x \in_{j_1} S(a)) \& (y \in_{j_2} S(b)))(x \in_{j_3} S(a)) \& (y \in_{j_4} S(b))\right]$$
$$= P\left[(\phi(S(a))(x) = j_1) \& (\phi(S(a))(x) = j_2)\right.$$
$$\left. \& (\phi(S(a))(x) = j_3) \& (\phi(S(a))(x) = j_4)\right], \quad (15.29)$$

for any $j_1, j_2, j_3, j_4 \in \{0, 1\}$ and all $x \in D_a$, $y \in D_c$ (recalling again, $D_a = D_c$, $D_b = D_d$).

Next, Sklar's copula theorem [11], [12], shows that the joint cdf of any collection of one-dimensional random variables, say $X_1, .., X_m$, is determined by the evaluation of an appropriately determined copula acting compositionally on the set of (marginal) cdf's of $X_1, ..., X_m$. Conversely, given the (marginal) cdf of each random variable X_j, any choice of copula acting as a composition on the entire set such cdf's produces a legitimate joint cdf corresponding to the entire set of random variables in the sense that each of its one-dimensional marginal cdf's matches the given cdf of the corresponding random variable. Thus, using obvious notation for the joint and marginal cdf's, for any choice of copula cop, and any outcomes x_j of X_j ($x_j \in \{0, 1\}$),

$$F(x_1, ..., x_m) = \text{cop}(F_1(x_1), ..., F_m(x_m)). \quad (15.30)$$

Furthermore, when each of the X_j are 0-1-valued random variables, using standard probability expansions, the probability for any possible joint 0-1-event outcome can be expressed as an appropriate sum and difference of probabilities of joint 0-outcome events of various differing numbers of arguments (necessarily less than m). For example, for $m = 2$, the joint (and marginal) 0-event outcomes are $P(X_1 = 0)$, $P(X_2 = 0)$, $P((X_1 = 0) \& (X_2 = 0))$ and the probabilities of the remaining non-0-outcome events are

$$P((X_1 = 1) \& (X_2 = 1)) = 1 - P(X_1 = 0) - P(X_2 = 0)$$

$$+P((X_1 = 0)\&(X_2 = 0)),$$
$$P((X_1 = 1)\&(X_2 = 0)) = P(X_2 = 0) - P((X_1 = 0)\&(X_2 = 0)),$$
$$P((X_1 = 0)\&(X_2 = 1)) = P(X_1 = 0) - P((X_1 = 0)\&(X_2 = 0)). \quad (15.31)$$

Moreover, the probability evaluation of any such joint 0-valued event coincides with a corresponding cdf evaluation (but not so for other non-0 joint events). Thus, one can use Sklar's copula theorem to show for cop determining the joint distribution of the X_j,

$$
\begin{aligned}
P\left(\&_{j \in J}(X_j = 0)\right) &= P\left(\&_{j \in J}(X_j \leq 0)\right) = \mathrm{cop}((P(X_j \leq 0))_{j \in J}) \\
&= \mathrm{cop}\,((P(X_j = 0))_{j \in J}) \quad (15.32)
\end{aligned}
$$

for any $\emptyset \neq J \subseteq \{1, ..., m\}$. Note that eq.(15.32) also shows a simple evaluation for the disjunctive probability

$$
\begin{aligned}
P\left(\bigvee_{j \in J}(X_j = 1)\right) &= 1 - P\left(\&_{j \in J}(X_j = 0)\right) \\
&= 1 - \mathrm{cop}((P(X_j = 0))_{j \in J}) = \mathrm{cocop}((P(X_j = 1))_{j \in J}). \quad (15.33)
\end{aligned}
$$

Further, the evaluations in eq.(15.31) can now be analogously expressed in terms of 1-values as

$$
\begin{aligned}
P((X_1 = 1)\&(X_2 = 1)) &= 1 - P(X_1 = 0) - P(X_2 = 0) \\
&\quad + \mathrm{cop}(P(X_1 = 0)), P(X_2 = 0)) \\
&= P(X_1 = 1) + P(X_2 = 1) - \mathrm{cocop}(P(X_1 = 1), P(X_2 = 1)), \\
P((X_1 = 1)\&(X_2 = 0)) &= P(X_2 = 0) - \mathrm{cop}(P(X_1 = 0), P(X_2 = 0)) \\
&= \mathrm{cocop}(P(X_1 = 1), P(X_2 = 1)) - P(X_2 = 1), \\
P((X_1 = 0)\&(X_2 = 1)) &= P(X_1 = 0) - \mathrm{cop}(P(X_1 = 0), P(X_2 = 0)) \\
&= \mathrm{cocop}(P(X_1 = 1), P(X_2 = 1) - P(X_1 = 1)). \quad (15.34)
\end{aligned}
$$

Hence, the joint behavior - in the form of the joint distribution - of any finite collection of 0-1-valued random variables is completely determined by the choice of a copula evaluated at all possible combinations of marginal probabilities of 0-outcome events. Dually, as seen in eq.(15.34), this can be rewritten as a choice of cocopula evaluated at all possible combinations of 1-outcome events. Thus, in the above example for $m = 2$, the values $P(X_1 = 0)$, $P(X_2 = 0)$, $\mathrm{cop}(P(X_1 = 0), P(X_2 = 0))$ completely specify the joint cdf of X_1, X_2, where cop can be chosen arbitrarily (or restricted appropriately, where desired) within the class of all copulas; dually, the values $P(X_1 = 1)$, $P(X_2 = 1)$, $\mathrm{cocop}(P(X_1 = 1), P(X_2 = 1))$ also completely determine the joint cdf of X_1, X_2. All of this is generalizable to more than two arguments. In particular, taking into account the results on the bijection between any joint collection of one-point coverage type random sets and the corresponding joint collection of 0-1-valued random variables given in eqs.(15.21)-(15.23), it follows that for the case of D_a and

D_c finite, any possible collection of joint random sets $S(a)$, $S(b)$, $S(c)$, $S(d)$ is uniquely determined – up to induced probability measure, as usual – by simply any choice of copula cop acting on all combinations of the probability for each of their marginal 0-outcome events

$$P(\phi(S(a))(x) = 0) = P(x \notin S(a)) = 1 - t(a)(x),$$
$$P(\phi(S(b))(y) = 0) = P(y \notin S(b)) = 1 - t(b)(y),$$
$$P(\phi(S(c))(x) = 0) = P(x \notin S(c)) = 1 - t(c)(x),$$
$$P(\phi(S(d))(y) = 0) = P(y \notin S(d)) = 1 - t(d)(y), \qquad (15.35)$$

over all possible $x \in D_a$, $y \in D_c$, or equivalently, any choice of cocopula acting upon all combination of the given fuzzy set membership functions $t(a)$, $t(b)$, $t(c)$, $t(d)$ at any arguments $x \in D_a$, $y \in D_c$:

$$P(\phi(S(a))(x) = 1) = P(x \in S(a)) = t(a)(x),$$
$$P(\phi(S(b))(y) = 1) = P(y \in S(b)) = t(b)(y),$$
$$P(\phi(S(c))(x) = 1) = P(x \in S(c)) = t(c)(x),$$
$$P(\phi(S(d))(y) = 1) = P(y \in S(d)) = t(d)(y). \qquad (15.36)$$

As cop or cocop is made to vary arbitrarily - with the above fuzzy set membership functions held fixed – the corresponding joint one-point coverage random sets $S(a)$, $S(b)$, $S(c)$, $S(d)$ vary over all possible solutions . (See again [8], Section 4 for more details.)

Combining eqs.(15.15), (15.20) and (15.32), using the definitions

$$X_1 = \phi(S(a))(x), \qquad X_2 = \phi(S(b))(y),$$
$$X_3 = \phi(S(c))(x), \qquad X_4 = \phi(S(d))(y), \qquad (15.37)$$

for, at first, $J = \{1, 2\}$, then $J = \{3, 4\}$, shows for

$$\bar{\vee} = \text{cocop}, \qquad (15.38)$$

where cocop (and equivalently, its DeMorgan dual cop) *can be arbitrary and used to determine jointly* $S(a)$, $S(b)$, $S(c)$, $S(d)$, as discussed above,

$$
\begin{aligned}
t(\alpha)(x, y) &= t(a)(x) \,\bar{\vee}\, t(b)(y) = P(x \in S(a)) \,\bar{\vee}\, P(y \in S(b)) \\
&= P((x \in S(a)) \vee (y \in S(b))), \\
t(\beta)(x, y) &= t(c)(x) \,\bar{\vee}\, t(d)(y) = P(x \in S(c)) \,\bar{\vee}\, P(y \in S(d)) \\
&= P((x \in S(c)) \vee (y \in S(d))), \qquad (15.39)
\end{aligned}
$$

for all $x \in D_a$, $y \in D_b$. Thus, such a choice for $\bar{\vee}$ produces a *homomorphic-like relation* in eq.(15.39), where, as usual, un-barred $\vee$ here refers to ordinary boolean disjunction. *Thus, it is natural to identify with α, β the corresponding underlying boolean events to eq.(15.2) as*

$$A(a, b; x, y; S) = (x \in S(a)) \vee (y \in S(b))),$$
$$B(c, d; x, y; S) = (x \in S(c)) \vee (y \in S(d))), \qquad (15.40)$$

for all $x \in D_a$, $y \in D_b$, whose probabilities via P give back eq.(15.39).

15.5   Use of One-Point Coverages with Probability Distance Functions

Consider now the distance expression given in eq.(15.18) fully evaluated by use of eq.(15.15). By use of eqs.(15.39) and (15.40), it is clear that this distance expression can be written now as

$$
\begin{aligned}
t(\delta)(x,y) &= |t(\alpha)(x,y) - t(\beta)(x,y)| \\
&= |P(A(a,b;x,y;S)) - P(B(c,d;x,y;S')| \\
&= |P(A(a,b;x,y;S))\&(B(c,d;x,y;S))') \\
&\quad -P((A(a,b;x,y;S))'\&B(c,d;x,y;S))|, \qquad (15.41)
\end{aligned}
$$

the same forms as for the naive probability distance function given in eq.(9.25). However, note that the bottom form of equation (15.41) need not be used to obtain the full evaluation. On the other hand, consider the more useful natural absolute probability distance, recalling the argument in Example 3: This certainly does require the probability evaluations $P((A(a,b;x,y;S))\&(B(c,d;x,y;S))')$ and $P((A(a,b;x,y;S))'\&B(c,d;x,y;S))$ as

$$
\begin{aligned}
d_{P,2}&((A(a,b;x,y;S)),(B(c,d;x,y;S))) \\
&= P((A(a,b;x,y;S))\&(B(c,d;x,y;S))') \\
&\quad +P((A(a,b;x,y;S))'\&B(c,d;x,y;S)). \qquad (15.42)
\end{aligned}
$$

If we now assume that α and β are formed in some sense independently, it is reasonable to have $S(a)$, $S(b)$ statistically independent of $S(c)$, $S(d)$ – though $S(a)$ and $S(b)$ need not be independent and similar remarks hold for $S(c)$, $S(d)$. In this case, the events $A(a,b;x,y;S)$, $B(c,d;x,y;S)$ defined in eq.(15.40) are P-statistically independent and eq.(15.42) simplifies to

$$
\begin{aligned}
d_{P,2}&(A(a,b;x,y;S)), B(c,d;x,y;S)) \\
&= P(A(a,b;x,y;S))P((B(c,d;x,y;S))') \\
&\quad +P((A(a,b;x,y;S))')P(B(c,d;x,y;S)) \\
&= t(\alpha)(x,y)(1 - t(\beta)(x,y)) + (1 - t(\alpha)(x,y))t(\beta)(x,y) \\
&= t(\alpha)(x,y) + t(\beta)(x,y) - 2t(\alpha)(x,y)t(\beta)(x,y), \qquad (15.43)
\end{aligned}
$$

a result identifiable with eq.(9.27) for independent a and b. Note that because α and β are independent, even when they have identical marginal forms, in general the distance in eq.(15.43) remains positive. In fact, all of the remarks in Example 3 relevant to the comparison of the two probability distances are fully applicable here.

Consider next the situation where α and β are not necessarily independent, i.e., the pairs $(S(a), S(b))$ and $(S(c), S(d))$ may not be P-statistically independent. Then, for this more general case, we need to obtain explicitly the probabilities

$$
P((A(a,b;x,y;S))\&(B(c,d;x,y;S))') \text{ and } P((A(a,b;x,y;S))'\&B(c,d;x,y;S)).
$$

Using eqs.(15.34), (15.37), and analogues of eq.(15.34), after simplifying terms,

$$P((A(a, b; x, y; S))\&(B(c, d; x, y; S))')$$
$$= P(((X_1 = 1) \vee (X_2 = 1))\&(X_3 = 0)\&(X_4 = 0))$$
$$= cop(P(X_3 = 0), P(X_4 = 0))$$
$$\quad -cop(P(X_1 = 0), P(X_2 = 0), P(X_3 = 0)), P(X_4 = 0))$$
$$= cop(1 - t(c)(x), 1 - t(d)(y))$$
$$\quad -cop(1 - t(a)(x), 1 - t(b)(y), 1 - t(c)(x), 1 - t(d)(y))$$
$$= cocop(t(a)(x), t(b)(y), t(c)(x), t(d)(y))$$
$$\quad -cocop(t(c)(x), t(d)(y)), \tag{15.44}$$

and similarly,

$$P((A(a, b; x, y; S))'\&(B(c, d; x, y; S))$$
$$= cocop(t(a)(x), t(b)(y), t(c)(x), t(d)(y))$$
$$\quad -cocop(t(a)(x), t(b)(y)). \tag{15.45}$$

Hence, using eqs.(15.43)-(15.45),

$$d_{P,2}(A(a, b; x, y; S)), B(c, d; x, y; S))$$
$$= P(A(a, b; x, y; S))P((B(c, d; x, y; S))')$$
$$\quad +P((A(a, b; x, y; S))')P(B(c, d; x, y; S))$$
$$= 2cocop(t(a)(x), t(b)(y), t(c)(x), t(d)(y)) - cocop(t(a)(x), t(b)(y))$$
$$\quad -cocop(t(c)(x), t(d)(y)). \tag{15.46}$$

15.6   Incorporation of Fuzzy Logic Modifiers and Use of Relational Event Algebra in Example 4

A further refinement of the above problem can be carried out by using the standard fuzzy logic modifiers to interpret the natural language modifiers "very", "quite a", "extremely", "moderate" in eq.(15.1) by, e.g., appropriate exponentiation. (The reasoning here is that numerical exponentiation above unity produces lower membership values, while exponentiation below unity produces higher membership function values – all with respect to a moderate or neutral modification of an attribute in question.) Then, for all $x \in D_a$ and $y \in D_b$,

$$t(a)(x) = (t(g)(x))^2, \quad t(b)(y) = (t(d)(y))^{1.5}, \quad t(c)(x) = (t(g)(x))^3, \tag{15.47}$$

and $t(d)(y)$ remains the same ("moderate quantity ..."), where

$$g = \text{"moderately long" and } g \text{ corresponds to } f_g : D_g \rightarrow [0, 1]. \tag{15.48}$$

Then, applying the one-point coverage relations first to g and using eq.(15.15), yields the functional forms

$$t(\alpha)(x, y) = (P(x \in S_1(g)))^2 \; \bar{\vee} \; (P(y \in S_1(d)))^{1.5},$$
$$t(\beta)(x, y) = (P(x \in S_2(g)))^3 \; \bar{\vee} \; P(y \in S_2(d)) \qquad (15.49)$$

for all $x \in D_a$, $y \in D_b$, where now we use the S subscript notation 1 and 2 to distinguish between events formulated for α and those for β. (Recall the earlier discussion on the jointness of pairs of one-point coverage type random sets following eq.(15.29).) In this situation, the joint random sets $S_1(g)$, $S_1(d)$, $S_2(g)$, $S_2(d)$ – or equivalently, the joint 0-1 random variables

$$Y_1 = \phi(S_1(g))(x), \qquad Y_2 = \phi(S_1(d))(y),$$
$$Y_3 = \phi(S_2(g))(x), \qquad Y_4 = \phi(S_1(d))(y), \qquad (15.50)$$

noting marginally, Y_1 and Y_3 are identically distributed, as are Y_2 and Y_4 (but their joint distributions can be nontrivial):

$$P(Y_1 = 1) = P(Y_3 = 1) = t(g)(x),$$
$$P(Y_1 = 0) = P(Y_3 = 0) = 1 - t(g)(x),$$
$$P(Y_2 = 1) = P(Y_4 = 1) = t(d)(y),$$
$$P(Y_2 = 0) = P(Y_4 = 0) = 1 - t(d)(y), \qquad (15.51)$$

for all $x \in D_a$, $y \in D_b$, determine all relevant joint distributions involving eq.(15.49). Here, the homomorphic-like relations in eq.(15.39) still hold, but no longer can cop, nor corresponding cocop, be arbitrarily chosen – both must now be derivable from the choice of any copula or cocopula corresponding to the joint distribution of 0-1 random variables Y_1, Y_2, Y_3, Y_4 (for all $x \in D_a$, $y \in D_b$).

Consider next the exponential probabilities used in eq.(15.49). For any positive integer m, and any choice of boolean event, say C, the exponentiated probability $(P(C))^m$, naturally matches a corresponding m-factor cartesian product evaluation:

$$(P(C))^m = P_o(C^m), \qquad (15.52)$$

where, by definition, the *m-exponentiation of event* C, for any choice of event,

$$C^m = C \times \ldots \times C \qquad (m \;\; \text{factors}) \qquad (15.53)$$

and P_o is the m-product probability measure generated from probability space $(\Omega, \mathcal{B}, P)$ with $C \in \mathcal{B}$, with P_o corresponding to m statistically independent marginal probabilities, each identical to P. Thus, we have in eq.(15.49) the relations

$$P_o((x \in S_1(g))^2) = (P(x \in S_1(g)))^2,$$
$$P_o((x \in S_2(g))^3) = (P(x \in S_2(g)))^3,$$
$$Po((y \in S_2(d) \times \Omega \times \Omega) = P(y \in S_2(d)). \qquad (15.54)$$

Consider, in particular, the fractional exponentiation $(P(y \in S(d)))^{1.5}$ in light of the development of constant-probability events in Sections 10.3 and 13.3. Thus, for any collection of real numbers (not necessarily weights), say t_1, t_2, t_3,..., where $0 \leq t_j \leq 1$, $j = 1, 2, 3,...$, there is a corresponding collection of events $\theta(t_j)$, $j = 1, 2, 3, ...,$ (not necessarily disjoint) so that for any choice of P – within a large class of such probability measures

$$P_1(\theta(t_j)) = t_j, \qquad j = 1, 2, ..., \tag{15.55}$$

where P_1 is an appropriately chosen extension for each P.

Consider also any analytic function $h : [0, 1] \to [0, 1]$ (where the endpoints may be considered as limiting cases) with

$$h(s) = (1 - s)^m k(s), \tag{15.56}$$

where $k : [0, 1] \to [0, 1]$ is an analytic function such that h takes the form of a finite polynomial or infinite series, given, without loss of generality, as

$$h(s) = \sum_{j=0}^{\infty} (s^j (1 - s)^m t_j), \qquad \text{for all } s \in [0, 1], \tag{15.57}$$

where the coefficients t_j are all $\in [0, 1]$ and m is a positive integer. Then, there is a natural corresponding event-valued function, denoted also by h, when no ambiguity is present (we have already used such notation above to denote exponentiation of events), such that for any choice of event C – the $(C')^m$ broken up between $(C')^{m-1}$ and C' to insure disjointness –

$$h(C) = \bigvee_{j=0}^{\infty} ((C')^{m-1} \times C^j \times C' \times \theta(t_j)), \tag{15.58}$$

and for any choice of P in a large class of probability measures, slightly abusing notation, using the product probability measure form introduced earlier and the disjointness of terms,

$$\begin{aligned} P_1(h(C)) &= \sum_{j=0}^{\infty} (P_o(C^j) P_o((C')^m) P_1(\theta(t_j))) \\ &= \sum_{j=0}^{\infty} ((P(C))^j (P(C'))^m t_j) = h(P(C)). \end{aligned} \tag{15.59}$$

Also, noting the special role 0 and 1 play in probability, we may at times consider in place of the form in eq.(15.57) the more appropriate dual expansion h in terms of $s' = 1 - s$,

$$h(s') = \sum_{j=0}^{\infty} ((s')^j s^m t_j), \qquad \text{for all } s \in [0, 1], \tag{15.60}$$

with corresponding event-valued form

$$h(C') = \bigvee_{j=0}^{\infty} (C^{m-1} \times (C')^j \times C \times \theta(t_j)) \tag{15.61}$$

with compatibility eq.(15.59) now becoming

$$P_1(h(C')) = h(P(C')) = h(1 - P(C)), \tag{15.62}$$

for all choices of P in an appropriately large class. Thus, eqs.(15.52), (15.53) can be considered limiting cases of eqs.(15.60), (15.61).

   Returning to the problem at hand, the exponentiation $(P(x \in S(g)))^{1.5}$, it is clear that the associated numerical function $s^{1.5}$, for all $s \in [0, 1]$, as it stands, obviously takes only values in [0,1] and can be expanded as a power series about 1, equivalently as a function of s' as a power series about 0, in the form $(1 - s')^{1.5}$. However, the very first derivative of this function with respect to variable s' evaluated at 0 yields the negative value -1.5, violating the weighting conditions required for coefficients imposed via eq.(15.35). On the other hand, recall the basic identity, for any fixed real r, $0 < r < 1$,

$$s^r = (1 - s')^r = s(1 - s')^{-r'}, \qquad \text{for all } 0 \le s < 1, \tag{15.63}$$

where

$$s' = 1 - s, \qquad r' = 1 - r \tag{15.64}$$

and we have the following series with coefficients in [0,1]:

$$(1 - s')^{-r'} = \sum_{j=0}^{\infty} (s')^j t_{j,r'}, \tag{15.65}$$

where

$$t_{j,r'} = \begin{cases} 1 & \text{if } j = 0 \\ r'(r' + 1) \cdots (r' + j - 1)/j! & \text{if } j = 1, 2, \ldots. \end{cases} \tag{15.66}$$

Then, define for any $C \in \mathcal{B}$,

$$C^r = \bigvee_{j=0}^{\infty} (C')^j \times C \times \theta(t_{j,r'}). \tag{15.67}$$

Because the infinite disjunctive series in eq.(15.67) is a mutually disjoint one, using the product space properties of P_1, we obtain via $s = P(C)$ in eq.(15.65), and using the identity in eq.(15.63),

$$P_1(C^r) = (P(C))^r, \qquad \text{for all well-defined } P. \tag{15.68}$$

Similarly, if we consider any positive real number t, assuming without loss of generality that t itself is not a positive integer, denoting the greatest integer part of t , as $[[t]]$ and the fractional excess $t-[[t]]$ as $\{\{t\}\}$, we obtain

$$s^t = s^{[[t]]} s^{\{\{t\}\}}, \tag{15.69}$$

with algebraic analogue

$$C^t = C^{[[t]]} \times C^{\{\{t\}\}}, \tag{15.70}$$

where $C^{\{\{t\}\}}$ is obtained from eq.(15.67), letting $r = \{\{t\}\}$ and $C^{[[t]]}$ is obtained from eq.(15.53) letting $m = [[t]]$. More directly, eq.(15.70) is the same as

$$C^t = \bigvee_{j=0}^{\infty} \left(C^{[[t]]} \times ((C')^j \times C \times \theta(t_{j,\{\{t\}\}'}) \right). \tag{15.71}$$

As in eq.(15.68), we have the faithful evaluation

$$P_1(C^t) = (P(C))^t, \qquad \text{for all well-defined } P. \tag{15.72}$$

Back to the problem of interest, we now specialize the above general development in eqs.(15.63)-(15.72) to $r = 0.5$ and $t = 1.5$. Thus, we have any event $C \in \mathcal{B}$, the disjoint event series

$$C^{1.5} = \bigvee_{j=0}^{\infty} \left(C \times (C')^j \times C \times \theta(t_{j,0.5}) \right), \tag{15.73}$$

with probability evaluation

$$P_1(C^{1.5}) = (P(C))^{1.5}. \tag{15.74}$$

Since the coefficients $t_{j,0.5}$ in the above series are all dominated by unity, this series converges at least as fast as the ordinary power series and thus reasonably small finite truncation approximations for all event values $P(C)$ bounded away from zero are possible. These convergence rates can also be refined by noting the series given in eq.(15.57) is a particular member of a well-studied class of series [2].

In any case, note the complement of $C^{1.5}$, up to P_1-probability one, is simply

$$(C^{1.5})' = \bigvee_{j=0}^{\infty} \left(C \times (C')^j \times C \times (\theta(t_{j,0.5}))' \right) \vee C' \tag{15.75}$$

since

$$
\begin{aligned}
P_1(C^{1.5} \vee (C^{1.5})') &= P_1\left(\bigvee_{j=0}^{\infty} (C \times (C')^j \times C \times (\theta(t_{j,0.5}) \vee (\theta(t_{j,0.5}))')) \vee C' \right) \\
&= P_1\left(\bigvee_{j=0}^{\infty} (C \times (C')^j \times C) \right) + P(C') \\
&= (P(C))^2(1/(1 - P(C'))) + 1 - P(C) = 1. \tag{15.76}
\end{aligned}
$$

In a similar vein, note the exponential integer power complements

$$(C^2)' = (C \times C') \vee C', \quad (C^3)' = (C^2 \times C') \vee (C \times C') \vee C'. \tag{15.77}$$

Also, via eqs.(15.53) and (15.73), we have the natural orderings when $C > \emptyset$

$$\emptyset < C^3 < C^2 < C^{1.5} < C < \Omega. \tag{15.78}$$

The above extends to the general case where exponent 1.5 is replaced by arbitrary real s, t, with $0 < s < t$,

$$\emptyset < C^t < C^s. \tag{15.79}$$

Substituting eqs.(15.54) and (15.74) into eq.(15.49) shows for the choice of $\bar{V}=\text{cocop}_o$, say, determined through cop or cocop (which may be arbitrary) corresponding to the joint distribution of Y_1, Y_2, Y_3, Y_4, given in eqs.(15.50), (15.51),

$$
\begin{aligned}
t(\alpha)(x,y) &= (P(x \in S_1(g)))^2 \ \bar{V} \ (P(y \in S_1(d)))^{1.5} \\
&= P_1((x \in S_1(g))^2) \ \bar{V} \ P((y \in S_1(d))^{1.5}) \\
&= P_1((x \in S_1(g))^2 \ \vee \ (y \in S_1(d))^{1.5}), \\
t(\beta)(x,y) &= (P(x \in S_2(g)))^3 \ \bar{V} \ P(y \in S_2(d)) \\
&= P_1((x \in S_2(g))^3) \ \bar{V} \ P(y \in S_2(d)) \\
&= P_1((x \in S_2(g))^3) \ \vee \ (y \in S_2(d)))
\end{aligned} \tag{15.80}
$$

for all $x \in D_a$, $y \in D_b$. Thus, provided $\bar{V}$ is appropriately determined, as discussed above, it is natural to let here, for all $x \in D_a$, $y \in D_b$,

$$
\begin{aligned}
A(d,g;x,y;S_1) &= (x \in S_1(g))^2 \vee (y \in S_1(d))^{1.5} \\
&= ((x,x) \in (S_1(g))^2) \vee ((y,y,y,...) \in (S_1(d))^{1.5}), \\
B(d,g;x,y;S_2) &= (x \in S_2(g))^3 \vee (y \in S_2(d)) \\
&= ((x,x,x) \in (S_2(g))^3) \vee (y \in S_2(d)),
\end{aligned} \tag{15.81}
$$

since evaluation of A and B in eq.(15.81) yields back eq.(15.80). Note further, that to unify notation, we can rewrite the far right-hand side of eq.(15.81) as the following one-point coverage relations

$$
\begin{aligned}
A(d,g;x,y;S_1) &= ((\underline{x} \in (S_1(g))^2 \times \Omega_1)) \vee (\underline{y} \in (S_1(d))^{1.5}), \\
B(d,g;x,y;S_2) &= ((\underline{x} \in (S_2(g))^3 \times \Omega_1)) \vee (\underline{y} \in (S_2(d) \times \Omega_1))
\end{aligned} \tag{15.82}
$$

where

$$\underline{x} = (x,x,x,...), \quad \underline{y} = (y,y,y,...), \quad \text{for all } x \in D_a, y \in D_c, \tag{15.83}$$

where the infinite product probability space here is $(\Omega_1, \mathcal{B}_1, P_1)$. Thus, analogous to the previous relations developed between one-point coverages and 0-1 random variables, we can let the corresponding 0-1 random variables be

$$
\begin{aligned}
Z_1 &= \phi((S_1(g))^2 \times \Omega_1)(\underline{x}) = \phi(S_1(a))(x), \\
Z_2 &= \phi((S_1(d))^{1.5})(\underline{y}) = \phi(S_1(b))(y), \\
Z_3 &= \phi((S_1(g))^3 \times \Omega_1)(\underline{x}) = \phi(S_2(c))(x), \\
Z_4 &= \phi(S_2(d))(\underline{y}),
\end{aligned} \tag{15.84}
$$

where by the product probability structure of P_1, compatible with eqs.(15.47), (15.54), (15.74), (15.80), etc.,

$$P(Z_1 = 1) = (t(g)(x))^2 = t(a)(x),$$
$$P(Z_1 = 0) = 1 - (t(g)(x))^2 = 1 - t(a)(x),$$
$$P(Z_2 = 1) = (t(d)(y))^{1.5} = t(b)(y),$$
$$P(Z_2 = 0) = 1 - (t(d)(y))^{1.5} = 1 - t(b)(y),$$
$$P(Z_3 = 1) = (t(g)(x))^3 = t(c)(x),$$
$$P(Z_3 = 0) = 1 - (t(g)(x))^3 = 1 - t(c)(x),$$
$$P(Z_4 = 1) = t(d)(y), \quad P(Z_4 = 0) = 1 - t(d)(y). \qquad (15.85)$$

Note, in terms of Y_1, Y_2, Y_3, Y_4, in eqs.(15.50), (15.51), using the above equations,

$$(Z_1 = 0) \text{ iff } ((Y_1 = 1)^2)' = ((Y_1 = 1) \times (Y_1 = 0)) \vee (Y_1 = 0);$$
$$(Z_1 = 1) \text{ iff } (Y_1 = 1)^2;$$

$$(Z_2 = 0) \text{ iff } ((Y_2 = 1)^{1.5})' = \bigvee_{j=0}^{\infty} (G_j) \vee H;$$

$$(Z_2 = 1) \text{ iff } (Y_2 = 1)^{1.5} = \bigvee_{j=0}^{\infty} (I_j);$$

$$(Z_3 = 0) \text{ iff } ((Y_3 = 1)^3)' = ((Y_3 = 1) \times (Y_3 = 1) \times (Y_3 = 0)) \vee$$
$$((Y_3 = 1) \times (Y_3 = 0)) \vee (Y_3 = 0);$$
$$(Z_3 = 1) \text{ iff } (Y_3 = 1)^3;$$
$$Z_4 = 0 \text{ iff } (Y_4 = 0); \ Z_4 = 1 \text{ iff } (Y_4 = 1), \qquad (15.86)$$

where

$$G_j = (Y_2 = 1) \times (Y_2 = 0)^j \times (Y_2 = 1) \times (\theta(t_{j,0.5}))', \quad j = 1, 2, \ldots$$
$$H = (Y_2 = 0),$$
$$I_j = (Y_2 = 1) \times (Y_2 = 0)^j \times (Y_2 = 1) \times \theta(t_{j,0.5}), \qquad (15.87)$$

$j = 0, 1, 2, \ldots$ Then, by inspection of the forms in eq.(15.86), the following computation simplifies

$$(Z_1 = 0)\&(Z_3 = 0)\&(Z_4 = 0) \quad \text{iff } A_1 \vee \ldots \vee A_6, \qquad (15.88)$$

where

$$A_1 = ((Y_1 = 1)\&(Y_3 = 1)\&(Y_4 = 0)) \times ((Y_1 = 0)\&(Y_3 = 1)) \times (Y_3 = 0),$$
$$A_2 = ((Y_1 = 1)\&(Y_3 = 1)\&(Y_4 = 0)) \times ((Y_1 = 0)\&(Y_3 = 0)),$$
$$A_3 = ((Y_1 = 1)\&(Y_3 = 0)\&(Y_4 = 0)) \times (Y_1 = 0),$$
$$A_4 = ((Y_1 = 0)\&(Y_3 = 1)\&(Y_4 = 0)) \times (Y_3 = 1) \times (Y_3 = 0),$$
$$A_5 = ((Y_1 = 0)\&(Y_3 = 1)\&(Y_4 = 0)) \times (Y_3 = 0),$$
$$A_6 = (Y_1 = 0)\&(Y_3 = 0)\&(Y_4 = 0). \qquad (15.89)$$

Thus, using the above equations, noting he mutual disjointness of the terms A_j, $j = 1, ..., 6$, and that of the G_j and H , we can in principle, obtain the full evaluation

$$P(Z_1 = 0)\&(Z_2 = 0)\&(Z_3 = 0)\&(Z_4 = 0))$$
$$= \sum_{i=1}^{6} \left(\sum_{j=0}^{\infty} (P_1(A_i\&G_j) + P(A_i\&H)) \right), \qquad (15.90)$$

where each inner infinite sum can actually be expressed in closed form.

For example, for i=6,

$$P_1(A_6\&G_j)) \quad = \quad P((Y_1 = 0)\&(Y_2 = 1)\&(Y_3 = 0)\&(Y_4 = 0))$$
$$\cdot \sum_{j=1}^{\infty} (P(Y_2 = 1)(P(Y_2 = 0))^j (1 - t_{j,0.5})) . \quad (15.91)$$

Analogous to the computations in eqs.(15.34), (15.44), etc., the first term on the right-hand side of eq.(15.91) is simply

$$P((Y_1 = 0)\&(Y_2 = 1)\&(Y_3 = 0)\&(Y_4 = 0))$$
$$= \quad cocop(t(g)(x), t(d)(y), t(g)(x), t(d)(y))$$
$$-cocop(t(g)(x), t(g)(x), t(d)(y)), \qquad (15.92)$$

while the last term, an infinite series is actually always obtainable in closed form by first evaluating eq.(15.75)

$$P_1((C^{1.5})') \quad = \quad 1 - P_1(C^{1.5}) = 1 - (P(C))^{1.5}$$
$$= \quad P\left(\bigvee_{j=0}^{\infty} (C \times (C')^j \times C \times (\theta(t_{j,0.5}))') \vee C' \right)$$
$$= \quad \sum_{j=1}^{\infty} ((P(C))^2 (P(C))^j (1 - t_{j,0.5})) + P(C'), \quad (15.93)$$

and solving for the series divided by $P(C)$, with $C = (Y_2 = 1)$,

$$\sum_{j=1}^{\infty} (P(C)(P(C'))^j (1 - t_{j,0.5})) = (1 - (P(C))^{1.5} - P(C'))/P(C)$$
$$= (P(C) - (P(C))^{1.5})/P(C) = 1 - (t(d)(y))^{0.5} . \qquad (15.94)$$

Similarly, the remaining expressions in eq.(15.90) can be evaluated in closed form. (In fact, these are specifically evaluated following eq.(15.107) for the purpose of obtaining the natural absolute probability distance between the models for a and b.) Then, using eq.(15.85) and the basic properties of copulas discussed earlier,

$$P(Z_1 = 0)\&(Z_2 = 0)\&(Z_3 = 0)\&(Z_4 = 0))$$
$$= \text{cop}_0(P(Z_1 = 0), P(Z_2 = 0), P(Z_3 = 0), P(Z_4 = 0))$$
$$= \text{cop}_0 \left(1 - (t(g)(x))^2, 1 - (t(d)(y))^{1.5}, 1 - (t(g)(x))^3, 1 - t(d)(y)\right)$$
$$= \text{computable function of } (1 - t(g)(x), 1 - t(d)(y))$$

and cocop acting on them. $\hspace{4cm}$ (15.95)

Following this, one can substitute the variables

$$r = t(g(x)), \qquad s = t(d(y)), \hspace{3cm} (15.96)$$

in eq.(15.95), noting that $r = t(g)(x)$ and $s = t(d)(y)$ are obviously in general functionally independent and can vary arbitrarily over $[0,1]$. Thus, one can solve for $\text{cocop}_0(r^2, s^{1.5}, r^3, s)$ explicitly as a function of r,s and $\text{cocop}(r, s)$, acting upon them. By reformulating the problem so that the parameters associated with the Y_j are all functionally independent, without changing the nature of the random quantities involved, cocop_0 as a full four-argument function can be explicitly obtained, but for reasons of simplicity we omit the details here

15.7　Additional Analysis of Example 4

On the other hand, for the two marginal cases of pair Z_1, Z_2 relative to determining $\bar{V}$ for γ and pair Z_3, Z_4 for determining $\bar{V}$ for β in eq.(15.80), we can actually obtain the desired relations fully and in relatively simple form:

First, note from previous results and eqs.(15.85), (15.96), using additional subscript 1 to indicate the joint marginal cocopula of $\text{cocop}_{o,1}$ for the joint cdf of Z_1, Z_2, as opposed to $\text{cocop}_{o,2}$ for Z_3, Z_4,

$$P_1((Z_1 = 0)\&(Z_2 = 0)) = \text{cop}_{o,1}(P(Z_1 = 0), P(Z_2 = 0))$$
$$= 1 - \text{cocop}_{o,1}(r^2, s^{1.5}), \hspace{3cm} (15.97)$$

Next, let us compute directly, using eq.(15.86),

$$P_1((Z_1 = 0)\&(Z_2 = 0)) = P_1(\bigvee_{j=0}^{\infty} ((I \vee J) \times K_j) \vee L)$$

$$= [(P_1(I) + P_1(J)) \sum_{j=1}^{\infty}(P_1(K_j))] + P_1(L), \hspace{2cm} (15.98)$$

by the disjointness of terms and the product probability form for P_1, where

$$I = ((Y_1 = 1)\&(Y_2 = 1)) \times ((Y_1 = 0)\&(Y_2 = 0)),$$
$$J = ((Y_1 = 0)\&(Y_2 = 1)) \times (Y_2 = 0),$$
$$K_j = (Y_2 = 0)^{j-1} \times (Y_2 = 1) \times (\theta(t_{j,0.5}))',$$
$$L = [(Y_1 = 1)\&(Y_2 = 0)) \times (Y_1 = 0)] \vee [(Y_1 = 0)\&(Y_2 = 0)]. \hspace{0.8cm} (15.99)$$

Then, using eq.(15.34) and its analogues, in conjunction with eqs.(15.50), (15.51),

$$P_1(I) = [t(g)(x) + t(d)(y) - \text{cocop}(t(g))(x), t(d)(y)]$$
$$\cdot(1 - \text{cocop}(t(g)(x), t(d)(y))),$$
$$P_1(J) = [\text{cocop}(t(g)(x), t(d)(y)) - t(g)(x)](1 - t(d)(y)),$$
$$P_1(L) = [(\text{cocop}(t(g)(x), t(d)(y)) - t(d)(y)) \cdot (1 - t(d)(y))]$$
$$+[1 - \text{cocop}(t(g)(x), t(d)(y))] \tag{15.100}$$

and eq.(15.94)) shows

$$\sum_{j=1}^{\infty} P_1(K_j) = (1 - (t(d)(y))^{0.5} / (1 - t(d)(y)). \tag{15.101}$$

Then, eqs.(15.96)-(15.101) show, for all $0 \le r, s \le 1$,

$$1 - \text{cocop}_{0,1}(r^2, s^{1.5}) = f_1(r, s, \text{cocop}(r, s)), \tag{15.102}$$

where

$$f_1(r, s, \text{cocop}(r, s)) = ([[(r + s - \text{cocop}(r, s))(1 - \text{cocop}(r, s))]$$
$$+ [(\text{cocop}(r, s) - r)(1 - s)]) [(1 - s^{0.5})/(1 - s)]$$
$$+((\text{cocop}(r, s) - s)(1 - r)) + 1 - \text{cocop}(r, s), \tag{15.103}$$

with some further simplification possible. Finally, eqs.(15.102), (15.103) imply

$$\text{cocop}_{0,1}(r, s) = 1 - f_1(r^{1/2}, s^{2/3}, \text{cocop}(r^{1/2}, s^{s/3})), \tag{15.104}$$

for all $0 \le r, s \le 1$. Even more simply, we obtain $\text{cocop}_{0,2}$ analogously:

$$P_1((Z_3 = 0)\&(Z_4 = 0)) = \text{cop}_{0,2}(P(Z_3 = 0), P(Z_4 = 0))$$
$$= 1 - \text{cocop}_{0,2}(r^3, s), \tag{15.105}$$

Next, let us compute directly, using eq.(15.86),

$$P_1[(Z_3 = 0)\&(Z_4 = 0)] = P_1[((Y_3 = 1)\&(Y_4 = 0)) \times (Y_3 = 1) \times (Y_3 = 0)$$
$$\vee((Y_3 = 1)\&(Y_4 = 0)) \times (Y_3 = 0) \vee ((Y_3 = 0)\&(Y_4 = 0))]$$
$$= f_2(r, s, \text{cocop}(r, s))$$
$$= ((\text{cocop}(r, s) - s)r(1 - r)) + ((\text{cocop}(r, s) - s)(1 - r))$$
$$+1 - \text{cocop}(r, s), \tag{15.106}$$

with possible additional simplifications. Hence,

$$\text{cocop}_{0,2}(r, s) = 1 - f_2(r^{1/3}, s, \text{cocop}(r^{1/3}, s)), \text{ all } 0 \le r, s \le 1. \tag{15.107}$$

Thus, finally, eq.(15.107) allows fully the compatible choices of $\bar{V} = \text{cocop}_{o,1}$ for the top (α) part and $\text{cocop}_{0,2}$ for the bottom (β) part of eq.(15.80), thereby justifying the homomorphic-like form there.

Finally, let us consider the natural absolute probability distance between the a and b models for this case. It will be most convenient to consider the last form in eq.(9.27), so that apropos to eq.(15.80),

$$d_{P_1,2}(A(d,g;x,y;S_1), B(d,g;x,y;S_2)) = 2 - P_1(A(d,g;x,y;S_1))$$
$$-P_1(B(d,g;x,y;S_2)) - 2P_1((A(d,g;x,y;S_1))'\&(B(d,g;x,y;S_2))')$$
$$= 2 \quad t(\alpha)(\varpi,y) \quad t(\beta)(\varpi,y)$$
$$-2P(Z_1 = 0)\&(Z_2 = 0)\&(Z_3 = 0)\&(Z_4 = 0)). \tag{15.108}$$

Thus, all we need do to implement the above evaluation is obtain the last term as given in eq.(15.90). We have already considered the evaluation of one of the inner infinite sums for eq.(15.90), namely the case $i = 6$ given in eq.(15.91). For completeness, we present in the following list the set-up for complete closed form evaluations for all six inner infinite sums and the remaining term as well. Using the previously developed techniques, we then have: for $i = 1$,

$$\sum_{j=1}^{\infty} P_1(A_1 \& G_j) = P_1((Y_1 = 1)\&(Y_2 = 1)\&(Y_3 = 1)\&(Y_4 = 0))$$
$$\cdot P_1((Y_1 = 0)\&(Y_2 = 0)\&(Y_3 = 1))$$
$$\cdot [P_1((Y_2 = 1)\&(Y_3 = 0))(1 - t_{1,0.5}) + (P((Y_2 = 0)\&(Y_3 = 0))$$
$$\cdot \sum_{j=2}^{\infty} (P(Y_2 = 1)(P(Y_2 = 0))^{j-2}(1 - t_{j,0.5})))]; \tag{15.109}$$

for $i = 2$,

$$\sum_{j=1}^{\infty} P_1(A_2 \& G_j)) = P_1((Y_1 = 1)\&(Y_2 = 1)\&(Y_3 = 1)\&(Y_4 = 0))$$
$$\cdot P_1((Y_1 = 0)\&(Y_2 = 0)\&(Y_3 = 0))$$
$$\cdot \sum_{j=1}^{\infty} P(Y_2 = 1)(P(Y_2 = 0))^{j-1}(1 - t_{j,0.5}); \tag{15.110}$$

for $i = 3$,

$$\sum_{j=1}^{\infty} P_1(A_3 \& G_j) = P_1((Y_1 = 1)\&(Y_2 = 1)\&(Y_3 = 0)\&(Y_4 = 0))$$
$$\cdot P_1((Y_1 = 0)\&(Y_2 = 0))$$
$$\cdot \sum_{j=1}^{\infty} P(Y_2 = 1)(P(Y_2 = 0))^{j-1}(1 - t_{j,0.5}); \tag{15.111}$$

for $i = 4$,

$$\sum_{j=1}^{\infty} P_1(A_4 \& G_j) = P_1[(Y_1 = 0)\&(Y_2 = 1)\&(Y_3 = 1)\&(Y_4 = 0)]$$

$$\cdot P_1((Y_2 = 0)\&(Y_3 = 1))[(P_1((Y_2 = 1)\&(Y_3 = 0))(1 - t_{1,0.5}))$$
$$+P((Y_2 = 0)\&(Y_3 = 0))P(Y_2 = 1)$$
$$\cdot \sum_{j=2}^{\infty}(P(Y_2 = 0))^{j-2}(1 - t_{j,0.5})]; \tag{15.112}$$

for $i = 5$,

$$\sum_{j=1}^{\infty} P_1(A_5\&G_j) = P_1((Y_1 = 0)\&(Y_2 = 1)\&(Y_3 = 1)\&(Y_4 = 0))$$
$$\cdot P_1((Y_2 = 0)\&(Y_3 = 0))P(Y_2 = 1)$$
$$\cdot \sum_{j=1}^{\infty}(P(Y_2 = 0))^{j-1}(1 - t_{j,0.5}); \tag{15.113}$$

for $i = 6$,

$$\sum_{j=1}^{\infty} P_1(A_6\&G_j)) = P((Y_1 = 0)\&(Y_2 = 1)\&(Y_3 = 0)\&(Y_4 = 0))$$
$$\cdot \sum_{j=1}^{\infty} P(Y_2 = 1)(P(Y_2 = 0))^j(1 - t_{j,0.5}); \tag{15.114}$$

and the remaining expression

$$\sum_{i=1}^{6} P(A_i\&H) = P_1[(Y_1 = 1)\&(Y_2 = 0)\&(Y_3 = 1)\&(Y_4 = 0)]P(Y_3 = 0)$$
$$+P_1[(Y_1 = 1)\&(Y_2 = 0)\&(Y_3 = 1)\&(Y_4 = 0)]P[(Y_1 = 0)\&(Y_3 = 0)]$$
$$+P_1[((Y_1 = 1)\&(Y_2 = 0))\&(Y_3 = 0)\&(Y_4 = 0)]P(Y_1 = 0)$$
$$+P_1[((Y_1 = 0)\&(Y_2 = 0))\&(Y_3 = 1)\&(Y_4 = 0)]P(Y_3 = 1)P(Y_3 = 0)$$
$$+P_1[(Y_1 = 0)\&(Y_2 = 0)\&(Y_3 = 1)\&(Y_4 = 0)]P(Y_3 = 0)$$
$$+P_1[(Y_1 = 0)\&(Y_2 = 0)\&(Y_3 = 0)\&(Y_4 = 0)]. \tag{15.115}$$

Note also that the right-hand infinite series factors appearing in all six cases are actually of three types and can be evaluated in closed form, similar to that in eq.(15.94). Specifically, we have either

$$\sum_{j=1}^{\infty}(P(Y_2 = 1)(P(Y_2 = 0))^j(1 - t_{j,0.5})) = 1 - (t(d)(y))^{0.5}, \tag{15.116}$$

as in case $i = 6$ only; or

$$\sum_{j=1}^{\infty}(P(Y_2 = 1)(P(Y_2 = 0))^{j-1}(1 - t_{j,0.5})) = \frac{1 - (t(d)(y))^{0.5}}{1 - t(d)(y)}. \tag{15.117}$$

as in cases $i = 2, 3, 5$; or

$$\sum_{j=1}^{\infty} \left(P(Y_2 = 1)(P(Y_2 = 0))^{j-2}(1 - t_{j,0.5}) \right)$$

$$= \frac{(1 - t(d)(y))^{0.5} - \frac{1}{2}t(d)(y) + \frac{1}{2}(t(d)(y))^2}{(1 - t(d)(y))^2}, \qquad (15.118)$$

as in cases $i = 1, 4$.

Again, as in case $i = 6$ in eq.(15.92), the evaluation of all of the right-hand factors in cases $i = 1, 2, 3, 4, 5$ above, as well as the terms in eq.(15.115) can be similarly evaluated as straightforward (though, somewhat tedious) finite signed sums of cocopula and single term evaluations over $t(g)(x)$, $t(g)(y)$ in possibly repeated argument form. Once numerical evaluations are established for $d_{P_1,2}(A(d, g; x, y; S_1), B(d, g; x, y; S_2))$, following the above scheme, all similarity and hypotheses testing results discussed in Example 2 are applicable here as well, but in a functional or component-wise sense, as variables $x \in D_a$ and $y \in D_b$ vary. If it is decided that both models are sufficiently similar (for a reasonable range of values of x and y), then the same techniques for combination of information – such as using weighting events – or deduction with respect to the collection $\{A(d, g; x, y; S_1), B(d, g; x, y; S_2))\}$ is applicable here, analogous to that presented in Section 10.4 and in Example 3, but again, keeping in mind the basic index variables x, y.

Bibliography

[1] Bezdek, J., Ed. (1994) Special Issue, *IEEE Transactions on Fuzzy Systems*, **2**(1), 1994.

[2] Bromwich, T. J. (1959) *An Introduction to the Theory of Infinite Series, 2nd Edition*. MacMillan Co., New York, especially, pp. 292-293.

[3] Dall'aglio, G., Kotz, S., and Salinetti, G., (Eds.) (1991) *Advances in Probability Distributions with Given Marginals*, Kluwer Academic Publishers, Dordrecht, Holland.

[4] Dubois, D. and Prade, H. (1980) *Fuzzy Sets and Systems*, Academic Press, New York, especially, pp. 38-40.

[5] Frank, M. J. (1979) On the simultaneous associativity of $F(x,y)$ and $x + y - F(x,y)$, *Aequationes Mathematicae*, **19**, 194-226.

[6] Genest, C. (1987) Frank's family of bivariate distributions, *Biometrika*, **74**(3), 549-555.

[7] Genest, C. and MacKay, J. (1986) The joy of copulas: bivariate distributions with uniform marginals, *The American Statistician*, **40**(4), 280–283.

[8] Goodman, I. R. and Kramer, G. F. (1997) Extension of relational event algebra to random sets with applications to data fusion, *Proceedings of Workshop on Applications and Theory of Random Sets*, Institute for Mathematics and Its Applications (IMA), University of Minnesota, Minneapolis, August 22-24, 1996, Springer-Verlag, Berlin, (to appear).

[9] Laviolette, M., Seaman, J. W., Barrett, J. D., and Woodall, W. H. (1995) A probabilistic and statistical view of fuzzy methods, *Technometrics*, **37**(3), 249-292.

[10] McShane, E. J. and Botts, T. (1959) *Real Analysis*, D. Van Nostrand Co., Princeton, NJ.

[11] Schweizer, B. and Sklar, A. (1983) *Probabilistic Metric Spaces*, North-Holland, Amsterdam.

[12] Sklar, A. (1973) Random variables, joint distribution functions, and copulas, *Kybernetika*, 9, 449-460.

[13] Zadeh, L. A. (1977) A theory of approximate reasoning (AR), in J. E. Hayes, D. Michie and L.I. Mikulich (Eds.), *Machine Intelligence*, Elsevier, New York, pp. 149-194.

Chapter 16

Development of Relational Event Algebra Proper to Address Data Fusion Problems

Overview

For completeness, relational events representing weighted linear combinations of multiple event probabilities or polynomial or analytic functions of one probability variable, in Sections 16.1 and 16.2, which have been already treated in [2], [3], are again presented here. New corrected results concerning relational events representing quadratic polynomials in two event probability variables are given in Section 16.3. Finally, Section 16.4 provides new results on representing various functions of event probabilities, which up to now, were thought to be unrepresentable, including minimum and maximum, multiplication by integers exceeding unity, bounded sums, and absolute differences – and $d_{P,1}$. The device used to carry this out is to relax the requirement that the relational event in question be independent of any particular choice of probability evaluation. In particular, the constant-probability events (but, not the ordinary or other conditional events involved) of the relational events are allowed to be dependent on the probability measure chosen for evaluation. Finally, a number of open questions concerning conditional and relational event algebra are presented in Section 16.5.

16.1 Use of Relational Event Algebra and Probability Distances in Comparing Opinions of Two Experts' Weighted Combinations of Probabilities

Motivating Example 5

This example is related to the algebraic averaging of events considered in Section 9.3.. In this situation, two experts consider the overall probability of enemy attack tomorrow as separate weighted linear combinations of contributing probabilities. Each expert has his own bias as to the importance to be given to a particular type of probability. For simplicity, suppose that the relevant contributing events are:

$a = $ enemy is actually moving at least a thousand troops to area C,

$b = $ weather tomorrow will be dry and over 70 degrees F.,

$c = $ sea-state will be no more than level A,   (16.1)

noting that in general the above events may be neither disjoint nor independent. Suppose the experts yield the subjective formal probabilities

$$P(\alpha) = w_{11}P(a) + w_{12}P(b) + w_{13}P(c),$$
$$P(\beta) = w_{21}P(a) + w_{22}P(b) + w_{23}P(c),$$ (16.2)

where the w_{ij} are weights, i.e.,

$$0 \le w_{ij} \le 1, \qquad w_{i1} + w_{i2} + w_{i3} = 1, \quad i = 1, 2.$$ (16.3)

If events a, b, c were actually disjoint, then the total probability theorem is applicable and eq.(16.4) also holds:

$$w_{11} = P(\alpha|a), \qquad w_{12} = P(\alpha|b), \qquad w_{13} = P(\alpha|c),$$
$$w_{21} = P(\beta|a), \qquad w_{22} = P(\beta|b), \qquad w_{23} = P(\beta|c).$$ (16.4)

In this case, if we need to obtain $P(\alpha\beta)$ – such as for the computation of $d_{P,2}(\alpha, \beta)$ in eq.(14.8) – we note that, using standard probability techniques, we still cannot obtain $P(\alpha\beta)$ in terms of only knowing all of the joint probabilities of a, b, c and the w_{ij} : we must also be able to obtain joint probabilities of various combinations of α, β, with a, b, c. Obviously, when a, b, c are not mutually disjoint this problem still remains. Thus, we seek a way to obtain α and β explicitly in terms of a, b, c and events relating to the w_{ij}. By use again of constant-probability events, we can find events $\theta(w_{ij})$ and extensions a_1, b_1, c_1, of a, b, c, respectively, all lying in a sample space B_1, (associated with a product probability space (Ω_1, B_1, P_1)) extending the original sample space B (associated with probability space (Ω, B, P)) such that

$$P_1(\theta(w_{ij})) = w_{ij}, \qquad \text{for all } i, j,$$
$$P_1(a_1) = P(a), \quad P_1(b_1) = P(b), \quad P_1(c_1) = P(c),$$ (16.5)

essentially independent of the choice of P. Thus, for the disjoint case, if we let the natural corresponding events representing α, β be the algebraic averages

$$\alpha = \mathrm{av}(a, b, c; \underline{w}_1) = (a \times \theta(w_{11})) \vee (b \times \theta(w_{12})) \vee (c \times \theta(w_{13})),$$
$$\beta = \mathrm{av}(a, b, c; \underline{w}_2) = (a \times \theta(w_{21})) \vee (b \times \theta(w_{22})) \vee (c \times \theta(w_{23})), \qquad (16.6)$$

then clearly, by the basic properties of P_1, for all P of interest

$$P_1(a) = (P(a)w_{11}) + (P(b)w_{12}) + (P(c)w_{13}),$$
$$P_1(b) = (P(a)w_{21}) + (P(b)w_{22}) + (P(c)w_{23}), \qquad (16.7)$$

compatible with eq.(16.2). In turn, using the explicit algebraic forms for α, β in eq.(16.6), we can now obtain

$$\alpha\beta \;=\; (a \times (\theta(w_{11}) \,\&\, \theta(w_{21}))) \vee (b \times (\theta(w_{12}) \,\&\, \theta(w_{22})))$$
$$\vee (c \times (\theta(w_{13}) \& \theta(w_{23}))), \qquad (16.8)$$

where by straightforward reasoning it is seen that for all i, j, k, l, we must have (see Section 13.3)

$$\theta(w_{ij}) \,\&\, \theta(w_{kl}) = \theta(\min(w_{ij}, w_{kl})). \qquad (16.9)$$

Again, using disjointness of a, b, c, eqs.(16.8) and (16.9) finally show that

$$P_1(\alpha\beta) \;=\; (P(a)\min(w_{11}, w_{21})) + (P(b)\min(w_{12}, w_{22}))$$
$$+ (P(c)\min(w_{13}, w_{23})). \qquad (16.10)$$

Then, e.g.,

$$d_{P,2}(\alpha, \beta) = P_1(\alpha) + P_1(\beta) - 2P_1(\alpha\beta)$$
$$= w_{11}P(a) + w_{12}P(b) + w_{13}P(c) + w_{21}P(a)$$
$$+ w_{22}P(b) + w_{23}P(c) - 2[(\min(w_{11}, w_{21}))P(a)$$
$$+ (\min(w_{12}, w_{22}))P(b) + (\min(w_{13}, w_{23}))P(c)]$$
$$= |w_{11} - w_{21}|P(a) + |w_{12} - w_{22}|P(b) + |w_{13} - w_{23}|P(c). \qquad (16.11)$$

All of the above holds for the case of a, b, c mutually disjoint; but what about the more general situation when they are not disjoint ? Answer: Consider the probabilities of the relative atoms of a, b, c. Then, it is easy to see that the following numerical identity holds:

$$w_{11}P(a) + w_{12}P(b) + w_{13}P(c) = w_{11}[P(abc) + P(abc') + P(ab'c) + P(ab'c')]$$
$$+ w_{12}[P(abc) + P(a'bc) + P(abc') + P(a'bc')]$$
$$+ w_{13}[P(abc) + P(a'bc) + P(ab'c) + P(a'b'c)]$$
$$= (w_{11} + w_{12} + w_{13})P(abc) + (w_{11} + w_{12})P(abc')$$
$$+ (w_{11} + w_{13})P(ab'c) + (w_{12} + w_{13})P(a'bc)$$
$$+ w_{11}P(ab'c') + w_{12}P(a'bc') + w_{13}P(a'b'c), \qquad (16.12)$$

by gathering the coefficients of each relative atom. Similarly,

$$w_{21}P(a) + w_{22}P(b) + w_{23}P(c) = (w_{21} + w_{22} + w_{23})P(abc)$$
$$+(w_{21} + w_{22})P(abc') + (w_{21} + w_{23})P(ab'c) + (w_{22}$$
$$+w_{23})P(a'bc) + w_{21}P(ab'c') + w_{22}P(a'bc') + w_{23}P(a'b'c), \quad (16.13)$$

where the above identities are easily generalized to any number of arguments. Thus, if we replace in the probability evaluations in eqs.(16.12), (16.13) the relative atomic probabilities by the corresponding relative atoms weighted by the corresponding constant probability events, multiplications by cartesian products, and additions of the (disjoint) event probabilities by disjunctions, we obtain

$$
\begin{aligned}
\alpha \;=\; & abc \times \theta(w_{11} + w_{12} + w_{13}) \vee abc' \times \theta(w_{11} + w_{12}) \\
& \vee ab'c \times \theta(w_{11} + w_{13}) \vee a'bc \times \theta(w_{12} + w_{13}) \\
& \vee ab'c' \times \theta(w_{11}) \vee a'bc' \times \theta(w_{12}) \vee a'b'c \times \theta(w_{13}), \\
\beta \;=\; & abc \times \theta(w_{21} + w_{22} + w_{23}) \vee abc' \times \theta(w_{21} + w_{22}) \\
& \vee ab'c \times \theta(w_{21} + w_{23}) \vee a'bc \times \theta(w_{22} + w_{23}) \\
& \vee ab'c' \times \theta(w_{21}) \vee a'bc' \times \theta(w_{22}) \vee a'b'c \times \theta(w_{23}). \quad (16.14)
\end{aligned}
$$

Hence, eqs.(16.12)-(16.14) show that using the product probability structure of P_o, for all P of interest,

$$
\begin{aligned}
P_o(\alpha) &= w_{11}P(a) + w_{12}P(b) + w_{13}P(c), \\
P_o(\beta) &= w_{21}P(a) + w_{22}P(b) + w_{13}P(c), \quad (16.15)
\end{aligned}
$$

compatible with eq.(16.2). Thus, α and β given in eq.(16.14) are the natural underlying events for the numerical probability model given in eq.(16.2) when the events a, b, c are not necessarily mutually disjoint. Note that the constant-probability events in eq.(16.14) each represent numbers still in the unit interval, as required, but are no longer true weighting events, since obviously the total sum of their probability values exceeds unity in general. In turn, eq.(16.14) and use again of the min-conjunction property of constant-probability events shows

$$
\begin{aligned}
\alpha\beta \;=\; & abc \times \theta(min(w_{11} + w_{12} + w_{13}, w_{21} + w_{22} + w_{23})) \\
& \vee abc' \times \theta(\min(w_{11} + w_{12}, w_{21} + w_{22})) \\
& \vee ab'c \times \theta(\min(w_{11} + w_{13}, w_{21} + w_{23})) \\
& \vee a'bc \times \theta(\min(w_{12} + w_{13}, w_{22} + w_{23})) \\
& \vee ab'c' \times \theta(\min(w_{11}, w_{21})) \vee a'bc' \times \theta(\min(w_{12}, w_{22})) \\
& \vee a'b'c \times \theta(\min(w_{13}, w_{23})). \quad (16.16)
\end{aligned}
$$

Hence,

$$P_1(\alpha\beta) \;=\; P(abc)\min(w_{11} + w_{12} + w_{13}, w_{21} + w_{22} + w_{23})$$

$$+P(abc')\min(w_{11}+w_{12}, w_{21}+w_{22})$$
$$+P(ab'c)\min(w_{11}+w_{13}, w_{21}+w_{23})$$
$$+P(a'bc)\min(w_{12}+w_{13}, w_{22}+w_{23})$$
$$+P(ab'c')\min(w_{11}, w_{21})) + P(a'bc')\min(w_{12}, w_{22})$$
$$+P(a'b'c)\min(w_{13}, w_{23})), \tag{16.17}$$

and analogous to eq.(16.11), eq.(16.17) implies

$$
\begin{aligned}
d_{P,2}(\alpha,\beta) &= P_1(\alpha) + P_1(\beta) - 2P_1(\alpha\beta) \\
&= |(w_{11}+w_{12}+w_{13}) - (w_{21}+w_{22}+w_{23})|P(abc) \\
&\quad + |(w_{11}+w_{12}) - (w_{21}+w_{22})|P(abc') \\
&\quad + |(w_{11}+w_{13}) - (w_{21}+w_{23})|P(ab'c) \\
&\quad + |(w_{12}+w_{13}) - (w_{22}+w_{23})|P(a'bc) \\
&\quad + |w_{11} - w_{21}|P(ab'c') \\
&\quad + |w_{12} - w_{22}|P(a'bc') \\
&\quad + |w_{13} - w_{23}|P(a'b'c). \tag{16.18}
\end{aligned}
$$

Thus, the hypotheses tests discussed in Section 9.4 are applicable here using the evaluation in eq.(16.18). Note also that for a, b, c mutually disjoint, we have the only nonvacuous relative atoms as $ab'c' = a$, $a'bc' = b$, $a'b'c = c$ and all of the results in eqs.(16.12)-(16.18) reduce to the corresponding ones in eqs.(16.6)-(16.11).

Finally, it should be remarked that Example 5 can be made more complex, by replacing each unconditional statement a, b, c used above by corresponding PS conditional events, such as $(A|B)$, $(C|D)$, $(E|F)$, where, for simplicity, A, B, C, D, E, F are all unconditional events. Then, all of the above is applicable where now we invoke the PS calculus of operations (such as in Theorem 9 and the following Remark) to determine typical expressions replacing, e.g. $P(ab'c)$ by $P_o((A|B)\&(C|D)'\&(E|F))$, etc.

16.2   Comparing and Combining Polynomials or Analytic Functions of Probabilities with Weighted Coefficients Using Relational Event Algebra and Probability Distances

Motivating Example 6.

Previously in Example 4, eqs.(15.56)-(15.62), the case of

$$h(s) = \sum_{j=0}^{\infty} s^j (1 - s)^m t_j, \qquad \text{for all } s \text{ in } [0,1], \tag{16.19}$$

where the coefficients t_j are all in [0,1] and m is a positive integer was considered, yielding by direct replacement of real variable s by event variable C, multiplication by cartesian product and addition by disjoint disjunction, with appropriate placement,

$$h(C) = \bigvee_{j=0}^{\infty} (C')^{m-1} \times C^j \times C' \times \theta(t_j), \qquad \text{for all } C \in \mathcal{B}. \qquad (16.20)$$

h is indeed a relational event, since for all probabilities P of interest,

$$P_1(h(C)) = h(P(C)). \qquad (16.21)$$

Here, we consider h when $m = 0$, i.e., h takes the simple analytic or polynomial form (about zero)

$$h(s) = \sum_{j=0}^{\infty} s^j t_j, \qquad \text{for all } s \text{ in } [0,1], \qquad (16.22)$$

where the coefficients $t_j \in [0, 1]$, for all j. Thus, again, we could have in place of the two linear forms in three variables in Example 5, eq.(16.2), two polynomials or analytic functions in one variable, such as for the formal relations

$$P(\alpha) = h_1(P(a)), \qquad P(\beta) = h_2(P(a)), \qquad (16.23)$$

where,

$\alpha = $ partially hidden enemy build-up along a frontier exceeds x troops, according to expert 1, using modifier/connector h_1,

$\beta = $ partially hidden enemy build-up along a frontier exceeds x troops, according to expert 2, using modifier/connector h_2.   (16.24)

For example, it could be known that a relationship exists between the probability of

$a = $ partially hidden enemy build-up along a frontier exceeds x troops

and the probability of enemy attack tomorrow. Or, a similar situation could exist as in Example 4, where linguistic-based information produces a modifier, recalling that in Example 4, the modifiers were intensifiers or extensifiers in the form of exponentials, one of which 1.5 required an analytic function of the form in eq.(15.57), with the complementary argument replacements s' for s and C' for C. Here, we assume that the modifier yields a form as in eq.(16.22) or through its complementary argument replacements as in Example 4:

$$h_i(s) = \sum_{j=0}^{\infty} s^j t_{ij} \qquad \text{for all } s \in [0, 1], \, i = 1, 2, \qquad (16.25)$$

with coefficients $t_{ij} \in [0, 1]$, all i, j. Note of course that if we started out with any analytic function h_o with all nonnegative coefficients in its expansion, by normalizing through division by $h_o(1)$, we can always obtain the coefficients to be true weights. One convenient class of such functions is the exponential type $h_{\lambda_1,\lambda_2,\lambda_3}$, where

$$h_{\lambda_1,\lambda_2,\lambda_3}(s) = (\lambda_1 \exp(\lambda_2 s) + \lambda_3)/(\lambda_1 \exp(\lambda_2) + \lambda_3), \quad \text{all } s \in [0,1], \quad (16.26)$$

with λ_1, λ_2 any positive real constants and real constant λ_3 such that $\lambda_3 \geq -\lambda_1$; the limiting case $\lambda_3 = -\lambda_1$, yielding

$$h_{\lambda_1,\lambda_2}(s) = (\exp(\lambda_2 s) - 1)/(\exp(\lambda_2) - 1), \quad \text{all } s \in [0,1]. \quad (16.27)$$

We note first that, unlike the situation in Example 4, although we can substitute an event variable C for real number variable s, and as usual, replace multiplication by cartesian product, when we replace addition by disjunction, it is not disjoint, and in general we will not obtain the counterpart of eq.(15.59). However, if we see that this situation is analogous to that of the non-disjoint case of Example 5, where for the case of the four event variables, we have correspondingly successive powers of a: Ω, a, a^2, a^3 (i.e., $\Omega, a, a \times a, a \times a \times a$), where as usual we identify a with $a \times \Omega \times \Omega$, a^2 with $a^2 \times \Omega$, etc. In this case we seek all the nonvacuous relative atoms generated from the set of events $\{\Omega, a, a^2, a^3\}$ are:

$$\Omega a a^2 a^3 = a^3, \qquad \Omega a a^2 (a^3)' = a^2 \times a',$$
$$\Omega a (a^2)'(a^3)' = a \times a', \qquad \Omega a'(a^2)'(a^3)' = a', \qquad (16.28)$$

so that

$$\Omega = a^3 \vee a^2 \times a' \vee a \times a' \vee a', \quad a = a^3 \vee a^2 \times a' \vee a \times a',$$
$$a^2 = a^3 \vee a \times a' \vee a', \qquad a^3 = a^3. \qquad (16.29)$$

Hence, for h in eq.(16.22) being a polynomials of degree three, for all P of interest

$$
\begin{aligned}
h(P(a)) &= t_0 + t_1 P(a) + t_2 (P(a))^2 + t_3 (P(a))^3 \\
&= t_0 P_1(\Omega) + t_1 P_1(a) + t_2 P_1(a^2) + t_3 P_1(a^3) \\
&= t_0 \left[P_1(a^3) + P_1(a^2 \times a') + P_1(a \times a') + P_1(a') \right] \\
&\quad + t_1 \left[P_1(a^3) + P_1(a^2 \times a) + P_1(a \times a') \right] \\
&\quad + t_2 \left[P_1(a^3) + P_1(a^2 \times a') \right] + t_3 P_1(a^3) \\
&= (t_0 + t_1 + t_2 + t_3) P_1(a^3) + (t_0 + t_1 + t_2) P_1(a^2 \times a') \\
&\quad + (t_0 + t_1) P_1(a \times a') + t_3 P_1(a') \\
&= P_1(\alpha_3), \qquad (16.30)
\end{aligned}
$$

where by obvious inspection, we can let

$$
\begin{aligned}
\alpha_3 &= a^3 \times \theta(t_0 + t_1 + t_2 + t_3) \vee a^2 \times a' \times \theta(t_0 + t_2) \\
&\quad \vee a \times a' \times \theta(t_0 + t_1) \vee a' \times \theta(t_0), \qquad (16.31)
\end{aligned}
$$

noting by normalization $t_0 + t_1 + t_2 + t_3 = 1$, whence $\theta(t_0 + t_1 + t_2 + t_3) = \Omega_1$.

The general case readily follows from eq.(16.31), so that for h in eq.(16.22), we have for n-degree polynomials

$$h(P(a)) = P_1(\alpha_n), \tag{16.32}$$

where

$$\alpha_n = a^n \times \theta(t_0 \mid t_1 \mid \ \cdots \ \mid t_n) \ \lor \ u^{n-1} \land u' \land \theta(l_0 + l_1 + \ldots + l_{n-1})$$
$$\lor \cdots \lor \ a^2 \times a' \times \theta(t_0 + t_1 + t_2) \ \lor \ a \times a' \times \theta(t_0 + t_1) \ \lor a' \ \times \theta(t_0). \tag{16.33}$$

Thus, as $n \to \infty$, we have the full infinite series result for h in eq.(16.22), provided $P(a) < 1$:

$$h(P(a)) = P_1(\alpha_\infty), \tag{16.34}$$

where now

$$\alpha_\infty = \bigvee_{j=0}^{\infty} (a^j \times a' \times \theta(t_0 + t_1 + \cdots + t_j)) \tag{16.35}$$

thus reducing to a special case – corresponding to $m = 1$ – similar to Example 4, eqs.(15.57), (15.58), etc.

Returning back to the example at hand, let us then consider eqs. (16.23) and (16.25). Thus, eq.(16.35) furnishes the corresponding relational events are

$$\alpha = \bigvee_{j=0}^{\infty} \left(a^j \times a' \times \theta(t_{10} + t_{11} + \cdots + t_{1j}) \right),$$

$$\beta = \bigvee_{j=0}^{\infty} \left(a^j \times a' \times \theta(t_{20} + t_{21} + \cdots + t_{2j}) \right), \tag{16.36}$$

with the compatibility equations

$$P_1(\alpha) = \sum_{j=0}^{\infty} \left((P(a))^j P(a')(t_{10} + t_{11} + \cdots + t_{1j}) \right) = h_1(P(a)),$$

$$P_1(\beta) = \sum_{j=0}^{\infty} \left((P(a))^j P(a')(t_{20} + t_{21} + \cdots + t_{2j}) \right) = h_2(P(a)). \tag{16.37}$$

Hence, again using the mutual disjointness of terms and the min-conjunction property of constant-probability events (eq.(13.95)),

$$\alpha\beta = \bigvee_{j=0}^{\infty} \left(a^j \times a' \times \theta(\min(t_{10} + t_{11} + \cdots + t_{1j}, t_{20} + t_{21} + \cdots + t_{2j})) \right), \tag{16.38}$$

with

$$P_1(\alpha\beta) = \sum_{j=0}^{\infty} \left((P(a))^j P(a') \min(t_{10} + t_{11} + \cdots + t_{1j}, t_{20} + t_{21} + \cdots + t_{2j}) \right).$$
$$\tag{16.39}$$

Then, combining eqs.(16.37) and (16.38) with eq.(9.27) yields the probability distance computation

$$d_{P,2}(\alpha,\beta) = P(\alpha) + P(\beta) - 2P(\alpha\beta)$$

$$= \sum_{j=0}^{\infty} [(P(a))^j P(a')$$

$$\cdot |(t_{10} + t_{11} + \cdots + t_{1j}) - (t_{20} + t_{21} + \cdots + t_{2j})|]. \quad (16.40)$$

Notice the similarity of form in eq.(16.40) with the probability distance expressions in eq. (16.18): both being a sum of terms, each the product of the probability of a relative atom with a coefficient that is the absolute difference of corresponding weights or related functions from each of the two models.

16.3 Comparison of Models Whose Uncertainties Are Two Argument Quadratic Functions

Motivating Example 7.

The uncertainty of the models considered here is of the formal type

$$P(\alpha) = t_{10} + t_{11}P(a) + t_{12}P(b)$$
$$+ t_{13}(P(a))^2 + t_{14}(P(b))^2 + t_{15}P(a)P(b),$$
$$P(\beta) = t_{20} + t_{21}P(a) + t_{22}P(b)$$
$$+ t_{23}(P(a))^2 + t_{24}(P(b))^2 + t_{25}P(a)P(b), \quad (16.41)$$

where for each $i = 1, 2$, the constant coefficients t_{ij} are weights and a, b are contributing events (not necessarily disjoint) whose probabilities are known. Analogous to Example 6, we seek all the possibly nonvacuous relative atoms generated by $\{\Omega, a, a^2, b, b^2, a \times b\}$. From eq.(16.28), this is equivalent to finding the nonvacuous relative atoms of $\{a', a \times a', a^2, b', b \times b', b^2, a \times b\}$. By using the partitioning properties of the a and b terms, not including $a \times b$, this comes down to finding the nonvacuous terms among

$$\{(a \times b)\&\gamma, (a \times b)'\&\gamma : \gamma \in \{a'b', a'b \times b', a'b \times b, ab' \times a', ab \times a'b',$$
$$ab \times a'b, ab' \times a, ab \times ab', (ab)^2\}\}$$
$$= \{ab' \times a'b, ab \times a'b, ab' \times ab, (ab)^2, a'b', a'b \times b',$$
$$a'b \times b, ab' \times a'b', ab \times a'b', (ab')^2, ab \times ab'\}. \quad (16.42)$$

In turn, it follows that we have the following disjoint decompositions for each event in $\{\Omega, a, a^2, b, b^2, a \times b\}$, analogous to eq.(16.29):

$$\Omega_1 = ab' \times a'b \vee ab \times a'b \vee ab' \times ab \vee (ab)^2 \vee a'b' \vee a'b \times b'$$
$$\vee a'b \times b \vee ab' \times a'b' \vee ab \times a'b' \vee (ab')^2 \vee ab \times ab',$$

$$a = ab' \times a'b \vee ab \times a'b \vee ab' \times ab \vee (ab)^2 \vee ab' \times a'b' \vee ab \times a'b'$$
$$\vee (ab')^2 \vee ab \times ab',$$
$$a^2 = ab' \times ab \vee (ab)^2 \vee (ab')^2 \vee ab \times ab',$$
$$b = ab \times a'b \vee (ab)^2 \vee a'b \times b' \vee a'b \times b \vee ab \times a'b' \vee ab \times ab',$$
$$b^2 = ab \times a'b \vee (ab)^2 \vee a'b \times b,$$
$$a \times b = ab' \times a'b \vee ab \times a'b \vee ab' \times ab \vee (ab)^2. \tag{16.43}$$

Hence, using eqs.(16.41), (16.43), we can write, slightly abusing notation,

$$P(\alpha) = t_{10} + t_{11}P(a) + t_{12}P(b) + t_{13}(P(a))^2 + t_{14}(P(b))^2 + t_{15}P(a)P(b)$$
$$= t_{10}P_1(\Omega_1) + t_{11}P_1(a \times \Omega_1) + t_{12}P_1(b \times \Omega_1) + t_{13}P_1(a^2)$$
$$+ t_{14}P_1(b^2) + t_{15}P_1(a \times b)$$
$$= t_{10}P_1(ab' \times a'b \vee ab \times a'b \vee ab' \times ab \vee (ab)^2 \vee a'b' \vee a'b \times b'$$
$$\vee a'b \times b \vee ab' \times a'b' \vee ab \times a'b' \vee (ab')^2 \vee ab \times ab')$$
$$+ t_{11}P_1(ab' \times a'b \vee ab \times a'b \vee ab' \times ab \vee (ab)^2 \vee ab' \times a'b' \vee$$
$$ab \times a'b' \vee (ab')^2 \vee ab \times ab')$$
$$+ t_{12}P_1(ab \times a'b \vee (ab)^2 \vee a'b \times b' \vee a'b \times b \vee ab \times a'b' \vee ab \times ab')$$
$$+ t_{13}P_1(ab' \times ab \vee (ab)^2 \vee (ab')^2 \vee ab \times ab')$$
$$+ t_{14}P_1(ab \times a'b \vee (ab)^2 \vee a'b \times b)$$
$$+ t_{15}P_1(ab' \times a'b \vee ab \times a'b \vee ab' \times ab \vee (ab)^2)$$
$$= (t_{10} + t_{11} + t_{15})P_1(ab' \times a'b) + (t_{10} + t_{11} + t_{12} + t_{14} + t_{15})P_1(ab \times a'b)$$
$$+ (t_{10} + t_{11} + t_{13} + t_{15})P_1(ab' \times ab)$$
$$+ (t_{10} + t_{11} + t_{12} + t_{13} + t_{14} + t_{15})P_1((ab)^2)$$
$$+ t_{10}P_1(a'b') + (t_{10} + t_{12})P_1(a'b \times b') + (t_{10} + t_{12} + t_{14})P_1(a'b \times b)$$
$$+ (t_{10} + t_{11})P_1(ab' \times a'b') + (t_{10} + t_{12} + t_{12})P_1(ab \times a'b')$$
$$+ (t_{10} + t_{11} + t_{13})P_1((ab')^2) + (t_{10} + t_{11} + t_{12} + t_{13})P_1(ab \times ab'),$$
$$\tag{16.44}$$

by gathering appropriate coefficients, and using the constant-probability event and product structure of P_1, where

$$\alpha = ab' \times a'b \times \theta(t_{10} + t_{11} + t_{15}) \vee ab \times a'b \times \theta(t_{10} + t_{11} + t_{12} + t_{14} + t_{15})$$
$$\vee ab' \times ab \times \theta(t_{10} + t_{11} + t_{13} + t_{15})$$
$$\vee (ab)^2 \times \theta(t_{10} + t_{11} + t_{12} + t_{13} + t_{14} + t_{15})$$
$$\vee a'b' \times \theta(t_{10}) \vee a'b \times b' \times \theta(t_{10} + t_{12}) \vee a'b \times b \times \theta(t_{10} + t_{12} + t_{14})$$
$$\vee ab' \times a'b' \times \theta(t_{10} + t_{11}) \vee ab \times a'b' \times \theta(t_{10} + t_{11} + t_{12})$$
$$\vee (ab')^2 \times \theta(t_{10} + t_{11} + t_{13}) \vee ab \times ab' \times \theta(t_{10} + t_{11} + t_{12} + t_{13}), \tag{16.45}$$

noting that $t_{10} + t_{11} + t_{12} + t_{14} + t_{15} = 1$ and hence

$$\theta(t_{10} + t_{11} + t_{12} + t_{13} + t_{14} + t_{15}) = \Omega_1. \tag{16.46}$$

Similarly, by letting

$$
\begin{aligned}
\beta \;=\; & ab' \times a'b \times \theta(t_{20}+t_{21}+t_{25}) \;\vee\; ab \times a'b \times \theta(t_{20}+t_{21}+t_{22}+t_{24}+t_{25}) \\
& \vee \;\; ab' \times ab \times \theta(t_{20}+t_{21}+t_{23}+t_{25}) \\
& \vee \;\; (ab)^2 \times \theta(t_{20}+t_{21}+t_{22}+t_{23}+t_{24}+t_2) \\
& \vee \;\; a'b' \times \theta(t_{20}) \;\vee\; a'b \times b' \times \theta(t_{20}+t_{22}) \;\vee\; a'b \times b \times \theta(t_{20}+t_{22}+t_{24}) \\
& \vee \;\; ab' \times a'b' \times \theta(t_{20}+t_{21}) \;\vee\; ab \times a'b' \times \theta(t_{20}+t_{21}+t_{22}) \\
& \vee \;\; (ab')^2 \times \theta(t_{20}+t_{21}+t_{23}) \;\vee\; ab \times ab' \times \theta(t_{20}+t_{21}+t_{22}+t_{23}), \quad (16.47)
\end{aligned}
$$

$$
\begin{aligned}
P(\beta) \;=\;\; & t_{20}+t_{21}P(a)+t_{22}P(b)+t_{23}(P(a))^2+t_{24}(P(b))^2+t_{25}P(a)P(b) \\
=\;\; & P_1(\beta). \hspace{9cm} (16.48)
\end{aligned}
$$

Hence, α, β as given in eqs.(16.45), (16.46) are relational events corresponding to eq.(16.41). Thus, as before, we can now compute the conjunction probability, and in turn, a distance probability, such as $d_{P,2}(\alpha,\beta)$, analogous to the results in Example 6, yielding again the absolute difference of weight forms:

$$
\begin{aligned}
d_{P,2}(\alpha,\beta) = \; & P_1(ab' \times a'b)|(t_{10}+t_{11}+t_{15})-(t_{20}+t_{21}+t_{25})| \\
& +P_1(ab \times a'b)|(t_{10}+t_{11}+t_{12}+t_{14}+t_{15})-(t_{20}+t_{21}+t_{22}+t_{24}+t_{25})| \\
& +P_1(ab' \times ab)|(t_{10}+t_{11}+t_{13}+t_{15})-(t_{20}+t_{21}+t_{23}+t_{25})| \\
& +P_1((ab)^2)|(t_{10}+t_{11}+t_{12}+t_{13}+t_{14}+t_{15}) \\
& \hspace{2cm} -(t_{20}+t_{21}+t_{22}+t_{23}+t_{24}+t_{25})| \\
& +P_1(a'b')|t_{10}-t_{20}| + P_1(a'b \times b')|(t_{10}+t_{12})-(t_{20}+t_{22})| \\
& +P_1(a'b \times b')|(t_{10}+t_{12})-(t_{20}+t_{22})| \\
& +P_1(a'b \times b)|(t_{10}+t_{12}+t_{14})-(t_{20}+t_{22}+t_{24})| \\
& +P_1(ab' \times a'b')|(t_{10}+t_{11})-(t_{20}+t_{21})| \\
& +P_1(ab \times a'b)|(t_{10}+t_{11}+t_{12})-t_{20}+t_{21}+t_{22})| \\
& +P_1((ab')^2)|(t_{10}+t_{11}+t_{13})-(t_{20}+t_{21}+t_{23})| \\
& +P_1(ab \times ab')|(t_{10}+t_{11}+t_{12}+t_{13})-(t_{20}+t_{21}+t_{22}+t_{23})|, \quad (16.49)
\end{aligned}
$$

where

$$
P_1(ab' \times a'b) = P(ab')P(a'b), \; P_1((ab)^2) = (P(ab))^2, \hspace{2cm} (16.50)
$$

$$
|(t_{10}+t_{11}+t_{12}+t_{13}+t_{14}+t_{15})-(t_{20}+t_{21}+t_{22}+t_{23}+t_{24}+t_{25})| = 1-1 = 0. \; (16.51)
$$

and the $P_1((ab)^2)$ term drops out, etc. For more than two arguments and/or higher degree polynomials, similar results as above hold, except for the increased number of relative atomic terms involved.

16.4   General Relational Event Algebra Problem and a Modification for Relational Events Having Constant-Probability Event Coefficients Possibly Dependent upon Probabilities

So far, we have considered problems where numerical functions of probabilities were given and corresponding algebraic counterparts which when evaluated in an appropriate probability space yielded back the given numerical functions of the probabilities. These algebraic objects were actually events in an appropriately chosen boolean (or sigma-) algebra, being conditional events in the case of arithmetic divisions, and more generally, relational events, in the case of other numerical functions, such as weighted averages, or polynomials or analytic functions in the form of series with coefficients in the unit interval. The mathematical formulation of this, where we consider, for simplicity, as in all of the preceding examples, the case of two functions (the case for one function or any finite number exceeding two follows similarly) is: Given $f, g: [0,1]^m \rightarrow [0,1]$ and any probability space $(\Omega, \mathcal{B}, P)$ find a probability space $(\Omega_1, \mathcal{B}_1, P_1)$ extending $(\Omega, \mathcal{B}, P)$ (isomorphically, compatible with probability), with $(\Omega_1, \mathcal{B}_1)$ dependent on $(\Omega, \mathcal{B})$ and possibly f, g, but not on P, and functions $\alpha(f)$, $\alpha(g) : \mathcal{B}^n \rightarrow \mathcal{B}_1$, so that for any for events $a_1, ..., a_m \in \mathcal{B}$, subject to possibly some constraint R– usually involving relative ordering among them – the formal equations

$$P(\alpha) = f(P(a_1), ..., P(a_n)), \qquad P(\beta) = g(P(a_1), ..., P(a_n)), \qquad (16.52)$$

for all $a_1, ..., a_n \in \mathcal{B}$, subject to R, all well-defined P, can be replaced by probability-of-single-event equations, where $\alpha = \alpha(f)(a_1, ..., a_n)$ and $\beta = \beta(f)(a_1, ..., a_n)$ are actual (relational) events in $\mathcal{B}_1$ (the relational event algebra) – also, not dependent on P – and P is replaced by P_1 , so that eq.(16.52) now becomes

$$P_1(\alpha) = f(P(a_1), ..., P(a_n)), \qquad P_1(\beta) = g(P(a_1), ..., P(a_n)), \qquad (16.53)$$

for all $a_1, ..., a_n \in \mathcal{B}$ subject to R, all well-defined P.

In general (see the previous discussion on Lewis' theorem in Section 11.5), $(\Omega_1, \mathcal{B}_1, P_1)$ is a proper extension of $(\Omega, \mathcal{B}, P)$. Schematically, this can be summarized by the following homomorphic, or equivalently, commutative diagram:

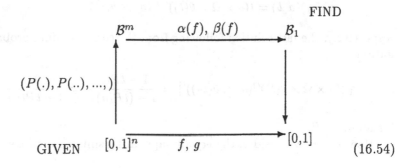

$$(16.54)$$

All functions in the above diagram commute.

Suppose, now we consider a few simple-appearing functions of probabilities for which we seek corresponding relational or conditional events. We know, of course, for functions such as

$f_1(P(a), P(b)) = P(a) + P(b) - P(ab)$, for all $a, b \in \mathcal{B}$,
$f_2(P(a), P(b)) = P(a) + P(b) - 2P(ab)$, for all $a, b \in \mathcal{B}$,
$f_3(P(a), P(b)) = P(b) - P(a)$, for all $a, b \in \mathcal{B}$, with $b \geq a$,
$f_4(P(a), P(b)) = P(a)/P(b)$, for all $a, b \in \mathcal{B}$, with $b \geq a$,
$f_5(P(a), P(b)) = P(a)P(b)$, for all $a, b \in \mathcal{B}$,
$f_6(P(a), P(b)) = w$, for all $a, b \in \mathcal{B}$, $0 \leq w \leq 1$ any fixed real number (weight),
$f_7(P(a), P(b)) = wP(a)$, for all $a, b \in \mathcal{B}$, $0 \leq w \leq 1$ any fixed real number,
$f_8(P(a), P(b)) = (P(a))^r$, for any $a \in \mathcal{B}$, r any fixed non-negative real number,
$f_9(P(a), P(b)) = wP(a) + (1-w)P(b)$, $0 \leq w \leq 1$ any fixed real number,
$f_{10}(P(a), P(b)) = \sum_{j=0}^{\infty}(P(a))^j t_j$, for any $a \in \mathcal{B}$, t_j constants, $0 \leq t_j \leq 1$, $j = 0, 1, 2, \ldots$,

$$f_{11}(P(a), P(b)) = 1/(1 + tP(a)), \quad t > 0, \tag{16.55}$$

we have, from either inspection of the form or use of relational or conditional event algebra from the previously developed examples, that the corresponding relational events (satisfying eq.(16.53) for the single function case) are:

$$\alpha(f_1)(a, b) = a \vee b, \qquad \alpha(f_2)(a, b) = a + b,$$
$$\alpha(f_3)(a, b) = b - a = a'b, \qquad \alpha(f_4)(a, b) = (a|b),$$
$$\alpha(f_5)(a, b) = \theta(w), \qquad \alpha(f_6)(a, b) = a \times \theta(w),$$
$$\alpha(f_7)(a, b) = a \times b \text{ or } b \times a \text{ or a symmetric weighted average}$$
$$\text{such as } (a \times b \times \theta(1/2)) \vee (b \times a \times \theta(1/2)),$$
$$\alpha(f_8)(a, b) = a^r,$$
$$\alpha(f_9)(a, b) = (ab \times \Omega_1) \vee (ab' \times \theta(w)) \vee (a'b \times \theta(1-w)),$$
$$\alpha(f_{10})(a, b) = \bigvee_{j=0}^{\infty}(a^j \times a' \times \theta(t_0 + \cdots + t_j)), \text{ provided } P(a) < 1$$

$$\alpha(f_{11})(a, b) = ((a \times \Omega \times \theta(t))' \mid (a^2 \times \theta(t^2))') \qquad (16.56)$$

respectively, the last solution following from the corresponding probability evaluation

$$P_1\left[(a \times \Omega \times \theta(t))'|(a^2 \times \theta(t_2))'\right] = \frac{1 - tP(a)}{1 - (tP(a))^2} = \frac{1}{1 + tP(a)} \qquad (16.57)$$

for all a, all P.

However, now consider, the equally-appearing simple functions

$$\begin{aligned}
f_{12}(P(a), P(b)) &= \min(P(a), P(b)), \\
f_{13}(P(a), P(b)) &= \max(P(a), P(b)), \\
f_{14}(P(a), P(b)) &= \min(2P(a), 1), \\
f_{15}(P(a), P(b)) &= \min(3P(a), 1), \\
f_{16}(P(a), P(b)) &= \min(P(a) + P(b), 1), \\
f_{17}(P(a), P(b)) &= \max(P(b) - P(a), 0), \\
f_{18}(P(a), P(b)) &= |P(b) - P(a)|.
\end{aligned} \qquad (16.58)$$

As opposed to the examples in eq.(16.56), with relational event solutions in eq.(16.57), there appears *no* way in general to construct corresponding relational events, as we have defined them, so. In addition, we note that all constant-probability events constructed so far, do not depend on any choice of a particular probability measure P and generating event c, so long as $1 > P(c) > 0$. However, this restriction may need to be relaxed when we seek relational event representation of certain functions of probabilities which do not appear amenable to the methods we have developed so far. That such events could be constructed dependent upon each P of choice poses no theoretical problem, by inspection of the constructions in Sections 10.3 and 13.3. We then construct a modified version of relational events as follows : We still construct an appropriate probability space $(\Omega_1, \mathcal{B}_1, P_1)$ extending $(\Omega, \mathcal{B}, P)$ with $(\Omega_1, \mathcal{B}_1)$ not depending on P, but for the choice of $\alpha(f)$, we attempt to modify it to be in a linear (if possible) form of the relative atoms with possible P-varying constant-probability event coefficients. Thus, for the two event variable case $f(P(a), P(b))$, we consider

$$\begin{aligned}
\alpha(f)(a, b, P) = \ & ab \times \theta(t_1(a, b, P)) \vee ab' \times \theta(t_2(a, b, P)) \\
& \vee a'b \times \theta(t_3(a, b, P)) \vee a'b' \times \theta(t_4(a, b, P)), \quad (16.59)
\end{aligned}$$

where, of course, the $t_j(a, b, P) \in [0, 1]$, so that the basic compatibility equation is satisfied :

$$P_1\left(\alpha(f)(a, b, P)\right) = f(P(a), P(b)) \qquad (16.60)$$

for all a, b subject to a possible constraint, and all P.

That such modified constant-probability events can be constructed follows from the construction of the previous type of constant-probability events. (See

comments at end of Section 13.3) More specifically, consider the first function
in eq.(16.58). Then, using basic properties of relative atomic probabilities,

$$
\begin{aligned}
\min(P(a), P(b)) &= \min(P(ab) + P(ab'), P(ab) + P(a'b)) \\
&= P(ab) + \min(P(ab'), P(a'b)),
\end{aligned}
\tag{16.61}
$$

where we want to find an appropriate choice of functions t_1, t_2 with ranges in
$[0,1]$, such that the t_1, t_2 are continuous functions of their arguments, and form
an event β disjoint from ab in the form

$$
\beta_{\min} = \beta_{\min}(a, b, P) = a'b \times \theta(t_1(a, b, P) \ \vee \ ab' \times \theta(t_2(a, b, P))
\tag{16.62}
$$

such that

$$
P_1(\beta_{\min}) = \min(P(ab'), P(a'b)),
\tag{16.63}
$$

and
   (i)   if $P(ab') = P(a'b)$, $t_1(a, b, P) = t_2(a, b, P) = 1/2$;
   (ii) if $P(ab') > P(a'b)$, then such a choice of t_1 and t_2 focuses more weight
toward $P(a'b)$ – and hence $P(b)$ – and increases as their relative separation, in
some sense, increases;
   (iii) if $P(ab') < P(a'b)$, then such a choice of t_1 and t_2 focuses

more weight toward $P(ab')$ – and hence $P(a)$ – and increases as

their relative separation, in some sense, increases.   (16.64)

Consider as a reasonable candidate for t_1, t_2 satisfying eqs.(16.62)-(16.64),
assuming $P(ab') > 0$ or $P(a'b) > 0$,

$$
\begin{aligned}
t_1(a, b, P) &= (P(ab')/(P(a'b) + P(ab'))) \min(P(ab')/P(a'b), 1), \\
t_2(a, b, P) &= (P(a'b)/(P(a'b) + P(ab'))) \min(P(a'b)/P(ab'), 1),
\end{aligned}
\tag{16.65}
$$

all $a, b \in \mathcal{B}$, all P, with the convention that

$$
t_1(a, b, P) = t_2(a, b, P) = 0, \quad \text{if } P(ab') = P(a'b) = 0.
\tag{16.66}
$$

By inspection of eq.(16.65), t_1 and t_2 satisfy eq.(16.63). Moreover, for example
for $P(a'b) < P(ab')$,

$$
\begin{aligned}
P_1(\beta) &= P(a'b)t_1(a, b, P) + P(ab')t_2(a, b, P) \\
&= P(a'b)[P(ab')/(P(a'b) + P(ab'))] \\
&\quad + P(ab')[P(a'b)/(P(a'b) + P(ab'))](P(a'b)/P(ab')) \\
&= [P(a'b)P(ab') + (P(a'b))^2]/(P(a'b) + P(ab')) \\
&= P(a'b) = \min(P(a'b), P(ab')),
\end{aligned}
\tag{16.67}
$$

as required.
   Hence, considering eq.(16.61), we can take

$$
\alpha(\min)(a, b, P) = ab \times \Omega_1 \ \vee \ \beta_{\min}(a, b, P),
\tag{16.68}
$$

where $\beta_{\min}(a, b, P)$ is given in eq.(16.62). It also follows easily that we can show for $\max(P(a), P(b))$ that the corresponding modified relational event is

$$\alpha(\max)(a, b, P) = (\alpha(\min)(a', b', P))' = ab \times \Omega_1 \vee \beta_{\max}, \qquad (16.69)$$

where now

$$\beta_{\max}(a, b, P) = a'b \times (\theta(t_1(a, b, P))' \vee ab' \times (\theta(t_2(a, b, P)))'. \qquad (16.70)$$

It also follows that, as a check, eqs.(16.68) and (16.69) show

$$\alpha(\max)(a, b, P) \vee (\alpha(\min)(a, b, P)) = a \vee b,$$
$$\alpha(\max)(a, b, P) \& (\alpha(\min)(a, b, P)) = ab, \qquad (16.71)$$

etc.

Next, consider in eq.(16.58) the function $f_{14}(a, b) = \min(2P(a), 1)$. An easily proven identity is

$$\min(2P(a), 1) = P(a) + P(a')\min(P(a)/P(a'), 1). \qquad (16.72)$$

Hence, by inspection of eq.(16.72), a reasonable form of the required modified relational event here is

$$\alpha(f_{14})(a, b, P) = a \times \Omega_1 \vee a' \times \theta(\min(P(a)/P(a'), 1)). \qquad (16.73)$$

Similarly, the identity

$$\min(3P(a), 1) = P(a) + P(a')\min(2P(a)/P(a'), 1) \qquad (16.74)$$

yields the modified relational event

$$\alpha(f_{15})(a, b, P) = a \times \Omega_1 \vee a' \times \theta(\min(2P(a)/P(a'), 1)). \qquad (16.75)$$

Next, consider the identity

$$P(a) + P(b) \leq 1 \quad \text{iff} \quad 2P(ab) + P(ab') + P(a'b) \leq 1$$
$$\text{iff} \quad P(ab) \leq P(a'b'). \qquad (16.76)$$

Then, by inspection of eq.(16.76), the corresponding relational event to f_{16} is

$$\alpha(f_{16})(a, b, P) = ab \times \Omega_1 \vee ab' \times \Omega_1 \vee a'b \times \Omega_1$$
$$\vee a'b' \times \theta(\min(P(ab)/P(a'b'), 1)). \qquad (16.77)$$

As a check, it can be shown, following simplifications, that analogous to the decomposition

$$\min(P(a) + P(b), 1)$$
$$= \min(P(ab') + P(a'b) + 2P(ab), P(ab') + P(a'b) + P(ab) + P(a'b'))$$
$$= P(a + b) + \min(\min(2P(ab), 1), P(a \Leftrightarrow b)), \qquad (16.78)$$

$$\alpha(f_{16})(a, b, P) = (a + b) \vee a(f_{12})(\alpha(f_{14})(a, b), (a \Leftrightarrow b) \times \Omega_1, P_1). \qquad (16.79)$$

Next, consider f_{17}. Note the identity

$$\max(P(b) - P(a), 0) = \max(P(a'b) - P(ab'), 0) \qquad (16.80)$$

and for any $1 \geq y \geq x > 0$, and functions $t_1 = t_1(x, y)$, $t_2 = t_2(x, y)$, the equivalence

$$y - x = t_1 y + t_2 x, \quad \text{for } 0 \leq t_1, t_2 \leq 1$$
$$\text{iff} \quad t_2 = \frac{y}{x}(1 - t_1) - 1, \quad \text{for } 0 \leq t_1, t_2 \leq 1$$
$$\text{iff} \quad t_2 = \frac{y}{x}(1 - t_1) - 1 \text{ and } \max(1 - 2\frac{x}{y}, 0) \leq t_1 \leq 1 - \frac{x}{y}. \qquad (16.81)$$

We also require that the t_i be such that they are both continuous functions of x, y through x/y and that assuming $1 \geq y \geq x > 0$,

(i) $t_1(x/y)$ is nonincreasing in x/y for $0 < x/y \leq 1$, with $\lim_{x/y \to 0} t_1(x/y) = 1$ and $t_1(1) = 0$;

(ii)   there is a point u_o, $0 \leq u_o \leq 1$, where $t_2(x/y)$ first is nondecreasing in x/y over $(0, u_o]$, with $\lim_{x/y \to 0} t_2(x/y) = 0$ and then, is nonincreasing in x/y over $[u_o, 1]$ with $t_2(1) = 0$ \hfill (16.82)

Choose any real constant λ, $0 \leq \lambda \leq 1$ and define

$$t_1(x/y, \lambda) = \begin{cases} \max(0, 1 - [(1 + \lambda(x/y))(x/y)]), & \text{if } 0 < x \leq y \leq 1 \text{ or } 0 = x < y \\ 0, & \text{if } 0 \leq x \leq y \leq 1 \text{ or } 0 = x = y \end{cases}$$
$$(16.83)$$

taking into account now all $0 \leq x, y \leq 1$. Note that for $1 \geq y \geq x > 0$, t_1 as defined in eq.(16.83) satisfies bottom(16.81) and (i) in eq.(16.82). Next, by lhs(16.81),

$$t_2(x/y, \lambda) = (y/x)(1 - t_1(x/y, \lambda)) - 1$$
$$= ((y/x)\min(1, (1 + \lambda(x/y))(x/y))) - 1$$
$$= \min((y/x) - 1, \lambda(x/y)), \quad \text{for } 1 \geq y \geq x > 0,$$

and thus define

$$t_2(x/y, \lambda) = \begin{cases} \min((y/x) - 1, \lambda(x/y)) & \text{if } 0 < x \leq y \leq 1 \text{ or } 0 = x < y \\ 0 & \text{if } 0 \leq y < x \leq 1 \text{ or } 0 = x = y \end{cases}$$
$$(16.84)$$

taking into account all $0 \leq x, y \leq 1$.

By inspection of eq.(16.81), $t_2(x/y, \lambda)$ satisfies (ii) in eq.(16.82), where $u_o(\lambda)$ satisfies via eq.(16.84),
$$(1/u_o(\lambda)) - 1 = \lambda u_o(\lambda), \text{ whence,}$$

$$u_o(\lambda) = (-1 + \sqrt{1 + 4\lambda})/(2\lambda), \qquad (16.85)$$

noting that as λ varies in choice from 0 to 1, correspondingly $u_o(\lambda)$ strictly decreases from 1 at $\lambda = 0$ down to $(-1 + \sqrt{5})/2 \approx 0.62$ at $\lambda = 1$. In fact for $\lambda = 0$,

$$t_1(x/y, 0) = 1 - (x/y), \qquad t_2(x/y, 0) = 0, \tag{16.86}$$

with corresponding trivial identity replacing top(16.81) as

$$y - x = t_1(x/y, 0)y + t_2(x/y, 0)x = 0 = (1 - (x/y))y. \tag{16.87}$$

Consider now the application of the development in eqs.(16.01)-(16.07) to eq.(16.80), where $y = P(a'b)$ and $x = P(ab')$. Thus, the corresponding modified relational event to f_{17} at any $a, b \in \mathcal{B}$ is

$$\begin{aligned}
\alpha(f_{17})(a, b, P) &= a'b \times \theta(t_1(P(ab')/P(a'b), \lambda)) \\
&\quad \vee ab' \times \theta(t_2(P(ab')/P(a'b), \lambda)). \tag{16.88}
\end{aligned}$$

The reason for introducing the constraints in eq.(16.82) is to have the weighting due to the constant-probability terms be compatible with the values of $P(ab')/P(a'b)$. For example, as $P(ab')/P(a'b)$ increases the coefficient term for $a'b$ should decrease, while that for ab' should increase, etc.

In turn, let

$$\begin{aligned}
\alpha(f_{18})(a, b, P) &= \alpha(f_{17})(a, b, P) \vee \alpha(f_{17})(b, a, P) \\
&= a'b \times \theta(\max[t_1(a, b, P), t_2(b, a, P)]) \\
&\quad \vee ab' \times \theta(\max[t_1(b, a, P), t_2(a, b, P)]), \tag{16.89}
\end{aligned}$$

and noting from the above constructions

$$\alpha(f_{17})(a, b, P) \& \alpha(f_{17})(b, a, P) = \emptyset_o, \tag{16.90}$$

$$\begin{aligned}
P_1(\alpha(f_{18})(a, b, P)) &= P_1(\alpha(f_{17})(a, b, P)) + P_1(\alpha(f_{17})(b, a, P)) \\
&= \max(P(b) - P(a), 0) + \max(P(a) - P(b), 0) \\
&= |P(a) - P(b)|. \tag{16.91}
\end{aligned}$$

Whether or not the naive distance function, represented in eq.(16.91) as the probability evaluation of a relation event in the modified sense as used here, though obviously a metric (see the Remark following Theorem 22) is a true probability distance function, i.e., for each fixed P, $\alpha(f_{18})(., .., P))$ is an algebraic metric in the obviously modified sense for each P is an open question. (Note also, because of the dependency on P, we cannot simply invoke Theorem 21(ii).) If not, it is of some interest to find a similar-structured function which is an algebraic metric (in the modified sense).

In summary, we need to be able to not only obtain reasonable forms of relational events representing given functions of probabilities – whether not dependent upon probabilities, if possible, as in the case of functions $f_1 - f_{11}$, or probability-dependent, as in the case of functions $f_{12} - f_{18}$, but also a calculus of operations and relations – especially conjunction or disjunction – in order to be able to compute desired probability distance functions for similarities or contrasts, as shown more explicitly in the previous motivating examples.

16.5   Concluding Remarks

A number of research issues remain in conditional event algebra, including:

1. Determine the full characterization of the class of boolean conditional event algebras in addition to PS. Thus, in light of Theorem 11, such a characterization necessarily involves omitting the special independence property requirement. (See the example in the remark following Theorem 10 for typical construction of non-PS boolean cea's satisfying modus ponens and other reasonable properties.) Are there such boolean cea's which have logical operations simpler than the PS ones ?

Apropos to the compatibility of conditional events and conditioning of random variables relative to PS, as shown in Section 12.3 (which is not applicable to the non-boolean cea's DGNW and AC) the following "local independence" property is always possessed by PS: For any joint random variables X, Y, W, Z with respect to a probability measure P, for any choice of events b, d which are each made to approach single points in n-dimensional space, the corresponding conditioned random variables $(X|Y \in b)$ and $(W|Z \in d)$ asymptotically are P_o-independent. (For more details, see [4], Sections 3.1 and 3.5.) Thus, it is natural to inquire whether the special independence property that PS possesses is too strong a property, and that boolean non-special independence cea's should be sought, in the spirit of the example constructed in Section 12.4.

2. Characterize in some convenient component-wise manner the partial ordering $\leq$ of events in $\mathcal{B}_o$ of the form $[\alpha|b]$, $\alpha \in \mathcal{B}_o$, $b \in \mathcal{B}$, analogous in some way to the characterization of $\leq$ between ordinary conditional events $(a|b) \in \mathcal{B}_o$, $a, b \in \mathcal{B}$. Such characterization can play a great simplifying role in the determination of deduction and enduction classes among $[\alpha|b]$-type events, extending the relatively closed-form results for ordinary conditionals in eqs. (12.28)-(12.31) and Theorem 14.

3. Despite the recursive forms available (see the remark following Theorem 9), the determination of probability evaluations of logical combinations of conditional events in PS require a much large number of elementary boolean operations than that utilized by the non-boolean cea's DGNW and AC. The growth rate of this number is of an order greater than the exponential of the number of arguments involved. Thus, there is a real need to determine approximation methods for practical implementations, in the same vein that eqs.(13.28)-(13.31) and Theorem 14 show that all six classes of deductions and enductions introduced in Section 9.3.2. relative to PS and simple conditionals, themselves depend on simple conditional event forms via appropriate corresponding DGNW or AC operations (imbedded within PS).

4. Determine the class of all probability metrics or, correspondingly (see Theorem 21), the class of all algebraic metrics. Similar issues remain for the class of all conditional algebraic or probability metrics, or even all unconditional algebraic (or probability) metrics. Comparisons and contrasts of various types of metrics need be considered, analogous to that provided here in Tables 1, 2 (Section 14.6).

5. Determine connections and applicability of the various cea's to non-

monotonic and related logics, especially developed for Artificial Intelligence and Expert Systems considerations. This requires extensive application of the various calculi of operations for the proposed cea's. A basic step toward this end for PS is provided in Goodman and Nguyen [4], Section 5.3, where a number of problems are considered, including the generalization of the basic syllogism, change of deductive assumptions in the form of the "penguin triangle" problem, and various non-monotonic preference axioms.

6. Until recently, despite a relatively large number of proposals for fuzzy logic extensions of conditioning, none have been consistent with extending conditional event algebra. However, a basic step toward such consistency has been initiated in [1] and [3], Section 5, utilizing one-point random set coverage representations of fuzzy sets. The long-range goal, then is to be able to extend relational event algebra in a similar sense.

Despite the presence of the six major issues raised above (and a good deal more), conditional event algebra is still a much more explored area than relational event algebra, which still remains in a stage of relative infancy. Unfortunately, at this stage, relational event algebra is still in an "art form" involving situation-specific techniques for the construction of the "most natural" corresponding relational events. It is expected that the philosophy of approach that relational event algebra invokes –namely, the search for the algebraic counterpart(s) for various given numeric-based functions of probabilities – will indeed be useful, especially in applications to similarity or contrast of various models, which up to now were only considered from a numerical viewpoint, as the six basic examples introduced in these sections illustrate. Some additional basic relational event algebra issues include:

7. Develop reasonably simple criteria which will yield conditions for unique relational event representation for given numerical functions of probabilities. Even for simple forms such as $P(a)P(b)$, one has to determine whether $a \times b$, $b \times a$, or even some weighted average of the two, is the most appropriate relational event representing the multiplication of the two probabilities.

8. A procedure is needed to represent analytic functions or infinite series – or polynomials – in probabilities – which, while having their requisite values being in the unit interval, have some or all coefficients not lying in the unit interval.

9. Criteria are needed to determine how to test which class of numerical functions of probabilities admit relational event counterparts (in the unmodified sense- not depending on the choice of particular probability measure involved) and which do not. In a similar, vein, it is of interest to determine criteria for which classes of numerical functions of probabilities admit even modified relational events and which do not.

Finally, it is the express hope of the authors that the relatively condensed presentation in these sections on conditional and relational event algebra will spark renewed interest in both the academic and applied analysis communities in the problems considered and the techniques utilized to treat them.

Bibliography

[1] Goodman, I. R. (1995) A new Approach to conditional fuzzy sets, *Proceedings Second Annual Joint Conference on Information Sciences*, September 28 - October 1, 1995, pp. 229-232. in P.P. Wang (ed.) *Advances in Machine Intelligence and Soft Computing*, Duke University, Durham, NC, pp. 121-131.

[2] Goodman, I. R. and Kramer, G. F. (1996) Applications of relational event algebra to the development of a decision aid in command and control, *Proceedings 1996 Command and Control Research and Technology Symposium*, Naval Postgraduate School, Monterey, CA, June 25-28, 1996, pp. 415-435.

[3] Goodman, I. R. and Kramer, G. F. (1997) Extension of relational event algebra to random sets with applications to data fusion, *Proceedings of Workshop on Applications and Theory of Random Sets*, Institute for Mathematics and Its Applications (IMA), University of Minnesota, Minneapolis, August 22-24, 1996, Springer-Verlag, Berlin, (to appear).

[4] Goodman, I. R. and Nguyen, H. T. (1995) Mathematical foundations of conditionals and their probabilistic assignments, *International Journal of Uncertainty, Fuzziness and Knowledge-Based Systems*, **3**(3), 247- 339.

Index

20. H. Bandemer and W. Näther: *Fuzzy Data Analysis*. 1992
ISBN 0-7923-1772-6
21. A.G. Sukharev: *Minimax Models in the Theory of Numerical Methods*. 1992
ISBN 0-7923-1821-8
22. J. Geweke (ed.): *Decision Making under Risk and Uncertainty*. New Models and Empirical Findings. 1992   ISBN 0-7923-1904-4
23. T. Kariya: *Quantitative Methods for Portfolio Analysis*. MTV Model Approach. 1993   ISBN 0-7923-2254-1
24. M.J. Panik: *Fundamentals of Convex Analysis*. Duality, Separation, Representation, and Resolution. 1993   ISBN 0-7923-2279-7
25. J.K. Sengupta: *Econometrics of Information and Efficiency*. 1993
ISBN 0-7923-2353-X
26. B.R. Munier (ed.): *Markets, Risk and Money*. Essays in Honor of Maurice Allais. 1995   ISBN 0-7923-2578-8
27. D. Denneberg: *Non-Additive Measure and Integral*. 1994
ISBN 0-7923-2840-X
28. V.L. Girko, *Statistical Analysis of Observations of Increasing Dimension*. 1995   ISBN 0-7923-2886-8
29. B.R. Munier and M.J. Machina (eds.): *Models and Experiments in Risk and Rationality*. 1994   ISBN 0-7923-3031-5
30. M. Grabisch, H.T. Nguyen and E.A. Walker: *Fundamentals of Uncertainty Calculi with Applications to Fuzzy Inference*. 1995   ISBN 0-7923-3175-3
31. D. Helbing: *Quantitative Sociodynamics*. Stochastic Methods and Models of Social Interaction Processes. 1995   ISBN 0-7923-3192-3
32. U. Höhle and E.P. Klement (eds.): *Non-Classical Logics and Their Applications to Fuzzy Subsets*. A Handbook of the Mathematical Foundations of Fuzzy Set Theory. 1995   ISBN 0-7923-3194-X
33. M. Wygralak: *Vaguely Defined Objects*. Representations, Fuzzy Sets and Nonclassical Cardinality Theory. 1996   ISBN 0-7923-3850-2
34. D. Bosq and H.T. Nguyen: *A Course in Stochastic Processes*. Stochastic Models and Statistical Inference. 1996   ISBN 0-7923-4087-6
35. R. Nau, E. Grønn, M. Machina and O. Bergland (eds.): *Economic and Environmental Risk and Uncertainty*. New Models and Methods. 1997
ISBN 0-7923-4556-8
36. M. Pirlot and Ph. Vincke: *Semiorders*. Properties, Representations, Applications. 1997   ISBN 0-7923-4617-3
37. I.R. Goodman, R.P.S. Mahler and H.T. Nguyen: *Mathematics of Data Fusion*. 1997   ISBN 0-7923-4674-2